VLADIMIR I. ARNOLD

Collected Works

Springer

Vladimir I. Arnold, lecturing in Syktyvkar, 1976.

Photo by courtesy of Yakov Eliashberg

VLADIMIR I. ARNOLD

Collected Works

VLADIMIR I. ARNOLD

Collected Works

VOLUME III

Singularity Theory 1972–1979

VLADIMIR I. ARNOLD

Collected Works

VOLUME III

Singularity Theory 1972–1979

Edited by

Alexander B. Givental
Boris A. Khesin
Mikhail B. Sevryuk
Victor A. Vassiliev
Oleg Ya. Viro

Springer

Vladimir I. Arnold

June 12, 1937 – June 3, 2010

Editors

Alexander B. Givental
Department of Mathematics
University of California
Berkeley, CA, USA

Boris A. Khesin
Department of Mathematics
University of Toronto
Toronto, ON, Canada

Victor A. Vassiliev
Steklov Mathematical Institute
Russian Academy of Sciences
Moscow, Russia

Oleg Ya. Viro
Institute for Mathematical Sciences
Stony Brook University
Stony Brook, NY, USA

Mikhail B. Sevryuk
V.L. Talroze Institute of Energy Problems of Chemical Physics
Russian Academy of Sciences
Moscow, Russia

ISBN 978-3-662-49612-1

Library of Congress Control Number: 2016943434

Printed on acid-free paper

Springer is part of Springer Science+Business Media (www.springer.com)

VLADIMIR I. ARNOLD

Collected Works

VLADIMIR I. ARNOLD

Collected Works

VOLUME III

Singularity Theory 1972–1979

Preface

This volume of the Collected Works contains papers written by V.I. Arnold mostly during the period from 1972 to 1979. While the papers collected here cover a variety of research directions, the main theme emerging in Arnold's work of this period is the development of singularity theory of smooth functions and mappings.

Among the articles specifically translated for this volume, the reader will find papers by V.I. Arnold on catastrophe theory and on A.N. Kolmogorov's school, his prefaces to Russian editions of several books related to singularity theory, a report on the first All-Union mathematical student competition in the USSR, as well as a hard-to-get preprint presenting notes by A.P. Shapiro of V.I. Arnold's lectures on bifurcations of discrete dynamical systems. The volume also contains a translation of the review by V.I. Arnold and Ya.B. Zeldovich of V.V. Beletsky's book on celestial mechanics.

While collecting materials for this volume the editors came across another review of the same book written by Ya.B. Zeldovich alone and included its translation into the volume for comparison.

Finally, we would like to reiterate the disclaimer appearing in the previous volumes that many reprinted translations of the 60s-80s may contain dramatic translation errors. While it was essentially impossible for us to edit the full body of work, as the very least we have provided correct versions of the titles in the Contents.

January 2016

Alexander Givental
Boris Khesin
Mikhail Sevryuk
Victor Vassiliev
Oleg Viro

Acknowledgements

The Editors thank the Göttingen State and University Library, the Mathematics Department Library of Heidelberg University, and the Mathematisches Forschungsinstitut Oberwolfach for providing the original articles for this edition. They are also grateful to Andrey Khesin for his help with the translation of several papers in this volume. The Editors also thank the Springer office in Heidelberg, and in particular Ruth Allewelt and Martin Peters, for their extensive help and tireless support with this project.

Contents

Books by V.I. Arnold written in 1972–1979 but not included in the "Collected Works"

1. Matematicheskie metody klassicheskoĭ mekhaniki (Mathematical Methods of Classical Mechanics) (in Russian), Nauka, Moscow, 1974, 432 pp. The subsequent Russian editions (by various publishing houses, essentially unrevised): 1979, 1989, 1999–2000, 2003. The first English translation: Mathematical Methods of Classical Mechanics. Translated from the Russian by K. Vogtmann and A. Weinstein. Graduate Texts in Mathematics, 60. Springer, New York–Heidelberg, 1978, x+462 pp. The subsequent English editions: 1989, 1995. There are also translations into French (1976), Polish (1981), Bulgarian (1985), German (1988), and many other languages.

2. Dopolnitel′nye glavy teorii obyknovennykh differentsial′nykh uravneniĭ (Supplementary Chapters of the Theory of Ordinary Differential Equations) (in Russian), Nauka, Moscow, 1978, 304 pp. The subsequent Russian editions under the title "Geometricheskie metody v teorii obyknovennykh differentsial′nykh uravneniĭ" (Geometrical Methods in the Theory of Ordinary Differential Equations) (by various publishing houses, essentially unrevised): 2000, 2002, 2012. The first English translation: Geometrical Methods in the Theory of Ordinary Differential Equations. Translated from the Russian by J. Szücs. Translation edited by M. Levi. Grundlehren der Mathematischen Wissenschaften, 250. Springer, New York–Berlin, 1983, xi+334 pp. The subsequent English edition: 1988. There are also translations into French (1980, 1984), German (1987), and many other languages.

MODES AND QUASIMODES

V. I. Arnol'd

The aim of this note is to direct attention to the fact that quasiclassical representations of the form of high-frequency characteristic oscillations of bounded regions can be found, contrary to common considerations of symmetry. These considerations are based on the study of the stratification of manifolds of ellipsoids in accordance with various symmetric groups. They are applied to oscillatory systems with a finite or infinite number of degrees of freedom, and in general to all problems in which quadratic forms depending on parameters occur.

The resolution of the above-indicated contradiction between quasiclassical asymptotics and considerations of symmetry and of the general situation is apparently found in the fact that the functions constructed according to the quasiclassical method are not modes (characteristic functions) and are not close to any modes, but are only quasimodes: with initial conditions given by these functions, the oscillations will resemble standing waves over a long period of time.

It should be noted that the appearance of quasimodes instead of approximations to real characteristic functions does not exclude the validity of quasiclassical approximations to the characteristic numbers.

§1. MANIFOLDS OF ELLIPSOIDS OF ROTATION

Let Q denote the set of all real quadratic forms in the Euclidean space $\mathbf{R}^N$. Q is itself a linear space (of dimension $N(N+1)/2$).

We classify the quadratic forms according to the multiplicities of their characteristic numbers. Let $\nu = (\nu_2, \nu_3, \ldots)$ be a set of integers; Q denotes the set of quadratic forms which have ν_2 characteristic numbers of multiplicity two, ν_3 of multiplicity three, and so on. It is easy to prove

1.1. LEMMA. Q_ν is a smooth semialgebraic* submanifold of the space of all quadratic forms Q; it has codimension

$$\operatorname{codim} Q_\nu = 2\nu_2 + 5\nu_3 + 9\nu_4 + \ldots = \sum \frac{1}{2}(i-1)(i+2)\nu_i.$$

in Q.

Note that if there is at least one multiple characteristic number, then codim $Q_\nu \geq 2$. (The simplest explanation of this is the fact that the discriminant of the characteristic equation of a symmetric matrix of order two is the sum of two squares.) In particular, the set of ellipsoids of rotation does not divide the manifold of all ellipsoids.†

Combining Lemma 1.1 with the "weak theorem of transversality" ([2], p. 35) we get

1.2. COROLLARY.‡ The set of ellipsoids whose semiaxes are all of different lengths is connected. In the space of one-parameter families of ellipsoids those families in which there are no ellipsoids of rotation (i.e., the lengths of the semiaxes are all different) form an everywhere dense set.

*A semialgebraic set is a finite union of sets, each of which is determined by a finite system of polynomial equations or inequalities in Cartesian coordinates.

†It is easy to see that 1.1 and 3.4 remain true if instead of all forms we consider only positive definite ones.

‡In essence, this is the Wigner-von Neumann theorem on the intersection of electron terms. See [3], §79.

Moscow State University. Translated from Funktsional'nyi Analiz i Ego Prilozheniya, Vol. 6, No. 2, pp. 12-20, April-June, 1972. Original article submitted December 8, 1971.

Originally publ. in: Funkts. Anal. Prilozh. 6(2), 12-20, © Russ. Acad. Sciences 1972
English transl.: Funct. Anal. Appl. 6(2), 94-101, © Consultants Bureau, New York 1972
Reprint by Springer-Verlag Berlin Heidelberg 2016 with kind permission.

In other words, if an ellipsoid of rotation is found in a one-parameter family of ellipsoids, it can be eliminated by an arbitrarily small deformation of the family.

1.3. Remark. In two-parameter families of the general kind ellipsoids of rotation ($\nu_2 = 1$) appear for isolated values of a parameter. Increasing the number of parameters, still more symmetric ellipsoids (which cannot be removed by a small change in the family) appear. In the following table all symmetries whose codimension k does not exceed 12 are presented (in a $k + l$-parameter family a symmetry of codimension k is observed on an l-dimensional submanifold):

k=codim Q_ν	2	3	4	5	6	7	8	9	10	11	12
$\{i_{\nu_i}\}$	2_1	—	2_2	3_1	2_3	$2_1 3_1$	2_4	$2_2 3_1$ 4_1	2_5 3_2	3_1 2_3	2_6 $2_1 3_2$

Here a symbol like $2_i 3_j$ corresponds to i different characteristic numbers of multiplicity two and j of multiplicity three.

From the table it can be seen, for example, that given an ellipsoid (more generally, a symmetric 2-tensor) at each point of a three-dimensional Euclidean space, in general, ellipsoids of rotation will be found on isolated lines, and spheres not at all.

In twelve-parameter families of the general kind ellipsoids with six pairs of equal axes are found at isolated values of the parameters, and also those with one pair and two triplets.

1.4. Remark. Near a manifold of ellipsoids with more symmetry the manifold Q_ν has a complicated singularity. As an example we describe below how the set A of ellipsoids with two equal axes approaches the set B of ellipsoids with three equal axes.

Through a point B' of B, construct the transversal cross section of dimension five (codimension of B). In the resulting five-dimensional space the manifold A (of codimension 2) cuts out a three-dimensional manifold A'. The closure of A' has the singular point B'.

Computations show that A' has the form of a double hollow cone over the real projective plane $\mathbf{R}P^2$ embedded in $\mathbf{R}^5$ by five spherical functions of the second degree (five linear combinations of the functions $x_i x_j$, independent on the sphere $x_1^2 + x_2^2 + x_3^2 = 1$, orthogonal to the identity).

1.5. Remark. The fundamental group of the space of all real quadratic forms with nonmultiple characteristic numbers in $\mathbf{R}^N$ is the same as that of the manifold of ellipsoids in $\mathbf{R}^N$ which are not ellipsoids of rotation.

For $N > 2$ it is natural to call this group a Clifford group, since it has a center of two elements $(1, -1)$ and $N-1$ generators $e_1, \ldots, e_{N-1}$ with the defining relations $e_i^2 = -1$, $e_i e_j = -e_j e_i$. The order of a Clifford group in $\mathbf{R}^N$ is equal to 2^N. For $N = 3$ the Hamilton quaternion group is obtained. The generator e_i is realized as follows in the space of ellipsoids: one rotates the ellipsoid 180° in the plane determined by the i-th and N-th principal axes.

§ 2. APPLICATIONS TO THE THEORY OF OSCILLATIONS

It follows from the results of § 1 that, given a family of oscillatory systems which depend in a "general way" on parameters, for two characteristic numbers to coincide it is necessary to change two parameters (and not one, as it appears at first glance).

By a family of the "general kind" we understand a family given by a transformation* of the space of parameters into the space of quadratic forms Q which is transversal† to the manifolds Q_ν of all possible ellipsoids of rotation.

*Here, for simplicity, it is assumed that the systems of the family are distinguished by their potential energies, while the kinetic energy is fixed. The family is then given by a transformation of the space of parameters into the space of quadratic forms: a potential energy is associated with each value of a parameter. The case when the kinetic energy also (or alone) depends on the parameters is treated similarly.

†A transformation f of a manifold X into a manifold Y is transversal to a submanifold Z in Y if at each point of the intersection of $f(X)$ and Z the sum of the space tangent to Z and the image of the space tangent to X consists of the entire space tangent to Y. Roughly speaking, this means that $f(X)$ and Z lie "in the common state" in Y.

2.1. Example. Consider the plane (diffeomorphic disc) membrane with boundary Γ. All such membranes form a family whose space of parameters is a functional space M of curves Γ. The potential energy associated with a curve Γ is the Dirichlet integral over the region bounded by the curve,

$$U = \frac{1}{2}\iint (\nabla u)^2 \, dx;$$

the kinetic energy is given by the equation $T = \frac{1}{2}\iint u^2 \, dx$.

2.2. Hypothesis of Transversality. The transformation $M \to Q$ corresponding to the family of all possible membranes is transversal to all submanifolds Q_ν.

Roughly speaking, this hypothesis means that any finite number of frequencies of the membrane can occur independently of one another, deforming the boundary of the membrane in the appropriate way.

(In order to formulate 2.2 accurately, it is necessary first to identify the configuration spaces of all membranes in M so that the kinetic energies become the same. This is done by a diffeomorphism of an arbitrary membrane onto a standard one which has a constant Jacobian.)

Other families of oscillatory systems where transversality is naturally expected are families of all possible Riemann metrics on a fixed smooth manifold.

The simplest model is the system given by the forms

$$U = \frac{1}{2}\int (\nabla u)^2 + b(x)\, u^2 \, dx, \qquad T = \frac{1}{2}\int u^2 \, dx$$

in the space of functions on the torus T^n; the case $n = 1$ (the Sturm-Liouville problem) is already included. The function b plays the role of the parameter.

The hypothesis of tranversality must be accepted without proof. From it* and Lemma 1.1 follows

2.3. COROLLARY. For an oscillation of membranes of the general kind, one characteristic oscillation corresponds to each characteristic frequency. Any two membranes of the general kind can be joined to one another by a continuous family of membranes such that none of the membranes of the family has a multiple spectrum. If a one-parameter family joining two membranes of the general kind contains a membrane with a multiple spectrum, then by means of an arbitrarily small change it is possible to obtain a family such that the spectra of all of its membranes become simple.

Thus from each characteristic oscillation of one membrane of the general kind it is possible to pass continuously to a specified characteristic oscillation of another without once encountering a multiple spectrum. Under such a continuous transformation, the k-th characteristic number, by magnitude, of the second membrane is obtained from the k-th characteristic number, by magnitude, of the first, regardless of the path of the transformation.

However, transformations of characteristic functions do depend on the path. Specifically, the sign of the resulting extension of the characteristic function depends on how the path goes around the set of membranes with multiple spectrum in the space of all membranes.

Of course, similar conclusions are obtained for other oscillatory systems, such as the family of Schrödinger equations with potentials of the general kind.

§3. INFLUENCE OF SYMMETRIES ON THE MULTIPLICITY OF THE SPECTRUM

A multiple spectrum is an exception in systems of the general kind, but it appears and is not removable by small changes in cases where the given system is symmetric and deformations preserve the symmetry.

Consider, for example, a round membrane. It is known that the majority of the characteristic numbers are of multiplicity two (the simple characteristic numbers correspond to characteristic functions which depend only on the radius). Now suppose that we change the form of the membrane, preserving the

*For applications even a weak hypothesis of transversality on a set of membranes of infinite codimension in the set of all membranes is sufficient.

third-order rotational symmetry (so that the membrane coincides with itself under a rotation of 120°). How will the spectrum change under deformations of the general kind which preserve the indicated symmetry?

3.1. Formalization. From the point of view of the theory of singularities of smooth transformations (see [2, 4]) we are considering the following problem. We are considering a transformation φ: $M \to Q \times O(N)$ which associates with each point x in the connected manifold of parameters M a pair consisting of a quadratic form q (the potential energy corresponding to the given value of the parameters) and an orthogonal transformation g of the configuration space which preserves q and satisfies the condition $g^3 = E$ (g gives the action of the group of symmetries of the problem on the configuration space).* We must investigate a point of the "common state" in the functional space Φ of transformations φ.

3.2. Eliminating the Deformation of the Representation. From the fact that the set of classes of the orthogonal representations of the group of symmetries (in our example, Z_3) is discrete, it follows that all of the transformations g(x) are similar in O (lie on one orbit G of the adjoint representation).

Therefore Φ is the space of transformations into the fiber space with basis G and a typical fiber F consisting of all quadratic forms remaining in place under a fixed transformation $g \in O$.

Thus the problem reduces to the investigation of points of the common state in the space of smooth transformations φ: $M \to F$, where M is the manifold of parameters and F is the space of all quadratic forms which remain unchanged under a fixed (not up to equivalence, but precisely fixed) representation of the group of symmetries in O(N).

Note that F is a linear subspace of the space Q of all quadratic forms. Therefore the theorem of transversality can be applied to transformations in F.

Note also that F inherits from Q a stratification† according to the multiplicities of the characteristic numbers. Our next goal is to investigate the resulting stratification of F.

3.3. Example. Consider the representation of $\mathbf{Z}_3$ (the group of rotations of an equilateral triangle) as transformations of the Euclidean space $\mathbf{R}^{2n}$, where the generator g rotates each of n two-dimensional orthogonal planes through 120° (so that the representation can be decomposed into n two-dimensional irreducible ones.‡

The investigation of the g-invariant quadratic forms is based on the following observation. The operator of rotation through 90°,

$$I = \frac{2g+E}{\sqrt{3}} : \mathbf{R}^{2n} \to \mathbf{R}^{2n}, \quad I^2 = -E,$$

gives $\mathbf{R}^{2n}$ the structure of the complex space $\mathbf{C}^n$. The ellipsoids which are invariant under g (or I) are in one-to-one correspondence with Hermitian scalar products in $\mathbf{C}^n$.

Hence it is clear that the number of axes of such an ellipsoid of some fixed length is even.

We will call the complex dimension of the largest complex subspace of $\mathbf{C}^n$ spanned by axes of a given length the complex multiplicity of an axis of the ellipsoid. An axis of complex multiplicity k is actually repeated 2k times.

Let H denote the set of all real quadratic forms in $\mathbf{R}^{2n}$ which are invariant under I (equivalently, under g). H is a linear subspace of dimension n^2 in the space of all quadratic forms in $\mathbf{R}^{2n}$.

Let $\mu = (\mu_2, \mu_3, \ldots)$ be a set of integers. Let H_μ denote the set of forms in H (i.e., of g-invariant quadratic forms on the Euclidean space $\mathbf{R}^{2n}$) which have μ_2 different characteristic numbers of complex multiplicity 2, μ_3 of complex multiplicity 3, etc. It is easy to prove

*In the general case of an arbitrary group of symmetries, instead of O one must consider O^k, and instead of one transformation g, k generators of the group of symmetries at once.

†Here and below a partition into a finite number of semialgebraic subsets is called a stratification; usually these subsets will be smooth submanifolds.

‡Here and below representations are orthogonal and real; equivalence and irreducibility are understood in the same O(N)-sense.

3.4. LEMMA. H_μ is a smooth real semialgebraic submanifold of H. The manifold H_μ has codimension

$$\operatorname{codim} H_\mu = 3\mu_2 + 8\mu_3 + 15\mu_4 + \ldots = \sum (i^2 - 1)\mu_i.$$

in H.

Combining this result with the theorem of transversality, we get

3.5. COROLLARY. In the space of smooth mappings φ: M → H of the manifold of parameters M into the space H of g-symmetric ellipsoids, the mappings for which ellipsoids with axes of complex multiplicity greater than 1 occur only on submanifolds of codimension 3 or higher in the space of parameters form an open everywhere dense set.

In particular, in one- and two-dimensional families of the general kind there are no ellipsoids with axes of complex multiplicity greater than 1.

In families with a larger number of parameters k, larger complex multiplicities can occur. Degeneracies of codimension not greater than 20 are exhausted by the following table (k = codim H_μ):

k	3	6	8	9	11	12	14	15	16	17	18	20
$\{i_{\mu_i}\}$	2_1	2_2	3_1	2_3	$2_1 3_1$	2_4	$2_2 3_1$	2_5 4_1	3_2	$2_3 3_1$	2_6 $2_1 4_1$	$2_4 3_1$

A symbol like $2_i 3_j$ corresponds to ellipsoids whose group of unitary symmetries is

$$(U(1))^{\mu_1}(U(2))^i (U(3))^j = (S^1)^{\mu_1+i+j}(S^3)^i (SU(3))^j.$$

For example, it can be seen from the table that ellipsoids with four axes of complex multiplicity two and one of multiplicity three can appear in a twenty-parameter family of the general kind, at isolated values of the parameters.

3.6. Remark. The singularity of the set A of ellipsoids with an axis of complex multiplicity two near the set B of those with an axis of complex multiplicity three is investigated as in the real case, § 1.

In a transversal cross-section R^8 there appears a five-dimensional manifold A' with an isolated singular point B' which has the form of a cone over the complex projective plane CP^2, embedded in R^8 by nine functions $z_i \bar{z}_j$ (from complex linear combinations of these it is possible to form eight real functions which are linearly independent on a sphere in C^3 and orthogonal to the identity).

3.7. Remark. In "physical" terms, Corollary 3.5 means that:

1) In systems of the general kind with rotational symmetry of the third order in which there are no completely symmetric oscillations, all characteristic numbers are of multiplicity two.

2) To each characteristic number correspond three modes which are carried into one another by a rotation through 120° and which add up to zero.

3) In order to make two characteristic numbers (of multiplicity two) coincide, while preserving symmetry, it is necessary to change three parameters; to make three coincide, eight parameters; and so on.

§ 4. THE CASE OF NONEQUIVALENT REPRESENTATIONS

It may happen that the representation of the group of symmetries in the configuration space decomposes into irreducible representations of a different kind. For example, in the case of membranes with rotational symmetry of order three, there are one-dimensional, as well as two-dimensional, representations of the group Z_3. The following lemma is easily proved and well known.

4.1. LEMMA. If a quadratic form is invariant under the orthogonal sum of some orthogonal representations $T_1 \oplus T_2 \oplus \ldots \oplus T_k$ of a compact group, divisible by pairwise distinct irreducible representations, then this quadratic form is the direct sum of the quadratic forms determined on the spaces of the representations T_i.

4.2. COROLLARY. A fixed space of transformations φ: $M \to Q(T)$ of a space of parameters M into a space of quadratic forms which are invariant under the orthogonal representation T of a compact group is the direct sum of spaces of transformations φ: $M \to Q(T_i)$, where $T = T_1 \oplus \ldots \oplus T_k$ is the decomposition of T into the orthogonal direct sum of representations which are divisible by distinct irreducible ones.

In other words, characteristic numbers corresponding to nonequivalent representations move independently of one another under deformations. In particular, the codimensions of degeneracy in families of the general kind can be computed by summing over the nonequivalent representations.

4.3. Example. Consider representations of Z_3 in R^{3k}. Such a representation can be decomposed into the direct sum of l one-dimensional and n two-dimensional representations (where $3k = l + 2n$). The set of representations consists of submanifolds of different dimensions discretely situated in O(3k). A calculation shows that the manifold of representations with $l = k$ has the highest dimension [if $l = k + 2s$, the dimension is less by $s(3s-1)$]. We can say that in the "most probable" representation of Z_3, the dimension of the space in which the representation is trivial is one-third the dimension of the space of the representation.

From Corollary 4.2 and the results of §§ 1-3, we get

4.4. COROLLARY. An oscillatory system of the "general kind" with group of symmetries Z_3, represented by the sum of l one-dimensional and n two-dimensional representations, has l simple characteristic frequencies and n of multiplicity two (for the "most probable" representation the simple frequencies are one-third of the total).

In general one-parameter families of such systems, at isolated values of the parameters one of the simple characteristic frequencies can pass through one of multiplicity two, but two simple frequencies cannot coincide, nor can two of multiplicity two (see [3], §79).

In two-parameter families it is possible for two simple frequencies to coincide with one another, and in three-parameter families, frequencies of multiplicity two may coincide. In order for i simple frequencies and j of multiplicity two to coincide at any point, $1/2(i-1)(i+2) + j^2$ parameters are needed.

§5. APPLICATIONS TO OSCILLATIONS OF SYMMETRIC MEMBRANES

We return to the problem of § 3 of the behavior of the frequencies of round membranes under deformations which preserve symmetry of order three.

Let M_G denote the space of all membranes which are invariant under transformations of a given compact group of motions G.

Let T be a representation of G in the configuration space of the membranes and let Q_G be the space of quadratic forms on this configuration space which are invariant under G.

5.1. Hypothesis of Transversality. The transformation φ: $M_G \to Q_G$ given by the Dirichlet integral is transversal to the stratification of Q_G according to the multiplicities of the characteristic numbers.

This hypothesis means that a deformation of the boundary of the membrane (preserving symmetry) is a deformation of a sufficiently general kind in the space of quadratic forms of the prescribed symmetry.

Combining 5.1 with the results of §§ 1-4, we get the following result.

5.2. THEOREM. If the hypothesis of transversality 5.1 is true, then:

1. For a membrane of the general kind, invariant under a rotation through 120°, one-third of the characteristic frequencies are simple and the corresponding characteristic functions are preserved under a rotation through 120°. The remaining characteristic frequencies are of multiplicity two; corresponding to a characteristic frequency of multiplicity two are three characteristic frequencies which add up to zero and which are translated into one another by a rotation through 120°.

2. In one-parameter families of the general kind of such symmetric membranes, a simple frequency can coincide with one of multiplicity two at isolated values of the parameters, but simple frequencies do not coincide with one another, nor do those of multiplicity two.

3. The number of parameters of families in which characteristic frequencies coincide in a more complicated way (not removably by means of a small change which preserves symmetry) is computed according to the formulas of par. 1.1, 3.4, and 4.3.

In particular, under a small general Z_3-symmetric deformation of a round membrane the three* characteristic numbers of multiplicity two (corresponding to characteristic functions with azimuthal part $\sin 3k\varphi$ and $\cos 3k\varphi$) split at once. Under further deformations a simple characteristic number and one of multiplicity two can pass through one another, however neither two simple characteristic numbers nor two of multiplicity two will coincide.

Problems in the theory of oscillations of other symmetric regions can be investigated similarly. It would probably be helpful to make up tables of bifurcational diagrams for all possible groups G of rotations in R^3 and of all possible decays of symmetry (pairs $G \supset G_1$) corresponding to degeneracies of small codimension.

§ 6. COMPARISON WITH QUASICLASSICAL RESULTS

In recent years the hope of understanding the asymptotic properties of high-frequency characteristic functions by using "quasiclassical" approximations has appeared ([6, 7, 8]). This hope is based not on mathematical theorems but on physical considerations. The example cited below† shows that the results obtained in this way contradict considerations of symmetry and the hypothesis of transversality, and therefore are probably false.

Consider a membrane of the general kind with rotational symmetry of order three, say one in the form of an equilateral triangle with rounded vertices. If the rounding-off curve is flat enough, the height of the triangle determines a stable periodic billiard ball trajectory. According to the representations of quasiclassical methods, there exists a series of high-frequency oscillations ("modes") whose characteristic functions are concentrated near a stable trajectory (i.e., near the height).

Rotating such a mode through 120° and 240°, we obtain two other modes of the same frequency. The sum of the characteristic functions corresponding to these three oscillations is different from zero (otherwise each of the modes would be concentrated near the center of the triangle).

This result contradicts that of §5, according to which there are no characteristic frequencies of multiplicity three for membranes of the general kind with groups of symmetries Z_3.

The results of §5 are based on the unproved hypothesis of transversality. The quasiclassical assertion singled out above has also never been proved rigorously for systems with non-separating variables. The contradiction shows that either the hypothesis of transversality or the quasiclassical conclusion (or both) is wrong.

The arguments in favor of the hypothesis of transversality seem all the more convincing since, evidently, the quasiclassical representation of the nature of the characteristic oscillations is false.

On the other hand, it is known that physics has successfully used these (evidently false) representations in calculations for lasers and other systems [9]. One possible explanation is the following.

The function u constructed by quasiclassical formulas, concentrated near a closed trajectory, is not a characteristic function (mode) and is not near any characteristic function; it is only quasicharacteristic in the sense that the quantity $\Delta u + \omega^2 u$ is small.

Consequently, if u is given as an initial condition, then over a long period of time (but not indefinitely) the oscillation will resemble a standing wave. Gradually, however, energy begins to be transferred and oscillations of another form, no longer concentrated near the height of the triangle, or concentrated near the second and third heights, are excited.

Thus, the function u determined by quasiclassical methods is not a mode but a quasimode, similar to the oscillation of one of two equal pendulums which are weakly connected. These modes are probably constructed differently; they cannot be concentrated anywhere (the whole membrane oscillates).

A knowledge of quasimodes, which are fairly easily constructed and which live a fairly long time, may be more useful in considering infinite, but difficult to construct real modes, from whose interference

*In this way the representation of Z_3 which we called the most probable in §4 is realized specifically. Such a coincidence is hardly accidental.

†This example was proposed by A. I. Shnirel'man during a discussion of the results of the preceding section.

quasimodes arise. Quasiclassical investigation of quasimodes ought to include a bound on the length of their lives.

The author is grateful to M. A. Isakovich, A. I. Shnirel'man, L. D. Faddeev, A. M. Veselova, and S. P. Novikov for useful discussions.

LITERATURE CITED

1. R. Perrin, Selection rules for the splitting of degenerate pairs of natural frequencies of thin circular rings, Acustica, 25, No. 2, 69-72 (1971).
2. R. Thom and H. Levine, Singularities of Differentiable Mappings [Russian translation], Mir, Moscow (1968).
3. L. D. Landau and E. M. Lifshitz, Quantum Mechanics. Non-relativistic Theory, Addison-Wesley (1958).
4. V. I. Arnol'd, Singularities of smooth mappings, Uspekhi Matem. Nauk, 23, No. 1, 3-44 (1968).
5. R. Thom, Les symetries brisees en physique macroscopique et la mécanique quantique, Publ. IHES (1970).
6. J. B. Keller and S. I. Rubinow, Asymptotic solution of eigenvalue problems, Ann. Phys., 9, No. 1, 24-75 (1960).
7. A. I. Baz', Ya. B. Zel'dovich, and A. M. Perelomov, Scattering, Reactions and Decay in Nonrelativistic Quantum Mechanics, Nauka, Moscow (1971). English translation, Israel Program for Scientific Translations, Ltd., Jerusalem (1969).
8. V. M. Babich, V. S. Buldyrev, and V. F. Lazutkin, Application of the methods of mathematical diffraction theory to the problem of the asymptotics of the characteristic numbers and functions of Laplace operators, Partial Differential Equations, Sibirsk. Otdel., Akad. Nauk SSSR, Moscow (1970).
9. V. P. Bykov and L. A. Vainshtein, High Power Electronics, 4, Nauka, Moscow (1965).

INTEGRALS OF RAPIDLY OSCILLATING FUNCTIONS AND SINGULARITIES OF PROJECTIONS OF LAGRANGIAN MANIFOLDS

V. I. Arnol'd

1. Integrals of Oscillating Functions. In various problems in analysis it is necessary to study the behavior of integrals of the form

$$I(h) = \int e^{\{i[S(x) - \langle p, x\rangle]/h\}} \varphi(x)\, dx \quad (x \in \mathbf{R}^n;\ p \in (\mathbf{R}^n)'), \tag{1}$$

as $h \to 0$, where S is a smooth real-valued function and φ is a finite C^∞-function. According to the stationary phase principle, those points x for which $\partial S/\partial x = p$ make the main contribution asymptotically. This equation determines an n-dimensional manifold M in the phase space $\{(x, p)\}$. A manifold of this form is a Lagrangian manifold (see [1]). The asymptotic character of the integral in (1) is related to the singularities of the smooth mapping (the projection onto the p-space)

$$\pi : M \to (\mathbf{R}^n)', \quad \pi(x, \partial S/\partial x) = \partial S/\partial x. \tag{2}$$

2. Projections of Lagrangian Manifolds. The classification of the singularities of smooth mappings was described in [2]. Projection mappings of Lagrangian manifolds have somewhat more singularities than the general mappings $M^n \to R^n$.

LEMMA. By means of an arbitrarily small perturbation, a Lagrangian manifold $M^n \subset R^{2n}$ can be brought into general position relative to the projection onto p-space so that the rank π falls to k only on a submanifold Σ^k of codimension $k(k + 1)/2$ in M.

COROLLARY. The singularities of projections of Lagrangian manifolds in general position for $n \le 3$ are exhausted by the following list (see [2]): Σ^1 (a fold, codim $\Sigma^1 = 1$); $\Sigma^{1,1}$ (a cusp, codim $\Sigma^{1,1} = 2$); $\Sigma^{1,1,1}$ (a "swallow tail," codim $\Sigma^{1,1,1} = 3$); $\Sigma^2_\pm$ (the junction of three cusps, two of which can be imaginary, codim $\Sigma^2_\pm = 3$).

Remark. It is not assumed above that M can be well projected onto x-space (i.e., that M is given by an equation $p = \partial S/\partial x$). However, all of the singularities listed can occur in this case, as the following examples show:

$$\begin{aligned} A_{l+1} &= \Sigma^{1_l}: S = \pm x_1^{l+2} + (x_l + x_1^l)^2 + \ldots + (x_2 + x_1^2)^2 + x_{l+1}^2 + \ldots + x_n^2, \\ D_4 &= \Sigma^2_\pm : S = x_1^3 \pm 3x_1x_2^2 + (x_1^2 + x_3)^2, \\ D_5 &= \Sigma^2_0 : S = \pm x_1^4 + x_1x_2^2 + (x_1^2 + x_3)^2 + (x_2^2 + x_4)^2. \end{aligned} \tag{3}$$

3. Exponent of Singularity. Together with (1), it is convenient to consider the more general integral

$$I(h) = \int \exp\{iF(x, \lambda)/h\}\, \varphi(x, \lambda)\, dx, \quad (x \in \mathbf{R}^n;\ \lambda \in \mathbf{R}^l), \tag{4}$$

depending on a parameter λ in addition to h. We introduce the exponent of singularity β of the function F (or of the germ of F at a point) as the lower bound of those numbers γ for which $I(h) = o(|h|^{(n/2)-\gamma})$ as $h \to 0$ for any smooth function φ and for any $\lambda \in \mathbf{R}^l$.

4. Examples. We can compute the following table of values of the exponent β for the simplest singularities:

Moscow State University. Translated from Funktsional'nyi Analiz i Ego Prilozheniya, Vol. 6, No. 3, pp. 61-62, July-September, 1972. Original article submitted April 4, 1972.

Originally publ. in: Funkts. Anal. Prilozh. 6(3), 61-62, © Russ. Acad. Sciences 1972
English transl.: Funct. Anal. Appl. 6(3), 222-224, © Consultants Bureau, New York 1972
Reprint by Springer-Verlag Berlin Heidelberg 2016 with kind permission.

Type of singularity	Σ^1	$\Sigma^{1,1}$	$\Sigma^{1,1,1}$	$\Sigma^{1,1,1,1}$	$\Sigma^2_{\pm}$	Σ^2_0	A_{l-1}	D_{l+1}	E_6	E_7	E_8
β	$1/6$	$1/4$	$3/10$	$1/3$	$1/3$	$3/8$	$\frac{1}{2}-\frac{1}{l}$	$\frac{1}{2}-\frac{1}{2l}$	$5/12$	$4/9$	$7/15$

For example, the integral (1) ordinarily behaves as follows in the three-dimensional case. At isolated "very singular" points in p-space ($\Sigma^2_{\pm}$), I is a quantity of order $h^{7/8}$; at other isolated points ($\Sigma^{1,1,1}$) it is of order $h^{6/5}$; on certain curves joining singular points ($\Sigma^{1,1}$) it is of order $h^{5/4}$; on certain surfaces (Σ^1) for which these curves are singularities the integral is of order $h^{4/3}$; and at other points it is of order $h^{3/2}$. If S, a function of three variables, also depends on a parameter $t \in R$, then for isolated values of t other singularities Σ^2_0 with I of order $h^{9/8}$ and $\Sigma^{1,1,1,1}$ with I of order $h^{7/6}$ may appear at isolated points p.

5. THEOREM. The exponent of singularity of almost all* functions $F(x, \lambda)$ depending on $l \leq 5$ parameters λ is not greater than the number β_l, where $\beta_1 = 1/6$, $\beta_2 = 1/4$, $\beta_3 = 1/3$, $\beta_4 = 3/8$, and $\beta_5 = 5/12$. Equality is attained, for example, for the following functions of three variables and for any functions sufficiently close to them:

$$\Sigma^1 l:\quad F = \pm x^{l+2} + \lambda_l x^l + \ldots + \lambda_1 x \pm y^2 \pm z^2 \qquad (l = 1, 2),$$
$$\Sigma^2_{\pm}:\quad F = x^3 \pm 3xy^2 + \lambda_3 x^2 + \lambda_2 x + \lambda_1 y,$$
$$\Sigma^2_0:\quad F = xy^2 \pm x^4 \pm z^2 + \lambda_4 y^2 + \lambda_3 x^2 + \lambda_2 y + \lambda_1 x,\ E_6.$$

The exponent of singularity of almost all functions $F = S - \langle p, x\rangle$ depending on $n \leq 5$ parameters p does not exceed β_n. Equality is attained, for example, for the functions S in (3) of types Σ^1, $\Sigma^{1,1}$, $\Sigma^2_{\pm}$, and Σ^2_0 and for any functions sufficiently close to them (when n = 5 for E_6).

6. LEMMA. In l-parameter families of functions ($l \leq 5$), the only degenerate critical points not removable by a small perturbation of the family which can occur are those in the neighborhood of which the function can be written in one of the following normal forms (in a suitable coordinate system $x_1, \ldots, x_n$):

$$\begin{aligned} \Sigma^{1m-1} = A_m:&\quad f = \pm x_1^{m+1} \pm x_2^2 \pm \ldots \pm x_n^2 + \text{const},\quad \operatorname{codim} A_m = m - 1 \leqslant l,\\ D_m:&\quad f = \pm x_1^{m-1} + x_1 x_2^2 \pm x_3^2 \pm \ldots \pm x_n^2 + \text{const},\quad \operatorname{codim} D_m = m - 1 \leqslant l,\\ E_6:&\quad f = \pm x_1^4 + x_2^3 \pm x_3^2 \pm \ldots \pm x_n^2 + \text{const},\quad \operatorname{codim} E_6 = 5.\end{aligned} \tag{5}$$

7. Definition. A function $F(x, \lambda)$ is called a versal deformation of the function $f(x) = F(x, 0)$ at the point $x = 0$ if for every smooth function $G = G(x, \mu)$ which is equal to $f(x) = G(x, 0)$ for $\mu = 0$ there exist a function $\lambda = \varphi(\mu)$ and a diffeomorphism of a neighborhood of zero $x \mapsto X(x, \mu)$, both smooth near $\mu = 0$, such that $G(x, \mu) \equiv F(X(x, \mu), \varphi(\mu))$ in a neighborhood of the point $x = 0$, $\mu = 0$, where $\varphi(0) = 0$ and $X(x, 0) = x$.

THEOREM. Versal deformations of the functions in (5) are given by the formulas

$$\begin{aligned} A_{m+1} = \Sigma^1 m:&\quad F = f + \lambda_m x_1^m + \ldots + \lambda_1 x + \lambda_0,\\ D_4 = \Sigma^2_{\pm}:&\quad F = f + \lambda_3 x_1^2 + \lambda_2 x_2 + \lambda_1 x_1 + \lambda_0,\\ D_5 = \Sigma^2_0:&\quad F = f + \lambda_4 x_2^2 + \lambda_3 x_1^2 + \lambda_2 x_2 + \lambda_1 x_1 + \lambda_0.\end{aligned} \tag{6}$$

It follows from this theorem that in order to study the integrals in (4) with functions F in general position, it suffices to consider on the special cases in (6) (if $l \leq 4$).

We can also deduce from (5) a classification scheme for singularities of projections of general Lagrangian manifolds for $n \leq 5$; for $n = 6$, "moduli" appear.

8. Remark. By Hironaka's theorem on the resolution of singularities, it follows, as I. N. Bernshtein has observed [3], that for any polynomial S the asymptotic form of the integral (1) is $h^{\varkappa} (\ln h)^{\nu}$. It seems probable that a bound for the integral in (4) of the form $|I| \leq ch^{(n/2)-\gamma}$ is uniform in λ near λ_0 for any $\lambda > \beta(F, \lambda_0)$ (and for $\gamma = \beta$ as well in the cases in (6)).

The author is grateful to V. P. Maslov, who asked him to compute the constants β_l, and also to I. N. Bernshtein and S. I. Gel'fand for useful discussions.

LITERATURE CITED

1. V. I. Arnol'd, Funktsional. Analiz i Ego Prilozhen., 1, No. 1, 1-14 (1967).

*All except the countable union of nowhere dense sets in the function space.

2. V. I. Arnol'd, Uspekhi Matem. Nauk, 23, No, 1, 3-44 (1968).
3. I. N. Bernstein, Funktsional. Analiz i Ego Prilozhen., 5, No. 2, 1-16 (1971).

To A. N. Kolmogorov on
his seventieth birthday

REMARKS ON THE STATIONARY PHASE METHOD AND COXETER NUMBERS

V. I. Arnol'd

We study integrals of rapidly oscillating functions. Such integrals tend to zero when the length of the oscillation wave tends to zero through wave fronts of constant form. The asymptotic decrease of the integral is determined by the character of the critical points of the function describing the front. If all of these critical points are non-degenerate (Morse), then the integral tends to zero like the wave length raised to the power of half the dimension of the space, and indeed this is the asymptotic behaviour of the integral for functions in general position.

However, if the integral depends on additional parameters, then for certain "caustic" parameter values there arise non-Morse critical points and the integral decreases slowly. The investigation of the asymptotic behaviour of the integral of an oscillating function in caustic cases can be regarded as a generalization of the theory of Airy functions; it is closely connected with Artin's braid theory, and the answer in the case of few parameters is expressed in terms of the Coxeter numbers of the Weyl groups of the series A, D, E, and in the case of many parameters it is expressed in terms of generalizations of them.

Contents

§1. Introduction

1.1. Trigonometric integrals. Let S: $\mathbf{R}^n \to \mathbf{R}$ be a smooth function. A *trigonometric integral with the function S* is an integral of the form

$$I(h) = \int_{\mathbf{R}^n} e^{iS(x)/h} \varphi(x)\, dx, \tag{1}$$

Originally publ. in: Usp. Mat. Nauk 28(5), 17-44, © Russ. Acad. Sciences 1973
English transl.: Russ. Math. Surv. 28(5), 19-48, © London Mathematical Society 1973
Reprint by Springer-Verlag Berlin Heidelberg 2016 with kind permission.

where h is a small real parameter and ϕ an infinitely differentiable function of compact support.

By the stationary phase principle, the principal contribution to the asymptotic behaviour of I as $h \to 0$ is given by neighbourhoods of the critical points of S. By Morse's lemma, in a neighbourhood of a non-degenerate critical point S can be bought by a diffeomorphism to the normal form $S = \pm x_1^2 \pm \dots \pm x_n^2 + \text{const}$. A trigonometric integral with such a function S and with $\phi \equiv 1$ can be computed explicitly: it is proportional to $h^{n/2}$.

If the support of ϕ is concentrated in a neighbourhood in which S reduces to the normal form, then clearly $|I(h)| \leqslant 0(h^{n/2})$ as $h \to 0$.

The asymptotic behaviour of a trigonometric integral with a function S for the worst ϕ is determined by the differentiable type of S (in other words, it is independent of the coordinate system): under a diffeomorphism $x = f(y)$ the trigonometric integral goes over to an integral of the same kind, but with ϕ multiplied by the Jacobian of f.

The local case when S is the germ of a smooth function at 0 can be treated in the same way. Thus, the determination of the asymptotic behaviour of trigonometric integrals is closely connected with the classification of critical points up to a diffeomorphism.

A function "in general position" has only non-degenerate critical points; hence a trigonometric integral with such a function decreases like $h^{n/2}$.

1.2. Trigonometric integrals depending on a parameter. In applications, however, we meet the situation when the trigonometric integral under discussion depends on some parameters; for example, we often come across integrals of the form

$$(2) \qquad I(h, p) = \int_{\mathbf{R}^n} e^{i[S(x) - \langle p, x \rangle]/h} \varphi(x)\, dx$$

(the so-called phase integrals, depending on an n-dimensional parameter $p \in \mathbf{R}^{n'}$), or the more general integrals

$$(3) \qquad I(h, \lambda) = \int_{\mathbf{R}^n} e^{iF(x, \lambda)/h} \varphi(x, \lambda)\, dx,$$

depending on an arbitrary number of parameters ($\lambda \in R^l$).

1.3. Estimates for trigonometric integrals. The asymptotic behaviour of such integrals depending on a small number of parameters was considered in [1] on the request of Maslov, who needed it for his investigation of focussing energy in crystals [2].

This clarified the astonishing connection [3] between the simplest degenerate critical points of the functions and the Weyl groups of the series A, D, E and their braid groups, which were introduced a little earlier by Brieskorn [4]. The asymptotic behaviour for these simplest degenerate critical points of the series A, D, E were expressed in terms of a certain integer N according to the formula

$$I(h) = O(h^{\frac{n}{2}-\beta}), \quad \text{where} \quad \beta = \frac{1}{2} - \frac{1}{N},$$

and

$$N(A_k) = k + 1, \quad N(D_k) = 2k - 2, \quad N(E_6) = 12, \quad N(E_7) = 18, \quad N(E_8) = 30.$$

I am grateful to B. Kostant for pointing out that N is the Coxeter number corresponding to the group generated by reflections.

At present there exist tables of normal forms to which functions reduce in the neighbourhood of a critical point such that all the cases not covered by these tables form a set of codimension 11. Thus, to complete the investigation of the asymptotic behaviour of the integrals (2), (3) with functions S and F in general position and depending on not more than 10 parameters, it remains simply to carry out the computations for reducing the functions to normal form. The answer looks like this:

1) *For every $\varepsilon > 0$ and for almost every*[1] *l-parameter family $F(x, \lambda)$ of trigonometric integrals* (3), *we have* $|I(h)| \leqslant O(h^{\frac{n}{2}-\beta_l-\varepsilon})$ *as* $h \to 0$ *for every value of* λ, *provided that* $n \geqslant 3$.

l	1	2	3	4	5	6	7	8	9	10, $n \geqslant 4$	$k(k+1)/2$, $n \geqslant k$
β_l	1/6	1/4	1/3	3/8	5/12	1/2	1/2	13/24	9/16	2/3	k/6?

2) *This estimate cannot be improved, in the sense that there exist l-parameter families such that the integral* (3) *decreases no faster than* $O(h^{\frac{n}{2}-\beta_l})$ *as* $h \to 0$ *for certain values of* λ: *this property of the family is preserved under small displacements.*

A similar answer is obtained for the integrals (2) with functions S in general position. See [3] for more details on the families of functions $S(x) - \langle p, x\rangle$ and their connection with Lagrange singularities.

The problem of the exact determination of the numbers β_l seems to be very difficult.

§2. Terminology

We need a whole series of concepts for our investigation of critical points of functions and the asymptotic behaviour of trigonometric integrals. This section is devoted to the definitions of these concepts.

---

[1] Excluding a thin set (that is, the union of countably many closed nowhere dense subsets) in the space of families.

2.1. The singularity index.

2.1.1. DEFINITION. *The singularity index* β *of the germ* s *of a function* S: $(\mathbf{R}^n, 0) \to \mathbf{R}$ *at the critical point* 0 is the infimum of the set of numbers γ for which the following estimate for the trigonometric integral (1) holds:

$$I(h) = O(h^{\frac{n}{2}-\gamma}) \text{ as } \quad h \to 0$$

for every representative S of the germ s and for all smooth ϕ with sufficiently small support (chosen in accordance with the representative s of S).

2.1.2. EXAMPLE. The singularity index of the germ of any function at a non-degenerate critical point is 0 (this case was discussed already by Fresnel).

2.1.3. The singularity index of the germ of the function x^3 of one variable at zero is 1/6 (this case was considered by Airy).

The function $x^3 + y^2$ of two variables has the same singularity index. In general, the definition of singularity index has been designed so that the addition of the squares of new variables does not alter it. Larger indices β correspond to "more complicated" singularities.

2.1.4. PROPOSITION. *The singularity index is an invariant of the differentiable type of the germ.*

We say that two germs s_1 and s_2 of function $\mathbf{R}^n \to \mathbf{R}$ are *differentiably equivalent* (or *have the same differentiable type*) if there exists a germ g of a diffeomorphism $(\mathbf{R}^n, 0) \to (\mathbf{R}^n, 0)$ such that $s_1 = s_2 \circ g$.

Thus, Proposition 2.1.4. means that the singularity indices of any two germs carried into each other by a smooth change of coordinates are the same.

The proof of 2.1.4. is obvious (see the Introduction).

2.1.5. PROPOSITION. *The singularity index of the germ of a polynomial at any point is a rational number.*

This proposition is non-trivial; a proof follows from recent papers of Bernshtein and Gel'fand [5] and Atiyah [6] and is based on Hironaka's theorem on resolution of singularities.

The results of these papers also give more exact information on the asymptotic behaviour of trigonometric integrals in the case where S is a polynomial. Such an integral has an asymptotic expansion with terms of the form $h^v \ln^\kappa h$; the indices v lie in finitely many rational arithmetic progressions. On the other hand, for general l-parameter families of smooth functions the germ of each function at every point is equivalent to that of a polynomial. To formulate this assertion more precisely, we use the following concept.

2.2. Multiplicity and the local ring of a critical point of a function.

2.2.1. DEFINITION. The *local ring* of a smooth function $S = S(x_1, \ldots, x_n)$ at the critical point $0 \in \mathbf{R}^n$ is the factor ring of the ring of germs of smooth functions of $x_1, \ldots, x_n$ at 0 by the ideal generated by the partial derivatives of S:

$$Q = C^\infty((x_1, \ldots, x_n))/(\partial S/\partial x_1, \ldots, \partial S/\partial x_n).$$

It is easy to see that Q does not depend on the choice of the coordinate system $(x_1, \ldots, x_n)$.

2.2.2. DEFINITION. The *multiplicity* μ of S at the critical point 0 is the dimension of Q as a linear space over the real field **R**: $\mu = \dim_{\mathbf{R}} Q$.

The critical point 0 of S is said to be of *finite multiplicity* if μ is finite.

2.2.3. EXAMPLE. The function $S = x^3$ of one variable x has a critical point of multiplicity 2 at the origin; under small displacements of S this critical point splits into two of multiplicity 1.

In exactly the same way, the function x^n has a critical point of multiplicity $\mu = n - 1$ at the origin.

2.2.4. REMARK. The multiplicity μ at the critical point 0 of a function S should not be confused with the multiplicity of the singular point 0 on the hyperplane $S = 0$, as defined in algebraic geometry.

2.2.5. DEFINITION. *The multiplicity at the critical point* 0 *of a holomorphic function* S: $(\mathbf{C}^n, 0) \to \mathbf{C}$ is defined as the dimension of the complex local ring Q: $\mu = \dim_{\mathbf{C}} Q$, $Q = H/(\partial S/\partial x_1, \ldots, \partial S/\partial x_n)$, where H is the ring of germs of functions of $x_1, \ldots, x_n$ holomorphic at 0.

2.2.6. REMARK. For a critical point of finite multiplicity, the rings of germs of smooth functions, C^∞ and H, in Definitions 2.2.1 and 2.2.5, can be replaced by the rings of formal power series or even the ring of truncated polynomials (factored by the ideal consisting of polynomials having a zero of sufficiently high order at 0).

All of this follows from the preparation theorem of Weierstrass and Malgrange (see [7]–[9]).

2.2.7. REMARK. If it is finite, the multiplicity μ may be defined also as the degree of the gradient mapping at the critical point, or as the degree of the following mapping of a small sphere with centre at the critical point onto itself: $z \to \|z\| \, (\partial S/\partial z)/\|\partial S/\partial z\|$. Concerning this see [10], [11].

2.3. **Sufficient jets.** We recall that the k-jet of a function at a point is defined, in terms of a fixed coordinate system, as the section of the Taylor series of the function at the point up to the terms of degree k inclusive.

2.3.1. DEFINITION. The k-jet of a function at a point is said to be *sufficient* if all functions having the given k-jet are differentiably equivalent to each other in the neighbourhood of the point.

It is clear that sufficiency is a property of the jet and not of the function.

2.3.2. EXAMPLE. The 2-jet of the function $\pm x_1^2 \pm \ldots \pm x_n^2$ of n variables at 0 is sufficient (Morse's Lemma).

2.3.3. EXAMPLE. The 3-jet of the function $x^2 y$ of two variables at zero is not sufficient, since the germs of $x^2 y$ and $x^2 y + y^4$ are not equivalent.

2.3.4. PROPOSITION. *Let* 0 *be a critical point of finite multiplicity of a smooth (real or complex) function. Then for sufficiently large k, the k-jet of this function at 0 is sufficient.*

In other words, *in a neighbourhood of a critical point the function is smoothly equivalent to its Taylor polynomial.*

The proof of Proposition 2.3.4 is by no means trivial. Apparently it was first proved by Tougeron [9]. There are other proofs (in the complex and real cases) in [8], [12], [13].

2.3.5. PROPOSITION. *The set of all ∞-jets of functions at critical points of infinite multiplicity has infinite codimension in the space of all jets of functions.*

In other words, the Taylor coefficients of a function at a critical point of infinite multiplicity are subject to an infinite number of independent (polynomial) conditions.

The proof of Proposition 2.3.5 is not difficult. Propositions 2.3.4, 2.3.5 and the transversality theorem imply that *every finite-parameter family of functions can be disturbed by arbitrarily small displacements in such a way that all functions in the disturbed family have critical points of finite multiplicity only.*

Thus, the investigation of critical points of functions that are not removable in finite-parameter families reduces to that of critical points of polynomials of finite multiplicity.

2.3.6. COROLLARY. *For almost every*[1)] *finite-parameter family of functions S the singularity index at every point and for every value of the parameter is a rational number.*

The proof follows immediately from Proposition 2.1.5, by virtue of Propositions 2.3.4 and 2.3.5.

2.4. Stabilization.

2.4.1. DEFINITION. The germs at zero of two smooth functions of m and n variables are said to be *stably equivalent* if there exist non-degenerate quadratic forms which yield on direct addition differentiably equivalent germs of smooth functions in the same number of variables.

2.4.2. EXAMPLE. The germ at zero of the function x^3 of one variable is stably equivalent to the germ at zero of the function $x^3 + y^2 + z^2$ of three variables.

2.4.3. PROPOSITION. *If germs of functions of the same number of variables are stably equivalent, they are equivalent* (see [8], [14]).

Thus, transition to classes of stably equivalent functions allows us to assume that all functions have "one and the same number of variables", and nothing is changed for the classification of functions of a fixed number of variables.

2.4.4. PROPOSITION. *Stably equivalent germs have the same singularity*

[1)] Excluding the union of finitely or countably many nowhere dense sets in the space of families of functions.

indices β, isomorphic local rings Q, and the same multiplicities μ of critical points.

The proofs are obvious.

2.5. Versal deformations. Versal deformations are the normal forms to which a family of functions can be brought in the neighbourhood of a parameter value corresponding to a degenerate critical point.

2.5.1. DEFINITION. A *deformation* of the germ of a smooth mapping $f\colon (\mathbf{R}^n, 0) \to \mathbf{R}^k$ is the germ of a smooth mapping $F\colon (\mathbf{R}^n \times \mathbf{R}^l, 0 \times 0) \to \mathbf{R}^k$ for which $F(., 0) = f$.

2.5.2. DEFINITION. By an *equivalence H of deformations F, G*: $(\mathbf{R}^n \times \mathbf{R}^l, 0 \times 0) \to \mathbf{R}$ of a function f we mean a deformation H: $(\mathbf{R}^n \times \mathbf{R}^l, 0 \times 0) \to \mathbf{R}^n$ of the identity mapping of $\mathbf{R}^n$ onto itself such that $F(x, \lambda) \equiv G(H(x, \lambda), \lambda)$.

2.5.3. DEFINITION. A deformation $K\colon (\mathbf{R}^n \times \mathbf{R}^k, 0 \times 0) \to \mathbf{R}$ of the germ of a function f at zero is said to be *induced from the deformation $F\colon (\mathbf{R}^n \times \mathbf{R}^l, 0 \times 0) \to \mathbf{R}$ under a change of parameter $\phi\colon (\mathbf{R}^k, 0) \to (\mathbf{R}^l, 0)$* if $K(x, \lambda) \equiv F(x, \phi(\lambda))$.

2.5.4. DEFINITION. A deformation F of the germ of a function f at 0 is said to be *versal* if every other deformation of f is equivalent to the deformation induced from F.

In other words, every deformation G can be expressed in terms of F,

$$G(x, \lambda) \equiv F(H(x, \lambda), \phi(\lambda)),$$

where H is a diffeomorphism depending smoothly on the parameter λ, ϕ is a smooth change of parameter, $\phi(0) = 0$, and $H(x, 0) = x$.

Differentiation of the preceding formula leads to the following concept.

2.5.5. DEFINITION. An l-parameter deformation F of the germ of a function f of n variables at 0 is said to be *infinitesimally versal* if every germ α of a smooth function of n variables has a decomposition

$$\alpha = \sum_{i=1}^{n} \frac{\partial f}{\partial x_i} h_i + \sum_{j=1}^{l} c_j g_j,$$

where the h_i are germs of smooth functions of n variables, the c_j are numbers, and the g_j are germs of the partial derivatives of F with respect to the parameters, taken at the value zero of these parameters.

2.5.6. PROPOSITION. *Every infinitesimally versal deformation of the germ of a function is versal.*

The proof of this "versality theorem" is by no means trivial: see [3], [15]–[19].

2.5.7. COROLLARY. *The germ of a function at a critical point of finite multiplicity μ has a μ-parameter versal deformation.*

It is sufficient to take this deformation of f:

$$F(x, \lambda) = f(x) + \Sigma\, \lambda_j g_j(x),$$

where $\{g_j\}$ is a set of μ functions giving a basis of the linear space Q of the local ring of f at zero after factorization.

2.5.8. EXAMPLE. The deformations $x^2 + \lambda_1$, $x^3 + \lambda_1 x + \lambda_2$, $x^4 + \lambda_1 x^2 + \lambda_2 x + \lambda_3$ are versal deformations of the functions x^2, x^3 and x^4, respectively, at zero.

The corresponding trigonometric integrals have been known in physics for a long time (see [20], [21], [22]).

2.6. Modality. Let G be a Lie group, X a smooth manifold, and $G \times X \to X$ an action of the group on the manifold. We consider any point x of X. The orbits of the action of G on X may form a finite stratification or a continuous family in the neighbourhood of x. We define the modality of a point x as the number of parameters necessary for number numbering the orbits in the neighbourhood of x.

2.6.1. DEFINITION. The *modality of a point $x \in X$ under the action of G on X* is the smallest number m such that a sufficiently small neighbourhood of x in X intersects at most finitely many at most m-parameter families of orbits of G.

2.6.2. EXAMPLE. A point is of modality 0 if and only if a sufficiently small neighbourhood of it intersects only finitely many orbits.

2.6.3. DEFINITION. The germ of a function f: $(\mathbf{R}^n, 0) \to (\mathbf{R}, 0)$ is said to be *m-modal* if some (and hence every) sufficient k-jet of the function at 0 is modal under the action of the group of k-jets of diffeomorphisms $(\mathbf{R}^n, 0) \to (\mathbf{R}^n, 0)$ on the manifold of k-jets of functions at the critical point 0 and with critical value 0.

2.6.4. EXAMPLE. The germ at zero of the function x^k of one variable is 0-modal since a sufficiently small neighbourhood of its k-jet intersects only finitely many orbits (namely the orbits of germs of the functions $\pm x^l$, where $l \leqslant k$).

So far, all the 0-modal and 1-modal germs have been calculated. It turns out that there are not too many of them, and indeed an analysis of the tables of 0-modal and 1-modal germs enables us to obtain the asymptotic behaviour of trigonometric integrals of general form with not too many parameters.

2.6.5. REMARK. *The modalities of stably equivalent germs are equal.* (The proof is obvious.)

In studying the modality of a germ of multiplicity μ we may replace the function space of germs by a μ-dimensional space, namely that with the versal deformations as basis, and examine how many parameters are needed to number the orbits in a neighbourhood of the parameter value zero. Another finite-dimensional space, and one that is even more suitable for this end, is a local transversal to the orbit in the space of k-jets of functions with the critical point 0 and critical value 0. Such a transversal is of dimension $\mu - 1$ and can be obtained by taking the vectors in a basis of the linear space $\mathfrak{m}^2/\mathfrak{m}(\partial f/\partial x)$ as tangent vectors to the transversal surface, where $\mathfrak{m}$ is a maximal ideal in the ring of germs of functions of x.

2.6.6. EXAMPLE. A convenient transversal to an orbit of the germ of x^4 is the system of germs $x^4 + \lambda_1 x^3 + \lambda_2 x^2$.

2.7. **The quadratic form of a singularity**. Some further topological invariants are connected with a critical point of a complex function. One of these is a manifold with boundary, the so-called local level manifold of the function. It can be defined as follows (see [23]).

Let f: $(\mathbf{C}^n, 0) \to (\mathbf{C}, 0)$ be a smooth function in a neighbourhood of 0, where 0 is an isolated critical point with the critical value 0. We fix a Hermitian metric in $\mathbf{C}^n$. We choose a number $\rho > 0$ so small that all spheres of radius $r \leqslant \rho$ centred at 0 intersect the singular set of the level set $f^{-1}(0)$ transversally. Fixing ρ, we choose a number $\varepsilon > 0$ so small that the sphere of radius ρ intersects transversally also all non-singular sets of the level sets $f^{-1}(z)$ for $|z| \leqslant \varepsilon$.

2.7.1. DEFINITION. *The local manifold at the level $z \neq 0$ of a function f at the critical point 0 is the set*

$$V_z = \{x \in \mathbf{C}^n \colon f(x) = z, \ |x| \leqslant \rho\},$$

that is, the intersection of the non-singular manifold of the level set $f^{-1}(z)$ of f with the ball of radius ρ and centre at the origin of coordinates.

It is easy to see that V_z is a smooth manifold with boundary, and that V_z has the real dimension $2n - 2$. Further, it is easy to check that the differentiable type of V_z does not depend on the choice of ρ and ε, provided only that these are small enough ($\rho < \rho_0(f)$, $\varepsilon < \varepsilon_0((\rho, f))$.

2.7.2. EXAMPLE. If $f(x) = x_1^2 + \ldots + x_n^2$, then V_z is diffeomorphic to the space of vectors of length $\leqslant 1$ in the tangent bundle of the sphere S^{n-1}. For example, for $n = 2$ V_z is an ordinary cylinder.

2.7.3. PROPOSITION (see 23). *The local level manifold V_z is homotopy equivalent to the union of as many spheres of dimension $n - 1$ as the multiplicity μ of the critical point*:

$$V_z \sim \underbrace{S^{n-1} \vee \ldots \vee S^{n-1}}_{\mu}.$$

Thus, the $(n - 1)$-dimensional homology group of the local level manifold is the μ-dimensional lattice $\mathbf{Z}^\mu$.

2.7.4. DEFINITION. By the *quadratic form* of the germ of a function at a critical point we mean the intersection index in the homology of mean dimension of the local level manifold of a function of $n \equiv 3 \bmod 4$ variables and stably equivalent to the given function:

$$H_{n-1}(V_z, \mathbf{Z}) \times H_{n-1}(V_z, \mathbf{Z}) \to \mathbf{Z}.$$

2.7.5. REMARK. The intersection index of $(n - 1)$-dimensional cycles gives a symmetric form for odd n and a skew-symmetric form for even n. For $n \equiv 3 \bmod 4$ the square of the generator in the homology group $H_{n-1}(V_z, \mathbf{Z}) \sim \mathbf{Z}$ is -2 for the function $f = x_1^2 + \ldots + x_n^2$.

The quadratic module (H_{n-1} together with the symmetric integral form) is determined by the stable equivalence class of f.

2.8. The monodromy group. The local level manifold of the germ of a function at a critical point serves as the fibre in certain fibrations. Here is the construction.

We fix some versal deformation F of the germ of f, say a μ-parameter one for definiteness. Let $\mathbf{C}^\mu$ be the space of parameters; we fix a Hermitian metric in it.

Take $\lambda \in \mathbf{C}^\mu$. Consider the set of zero level of $F(\cdot, \lambda)$:

$$V_\lambda = \{x \in \mathbf{C}^n: F(x, \lambda) = 0, \ |x| \leqslant \rho\}.$$

The set V_λ intersects the sphere S_ρ transversally if $|\lambda| \leqslant \varepsilon$, where ε is sufficiently small. However, it may have singularities inside this sphere.

2.8.1. DEFINITION. By the *bifurcation germ* Σ of the germ of f at $0 \in \mathbf{C}^n$ we mean the germ at $0 \in \mathbf{C}^\mu$ of the subset of a versal deformation consisting of those points λ of the parameter space for which 0 is a critical value of $F(\cdot, \lambda)$ in the ball $|x| \leqslant \rho$.

The intersection of a representative of the bifurcation germ with the ball $|\lambda| \leqslant \varepsilon$ for sufficiently small ε is called a *bifurcation diagram* Σ_ε. This complex hypersurface is not, in general, a smooth manifold. The topological type of the pair $(\Sigma_\varepsilon, D_\varepsilon)$ (where D_ε is the ball $|\lambda| \leqslant \varepsilon$) does not depend on ε, as long as ε is sufficiently small; neither does it depend on the special choice of the μ-parameter versal deformation. In particular, the complement $D_\varepsilon - \Sigma_\varepsilon$ is a base space for a fibration with fibre V_λ and projection $(x, \lambda) \to \lambda$.

2.8.2. EXAMPLE. For the germ at zero of the function x^{k+1} of a single variable we can take as a versal deformation the polynomial.

$$F(x, \lambda) = x^{k+1} + \lambda_1 x^{k-1} + \lambda_2 x^{k-2} + \ldots + \lambda_k \quad (\mu = k).$$

A bifurcation diagram is obtained by choosing the coefficients in such a way that the polynomial $F(\cdot, \lambda)$ has multiple roots. The equation of the bifurcation diagram then has the form $\Delta(\lambda_1, \ldots \lambda_k) = 0$, where Δ is the discriminant. This hypersurface is called a *swallow-tail*, and its complement is the space $K(\pi, 1)$ for the Artin braid group with $k + 1$ threads.

2.8.3. DEFINITION. *The braid group* of the germ of a function f at a critical point is the fundamental group $\pi_1(D_\varepsilon - \Sigma_\varepsilon)$ of the complement of the bifurcation diagram.

Since the fundamental group of the base space of a fibration acts on the homology of the fibre, we get a representation

$$\pi_1(D_\varepsilon - \Sigma_\varepsilon) \to \operatorname{Aut} H_{n-1}(V_\lambda, \mathbf{Z})$$

of the braid group in the group of automorphisms of the quadratic module.

This representation is called the *monodromy representation* and its image the *monodromy group of the germ.*

Similar actions arise for the homology with closed supports and other groups connected with the fibre. In defining monodromy it is useful to bear in mind that every element of the fundamental group of the base space gives rise to a diffeomorphism of the manifold V_λ with boundary, *which is the identity in some neighbourhood of the boundary.*

2.9. Classical monodromy. The monodromy group of the germ of a function has a significant subgroup, closely connected with the asymptotic behaviour of trigonometric integrals.

We incorporate a function f: $(\mathbf{C}^n, 0) \to)\mathbf{C}, 0)$ with isolated critical point 0 in a one-parameter system of functions $F(x, \lambda) = f(x) - \lambda$.

2.9.1. DEFINITION. The *classical monodromy* of the germ of f is the automorphism

$$h\colon H_{n-1}(V_\lambda, \mathbf{Z}) \to H_{n-1}(V_\lambda, \mathbf{Z})$$

of the homology of the local level manifold induced by a small rotation of the parameter λ around 0 ($\lambda(t) = e^{it}\lambda(0)$, $0 \leqslant t \leqslant 2\pi$).

The one-parameter deformation F of f indicated above is induced from a versal deformation (as is every deformation of f). The rotation of λ about 0 carries it into a certain remarkable element of the braid group of the germ.

This element can also be defined in terms of a special system of generators of the braid group, which is constructed as follows.

We fix a sphere of sufficiently small radius in the space $\mathbf{C}^\mu$ of parameters of the versal deformation. Almost every line sufficiently near to 0 intersects the bifurcation diagram transversally within the sphere in exactly μ points near 0.

2.9.2. PROPOSITION (see [24], [25], [26]). *The fundamental group of the complement of the bifurcation diagram in the sphere is generated by μ paths lying in a line L as above and encircling the μ points of intersection with the bifurcation diagram (generators of $\pi_1(L - \Sigma)$).*

The classical monodromy is generated by the product of these generators, that is, by a path on the complex line encircling each of the μ points of the bifurcation diagram once.

On the other hand, having fixed the initial value of λ on a chosen line in general position, we can *approach* each of the μ bifurcation values of the parameter instead of encircling them. The local level set corresponding to the bifurcation value of the parameter passes through a non-degenerate critical point. For values of the parameter near to a bifurcation value, the part of V_λ near such a critical point is diffeomorphic to the tangent bundle of a sphere (see Example 2.7.2). In this way a differentiable sphere (zero section of this bundle) is defined in V_λ, a so-called *vanishing cycle*, or *cycle vanishing at the given critical point.*

On the arc along which we proceeded from the initial point λ to the neighbourhood of a bifurcation value of the parameter, the sets V_λ form a smooth fibration. Therefore we can define a vanishing cycle in every fibre, among them the initial one.

The sphere S^{n-1} obtained in the initial fibre V_λ is determined up to a smooth isotopy leaving the boundary of V_λ fixed, and depends on the homotopy class of paths joining the initial point to the bifurcation value.

The standard basis of the fundamental group of the complex line with μ points removed thus determines a set of vanishing cycles in the homology group group of mean dimension of the local manifold V_λ (unique apart from the sign).

2.9.3. PROPOSITION. (Tyurina, Lamotke). *The μ vanishing cycles constructed above generate $H_{n-1}(V_\lambda, \mathbf{Z})$; moreover, they can be realized by embeddings of spheres $S^{n-1} \subset V_\lambda$ such that V_λ is homotopy equivalent to the union of these spheres.*

The basis of vanishing cycles just indicated is very convenient for calculation with vanishing cycles.

2.9.4. PROPOSITION (Picard, Lefschetz). *Consider any path joining the initial parameter value λ to a bifurcation value as described above, let $e \in H_{n-1}(V_\lambda, \mathbf{Z})$ be the corresponding vanishing cycle, and let γ be an element of the braid group of the singularity corresponding to the circuit around the bifurcation value along the indicated path, encircling the bifurcation value and returning along the old path. Then the action γ_* of γ on the homology of the local level manifold is described by the formula*

$$\gamma_* c = c + (c, e)e,$$

where

$$c \in H_{n-1}(V_\lambda, \mathbf{Z}), \qquad n \equiv 3 (\mathrm{mod}\ 4).$$

This formula is valid also when c is a cycle with closed support.

2.9.5. PROPOSITION. *The classical monodromy of a singularity is the action of the product of the standard generators of the braid group of the singularity (if the standard non-intersecting generators of $\pi_1(L - \Sigma)$ go around μ bifurcation values "anticlockwise", then their product is also "anticlockwise", in the order of the arguments of the starting directions of the generators).*

The proof is obvious, because this is the way in which the path is expressed in terms of μ generators encircles all the bifurcation points once each.

§3. Classification theorems

In this section we list the normal forms to which functions can be brought in the neighbourhood of not too complicated critical points. In particular, we list

1) all 0-modal germs,
2) all 1-modal germs,
3) all elliptical germs (with a definite quadratic form),
4) all parabolic germs (with a semidefinite quadratic form),
5) all hyperbolic germs (with a quadratic form of positive index of inertia 1).

The more than 1-modal germs form a set whose codimension (in the space of germs of functions of n variables at the critical point 0 with zero critical value) is equal to

10 for $n \geqslant 4$, 11 for $n = 3$, 12 for $n = 2$, ∞ for $n = 1$.

We also list all singularities with a critical point of multiplicity $\mu < 14$ (*the critical values are everywhere assumed to be zero*).

These classification theorems reduce the determination of the asymptotic behaviour of trigonometric integrals with functions depending, in general, on not too many parameters to the investigation of a finite number of normal forms.

3.1. 0-modal singularities. We consider the following functions:

$$\begin{aligned} A_k:&\quad f(x) = \pm x^{k+1}, \quad k \geqslant 1,\\ D_k:&\quad f(x, y) = x^2 y \pm y^{k-1}, \quad k \geqslant 4,\\ E_6:&\quad f(x, y) = x^3 \pm y^4,\\ E_7:&\quad f(x, y) = x^3 + xy^3,\\ E_8:&\quad f(x, y) = x^3 + y^5. \end{aligned}$$

3.1.1. THEOREM ([13]). *Every 0-modal germ is stably equivalent to one of the germs of A, D, or E at zero; these germs are themselves 0-modal.*

3.1.2. THEOREM ([13]). *The braid group of a 0-modal germ is a generalized Artin braid group (as defined by Brieskorn* [14], [17]) *of the corresponding Weyl groups of A_k, D_k and E_k.*

Brieskorn's definition goes like this. We consider a finite group W acting on $\mathbf{R}^n$ and generated by reflections in planes passing through the origin of hyperplanes (mirrors). We complexify $\mathbf{R}^n$ and the mirrors; then W acts also on the complex space $\mathbf{C}^n$ so obtained.

3.1.3. DEFINITION. *The complex fundamental domain* of W is the complement in $\mathbf{C}^n$ of the complex mirrors factored by the action of W.

3.1.4. DEFINITION. The *Artin group of W* is the fundamental group of the complex fundamental domain.

3.1.5. THEOREM ([3], [28]). *The complement to the bifurcation diagram of a germ of type A, D or E is homeomorphic to the complex fundamental domain of the corresponding Coxeter group.*

The Coxeter groups of the series A, D, E can be defined as follows. We consider graphs with k points (the so-called Dynkin diagrams):

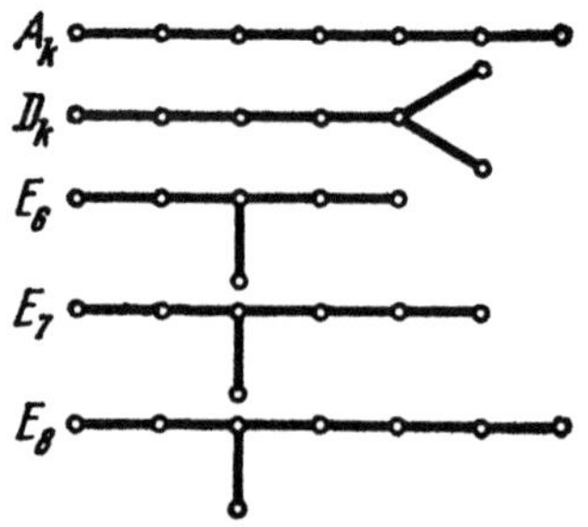

To each graph we assign the lattice $\mathbf{Z}^k$ whose generators $e_1, \ldots, e_k$ correspond to the vertices of the graph. We define an integral bilinear form on the lattice as follows:

1) the square (e_i, e_i) of each generator is -2.

2) the product (e_i, e_j) of two generators is 1 or 0 according as the corresponding points of the graph are joined by an edge or not.

3.1.6. DEFINITION. The *Coxeter group* $A_k(D_k, E_k)$ is a group of linear transformations of the linear space spanned by the lattice that preserve the relevant form, and is generated by k reflections acting according to the formula

$$c \mapsto c + (c, e)e, \quad e = e_1, \ldots, e_k.$$

It is easy to check that this formula gives the reflection in the mirror orthogonal to the vector e, since $(e, e) = -2$. The quadratic forms corresponding to these graphs are negative definite.

3.1.7. THEOREM ([3], [28], [29]). *The quadratic forms of the singularities of type A, D, and E are equivalent to the forms constructed above, and a system of vanishing cycles serves as the basis e_k. The monodromy of each of the singularities of type A, D and E is a natural homomorphism of the Artin group into the Coxeter group.*

3.1.8. THEOREM ([3]). *The complement to the bifurcation diagram of singularities of type A, D and E is the space $K(\pi, 1)$ for the Artin group of the corresponding Coxeter group.*

The proof is fairly simple in the cases A and D, while for the case E we have to use a theorem of Deligne [30].

Consider now a product of the generating reflections of the Coxeter group, a so-called *Coxeter element.* Coxeter elements for different orders of the product of the generators are conjugate, and the order of a Coxeter element is called the *Coxeter number.* This invariant can be defined in many other ways (for instance, as the highest degree of a generator in the ring of polynomial invariants of the action of the group). It is not hard to check that the Coxeter numbers of the groups A_k, D_k, E_k are $k + 1$, $2k - 2$, 12, 18, 30.

3.1.9. THEOREM ([1]). *The singularity index of functions of type A_k, D_k or E_k is given by the formula $\beta = \frac{1}{2} - \frac{1}{N}$, where N is the corresponding*

Coxeter number.

The classical monodromy of singularities of types A_k, D_k or E_k is of finite order equal to the corresponding Cexeter number. The connection between N and β is not surprising, since after an N-fold circuit h around the origin the trigonometric integral returns to its original value. However, the only proofs of Theorem 3.1.9 known to me are not based on this fact, but on a direct calculation for each individual case.

3.1.10. THE CLASSIFICATION THEOREM ([3]). *Every germ of a function at a critical point of finite multiplicity is stably equivalent either to one of the germs of type A_k, D_k, E_k, or else to a germ in one of the following three sets:*

P: germs of functions of three or more variables with a 2-jet zero,

X: germs of functions of two variables with a 3-jet zero,

J: germs of functions of two variables with the 5-jet $x^3 + axy^4$.

The codimensions of the sets of germs of the types A_k, D_k, E_k (in the manifold of jets of functions at the critical point 0 with the critical value 0) is $k - 1$, and the codimensions of P, X and J are 6, 7 and 8, respectively.

Theorems 3.1.9, 3.1.10 and Thom's transversality theorem [31] immediately give the next result:

3.1.11. COROLLARY ([1]). *For $l < 6$ every l-parameter family of functions can be brought by a small displacement to general position in such a way that the germ of every function of the family at every critical point is stably equivalent to one of the germs of A_k, D_k, E_k* (+ const), $k \leqslant 6$.

The singularities of these types are not removable by small displacements. In particular, the only irremovable singularities in one-parameter families are of type A_2 (that is, x^3), in two-parameter families only A_3, etc.:

Number of parameters	0	1	2	3	4	5
Singularities	A_1	A_2	A_3	A_4, D_4	A_5, D_5	A_6, D_6, E_6

Thus, all singularities of codimension <6 are 0-modal.

Combining Corollary 3.1.11 and Theorem 3.1.9 we obtain:

3.1.12. COROLLARY ([1]). *The following singularities have the largest singularity index β among singularities of codimension l:*

l	1	2	3	4	5
Singularity	A_2	A_3	D_4	D_5	E_6
β_l	1/6	1/4	1/3	3/8	5/12

3.1.13. DEFINITION. We say that a set B in the space of jets *borders* a set C, and we write $B \leftarrow C$, if every neighbourhood of every point of C contains a point of B.

3.1.14. THEOREM ([3]). *All borders of the orbits of 0-modal germs are contained in the following diagram*:

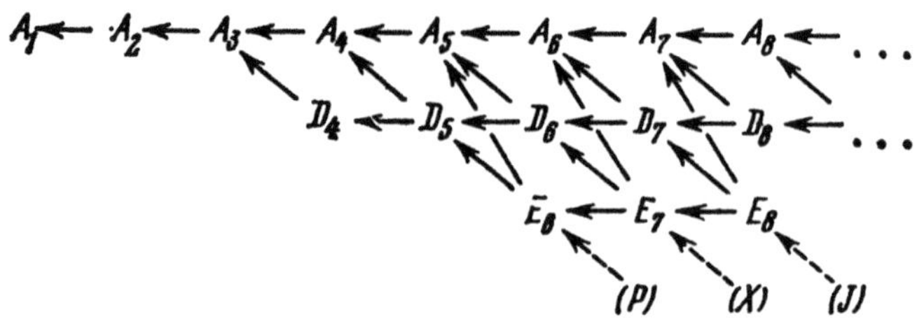

3.1.15. DEFINITION. The germ of a function at an isolated critical point is said to be *elliptic* if the quadratic form (given by the intersection index in the homology of mean dimension of the local level manifold of a function of $n \equiv 3 \pmod 4$ variables stably equivalent to the given function) is negative definite.

3.1.16. THEOREM. *All elliptic singularities occur among the* A_k, $k \geqslant 1$, D_k, $k \geqslant 4$, E_6, E_7 and E_8.

This theorem was proved by Tyurina in [32] on the basis of arguments about the semicontinuity of indices of inertia.

Thus, the 0-modal singularities are exactly the elliptic singularities. Unfortunately, there is no direct proof ot this fact: the result is obtained by matching lists of singularities of various types and obtained in different ways.

3.2. **Parabolic and hyperbolic singularities.** To decrease the number of variants we restrict ourselves from now on to a complex classification.

3.2.1. DEFINITION. A germ of a function at an isolated critical point is said to be *parabolic* if the corresponding quadratic form is semidefinite. From the classification of unimodal singularities given below we obtain the next result:

3.2.2. THEOREM ([33]). *Every parabolic germ is stably equivalent to one of the germs of the following three one-parameter families*:

$$P_8\colon\ f(x, y, z) = x^2z + y^3 + \varepsilon y^2 z + az^3, \quad a(4\varepsilon + 27a) \neq 0, \quad \varepsilon^3 = \varepsilon,$$
$$X_9\colon\ f(x, y) = x^4 + \varepsilon x^2y^2 + ay^4, \quad a(4a - \varepsilon^2) \neq 0, \quad \varepsilon^3 = \varepsilon,$$
$$J_{10}\colon\ f(x, y) = x^3 + \varepsilon x^2y^2 + ay^6, \quad a(4\varepsilon + 27a) \neq 0, \quad \varepsilon^3 = \varepsilon.$$

The germs of these families are parabolic.

This was stated by Milnor as a conjecture based on the paper [34] of Wagreich, in which these germs arose in another connection. The Corollaries 3.2.3, 3.2.4 and 3.2.6 below were also stated by Milnor.

3.2.3. COROLLARY. *The quadratic forms of the parabolic germs* P_8, X_9, J_{10} *are obtained from the quadratic forms of* E_6, E_7, E_8 *respectively, by adding a zero quadratic form in two variables. In particular, the dimensions of the zero spaces of the quadratic form of a parabolic germ is equal to* 2.

3.2.4. COROLLARY. *The monodromy group of a parabolic germ has an abelian subgroup* $\mathbf{Z}^{12}$ *(or* $\mathbf{Z}^{14}$, $\mathbf{Z}^{16}$, *respectively) whose factor group is isomorphic to the Coxeter group for* E_6, E_7, *or* E_8.

Both corollaries can easily be obtained from a consideration of the Dynkin diagram of the vanishing cycles found by Gabrielov.

3.2.5. THEOREM ([35]). *There is a basis of vanishing cycles of a parabolic germ such that the monodromy group is generated by reflections in the orthogonal complements to the generators, and the Dynkin diagram has the elevant form*

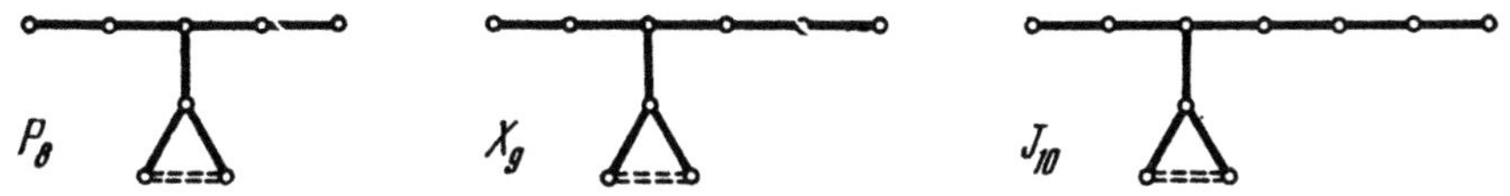

where the double dotted line indicates an intersection index +2.

3.2.6. COROLLARY ([33]). *Parabolic germs are defined by the following property: in a sufficiently small neighbourhood of a sufficient k-jet there are only finitely many orbits of dimension higher than that of the given jet (and infinitely many of the same dimension).*

3.2.7. COROLLARY. *The singularity index* β *of all parabolic germs* P_8, X_9, J_{10} *is* 1/2.

3.2.8. DEFINITION. The germ of a function at an isolated critical point is said to be *hyperbolic* if the positive index of inertia of the corresponding quadratic form (in the homology H_{n-1}, $n \equiv 3 \bmod 4$) is 1.

The classification of unimodal singularities to be given later yields:

3.2.9. THEOREM ([33]). *Every hyperbolic germ is stably equivalent to one of the germs in the following one-parameter families*:

$$T_{k,l,m}\colon\ f(x, y, z) = axyz + x^k + y^l + z^m, \qquad a \neq 0,$$

where

$$\frac{1}{k} + \frac{1}{l} + \frac{1}{m} < 1.$$

The germs in these families are hyperbolic,

$$\mu = k + l + m - 1.$$

The quadratic forms and the monodromy groups of these germs were found by Gabrielov when he proved the following theorem.

3.2.10. THEOREM. *For every hyperbolic germ there exists a basis of vanishing cycles such that the monodromy group is generated by reflections in the orthogonal complements to the generators, and the Dynkin diagram*

has the form

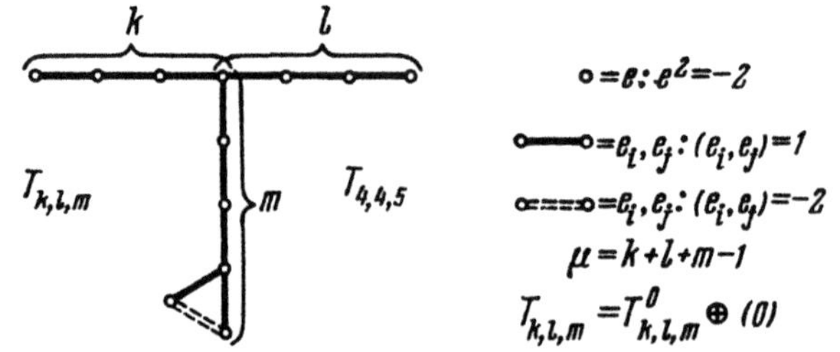

3.2.11. COROLLARY. *The quadratic form of a hyperbolic germ is the direct sum of a zero form in one variable and the form* $T^0_{k,l,m}$ *with indices of inertia* $\mu_+ = 1$, $\mu_- = \mu - 2$ *given by a T-shaped Dynkin diagram (the left lower vertex is omitted from the preceding diagram)*:

$$T_{k,\,l,\,m} = T^0_{k,\,l,\,m} \oplus (0).$$

3.2.12. REMARK. The sum $\mu_+ + \mu_0$ of the number of positive and zero squares in the canonical form of the quadratic form turns out to be even in all examples known to me.

3.3. **Unimodal germs.** We call germs of modality 1 unimodal (see Definition 2.6.3).

3.3.1. THEOREM ([33]). *Every unimodal germ is either parabolic, or hyperbolic, or else stably equivalent to a germ in one of the following 14 exceptional one-parameter families (throughout, the parameter a runs over* $\mathbb{C}$):

$$\begin{aligned}
Q_{10}&: \; x^2z + y^3 + ayz^3 + z^4,\\
Q_{11}&: \; x^2z + y^3 + yz^3 + az^5,\\
Q_{12}&: \; x^2z + y^3 + ayz^4 + z^5,\\
S_{11}&: \; x^2z + yz^2 + y^4 + ay^3z,\\
S_{12}&: \; x^2z + yz^2 + xy^3 + ay^5,\\
U_{12}&: \; x^3 + y^3 + z^4 + 3axyz^2,\\
Z_{11}&: \; x^3y + 3axy^4 + y^5,\\
Z_{12}&: \; x^3y + 3xy^4 + ay^6,\\
Z_{13}&: \; x^3y + 3axy^5 + y^6,\\
W_{12}&: \; x^4 + y^5 + 2ax^2y^3,\\
W_{13}&: \; x^4 + 4xy^4 + ay^6,\\
K_{12}&: \; x^3 + axy^5 + y^7,\\
K_{13}&: \; x^3 + axy^5 + y^8,\\
K_{14}&: \; x^3 + xy^6 + ay^8.
\end{aligned}$$

All germs of the types listed (parabolic, hyperbolic and exceptional) are 1-modal.

The quadratic forms and monodromy groups of the exceptional germs were calculated by Gabrielov.[1)]

3.3.2. THEOREM. *For every exceptional germ there exists a basis of vanishing cycles such that the Dynkin diagram takes the form*

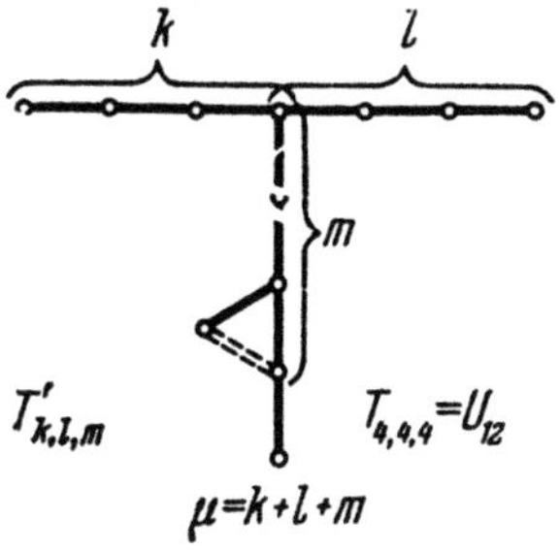

$$
\begin{array}{lll}
Q_{10}\sim(3,3,4), & Q_{11}\sim(3,3,5), & Q_{12}\sim(3,3,6),\\
Z_{11}\sim(2,4,5), & Z_{12}\sim(2,4,6), & Z_{13}\sim(2,4,7),\\
S_{11}\sim(3,4,4), & S_{12}\sim(3,4,5), & U_{12}\sim(4,4,4),\\
W_{12}\sim(2,5,5), & W_{13}\sim(2,5,6), & K_{12}\sim(2,3,7),\\
K_{13}\sim(2,3,8), & K_{14}\sim(2,3,9). &
\end{array}
$$

It follows, in particular, that the quadratic form of an exceptional germ is non-degenerate and has the positive index of inertia 2:

$$T'_{k,\,l,\,m} = T^0_{k,\,l,\,m} \oplus \begin{pmatrix} 0 & 1\\ 1 & 0\end{pmatrix}.$$

The proofs of Theorems **3.2.2**, **3.2.9** and **3.3.1** are based on a fairly cumbersome calculation of the orbits of the action of the group of diffeomorphisms on the space of jets of functions. To get the stated theorems it proved necessary to carry the calculations a little further so as to include the simplest non-unimodal germs.

According to the classification theorem **3.1.10** we have to deal with three cases:

P: germs of functions of three or more variables with a zero 2-jet,

X: germs of functions of two variables with a zero 3-jet,

J: germs of functions of two variables with a 5-jet $x^3 + axy^4$.

3.3.3. THEOREM. *Every germ of a function of two variables with a 5-jet $x^3 + axy^4$ lies in one of the following classes:*

$$
\begin{array}{l}
J^8_{10} \leftarrow J^9_{11} \leftarrow J^{10}_{12} \leftarrow \dots\\
\qquad\quad \nwarrow\\
\qquad\qquad K^{10}_{12} \leftarrow K^{11}_{13} \leftarrow K^{12}_{14} \leftarrow K^{13}_{16} \leftarrow K^{14}
\end{array}
$$

Here the letters denote classes of germs equivalent to germs in the following list:

J^8_{10}: $x^3 + \varepsilon x^2y^2 + ay^6$, $a(4\varepsilon + 27a) \neq 0$, $\varepsilon^3 = \varepsilon$,

J^{p+2}_{k+4}: $x^3 + \varepsilon x^2y^2 + ay^p$, $a \neq 0$, $\varepsilon^2 = 1$, $p > 6$,

K^{10}_{12}: $x^3 + axy^5 + y^7$, $a \in \mathbf{C}$,

K^{11}_{13}: $x^3 + xy^5 + ay^8$, $a \in \mathbf{C}$,

K^{12}_{14}: $x^3 + axy^6 + y^8$, $a \in \mathbf{C}$,

K^{13}_{16}: $x^3 + xy^6 + ay^9 + by^{10}$, $27a^2 + 4 \neq 0$.

---

1) A special technique was developed for the computation of the quadratic forms. A sketch of the level lines of the real form of the Morse function obtained by a disturbance of a degenerate critical point turned out to be very useful for this. Using this method, S. M. Husein–Zade has obtained the Dynkin diagrams of many singularities of the form $x^2 + f(y, z)$.

The letters a and b denote parameters. In the neighbourhood of a general point of the line (plane) of parameters the same values of the parameter may give rise to non-equivalent germs. Further, $\mathbf{K}^{14}$ *denotes a certain algebraic submanifold of codimension* 14 *(in the manifold of jets of functions at the critical point* 0 *with the critical value* 0*). In general, the upper index refers to the codimension of the set formed by orbits of the given type in the relevant manifold of jets. The lower index denotes the multiplicity* μ *of the critical point.*

The polynomials listed give sufficient k-jets. The singularity of J_{10}^8 *is parabolic, while that of* J_{p+4} *for* $p > 6$ *is hyperbolic (and is equivalent to* $T_{2,3,p}$ *in Theorem* 3.2.9*).*

The arrows in the diagram above denote borderings (see Definition 3.1.13). Not all of these are indicated, but only the most important ones. In particular, the theorem asserts that *every germ of type J not occurring in the above list, that is, a germ in* $\mathbf{K}^{14}$*, can be transformed by an arbitrarily small disturbance of a finite jet to a germ equivalent to one of the germs* $\mathbf{K}_{16}^{13}$. For any such germ $\mu \geqslant 16$, and the modality is $\geqslant 2$. By a comparison with Theorem 3.1.10 we came to the next result:

3.3.4. COROLLARY. *The germs in the above list are unimodal, apart from* $\mathbf{K}_{16}^{13}$*. The germs* $\mathbf{K}_{16}^{13}$ *are bimodal. All the remaining germs with a 5-jet* $x^3 + axy^4$ *(that is, germs in* $\mathbf{K}^{14}$*) have modality at least* 2.

We consider now functions of two variables of type X (with a zero 2-jet).

3.3.5. THEOREM. *Every germ of a function of two variables with zero 2-jet lies in one of the following classes:*

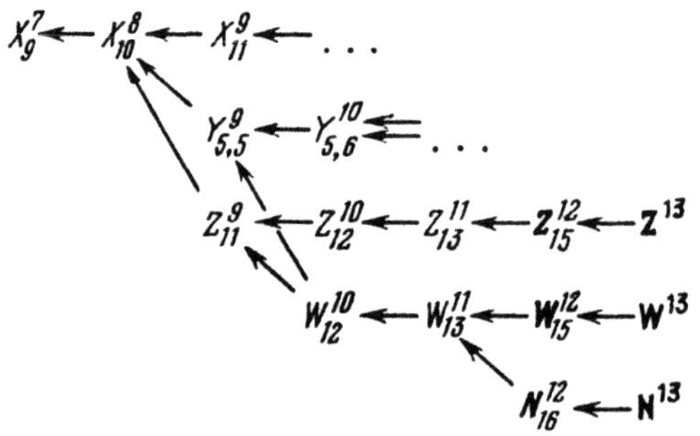

Here classes of germs are denoted by letters corresponding to the germs in the following list:

X_9^7: $x^4 + \varepsilon x^2y^2 + ay^4$, $a(4a - \varepsilon^2) \neq 0$, $\varepsilon^3 = \varepsilon$,
X_{p+5}^{p+3}: $x^4 + x^2y^2 + ay^p$, $a \neq 0$, $p > 4$,
$Y_{p,q}^{p+q-1}$: $x^2y^2 + x^p + ay^q$, $5 \leqslant p \leqslant q$, $a \neq 0$, $\mu = p + q + 1$,
Z_{11}^9: $x^3y + 3axy^4 + y^5$, $a \in \mathbf{C}$,
Z_{12}^{10}: $x^3y + 3xy^4 + ay^6$, $a \in \mathbf{C}$,
Z_{13}^{11}: $x^3y + 3axy^5 + y^6$, $a \in \mathbf{C}$,
Z_{15}^{12}: $x^3y + 3xy^5 + ay^7 + by^8$, $a^2 + 4 \neq 0$,

W_{12}^{10}: $x^4 + y^5 + 2ax^2y^3, \quad a \in \mathrm{C},$
W_{13}^{11}: $x^4 + 4xy^4 + ay^6, \quad a \in \mathrm{C},$
W_{15}^{12}: $x^4 + 2x^2y^3 + ay^6 + by^7, \quad a^2 \neq a,$
N_{16}^{12}. $x^4y + ax^3y^2 + bx^2y^3 + xy^4 + cx^3y^3, \quad \Delta(a,b) \neq 0.$

The arrows denote borderings (again, not all of them are indicated). In particular, arrows from $\mathbf{Z}^{13}$, $\mathbf{W}^{13}$ and $\mathbf{N}^{13}$ mean that 1) *all germs of functions of two variables with a zero 2-jet and not equivalent to germs in the list form a subset of codimension* 13 (*in the space of jets of functions with the critical value* 0). 2) *in every neighbourhood of a jet in this subset there is a jet of type* $\mathbf{Z}_{15}^{12}$, $\mathbf{W}_{15}^{12}$ *or* $\mathbf{N}_{16}^{12}$. *The above polynomials give sufficient jets. The singularity* X_9^7 *is parabolic,* X_{p+5}^{p+3} $(p > 4)$ *and* $Y_{p,q}$, $5 \leqslant p$ $5 \leqslant p \leqslant q$, *are hyperbolic* (*they are equivalent to* $T_{2,4,p}$ *and* $T_{2,p,q}$, *respectively, of Theorem* 3.2.9).

We note that small displacements of a germ of class X can produce only germs of classes A, D, E and J in Theorem 3.1.10.

3.3.6. COROLLARY. *The germs of the above list are unimodal, apart from* $\mathbf{Z}_{15}^{12}$, $\mathbf{W}_{15}^{12}$ *and* $\mathbf{N}_{16}^{12}$. *The germs* $\mathbf{Z}_{15}^{12}$ *and* $\mathbf{W}_{15}^{12}$ *are bimodal, and* N_{16}^{12} *are trimodal. The remaining germs, which are stably equivalent to germs of two variables with a zero 2-jet* (*that is, to germs in* $\mathbf{Z}^{13}$, $\mathbf{W}^{13}$ *or* $\mathbf{N}^{13}$), *have modality not less than* 2.

We turn finally to functions of type P (the corank of their second differential is 3 or more).

3.3.7. THEOREM. *Every germ of a function at a critical point at which the second differential has corank* 3 *or more is stably equivalent to a germ in one of the following classes*:

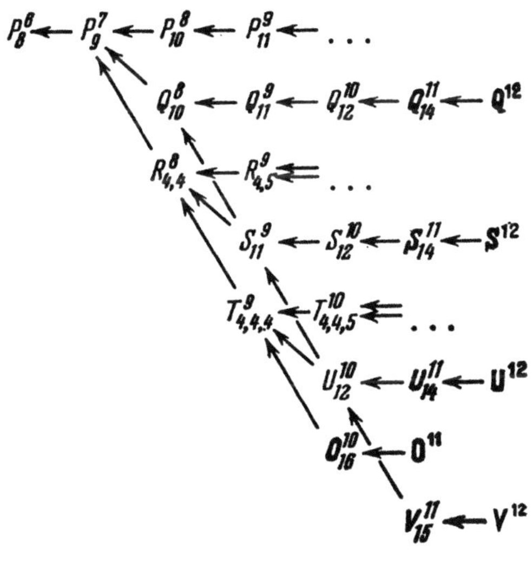

The letters denote germs in the following list:

P_8^6: $x^2z + y^3 + \varepsilon y^2 z + az^3$, $a\,(4\varepsilon + 27a) \neq 0$, $\varepsilon^3 = \varepsilon$,
P_{p+5}^{p+3}: $x^2z + y^3 + y^2z + az^p$, $a \neq 0$, $p > 3$,
Q_{10}^8: $x^2z + y^3 + ayz^3 + z^4$, $a \in \mathbf{C}$,
Q_{11}^9: $x^2z + y^3 + yz^3 + az^5$, $a \in \mathbf{C}$,
Q_{12}^{10}: $x^2z + y^3 + ayz^4 + z^5$, $a \in \mathbf{C}$,
Q_{14}^{11}: $x^2z + y^3 + yz^4 + az^6 + bz^7$, $27a^2 + 4 \neq 0$,
$R_{p,q}^{p+q}$: $x^3 + 3xyz + y^p + az^q$, $4 \leqslant p \leqslant q$, $\mu = p + q + 2$, $a \neq 0$,
S_{11}^9: $x^2z + yz^2 + y^4 + ay^3z$, $a \in \mathbf{C}$,
S_{12}^{10}: $x^2z + yz^2 + xy^3 + ay^5$, $a \in \mathbf{C}$,
S_{14}^{11}: $x^2z + yz^2 + y^3z + ay^5 + by^6$, $a \neq 0$, $a \neq 1/4$,
$T_{p,q,r}^{p+q+r-3}$: $axyz + x^p + y^q + z^r$, $4 \leqslant p \leqslant q \leqslant r$, $\mu = p + q + r - 1$, $a \neq 0$,
U_{12}^{10}: $x^3 + y^3 + z^4 + 3axyz^2$, $a \in \mathbf{C}$,
U_{14}^{11}: $x^3 + y^3 + 3xz^3 + 3ayz^3 + 9byz^4$, $a^3 \neq 1$,
V_{15}^{11}: $x^2y + z^4 + y^2z^2 + ay^3z + by^4 + cy^4z$, $a(12b + 1)\Delta(a, b) \neq 0$,
O_{16}^{10}: $x^3 + y^3 + z^3 + u^3 + (ax + by + cz + du)^3 + exyzu$, $\Delta(a, b, c, d) \neq 0$,

where Δ is the discriminant.

The notation is as in the preceding theorem. In particular, the theorem asserts that *germs not equivalent to one in the list form the set* $\mathbf{Q}^{12} + \mathbf{S}^{12} + \mathbf{U}^{12} + \mathbf{V}^{12} + \mathbf{O}^{11}$ *of codimension* 11 *and can be brought by sufficiently small displacements to germs stably equivalent to germs of type* Q_{14}^{11}, S_{14}^{11}, U_{15}^{11}, V_{15}^{11} *or* O_{16}^{10}.

The singularity of P_8^6 is parabolic, those of P, R, and T are hyperbolic $(P_{p+5} \sim T_{3,3,p},\ R_{p,q} \sim T_{3,p,q})$.

A comparison with Theorems 3.3.3, 3.3.5 and 3.3.7 gives the next result:

3.3.8. COROLLARY. *The germs in the above list are unimodal, apart from Q_{14}^{11}, S_{14}^{11}, U_{14}^{11} (which are bimodal), V_{15}^{11} (trimodal) and O_{16}^{10} (5-modal). The remaining germs, which are stably equivalent to germs of functions of more than 2 variables with a zero 2-jet (that is, to germs from $\mathbf{Q}^{12}$, $\mathbf{S}^{12}$, $\mathbf{U}^{12}$, $\mathbf{V}^{12}$ and $\mathbf{O}^{11}$), have modality not less than 2.*

3.3.9. COROLLARY. *The following table is a classification of critical points of codimension $l < 10$:*

l	0	1	2	3	4	5	6	7	8	9	10
	A_1	A_2	A_3	A_4	A_5	A_6	A_7	A_8	A_9	A_{10}	A_{11}
				D_4	D_5	D_6	D_7	D_8	D_9	D_{10}	D_{11}
						E_6	E_7	E_8	J_{10}	J_{11}	J_{12}
											K_{12}
								X_9	X_{10}	X_{11}	X_{12}
										$Y_{5,5}$	$Y_{5,6}$
										Z_{11}	Z_{12}
											W_{12}
							P_8	P_9	P_{10}	P_{11}	P_{12}
									Q_{10}	Q_{11}	Q_{12}

l	0	1	2	3	4	5	6	7	8	9	10
									$R_{4,4}$	$R_{4,5}$	$R_{5,5}$
											$R_{5,6}$
										S_{11}	S_{12}
										$T_{4,4,4}$	$T_{4,4,5}$
											U_{12}
											O_{16}

in the sense that the set of all germs not stably equivalent to germs in this list has codimension 11 *(in the set of jets of functions with the critical point* 0 *and the critical value* 0*). For* $s < 11$ *every s-parameter family of functions can be brought by a sufficiently small displacement to "general position" in such a way that the germs of all functions of the family at all critical points are stably equivalent to germs in the columns in the preceding table for which* $l \leqslant s$ *(up to additive constants).*

From these theorems we can extract a list of all singularities with $\mu < 14$. Concerning these theorems, we note that we have not considered the question whether different singularities in our table may not be topologically equivalent to each other. In the majority of cases, we can easily deduce that they are non-equivalent merely from the values of these invariants. Doubt remains only for certain members of the three-index series $T_{k,l,m}$ (it is not clear whether k, l, m can be recovered from the quadratic forms). We note too that the parameter a in the normal form of non-parabolic unimodal singularities can be replaced by 0 for 1, if we are not classifying functions but hypersurfaces (or functions up to a change of coordinates in preimage and image). Notwithstanding the disappearance of the "Modulus" a, the singularities remain unimodal also under the new equivalence relations (because they are close to parabolic singularities).

§4. Singularity indices

The classification theorems in the preceding sections reduce the investigation of the asymptotic behaviour of trigonometric integrals with functions in general position and depending, in general, on not too many parameters to a discussion of explicitly given polynomial cases. The calculation of the singularity index in each of these special cases reduces in principle to a finite number of arithmetical steps.

The calculations can be carried out without difficulty in the cases where the polynomial that determines the normal form is homogeneous or weighted homogeneous. From the results of the calculation for the weighted homogeneous case we can deduce the values of the singularity indices for all singularities of modality $\leqslant 1$, on the basis of a few propositions about the semicontinuity of the singularity index. The results so obtained can also be checked independently of the unproved assumptions. The author is grateful to A. Ya. Povzner for doing the checking.

4.1. Conjectures concerning the semicontinuity of the singularity index. Let $S\colon \mathbf{R}^n \to \mathbf{R}$ be a polynomial with the critical point 0, and let β be the singularity index of S at zero (see Definition 2.1.1). Consider any polynomial deformation $F\colon \mathbf{R}^n \times \mathbf{R}^l \to \mathbf{R}$ of the germ of S at zero.

4.1.1. PROPOSITION ([1]). *For every $\varepsilon > 0$ the trigonometric integral with a function $F(\cdot, \lambda)$ satisfies the inequality*

$$\left| \int\limits_{\mathbf{R}^n} e^{iF(x,\lambda)/h} \varphi(x, \lambda)\, dx \right| \leqslant C(\varphi, \varepsilon)\, h^{\frac{n}{2}-\beta-\varepsilon}$$

for every C^∞-function ϕ concentrated in a sufficiently small neighbourhood of the origin in $\mathbf{R}^n \times \mathbf{R}^l$.

If the critical point is of finite multiplicity, then it is enough to consider just one versal μ-parameter deformation F.

4.1.2. EXAMPLE. *For singularities of the series A_k, Proposition 4.1.1 takes the form of an estimate uniform in $|\lambda| \leqslant \delta$, $|h| \leqslant h_0$:*

$$\left| \int\limits_{-\infty}^{\infty} e^{\frac{i}{h}(x^{k+1}+\lambda_1 x^{k-1}+\ldots+\lambda_k)} \cdot \varphi\, dx \right| \leqslant C(\varphi, \delta)\, h^{\frac{1}{k+1}-\varepsilon}$$

As A. A. Karatsuba has informed me, an estimate of this form (even for $\varepsilon = 0$) has been established by I. M. Vinogradov [36] (and with different methods, by Hua Lo Ken [37]). In the general form, 4.1.1 remains unproved.

4.1.3. PROPOSITION. *The singularity index β is semicontinuous in the sense that the singularity indices of all jets of functions at the critical point lying in a sufficiently small neighbourhood of a sufficient jet s does not exceed the singularity index $\beta(s)$ of s.*

In particular, when the critical point is split up into simpler ones, the singularity index of each one of the new points is not greater than that of the original point. Proposition 4.1.3 follows from 4.1.1; it has not been proved, in general.

4.1.4. PROPOSITION. *Suppose that we are given a family of functions depending smoothly on some parameters and having 0 as a critical point for all values of the parameters. Assume further that the multiplicity of the critical point 0 does not change under a change of parameters. Then the singularity indices at 0 of all functions of the family are the same.*

This proposition also remains unproved, it has been proved only that the quadratic form of the singularity does not change along the family (Tyurina [32]), nor (for $n \neq 3$) does the complex topological type of the germ (Lê Dũng Tráng and Ramanujam [38]). In this connection it is natural to conjecture that the singularity index is a topological invariant of the complex germ. It would follow, in particular, that *the singularity indices of all real forms with one and the same set of complex germs are identical.*

4.2. **The rationality of the numbers β_l.** The symbol β_l denotes the singularity index of the "strongest" singularity not removable in l-parameter families.

4.2.1. DEFINITION ([1]). Let $l \geqslant 0$ be an integer. Then β_l denotes the greatest lower bound of the numbers γ having the property that every γ-parameter family of functions can be brought by a small displacement to a family for which the estimate

$$\left|\int_{\mathbf{R}^n} e^{\frac{i}{h}F(x,\lambda)}\varphi(x,\lambda)\,dx\right| \leqslant C(\varphi)\,h^{\frac{n}{2}-\gamma}$$

holds uniformly in the parameter $\lambda \in \mathbf{R}^l$, for all sufficiently small $h > 0$ and all ϕ with compact support.

4.2.2. EXAMPLE. By Morses's lemma, $\beta_0 = 0$. Further, $\beta_1 = 1/6$, as can be seen by considering Airy integrals (in which F -- $x^3 + \lambda_1 x + \lambda_2$).

4.2.3. THEOREM. *It follows from* 4.1.1 *and* 4.1.-, *that all the β_l are rational.*

PROOF. It is not hard to extract a semi-algebraic set of codimension more than l from the space of jets of functions with the critical point 0 and the critical value 0 such that the complement consists of jets that are sufficient everywhere and splits into finitely many smooth manifolds along each of which the multiplicity μ is constant. Along each of these manifolds the singularity index β is also constant (Proposition 4.1.4). By 2.1.5 this index is rational. Since β_l is the maximum of these numbers, it is itself rational.

4.3. **Singularity indices of quasi-homogeneous singularities.** A polynomial $S(x_1, \ldots, x_n)$ is said to be *quasi-homogeneous with weights* $\alpha_1, \ldots, \alpha_n$ if $S(\lambda^{\alpha_1}x_1, \ldots, \lambda^{\alpha_n}x_n) \equiv \lambda S(x_1, \ldots, x_n)$, $0 < \alpha_k \leqslant 1/2$.

For example, $x^a + y^b + z^c$ has weights $1/a$, $1/b$, $1/c$, and a homogeneous polynomial of degree a has weight $1/a$ in each argument. See Saito [39] for quasi-homogeneous functions, where it is proved, in particular, that the germ of a function at a critical point is contained in the ideal generated by its partial derivatives if and only if the germ is equivalent to a quasi-homogeneous germ.

4.3.1. PROPOSITION. *A quasi-homogeneous polynomial with weights $\alpha_1, \ldots, \alpha_n$ having* 0 *as an isolated critical point has the singularity index*

$$\beta = \frac{n}{2} - \sum_{k=1}^{n} \alpha_k = \sum_{k=1}^{n}\left(\frac{1}{2} - \alpha_k\right) \text{ there.}$$

We remark here that the normal forms of all 0-modal (that is, elliptic) singularities and all parabolic singularities are quasi-homogeneous.

Computing β from the preceding formula, we get in the elliptic case of Theorem 3.1.9: $\beta = \dfrac{1}{2} - \dfrac{1}{N}$, where N is the Coxeter number, and in the parabolic case $\beta = 1/2$, so that here we must count the Coxeter number as being ∞.

Hyperbolic singularities are not homogeneous, and we consider their singularity indices β separately. As regards the remaining families of singularities listed in §3, many of them have quasi-homogeneous representatives, and this together with Proposition **4.1.4** enables us to compute β for every family. In particular, this can be done for all the unimodal exceptional singularities listed in §3. In these cases the singularity index exceeds 1/2 by the reciprocal of a natural number: $\beta = \frac{1}{2} - \frac{1}{N}$, $N \in \mathbf{Z}$, $N < 0$, that is, the "Coxeter number" defined by the same formula as in the elliptic case (and, in general, rational) turns out to be a *negative integer* for these singularities.

4.3.2. PROPOSITION. *The weights and singularity indices of the singularities listed in* §3 *are given by the following table*:

Type	Q_{10}	Q_{11}	Q_{12}	Q_{14}	Z_{11}	Z_{12}	Z_{13}	Z_{15}	S_{11}	S_{12}
$\lvert N\rvert \alpha_k$	9,8,6	7,6,4	6,5,3	5,4,2	15,8,6	11,6,4	9,5,3	7,4,2	6,5,4	5,4,3
$\lvert N\rvert$	24	18	15	12	30	22	18	14	16	13
β	$\frac{13}{24}$	$\frac{5}{9}$	$\frac{17}{30}$	$\frac{7}{12}$	$\frac{8}{15}$	$\frac{6}{11}$	$\frac{5}{9}$	$\frac{4}{7}$	$\frac{9}{16}$	$\frac{15}{26}$

U_{12}	S_{14}	U_{14}	V_{15}	W_{12}	W_{13}	W_{15}	N_{16}	K_{12}	K_{13}	K_{14}	K_{16}
4,4,3	4,3,2	3,3,2	3,2,2	10,5,4	8,4,3	6,3,2	5,2,2	21,14,6	15,10,4	12,8,3	9,6,2
12	10	9	8	20	16	12	10	42	30	24	18
$\frac{7}{12}$	$\frac{3}{5}$	$\frac{11}{18}$	$\frac{5}{8}$	$\frac{11}{20}$	$\frac{9}{16}$	$\frac{7}{12}$	$\frac{3}{5}$	$\frac{11}{21}$	$\frac{8}{15}$	$\frac{13}{24}$	$\frac{5}{9}$

The correctness of β can be checked by using finite computations that do not use the unproved general propositions.

4.3.3. REMARK. The singularity in $K_{12}(x^3 + y^7 + z^2)$ is connected with a regular 56-hedron of genus 3 with triangular faces, having 84 edges meeting 7 at a time in each of 24 vertices (and with congruence subgroup mod 7) in much the same way as the singularity $E_8(x^3 + y^5 + z^2)$ is connected with the icosahedron with congruence subgroup mod 5 (see [40]).

4.4. Singularity indices of hyperbolic germs. Hyperbolic germs are stably equivalent to the germs $T_{k,l,m}$: $axyz + x^k + y^l + z^m$, $a \neq 0$, in the complex domain (see §3). We denote the function xyz of three variables by

T_∞. The semi-continuity hypothesis in **4.1.1** gives

4.4.1. PROPOSITION. *For every hyperbolic germ,* $\beta(T_{k,l,m}) \leqslant \beta(T_\infty)$.

This proposition can be checked using calculations not running into the indicated general hypothesis. On the other hand, the following fact can be verified directly:

4.4.2. PROPOSITION. $\beta(T_\infty) = 1/2$.

Finally, from the semi-continuity hypothesis of **4.1.1** and the fact that $\beta = 1/2$ for parabolic singularities (see §3) we deduce:

4.4.3. PROPOSITION. *The singularity index of every non-elliptic singularity is at least* $1/2$.

Combining the last three propositions we obtain the following one (which can likewise be verified independently of the conjecture **4.1.1**):

4.4.4. COROLLARY. *The singularity index of every hyperbolic germ is* $1/2$.

4.5. The strongest singularities. Combining the values of the singularity index with the list of the singularities in Corollary 3.3.9, we can give for every $l \leqslant 10$ all singularity indices for singularities in l-parameter families in general position on submanifolds of the space of parameters. Selecting for each l the singularity in 3.3.9 with the largest singularity index, we came to the following values for β_l:

l	0	1	2	3	4	5	6	7	8	9	10, $n=3$	11, $n=3$	10, $n \geqslant 4$
Type	A_1	A_2	A_3	D_4	D_5	E_6	P_8	X_9, P_9	Q_{10}	S_{11}	U_{12}	V_{15}	O_{16}
β_l	0	$1/6$	$1/4$	$1/3$	$3/8$	$5/12$	$1/2$	$1/2$	$13/24$	$9/16$	$7/12$	$5/8$	$2/3$
N	+2	+3	+4	+6	+8	+12	∞	∞	−24	−16	−12	−8	−6

This reduced table contains the maximal singularity indices in general l-parameter families of functions of $n \geqslant 4$ variables for $l \leqslant 10$ and functions of 3 variables for $l \leqslant 11$. For functions of two variables the table can be extended to the case $l = 12$:

l	0	1	2	3	4	5	6	7	8	9	10	11	12
Type	A_1	A_2	A_3	D_4	D_5	E_6	E_7	X_9	X_{10}, J_{10}	Z_{11}	W_{12}	W_{13}	N_{16}
β_l	0	$1/6$	$1/4$	$1/3$	$3/8$	$5/12$	$4/9$	$1/2$	$1/2$	$8/15$	$11/20$	$9/16$	$3/5$
N	+2	+3	+4	+6	+8	+12	+18	∞	∞	−30	−20	−16	−10

The calculation of the exact value of β_l for all l would be extremely difficult. On the basis of experimental evidence, I believe that one can guess the asymptotic behaviour for large l.

CONJECTURE. *A non-degenerate cubic form in k variables is the singularity with maximum singularity index β for its codimension (that is, for $l = k(k+1)/2$). In other words, $\beta_{k(k+1)/2} = k/6$.*

We remark that the estimate $\beta_{k(k+1)/2} \geqslant k/6$ already follows from the semi-continuity conjecture. Note also that a sum of cubes appears to be the "strongest" among singularities with the same codimension not only in the sense of the singularity index β, but also of the other parameters (μ and m).

Singularities of the type of a sum of k cubes (for $k = 1$ this is A_2, for $k = 2$ it is D_4, for $k = 3$ it is P_8, and for $k = 4$ it is O_{16}) have multiplicity $\mu = 2^k$, modality $m = 2^k - \dfrac{k(k+1)}{2} - 1$, and they form a manifold of codimension $l = k(k+1)/2$ (in the space of jets of functions with the critical point 0 and the critical value 0). Thus, we can guess that the maximum multiplicity μ_l of singularities occurring in general l-parameter families and the maximum modality m_l both grow like $2^{\sqrt{(2l)}}$ as $l \to \infty$, while β_l grows like $\sqrt{(2l)}/6$.

References

[1] V. I. Arnol'd, Integrals of quickly oscillating functions and singularities of projections of Lagrange manifolds, Funktsional. Anal. i Prilozhen. 6:3 (1972), 61–62. MR

[2] V. P. Maslov, Focussing energy in a crystal lattice, Uspekhi Mat. Nauk 27:6 (1972), 224.

[3] V. I. Arnol'd, Normal forms of functions near degenerate critical points, the Weyl groups A_k, D_k, E_k, and Lagrange singularities, Funktsional. Anal. i Prilozhen. 6:4 (1972), 3–25. MR

[4] E. Brieskorn, Groupes des tresses, Séminaire Bourbaki 24:401, November 1971.

[5] I. N. Bernshtein and S. I. Gel'fand, Meromorphy of the function P^λ, Funktsional. Anal. i Prilozhen. 3:1 (1969), 84–86. MR 40 #723.

[6] M. F. Atiyah, Resolution of singularities and division of distributions, Comm. Pure. Appl. Math. 23 (1970), 145–150. MR 41 #815.

[7] B. Malgrange, Ideals of differentiable functions, Oxford Univ. Press, London 1967. MR 35 #3446.
Translation: *Idealy differentsiruemykh funktsii*, Izdat. "Mir", Moscow 1968.

[8] V. I. Arnol'd, Singularities of smooth mappings, Uspekhi. Mat. Nauk. 23:1 (1968), 3–44. MR 37 #2243.
= Russian Math. Surveys 23 (1968), 1–43.

[9] J. C. Tougeron, Idéaux de fonctions differentiables. I, Ann. Inst. Fourier (Grenoble) 18 (1968), fasc.1, 177–240. MR 39 #2171.

[10] V. P. Palamodov, On the multiplicity of a holomorphic transformation, Funktsional. Anal. i. Prilozhen. 13 (1967), 54–65. MR 38 #4720.

[11] V. P. Palamodov, Remarks on differentiable maps of finite multiplicity, Funktsional. Anal. i Prilozhen. **6**:2 (1972), 52–61. MR **45** #7732.
[12] A. M. Samoĭlenko, The equivalence of a smooth function to a Taylor polynomial in the vicinity of a critical point of finite type, Funktsional. Anal. i Prilozhen. **2**:4 (1968), 63–69. MR **42** #5132.
[13] M. Artin, On the solution of analytic equations, Invent. Math. **5** (1968), 277–291. MR **38** #344.
[14] A. Weinstein, Singularities of families of functions, Differential-geometrie im Grossen, Oberwolfach **4** (1971).
[15] John Mather, Structural stability of mappings, Singularities of differentiable maps, 216–267. Izdat "Mir", Moscow 1968 (in Russian). MR **40** #4966.
[16] F. Latour, Stabilité des champs d'applications différentiables: Généralisation d'un théorème de J. Mather, C.R. Acad. Sci. (Paris) Sér. A **268** (1969), 1331–1334. MR **39** #7617.
[17] J. Guckenheimer, Catastrophes and partial derivative equations, Princeton 1972, 1–29.
[18] V. M. Zakalyukin, A versality theorem, Funktsional. Anal. i Prilozhen. 7:2 (1973), 28–32. MR
[19] G. N. Tyurina, Locally semi-universal flat deformations of isolated singularities in complex spaces, Izv. Akad. Nauk. SSSR Ser. Mat. **33** (1969), 1026–1058. MR **40** # 5903.
= Math. USSR–Izv.
[20] A. Fresnel, Mémoire sur la diffraction de la lumière, Mém. de l'Acad. des sciences **5** (1818), 339–353.
[21] G. B. Airy, Intensity of light in a neighbourhood of a caustic, Trans. Cambridge Philos. Soc. **6** (1838), 379–403.
[22] T. Pearcey, The structure of an electromagnetic field in the neighbourhood of a cusp of a caustic, Philos. Mag. **37** (1946), 311–315.
[23] J. Milnor, Singular points of complex hypersurfaces, Ann. of Maths. Studies No. 61, Princeton University Press, Princeton, New Jersey 1968. MR **39** #969.
Translation: *Osobye tochki kompleksnykh giperpoverkhnostei*, Izdat. "Mir", Moscow 1971.
[24] O. Zariski, On the Poincaré group of a projective hypersurface, Ann. of Math. (2) **38** (1937), 131–141.
[25] A. N. Varchenko, Theorems on the topological equisingularity of systems of algebraic varieties and systems of polynomial mappings, Izv. Akad. Nauk. SSSR Ser. Mat. **36** (1972), 957–1019.
= Maths. USSR–Izv.
[26] H. Hamm and Lê Dũng Tráng, Un théorème du type de Lefschetz, C.R. Acad. Sci. (Paris) Sér. A **272** (1971), 946–949. MR **43** #8094.
[27] E. Brieskorn, Die Fundamentalgruppe des Raumes der regulären Orbits einer endlichen komplexen Spiegelungsgruppe, Invent. Math. **12** (1971), 57–61.
[28] E. Brieskorn, Singular elements of semi-simple algebraic groups, Proc. Internat. Congress Mathematicians, Nice 1970.
[29] G. N. Tyurina, Resolution of singularities of flat deformations of double rational points, Funktsional. Anal. i Prilozhen. **4**:1 (1970), 77–83. MR **42** #2031.
[30] P. Deligne, Les immeubles des groupes des tresses généralisés, Invent. Math. **17** (1972), 273–302.

[31] René Thom and Harold I. Levine, Singularities of differentiable mappings, Singularities of differentiable maps, 9–101, Izdat. "Mir", Moscow 1968 (in Russian). MR **40** # 896.
[32] G. N. Tyurina, The topological properties of isolated singularities of complex spaces of codimension one, Izv. Akad. Nauk. SSSR Ser. Mat **32** (1968), 605–620. MR **37** # 3053.
[33] V. I. Arnol'd, Classification of unimodal critical points of functions, Funktsional. Anal. i Prilozhen. 7:3 (1973), 75–76.
[34] P. Wagreich, Singularities of complex surfaces with solvable fundamental group, Topology **11** (1972), 51–72.
[35] A. M. Gabrielov, Intersection matrices for certain singularities, Funktsional. Anal. i Prilozhen. 7:3 (1973), 18–32.
[36] I. M. Vinogradov, *Metod trigonometricheskikh summ v teorii chisel* (The method of trigonometric sums in number theory), Izdat. "Nauka", Moscow 1971.
[37] L. K. Hua, On the number of solutions of Tarry's problem, Acta Scientica Sinica **1** (1953), 1–76.
[38] Lê Dũng Tráng and C. P. Ramanujan, The invariance of Milnor's number implies the invariance of the topological type, École Polytechnique, Paris 1973.
[39] K. Saito, Quasihomogene isolierte Singularitäten von Hyperflächen, Invent. Math. **14** (1971), 123–142.
[40] F. Klein, Vorlesungen über die Entwicklung der Mathematik im 19. Jahrhundert, Springer–Verlag, Berlin 1926.
Translation: *Lektsii o razvitii matematiki v xix stoletii*, ONTI, Moscow 1937.

Received 6 June 1973

Translated by J. Wiegold

NORMAL FORMS FOR FUNCTIONS NEAR DEGENERATE CRITICAL POINTS, THE WEYL GROUPS OF A_k, D_k, E_k AND LAGRANGIAN SINGULARITIES

V. I. Arnol'd

There exist interesting connections between the simplest degenerate critical points of functions and the simple Lie algebras A_k, D_k, E_k (or at least their Weyl groups). In the present paper it is shown that critical points which are simple (without moduli) are classified by the series A_k, D_k, E_k. All degeneracies of codimension not greater than 5 turn out to be simple. Hence this paper also contains the classification of all degenerate critical points of codimension not greater than 5.

Critical points of functions are closely connected with singularities of projections of lagrangian manifolds. At the end of the paper, the classification of simple singularities of lagrangian maps is given. In particular, a table is computed of normal forms of projections of lagrangian manifolds in general position up to dimension 6, where moduli first appear.

§1. Introduction

The behavior of a function near a nondegenerate critical point is determined by Morse's lemma: an appropriate choice of coordinate functions will lead to the morsian normal form $-x_1^2 - \ldots - x_\nu^2 + x_{\nu+1}^2 + \ldots + x_n^2 + C$. It is also known that near an isolated critical point, even if it is degenerate, an analytic function is transformed into its Taylor polynomial by an appropriate analytic change of variables (see, e.g., [1], [2]).

In this sense, a function near an isolated critical point can always be reduced to polynomial normal form. However, there is an essential difference between this polynomial normal form and the morsian one: the coefficients of the Taylor polynomial are continuous parameters for the polynomial normal forms, while the morsian normal forms are determined by a discrete parameter (the index of inertia ν).

It turns out, that for the simplest degeneracies one can give normal forms similar to the morsian ones, i.e., which do not involve continuous parameters. In the present paper the classification of degeneracies of codimension less than 6 is given. We shall give a finite list of normal forms such that in typical $l < 6$-parameter families of functions no singularities occur other than those enumerated.

Carrying out an analogous classification of all critical points of codimension 6 is impossible, since here the normal forms must inevitably contain parameters ("moduli"). However, there are some special degeneracies (we shall call them simple) of any codimension, near which there are no moduli (for precise definitions see § 2). The basic result of this paper is the classification of all simple critical points.

A complete list of normal forms of a function in the neighborhood of a simple critical point appears as follows (we assume that the critical value is equal to zero):

$$\begin{array}{lll} A_k: f = \pm x_1^{k+1} \pm x_2^2 + Q, & k \geqslant 1, & \operatorname{codim} A_k = k-1, \\ D_k: f = x_1^2 x_2 \pm x_2^{k-1} + Q, & k \geqslant 4, & \operatorname{codim} D_k = k-1, \\ E_6: f = x_1^3 \pm x_2^4 + Q, & & \operatorname{codim} E_6 = 5, \\ E_7: f = x_1^3 + x_1 x_2^3 + Q, & & \operatorname{codim} E_7 = 6, \\ E_8: f = x_1^3 + x_2^5 + Q, & & \operatorname{codim} E_8 = 7, \end{array} \qquad (1.1)$$

Moscow State University. Translated from Funktsional'nyi Analiz i Ego Prilozheniya, Vol. 6, No. 4, pp. 3-25, October-December, 1972. Original article submitted June 21, 1972.

Originally publ. in: Funkts. Anal. Prilozh. 6(4), 3-25, © Russ. Acad. Sciences 1972
English transl.: Funct. Anal. Appl. 6(4), 254-272, © Consultants Bureau, New York 1972
Reprint by Springer-Verlag Berlin Heidelberg 2016 with kind permission.

where Q denotes the standard quadratic form

$$Q = -x_3^2 - \dots - x_s^2 + x_{s+1}^2 + \dots + x_n^2.$$

The connection between these series and the series of simple Lie groups A_k (linear), D_k (orthogonal) and E_k is discussed in §9.

All singularities of codimension $l < 6$ are simple. They are indicated exhaustively in the following table:

l	1	2	3	4	5
Singularity	A_2	A_3	A_4, D_4	A_5, D_5	A_6, D_6, E_6

(1.2)

In codimension $l \geq 6$ with $n \geq 3$ variables, along with the singularities A_k, D_k, E_k one meets singularities of other types: their normal forms inevitably contain parameters. For $n = 2$, the parameters show up beginning with codimension 7.

§2. Formulation of Results

We shall consider smooth real functions with critical point $0 \in \mathbf{R}^n$ and critical value 0 (the complex, R-analytic and formal cases differ with slight simplifications). The group of germs at 0 of local diffeomorphisms of the space $\mathbf{R}^n$ which leave 0 fixed, acts on the space of germs of our functions at 0. We are interested in the orbits of this action. Orbits "in general position" have codimension 0: they consist of functions with nondegenerate critical points. The remaining orbits have positive codimension.

Orbits of positive codimension can form discrete "stratifications" or continuous families. In the first case we shall call the orbits simple. In order to give the exact definition, we shall first consider the action $G \times M \to M$ of a finite-dimensional Lie group G on a finite-dimensional manifold M.

Definition 2.1. An orbit W of the action of G on M abuts an orbit V at some (and then any) point $v \in V$, if any neighborhood of the point v in M meets W.

Definition 2.2. An orbit V is called simple, if a sufficiently small neighborhood of one (and then any) of its points v meets only a finite number of orbits.

Remark 2.3. If the action of G on M is algebraic, then a neighborhood of a point v meets either a finite number of orbits, or a continuous family.

Remark 2.4. If in definitions 2.1 and 2.2 one replaces neighborhoods by local transversals, then these definitions will apply to orbits of finite codimension even in the infinite-dimensional case.

We consider as M the manifold of r-jets at the point 0 of functions with critical point $0 \in \mathbf{R}^n$ and critical value 0. On this manifold the group of r-jets of diffeomorphisms $\mathbf{R}^n \to \mathbf{R}^n$ which leave 0 fixed, acts algebraically.

Definition 2.5. A germ of a function f with critical point $0 \in \mathbf{R}^n$ and critical value 0 will be called simple if for sufficiently large r, 1) its orbit in the space of r-jets is simple and 2) the number of abutting orbits in the space of r-jets remains bounded as $r \to \infty$.

Example 2.6. The germ of the function of one variable $v(x) = x^k$ is simple, since its orbit in the space of r-jets for $r \geq k$ abuts only the orbits of the germs $\pm x^m$, where $m < k$.

Example 2.7. The germ of the function $v(x_1, x_2) = x_1 x_2(x_1^2 - x_2^2)$ is not simple, since among the nearby functions $x_1(x_1^2 - x_2^2)(x_2 + tx_1)$ (where t is small) an infinite number belong to different orbits. Thus the cross-ratio of the four tangents to the branches of the zero level lines is an invariant diffeomorphism.

Example 2.8. The germ of the function of one variable $v(x) \equiv 0$ is not simple, although its r-jet for any r is simple.

The critical point 0 in this example is not isolated. A phenomenon, similar to what has been observed in this example is possible only for critical points of infinite multiplicity (in the analytic case not isolated). In fact, if a critical point is of finite multiplicity (in the analytic case isolated), sufficiently long r-jets define the germ up to diffeomorphisms, hence under passage to longer jets the number of abutting orbits is unchanged.

Definition 2.9. An r-jet of a function at 0 is called sufficient, if any 2 germs f_1, f_2 with this r-jet are carried into one another by a local diffeomorphism g which leaves 0 fixed: $f_1(x) = f_2(g(x))$.

The set of germs which do not have a sufficient jet (i.e., the set of germs with singularities of infinite multiplicity), has infinite codimension in the space of all germs.

Disregarding these completely unusual cases allows us to reduce the analytic problem of the classification of simple germs to an algebraic one (on orbits in the space of jets) and has no influence on the decisive results which are the following.

THEOREM 2.10. Each simple germ can be described in an appropriate coordinate system as one of the normal forms $A_k (k \geq 1)$, $D_k (k \geq 4)$, $E_k (6 \leq k \leq 8)$ of the list (1.1).

THEOREM 2.11. The set of all nonsimple germs has codimension 6 in the space of germs of functions of $n > 2$ variables with critical point 0 and critical value 0 (if it is desired, in the space of r-jets where $r \geq 3$). For $n = 2$ the codimension of the set of all nonsimple germs (or r-jets, where $r \geq 4$) is equal to 7, and for $n = 1$ it is infinite.

COROLLARY 2.12. The classification of degeneracies of codimension $l < 6$ is given by table (1.2). For example, in a typical five-parameter family of functions with separated values of the parameters, critical points of types A_6, D_6, E_6 occur, on some union of their curves types A_5, D_5, on some union of these curved surfaces types A_4 and D_4; the singularity A_3 occurs on a three-dimensional and A_2 on a four-dimensional submanifold of the parameter space. For the remaining values of the parameters, critical points are morsian (nondegenerate).

Remark 2.13. The results formulated are equally valid in C^∞ and $\mathbf{R}$- or $\mathbf{C}$-analytic situations and also in the framework of formal power series. In the complex case the number of normal forms is naturally lower, replacing all minus signs in (1.1) by plus signs. For even k the singularities A_k with x_1^{k+1} are equivalent even in the real domain (i.e., they belong to the same orbit). With this exception the remaining singularities in the list (1.1) are pairwise nonequivalent.

Remark 2.14. Further consequences of theorems 2.10 and 2.11 are mentioned in §§ 8-11. Among other things, from the results cited there it follows that in a typical family of functions of the special form $F(x, y) = S(x) - \langle x, y\rangle$ $(x \in \mathbf{R}^n, y \in \mathbf{R}^n)$, depending on n variables x and n parameters y, there occur exactly the same degenerate critical points as in a typical family F of general form which depends arbitrarily on n parameters.

The proof of Theorems 2.10 and 2.11 is given in §§ 3-7.

§ 3. Sufficient Jets

The use of the simple lemmas given below greatly shortens the calculations in the proof of the classification theorem.

We consider the ring of germs of smooth functions at the point $0 \in \mathbf{R}^n$. We denote by $\mathfrak{m}$ the maximal ideal of this ring, i.e., the collection of germs of functions equal to zero at the point 0. Thus, the notation $f \in \mathfrak{m}^s$ denotes that f has a zero of order s at 0.

LEMMA 3.1. Let f be a germ of a function at the point $0 \in \mathbf{R}^n$. We fix a natural number r and assume that f satisfies the following condition:

$$\forall \alpha \in \mathfrak{m}^{r+1}\ \exists h_i \in \mathfrak{m}^2:\ \alpha = \sum_{i=1}^{n} \frac{\partial f}{\partial x_i} h_i \mod \mathfrak{m}^{r+2}. \tag{3.1$_r$}$$

Then any function f' for which $f' = f \mod \mathfrak{m}^{r+1}$ also satisfies $(3.1)_r$.

Proof. $\sum_{i=1}^{n} \frac{\partial (f'-f)}{\partial x_i} h_i \in \mathfrak{m}^{r+2}$, hence it suffices to take $h_i' = h_i$.

Remark. Thus, $(3.1)_r$ is a restriction on r-jets, not on germs.

LEMMA 3.2. Let the r-jet of the function f at 0 satisfy $(3.1)_r$. Then the r-jet of the function f at 0 is sufficient (see definition 2.9).

Proof. 1°. Let J^s be the space of s-jets of functions at the point 0 in the space $\mathbf{R}^n$. Let $m \geq 1$. We consider in the jet space J^{r+m} the submanifold consisting of all r + m-jets, whose r-jet coincides with the r-jet of the function f. We denote this submanifold by F_m; thus, $F_m \subset J^{r+m}$. To prove Lemma 3.2, we shall show first of all the following

Assertion A_m. Under the condition $(3.1)_r$ the orbit of the jet of the function f in J^{r+m} contains F_m for any $m \geq 1$.

2°. We shall prove assertion A_1. The action $G \times J \to J$ of the Lie group of s-jets of diffeomorphisms leaving 0 fixed, on the manifold of jets $J = J^s$ gives a map of tangent spaces (derived from the action on the first argument) $T_eG \to T_jJ$ (e is the unit of the group, j is the jet of the function f). In coordinate notation, this map associates with the s-jet of the vector field $h = \sum h_i \partial/\partial x_i$ at zero the s-jet of the function $\sum h_i \partial f/\partial x_i$ at zero. Hence, from condition $(3.1)_r$ it follows that the tangent space to the orbit of the r + 1-jet of the function f contains the tangent space to F_1 at the point which is the r + 1-jet of the function f.

But by Lemma 3.1, condition $(3.1)_r$ is also satisfied for all $f \in F_1$. Consequently, the tangent space to the orbit of each point of F_1 contains the entire tangent space to F_1 at this point.

Whence it follows that each curve in F_1 is tangent to the orbit of each of its points, and so belongs entirely to an orbit. Since the space F_1 is arcwise connected (it is diffeomorphic to euclidean space), F_1 belongs entirely to an orbit. A_1 is proved.

3°. We shall show that $(3.1)_r \Rightarrow (3.1)_{r+1}$. Let $\alpha \in \mathfrak{m}^{r+2}$. Then $\alpha = \sum x_i \alpha_i$ $(\alpha_i \in \mathfrak{m}^{r+1})$ (Hadamard's lemma). We represent α_i in the form $\alpha_i = \sum_j h_{ij} \cdot \partial f/\partial x_j \bmod \mathfrak{m}^{r+2}, h_{ij} \in \mathfrak{m}^2$, by virtue of $(3.1)_r$. We set $h_j = \sum_i x_i h_{ij}$. Then $\alpha = \sum h_j \cdot \partial f/\partial x_j \bmod \mathfrak{m}^{r+3} (h_j \in \mathfrak{m}^3 \subset \mathfrak{m}^2)$, which is what was needed.

Applying what was proved s − r times, we get $(3.1)_r \Rightarrow (3.1)_s$ for any $s \geq r$.

4°. Let f_1 and f_2 be two functions whose r-jets coincide with the r-jet of the function f. We shall show that their s-jets belong to one orbit in J^s for $s \geq r$. This was already proved in Par. 2° for s = r + 1. Let the k-jets of the functions f_1 and f_2 belong to one orbit in J^k, we shall show that their k + 1-jets belong to one orbit in J^{k+1}. We make a diffeomorphism which carries the k-jet of the function f_2 into the k-jet of the function f_1. The function f_2 is carried into a function f_3, whose k + 1-jet lies in the same orbit in J^{k+1} as the k + 1-jet of the function f_2. But the k-jets of the functions f_1 and f_3 coincide and satisfy $(3.1)_k$ by virtue of 3°. From A_1 (proved in 2°) it follows that the k + 1-jet of the function f_3 belongs to the orbit of the k + 1-jet of the function f_1 in J^{k+1}. This means that the k + 1-jets of f_1 and f_2 lie in one orbit in J^{k+1}, and A_m is proved for any $m \geq 1$.

5°. For sufficiently large s the s-jet of the function f is sufficient (see [2] or [1] and [3]). Whence and from A_m which was proved in 4° it follows that the r-jet is already sufficient.

In fact, let f_4 be another function with the same r-jet as f. There exists a diffeomorphism which carries f_4 into a function f_5 with the same s-jet as $f (A_{s-r})$. Since the s-jet of the function f is sufficient, there exists a diffeomorphism carrying f_5 into f. Thus, there exists a diffeomorphism carrying f_4 into f, and Lemma 3.2 is proved.

Example 3.3. We consider the function $x^3 + y^4$ of the two variables x, y. Its 4-jet at 0 is sufficient by Lemma 3.2. In fact, any monomial of degree 5 in x, y can be represented either in the form $3x^2h_1$ or in the form $4y^3h_2$, where h_1 is a monomial of degree 3 and h_2 is a monomial of degree 2.

Remark 3.4. To replace $h \in \mathfrak{m}^2$ by $h \in \mathfrak{m}$ in lemmas 3.1 and 3.2 is impossible. In fact, let $f(x) = x^2$, r = 1. Then $\forall \alpha \in \mathfrak{m}^2 \; \exists h \in \mathfrak{m}$: $\alpha = 2xh \bmod \mathfrak{m}^3$, however, the 1-jet of the function f at zero is not sufficient. In the same example the function $f' \equiv 0$ satisfies the condition $f = f' \bmod \mathfrak{m}^2$, but does not admit a decomposition $\alpha = h\partial f/\partial x$, although f admits such a decomposition.

Remark 3.5. The assertions and proofs of lemmas 3.1 and 3.2 are still valid in the R- or C-analytic and R- or C- formal cases.

§4. Classification of Simple Germs: the Series A_k

Let f be a germ of a smooth function at the point $0 \in \mathbf{R}^n$; we shall assume that 0 is a critical point $(df|_0 = 0)$ and that $f(0) = 0$. Let ρ be the rank of the second differential $d^2f|_0$, $\mathfrak{m}$ the maximal ideal of the ring of germs so that germs in $\mathfrak{m}^k$ have zeroes of order k at 0.

LEMMA 4.1. In a neighborhood of the point 0 there exists a smooth system of coordinates x, y in which f can be written in the form

$$f = \varphi(x) + Q(y) \qquad (\varphi \in \mathfrak{m}^3),$$

where Q is a nondegenerate quadratic form ($\dim\{y\} = \rho$, $\dim\{x\} = n - \rho$).

Proof. First we reduce 2-jets to canonical form. By the theorem of Jacobi on quadratic forms, there exists a system of coordinates x, z($\dim\{z\} = \rho$), in which the 2-jet of the function f takes the form

$$(j_2 f)(x, z) = \sum \varepsilon_i z_i^2 \quad (i = 1, \ldots, \rho,\ \varepsilon_i = \pm 1). \tag{4.1}$$

We consider the restriction f_t of the function f to the plane x = t. For small $|t|$ the function f_t has a unique critical point near z = 0; this critical point depends smoothly on t, and at it $|z| = o(|t|)$ all of this follows from the nondegeneracy of the critical point z = 0 of the function f_0 and the implicit function theorem.

We denote by $\varphi(t)$ the value of the function f at this critical point; from formula (4.1), since $|z| = o(|t|)$ it follows that $|\varphi(t)| = O(|t|^3)$.

The difference $g(x, z) = f(x, z) - \varphi(x)$ can be considered as a family of functions of z depending smoothly on the parameter x, with nondegenerate critical point which depends smoothly on x and with critical value zero.

The generalized lemma of Morse asserts that such a family g can be reduced to the form $g(x, z) = \pm y_1^2 \pm \ldots \pm y_e^2$ by a smooth change of variables $y_i = y_i(x, z)$ (this lemma is proved in the same way as the ordinary lemma of Morse in which there is no parameter x; see, e.g., [4]). Now

$$f = \varphi(x) \pm y_1^2 \pm \cdots \pm y_\rho^2 \qquad (\varphi \in \mathfrak{m}^3),$$

which is what had to be proved.

The number of deficient squares $n - \rho = \dim\{x\}$ will be called the corank of the function f at zero. Thus, the corank of a function at a nondegenerate critical point is equal to zero.

LEMMA 4.2. The corank of a simple germ does not exceed two.

Proof. We consider the cubical Taylor polynomial of the function f at a critical point of corank k. The restriction to the null space of the second differential defines on this k-dimensional linear subspace of the tangent space a cubical form which is independent of the coordinate system.

The action of the group of diffeomorphisms on the space J^3 (3-jets of functions) induces an action of the linear group $GL(\mathbf{R}^k)$ in the form of a linear substitution in the cubical forms on $\mathbf{R}^k$. If the cubical forms of the functions f and g lie in different orbits of the action of $GL(\mathbf{R}^k)$ on 3-jets, these functions lie in different orbits of the action of the group of diffeomorphisms on the space of jets J^3.

But the dimension of the space E of cubical forms in $\mathbf{R}^k$ is equal to C_{k+2}^3, and the dimension of the group $GL(\mathbf{R}^k)$ is equal to k^2. For $k \geq 3$ we have $C_{k+2}^3 > k^2$. Consequently the dimension of all orbits of $GL(\mathbf{R}^k)$ in E is less than the dimension of E. Hence the orbits form a continuous family in E for $k \geq 3$. This means that the orbits of the group of diffeomorphisms in J^3 also form a continuous family near the 3-jet of the function f.

Thus, a germ of corank $k \geq 3$ cannot be simple, and Lemma 4.2 is proved.

It remains for us to classify simple germs of coranks 1 and 2.

LEMMA 4.3. A simple germ of corank 1 can be reduced to one of the normal forms of type A_k with some $k \geq 2$,

$$f = \pm x^{k+1} + Q, \quad Q = \sum_{i=1}^{n-1} \varepsilon_i y_i^2 \qquad (\varepsilon_i = \pm 1). \tag{4.2}$$

Proof. By virtue of Lemma 4.1, in some coordinate system x, y we have $f = \varphi(x) + Q(y)$, where $\varphi \in \mathfrak{m}^3$. We investigate the function φ of one variable x. If all derivatives of φ at zero are equal to 0, then the germ f is not simple (infinite number of abutting orbits $-x^k + Q$). Let, for some $k \geq 2$, all derivatives of φ at 0 up to order k inclusive be equal to zero, and the derivative of order k + 1 be different from zero.

Then f reduces to normal form (4.2). In fact, by virtue of Lemma 3.2 the k + 1-jet

$$c^{k+1} \pm y_1^2 \pm \cdots \pm y_{n-1}^2 \qquad (c \neq 0) \tag{4.3}$$

is sufficient: this follows from the factorization of all monomials in x, y of degree k + 2:

$$x^{k+2} = (k+1)\, c x^k h_0, \quad y_i g_\alpha (x, y) = \pm 2 y_i h_{i\alpha} \qquad (h_0, h_{i\alpha} \in \mathfrak{m}^2).$$

Thus, we can write f in the form (4.3), and hence, (after substituting $x' = |c|^{1/k+1} x$), also in the form (4.2). Lemma 4.3 is proved.

It is not difficult to verify that all germs (4.2) are simple (see below §8).

Remark 4.4. Actually, we have proved more than was formulated in Lemma 4.3. Namely, we have proved that every germ of corank 1 either reduces to one of the normal forms (4.2) or has a critical point of infinite multiplicity, and hence, belongs to a set of infinite codimension in the space of germs.

§5. Classification of Simple Germs of Corank 2: the Series D_k

Lemma 4.1 reduces the study of germs of corank k to the case of functions of k variables: the reduction is brought about in the general case just as it was done for the case k = 1 in the proof of lemma 4.3. Hence for the classification of simple germs of corank 2 it suffices to deal with functions of two variables. (However, the calculation carried out below could also be done for a larger number of variables, without using the above-cited reduction.)

A germ of a function of two variables has corank 2 if its 2-jet is zero. In this case the 3-jet determines a cubical form on the tangent plane. It is easy to prove

LEMMA 5.1. A cubical form on the plane R^2 is reduced by a linear change of variables to one of the following types: $x^2y \pm y^3$, x^2y, x^3, 0.

The proof depends on the fact that three different points in a projective line can be transformed into any other three by a projective transformation.

LEMMA 5.2. If the germ of a function of two variables of corank 2 is simple, then its cubical form is different from 0. Moreover, if a germ of corank 2 of a function of any number of variables has zero cubical form on the null plane of the second differential, then it is not simple.

Proof. Under the given hypotheses the 4-jet can be reduced to the form $\varphi(x_1, x_2) \pm x_3^2 \pm \ldots \pm x_n^2$ mod $\mathfrak{m}^5$, where φ is a homogeneous polynomial of degree 4. This polynomial gives a form of degree 4 on the null plane of the second differential, which does not depend on the coordinate system. The cross-ratio, defined on zeros of φ, is an invariant action of the diffeomorphisms on the 4-jets. The existence of a continuous family of orbits distinguishing values of this invariant is an obstruction to the simplicity of the germ.

LEMMA 5.3. The 3-jet of the function $x^2y \pm y^3$ at 0 is sufficient.

Proof. For calculation it is more convenient to take the (obviously equivalent) form $f = x^2y \pm (y^3/3)$. Then $\partial f/\partial y = x^2 \pm y^2$, $\partial f/\partial x = 2xy$, and Lemma 5.3 follows from Lemma 3.2 and the decompositions

$$y^4 = (x^2 \pm y^2)(\pm y^2) \mp 2xy\,(xy/2), \quad y^3x = 2xy\,(y^2/2),$$
$$x^2y^2 = 2xy\,(xy/2), \quad yx^3 = 2xy\,(x^2/2), \quad x^4 = (x^2 \pm y^2)x^2 \mp 2xy\,(xy/2).$$

From Lemma 5.3 and 5.1 it follows that if the cubical form is not degenerate, then the function can be reduced to the normal form

$$D_4 : f = x^2y \pm y^3.$$

It is not hard to verify that both germs of type D_4 are simple (see §8).

We now consider the case of a cubical form of type x^2y.

LEMMA 5.4. A simple germ of a function of two variables of corank 2 with cubical form x^2y can be reduced to one of the normal forms

$$D_k : f = x^2y \pm y^{k-1} \ (k \geqslant 5).$$

Proof. We assume that the s-jet of the function f in some coordinate system has the form $x^2y + \varphi_s(x, y)$, where φ_s is a homogeneous polynomial of degree s. (For s = 4 there is automatically such a coordinate system). We represent φ_s in the form

$$\varphi_s = ay^s + 2bxy^{s-1} + x^2\psi(x, y), \text{ where } \psi \in \mathfrak{m}^{s-2}.$$

Substituting $x - by^{s-2} = x_1$, $y - \psi(x, y) = y_1$ reduces f to the form $f = x_1^2y_1 + ay_1^s \mod \mathfrak{m}^{s+1}$ (because $\mathfrak{m}^{2s-3} \subset \mathfrak{m}^{s+1}$ for s > 3). Next, two cases are possible: $a = 0$ and $a \neq 0$. If $a = 0$, then the s-jet of the function f is reduced to the form x^2y. Hence the s + 1-jet has the form $x_1^2y_1 + \varphi_{s+1}(x_1, y_1)$, and we can repeat the previous argument. Going on, we will either get $a = 0$ for all s = 4, 5, . . ., or at some step we get $a \neq 0$.

If $a \neq 0$ never occurs, then the singularity has infinite codimension and the germ of f at 0 is not simple (for example, its orbit in J^s abuts all orbits of $x^2y + x^k$ for $k \leq s$).

Now if for some s one has $a \neq 0$, then the germ of the function f at zero will be equivalent to the germ of $x^2y + ay^s$. For the proof, we verify that the s-jet $x^2y + (ay^s/s)$ ($s \geq 4$, $a \neq 0$) is sufficient. By Lemma 3.2 this follows from the following decompositions:

$$x^\alpha y^\beta = 2xyh(x, y), \quad h \in \mathfrak{m}^2 \text{ for } \alpha + \beta = s + 1 > 3 \quad (\alpha > 0,\ \beta > 0),$$

$$x^{s+1} = (x^2 + ay^{s-1})x^{s-1} + 2xy(-x^{s-2}y^{s-2}/2a) \quad (x^{s-1} \in \mathfrak{m}^2,\ (xy)^{s-2} \in \mathfrak{m}^2),$$

$$y^{s+1} = (x^2 + ay^{s-1})(y^2/a) + 2xy(-xy/2) \quad (y^2 \in \mathfrak{m}^2,\ (-xy/2) \in \mathfrak{m}^2).$$

In order to convert $x^2y + (ay^s/s)$ into $x^2y \pm y^s$, it suffices to substitute $x_1 = px$, $y_1 = qy$. Lemma 5.4 is proved.

Remark 5.5. Actually, more was proved than Lemma 5.4. Namely, we have proved that every germ of corank 2 with cubical form x^2y either reduces to one of the normal forms $D_k(f = x_1^2x_2 \pm x_2^{k-1} \pm x_3^2 \pm \ldots \pm x_n^2)$, or has a critical point of infinite multiplicity, and hence belongs to a set of infinite codimension in the space of germs.

It is not hard to verify that the germs D_k, $k \geq 4$, are simple (see § 8).

§6. The Germs E_6, E_7, E_8

We shall look for simple germs of functions of two variables of corank 2 with cubical form x^3.

LEMMA 6.1. A simple germ of a function of two variables of corank 2 with cubical form x^3 reduces to one of the following normal forms:

$$E_6 : f = x^3 \pm y^4,\ E_7 : f = x^3 + xy^3,\ E_8 : f = x^3 + y^5.$$

Proof. 1°. We write the 4-jet of the function considered in the form

$$x^3 + ay^4 + bxy^3 + 3x^2\varphi(x, y) \mod \mathfrak{m}^5, \text{where } \varphi \in \mathfrak{m}^2.$$

Substituting $x + \varphi(x, y) = x_1$ turns the 4-jet into $x_1^3 + ay^4 + bx_1y^3 \mod \mathfrak{m}^5$.

2°. We assume that $a \neq 0$. Then one can reduce the 4-jet to the form $x_1^3 \pm y_1^4 + 4cx_1y_1^3 \mod \mathfrak{m}^5$ by substituting $y = py_1$. After this we set $y_2 = y_1 \pm cx_1$. Then the 4-jet assumes the form $x_1^3 \pm y_2^4 + 3x_1^2\psi \mod \mathfrak{m}^5$, where $\varphi \in \mathfrak{m}^2$. Now, substituting $x_1 + \psi = x_2$ reduces the 4-jet to the form $x_2^3 \pm y_2^4 \mod \mathfrak{m}^5$. By Lemma 3.2 this jet is sufficient (see example 3.3). Thus, in the case $a \neq 0$ the germ is reduced to the normal form E_6.

3°. We assume that $a = 0$, $b \neq 0$. Then the 4-jet has the form $x_1^3 + bx_1y^3$, and it can be reduced to the form $x_1^3 + x_1y_1^3 \mod \mathfrak{m}^5$ by substituting $y = py_1$.

We shall show that the 4-jet $x^3 + xy^3$ is sufficient. First, we shall show that the orbit of any 5-jet $x^3 + xy^3 + \varphi \mod \mathfrak{m}^6$ ($\varphi \in \mathfrak{m}^5$) is open in the space of 5-jets with 4-jet $x^3 + xy^3$. For this it suffices to represent any monomial 5-jet $x^\alpha y^\beta(\alpha + \beta = 5)$ in the form

$$x^\alpha y^\beta = (3x^2 + y^3 + A)h_1 + (3xy^2 + B)h_2 \mod \mathfrak{m}^6, \tag{6.1}$$

where $A = \partial\varphi/\partial x$, $B = \partial\varphi/\partial y \in \mathfrak{m}^4$, $h \in \mathfrak{m}^2$. For the decompositions of xy^4, x^2y^3 and x^3y^2 we take $h_1 = 0$, $h_2 = x^{\alpha-1}y^{\beta-2}/3$. Further,

$$x^5 = (3x^2 + y^3 + A)(x^3/3) + (3xy^2 + B)(-x^2y/3) \mod \mathfrak{m}^6.$$

260

Thus, for all monomials divisible by x, the decomposition (6.1) is achieved. Finally,

$$y^5 = (3x^2 + y^2 + A)y^2 + (3xy^2 + B)(-x) + Bx \bmod \mathfrak{m}^6;$$

Here $Bx \bmod \mathfrak{m}^6$ is a monomial of degree 5, which is divisible by x. The decomposition (6.1) with left side Bx exists by what was proved. This means it also exists for y^5.

Thus, the orbit of the 5-jet $x^3 + xy^3 + \varphi(\varphi \in \mathfrak{m}^5)$ is open in the space of 5-jets with 4-jet $x^3 + xy^3$ and hence contains this entire space. Consequently, any 5-jet with 4-jet $x^3 + xy^3 \bmod \mathfrak{m}^5$ can be reduced to the form $x^3 + xy^3 \bmod \mathfrak{m}^6$.

4°. We shall show that the 4-jet of $x^3 + xy^3 \bmod \mathfrak{m}^6$ is sufficient. Our computation in Par. 3° shows for any $\gamma \in \mathfrak{m}^5$ there exists a decomposition of the form (6.1)

$$\gamma = (3x^2 + y^3)h_1 + (3xy^2)h_2 \bmod \mathfrak{m}^6, \text{where } h_i \in \mathfrak{m}.$$

Consequently, for $\delta \in \mathfrak{m}^6$ there exists a decomposition

$$\delta = (3x^2 + y^3)h_1 + (3xy^2)h_2 \bmod \mathfrak{m}^7, \text{ where } h_i \in \mathfrak{m}^2.$$

Thus, the 5-jet of $x^3 + xy^3$ is sufficient by Lemma 3.2.

But by virtue of 3°, every 5-jet with 4-jet $x^3 + xy^3 \bmod \mathfrak{m}^5$ can be transformed into the form $x^3 + xy^3 \bmod \mathfrak{m}^6$. Thus, the 4-jet of $x^3 + xy^3 \bmod \mathfrak{m}^5$ is already sufficient.

We have thus proved that if in 1°, $a = 0$ and $b \neq 0$, then the function can be transformed to the normal form E_7.

5°. We assume that $a = 0$, $b = 0$. Then the 4-jet has the form $x^3 \bmod \mathfrak{m}^5$. We write the 5-jet of our function in the form

$$x^3 + a'y^5 + b'xy^4 + 3x^2\varphi(x, y) \bmod \mathfrak{m}^6, \text{where } \varphi \in \mathfrak{m}^3.$$

Substituting $x + \varphi(x, y) = x_1$ transforms the 5-jet into $x_1^3 + a'y^5 + b'x_1y^4 \bmod \mathfrak{m}^6$.

6°. We assume that $a' \neq 0$. Then one can reduce the 5-jet to the form $x_1^3 + y_1^5 + 5c'x_1y_1^4 \bmod \mathfrak{m}^6$ by substituting $y = py_1$. After this we set $y_2 = y_1 \pm c'x_1$. Then the 5-jet assumes the form $x_1^3 + y_2^5 + 3x_1^2\psi \bmod \mathfrak{m}^6$, where $\psi \in \mathfrak{m}^3$. Now substituting $x_1 + \psi = x_2$ transforms the 5-jet to the form $x_2^3 + y_2^5 \bmod \mathfrak{m}^6$.

By Lemma 3.2, the 5-jet $x^3 + y^5 \bmod \mathfrak{m}^6$ is sufficient (the monomial $x^\alpha y^\beta$ for $\alpha + \beta = 6$ is divisible either by x^2 or by y^4). Thus, in the case $a' \neq 0$ the function reduces to the normal form E_8.

7°. We assume $a' = 0$. Then the 5-jet has the form $x^3 + b'xy^4 \bmod \mathfrak{m}^6$. We shall show that this jet is not sufficient. Moreover, the partition of the space of 6-jets with this 5-jet into orbits is continuous (at least for $b' \neq 0$) in the neighborhood of any such 6-jet. Thus we will show that for simple jets $a' \neq 0$.

8°. We shall show that our 6-jet for $b' \neq 0$ can be reduced to the form $x^3 \pm xy^4 + \lambda y^6 \bmod \mathfrak{m}^7$. In fact, we shall show that the coefficient b' in the 5-jet can be made equal to ± 1 by an axial dilatation. After this we shall show that the tangent plane to the orbit of any 6-jet $x^3 \pm xy^4 + \varphi \bmod \mathfrak{m}^7$ (where $\varphi \in \mathfrak{m}^6$) contains the entire space of monomials of degree 6 which are divisible by x ($\bmod \mathfrak{m}^7$). For this, we shall give a decomposition of the monomial $x^\alpha y^\beta$ ($\alpha + \beta = 6$, $\alpha \geq 1$):

$$x^\alpha y^\beta = (3x^2 \pm y^4 + A)h_1 + (\pm 4y^3 + B)h_2 \bmod \mathfrak{m}^7,$$

where $A = \partial\varphi/\partial x$, $B = \partial\varphi/\partial y$, A and $B \in \mathfrak{m}^5$. Namely, for $1 \leq \alpha \leq 3$ we set $h_1 = 0$, $h_2 = \pm x^{\alpha-1}y^{\beta-3}/4 \in \mathfrak{m}^2$, and for $4 \leq \alpha \leq 6$ we set $h_1 = x^{\alpha-2}y^\beta/3 \in \mathfrak{m}^4$, $h_2 = 0$.

Consequently, the orbit of the 6-jet $x^3 \pm xy^4 + \varphi \bmod \mathfrak{m}^7$ contains a representative with $\varphi = \lambda y^6$, which is what was asserted.

9°. We consider the family of functions depending on a parameter λ, $f_\lambda = x^3 \pm xy^4 + \lambda y^6$. We shall show that the orbit of the germ of the function f_λ at zero (and even the orbit of its 6-jet) varies continuously with λ.

With this aim, we note that the zeros of f_λ form 3 parabolas $x = t_iy^2$, where t_i are the roots of the equation $t^3 \pm t + \lambda = 0$. We shall show that the ratio $(t_3 - t_1)/(t_2 - t_1)$ (definite to within the order of the roots) is an invariant of the system of three tangent parabolas with respect to diffeomorphisms (it is sufficient to consider the 2-jets of the parabolas and the 1-jets of the diffeomorphisms).

We shall verify that the ordered triple of parabolas $x = 0$, $x = y^2$, $x = ky^2$ ($k \neq 0, 1$) is never transformed by a diffeomorphism of the plane (near $x = y = 0$) into another triple of the same form with different k (one parabola can be transformed by the diffeomorphism into $x = 0$, and the second into $x = y^2$).

A diffeomorphism which carries the first triple into the second leaves the y axis fixed and hence has the form

$$x' = x(a_{11} + u(x, y)), \quad y' = a_{21}x + a_{22}y + v(x, y), \text{where } u \in \mathfrak{m}, \ v \in \mathfrak{m}^2.$$

In order that the parabola $x = y^2$ remain in place, one must have $a_{11} = a_{22}^2$. But then the image of the parabola $x = ky^2$ will also be $x' = ky'^2 \mod y'^3$, which cannot be a parabola $x' = k'y'^2$ for $k' \neq k$.

10°. We return to the 6-jet of 8°. The number $(t_3 - t_1)/(t_2 - t_1)$, constructed from this jet, varies continuously with λ. Whence it follows that the 6-jets of the functions f_λ belong to different orbits (which vary continuously with λ). This means that not one of the germs with $a' = 0$ is simple, and Lemma 6.1 is proved.

One can verify (see §8), that the germs E_6, E_7, E_8 are simple.

Remark 6.2. Actually, we have proved more than Lemma 6.1. Namely, we have proved that every germ of corank 2 with cubical form x^3 either reduces to one of the forms

$$E_6\colon\ f = x_1^3 \pm x_2^4 \pm x_3^2 \pm \ldots \pm x_n^2, \quad E_7\colon\ f = x_1^3 + x_1x_2^3 \pm x_3^2 \pm \ldots \pm x_n^2,$$

$$E_8\colon\ f = x_1^3 + x_2^5 \pm x_3^2 \pm \ldots \pm x_n^2,$$

or belongs to the set of codimension 8, which consists of germs whose 5-jets reduce to the form

$$x_1^3 \pm x_1x_2^4 \pm x_3^2 \pm \ldots \pm x_n^2 \mod \mathfrak{m}^6.$$

All germs of this last type are not simple.

The last part of the assertion (that the germs with 5-jet $x_1^3 \pm x_1x_2^4 \pm x_3^2 \pm \ldots \pm x_n^2$ are not simple) for $n > 2$ does not follow formally from Lemma 6.1, but was also proved.

We shall show, for example, that the tangent plane to the orbit of the 6-jet $x_1^3 \pm x_1x_2^4 + \lambda x_2^6 \pm x_3^2 \pm \ldots \pm x_n^2$ (and every 6-jet with the given 5-jet reduces to this form) does not contain the direction x_2^6, i.e., that the equation relative to $h_i \in \mathfrak{m}$

$$(3x_1^2 \pm x_2^4)h_1 + (6\lambda x_2^5 \pm 4x_1x_2^3)h_2 \pm 2x_3h_3 \pm \ldots \pm 2x_nh_n = x_2^6 \mod \mathfrak{m}^7$$

is unsolvable. In fact, we set $\varepsilon = \pm 1$, $x_3 = \ldots = x_n = 0$, and we introduce the notation $h_\varepsilon(x_2) = h_2(\varepsilon\sqrt{\mp 1/3}\,x_2^2, x_2, 0, \ldots, 0)$. Then we get

$$(6\lambda \pm 4\varepsilon\sqrt{\mp 1/3})\,h^\varepsilon x_2^5 = x_2^6 \mod x_2^7.$$

Consequently,

$$h^\varepsilon(x_2) = cx_2 \mod x_2^2, \quad c = 6\lambda + 4\sqrt{\mp 1/3} = 6\lambda - 4\sqrt{\mp 1/3},$$

which is impossible for such a λ.

Thus, the orbits of the 6-jets $x_1^3 \pm x_1, x_2^4 + \lambda x_2^6 \pm x_3^2 \pm \ldots \pm x_n^2 \mod \mathfrak{m}^7$ vary continuously with λ, so no germ with such a 6-jet is simple.

Remark 6.3. The passage from the classification of germs of functions of k variables to germs of corank k of functions of any number of variables can also be justified with the help of any of the following assertions.

Proposition 6.4. The map which associates with the germ of φ at zero the germ of the function

$$f(x, y) = \varphi(x) + Q(y), \text{where } Q = \pm y_1^2 \pm \ldots \pm y_n^2 \quad (x \in \mathbb{R}^k, y \in \mathbb{R}^n, \varphi \in \mathfrak{m}^3),$$

at the point (0, 0) is transversal to the orbit of the germ of f at this point.

Proposition 6.5. If the germ of the functions $f_1(x, y) = \varphi_1(x) + Q(y)$ and $f_2(x, y) = \varphi_2(x) + Q(y)$ and (where $\varphi_1 \in \mathfrak{m}^3$, $\varphi_2 \in \mathfrak{m}^3$) at the point (0, 0) are smoothly equivalent and of finite multiplicity, then the germs of the functions φ_1 and φ_2 are also equivalent.

The proofs are not given since they are long and it is not clear if the requirement of finite multiplicity is essential. We shall not use propositions 6.4 and 6.5 in what follows.

§ 7. Classification of Singularities up to Codimension 6

Theorem 2.10 follows immediately from the lemmas proved above in §§ 4-6. As a matter of fact, these lemmas prove somewhat more, namely, from remarks 4.4, 5.5, and 6.2 follow the

THEOREM 7.1. Any germ of a function at a critical point* can either

1) be described in one of the forms A_k, D_k, E_6, E_7, E_8, or

2) belongs to the set of codimension 6, formed by all germs of corank greater than 2, or

3) belongs to the set of codimension 7, formed by those germs of corank 2, whose 3-jet reduces to the form $\pm x_3^2 \pm \ldots \pm x_n^2$, or

4) belongs to the set of codimension 8, formed by all germs of corank 2 with 5-jet which reduces to the form $x_1^3 \pm x_1x_2^4 \pm x_3^2 \pm \ldots \pm x_n^2$, or

5) belongs to the set of infinite codimension, formed by all germs of corank 1 with critical point of infinite multiplicity.

In case 1) the germ is simple, and in the remaining cases it is not.

Each of the sets described in 2)-5), has, for $n \geq 3$, codimension not less than 6 (for $n = 1$, not less than 7, for $n = 2$, infinite codimension). Hence from Theorem 7.1 follows Theorem 2.10 as well as Theorem 2.11 and Corollary 2.12.

§ 8. Versal Deformations of Functions

We recall the definition of versal deformations of functions (for more details on versal deformations (see [5], [6]).

Let f be the germ of a smooth† function f: $\mathbf{R}^n \to \mathbf{R}$ at the point 0. A deformation F of the function f is a germ of a smooth function F: $\mathbf{R}^n \times \mathbf{R}^l \to \mathbf{R}$ at the point (0, 0), for which $F(x, 0) = f(x)$. The space $\mathbf{R}^l$ of the second argument of F is called the base of the deformation, and its elements are called the parameters of the deformation. The dimensions of the bases of different deformations of one function f can be distinct.

Definition 8.1. A deformation F of the germ of a function f is called versal, if every other deformation of the function f is equivalent to one induced from F.

This means that for any function G: $\mathbf{R}^n \times \mathbf{R}^n \to \mathbf{R}$ which is smooth near the point (0, 0), such that $G(y, 0) = f(y)$, there exist 1) a map ψ: $\mathbf{R}^m \to \mathbf{R}^l$ which is smooth near 0 and $\psi(0) = 0$ (change of parameter), 2) a diffeomorphism ($x = X(y, \mu)$, where X: $\mathbf{R}^n \times \mathbf{R}^m \to \mathbf{R}^n$, $X(0, 0) = 0$, $\det (D_1X)(0, 0) \neq 0$) which depends smoothly on the parameter $\mu \in \mathbf{R}^l$, which turns into the identity transformation for $\mu = 0$ ($X(y, 0) = y$), such that

$$G(y, \mu) \equiv F(X(y, \mu), \psi(\mu)).$$

Differentiating this relation, we arrive at the following definition.

Definition 8.2. A deformation F: $\mathbf{R}^n \times \mathbf{R}^l \to \mathbf{R}$ of the germ of a function f: $\mathbf{R}^n \to \mathbf{R}$ at zero is called infinitesimally versal if every germ at 0 of a function α: $\mathbf{R}^n \to \mathbf{R}$ can be represented in the form

$$\alpha(x) = \sum_{i=1}^{n} \frac{\partial f}{\partial x_i} h_i(x) + \sum_{j=1}^{l} c_j \varphi_j(x),$$

where $\varphi_j = \left.\frac{\partial F}{\partial \lambda_j}\right|_{\lambda=0}$ $(\lambda = (\lambda_1, \ldots, \lambda_l) \in \mathbf{R}^l)$, h_i are smooth functions, c_i are numbers.

In other words, a deformation of the germ of a function at a critical point is infinitesimally versal if the germ of its derivative with respect to the parameter generates the local ring of gradient maps at the critical point.

*With critical value zero; the codimension is also in the space of germs with critical value zero.

†There exist C^∞, R-, and C-analytic and formal variants of the following definitions; the results carry over to all these cases.

THEOREM 8.3. Every infinitesimally versal deformation is versal.

The proof is too long to give it here; it can be carried out by any of a series of standard schemes (see, e.g., [1], [7]).

Choosing generators of the local ring for germs of the functions A_k, D_k, E_k, we get the basic result of the present paragraph.

COROLLARY 8.4. As versal deformations for the germs of functions A_k, D_k, and E_k [see formulas (1.1)] one can take the following k-parameter deformations:

$$A_k\colon\ F(x,\lambda)=\pm x_1^{k+1}\pm x_2^2+Q+\lambda_{k-1}x_1^{k-1}+\lambda_{k-2}x_1^{k-2}+\ldots+\lambda_1x_1+\lambda_0,$$
$$D_k\colon\ F(x,\lambda)=x_1^2x_2\pm x_2^{k-1}+Q+\lambda_{k-1}x_1+\lambda_{k-2}x_2^{k-2}+\ldots+\lambda_1x_2+\lambda_0,$$
$$E_6\colon\ F(x,\lambda)=x_1^3\pm x_2^4+Q+\lambda_5x_1x_2^2+\lambda_4x_1x_2+\lambda_3x_2^2+\lambda_2x_2+\lambda_1x_1+\lambda_0,$$
$$E_7\colon\ F(x,\lambda)=x_1^3+x_1x_2^3+Q+\lambda_6x_1x_2+\lambda_5x_2^4+\lambda_4x_2^3+\lambda_3x_2^2+\lambda_2x_2+\lambda_1x_1+\lambda_0,$$
$$E_8\colon\ F(x,\lambda)=x_1^3+x_2^5+Q+\lambda_7x_1x_2^3+\lambda_6x_1x_2^2+\lambda_5x_1x_2+$$
$$+\lambda_4x_2^3+\lambda_3x_2^2+\lambda_2x_2+\lambda_1x_1+\lambda_0.$$

Here Q is the standard quadratic form

$$-x_3^2-\ldots-x_s^2+x_{s+1}^2+\ldots+x_n^2.$$

COROLLARY 8.5. As transversals to the orbits of the simple germs f in the space of r-jets of functions with critical point 0 and critical value zero, one can take the linear family with $k-1$ parameters $\varepsilon_2,\ldots,\varepsilon_k$ of the form

$$A_k:\ \pm x_1^{k+1}\pm x_2^2+Q+\varepsilon_2x_1^2+\varepsilon_3x_1^3+\ldots+\varepsilon_kx_1^k \qquad (r\geqslant k+1),$$
$$D_k:\ x_1^2x_2\pm x_2^{k-1}+Q+\varepsilon_2x_2^2+\ldots+\varepsilon_{k-2}x_2^{k-2}+\varepsilon_{k-1}x_1x_2+\varepsilon_kx_1^2 \qquad (r\geqslant k-1),$$
$$E_6:x_1^3\pm x_2^4+Q+\varepsilon_2x_2^2+\varepsilon_3x_2^3+\varepsilon_4x_1^2+\varepsilon_5x_1x_2+\varepsilon_6x_1x_2^2 \qquad (r\geqslant 4),$$
$$E_7:\ x_1^3+x_1x_2^3+Q+\varepsilon_2x_2^2+\varepsilon_3x_2^3+\varepsilon_4x_2^4+\varepsilon_5x_1^2+\varepsilon_6x_1x_2+\varepsilon_7x_1x_2^2 \qquad (r\geqslant 4),$$
$$E_8:\ x_1^3+x_2^5+Q+\varepsilon_2x_2^2+\varepsilon_3x_2^3+\varepsilon_4x_2^4+\varepsilon_5x_1^2+\varepsilon_6x_1x_2+\varepsilon_7x_1x_2^2+\varepsilon_8x_1x_2^3 \qquad (r\geqslant 5).$$

From these formulas, in particular, follows

COROLLARY 8.6. The germs of types A_k, D_k, E_6, E_7, E_8 are simple.

In fact, by direct calculation it is easy to see that for all ε the germs indicated in Corollary 8.5 belong to the orbits of types A_l, D_l, E_l $(l \leq k)$, which (for fixed k and n) are finite in number. The formulas of Corollary 8.5 also clearly permit one to enumerate all orbits which abut orbits of the given simple germ.

COROLLARY 8.7. The diagram of abutting of simple germs has the form (in the complex case)

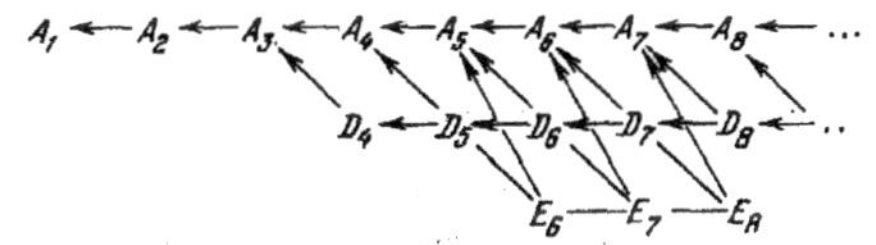

Here $P \leftarrow S$ denotes that the orbit P abuts the orbit S; the complete set of orbits abutting S is obtained from S by moving along arrows in the diagram.

It is easy to foresee the result of Corollary 8.7 heuristically, by considering the inclusions of the Dynkin diagrams of A_k, D_k, E_k in one another (see §9). However, the proof requires some calculation.

First of all, from dimensional considerations it is clear that S_k can abut only P_l with $l < k$ (P and S run through the values A, D, E). Further, from the semicontinuity of the rank and multiplicity, it follows that A_k can abut only A_l, and that D_k can abut only D_l and A_l. Thus, only the abutments indicated in the diagram are possible.

In order to prove that the abutment $P_{k-1} \leftarrow S_k$ is realized, it is sufficient to indicate a curve in the space of jets $(t \to f_t)$, for which f_0 has type S_k, and f_t for $t \neq 0$ has type P_{k-1}.

As such a curve, one can take, for example, the intersection of the transversals to S_k, indicated in Corollary 8.5, with orbit P_{k-1}.

The calculations (rather tedious, especially for $A_7 \leftarrow E_8$) give the following curves:

$$\begin{aligned}
&A_{k-1} \leftarrow A_k\colon\ f_t = x_1^{k+1} + t x_1^k + x_2^2 + Q,\\
&D_{k-1} \leftarrow D_k\colon\ f_t = x_1^2 x_2 + x_2^{k-1} + t x_2^{k-2} + Q,\\
&A_{k-1} \leftarrow D_k\colon\ f_t = x_1^2 x_2 - x_2^{k-1} - [t x_2^{k-2} + t^2 x_2^{k-3} + \ldots + t^{k-3} x_2^2] + 2\sqrt{t^{k-2}}\, x_1 x_2 - t x_1^2 + Q,\\
&D_5 \leftarrow E_6\colon\ f_t = (x_1 - t x_2)^2 (x_1 + 2t x_2) + x_2^4 + Q,\\
&A_5 \leftarrow E_6\colon\ f_t = x_1^3 + (x_2^2 + t x_1)^2 + Q,\\
&D_6 \leftarrow E_7\colon\ f_t = (x_1 - t x_2)^2 (x_1 + 2t x_2) + x_2^3 (x_1 - t x_2) + Q,\\
&A_6 \leftarrow E_7\colon\ f_t = 16 t^3 x_1^2 + x_1^3 - 4t x_1^2 x_2 - 8 t^2 x_1 x_2^2 + x_2^3 (x_1 + t x_2) + Q,\\
&D_7 \leftarrow E_8\colon\ f_t = (x_1 - 3t x_2)^2 (x_1 + 6t x_2) - 18 t^3 x_2^4 + 6t x_2^3 x_1 + x_2^5 + Q,\\
&A_7 \leftarrow E_8\colon\ f_t = t^5 (x_1 - t^2 x_2)^2 - x_1 (5t^4 x_2^2 + 4t x_2^3) + x_1^3 + 4t^6 x_2^3 + 5t^3 x_2^4 + x_2^5 + Q,\\
&E_6 \leftarrow E_7\colon f_t = x_1^3 + x_1 x_2^3 + t x_2^4 + Q,\\
&E_7 \leftarrow E_8\colon f_t = x_1^3 + x_2^5 + t x_1 x_2^3 + Q.
\end{aligned}$$

We shall show, for example, that the function f_t indicated in the line $A_5 \leftarrow E_6$, for $t \neq 0$, belongs to A_5. In fact, for $t \neq 0$ the corank of the germ of f_t at 0 is equal to 1, so f_t has type A_k with some $k \leq \infty$. In order to find k it suffices to consider the function $\varphi_t(x_2) = f_t(x_1(x_2), x_2)$, where the function $x_1(x_2)$ is defined by the condition $\left.\frac{\partial f_t}{\partial x_1}\right|_{x_1(x_2),\, x_2} = 0$; the order of zero of the function φ_t is then equal to $k + 1$. We find successively

$$3x_1^2 + 2t(x_2^2 + t x_1) = 0, \quad x_1(x_2) = -(x_2^2/t) + o(x_2^2),$$
$$(x_2^2 + t x_1(x_2))^2 = 9x_1^4/4t^2 = o(x_2^6), \quad \varphi_t = -x_2^6/t^3 + o(x_2^6).$$

Thus, $k + 1 = 6$, i.e., f_t has type A_5.

§9. Bifurcation Diagrams, Braids, and the Weyl Group

We consider the bases of the versal deformations indicated in Corollary 8.4. These bases are naturally stratified according to the singularities of the zeros of the level surfaces of the function $F(\cdot, \lambda)$. The set of all values of the parameter $\lambda \in C^k$ for which the function $F(\cdot, \lambda)$ has zero as a critical value, forms a hypersurface Σ in C^k (the bifurcation diagram), which can be called the generalized swallow's tail of the corresponding singularity (A_k, D_k or E_k); the ordinary swallow's tail in C^3 is obtained for the singularity A_3.

In this paragraph we consider (without proofs) more or less new propositions about bifurcation diagrams of germs of type A_k, D_k, E_k.

PROPOSITION 9.1. The complement in C^k to the bifurcation diagram Σ of a germ of type A_k, D_k or E_k is an Eilenberg-MacLane space $K(\pi, 1)$: $\pi_i(C^k - \Sigma) = 0$ for $i \geq 2$.

In order to describe the fundamental group of this complement, we recall the construction of E. Brieskorn of the group of braids of the group generated by a map (see [8], [9], [10]).

Let Γ be a finite group generated by a map of $\mathbf{R}^m$ into some hyperplane.

The complexification of the action of Γ on $\mathbf{R}^m$ gives a finite group in C^m. Let X be the domain in C^m formed by points of "general type" (with orbits of the smallest number of points). The domain X is obtained from C^k by throwing out a certain number of hyperplanes ("diagonals").

Definition 9.2. The braid group of the group Γ is the fundamental group of the quotient space X/Γ: $B(\Gamma) = \pi_1(X/\Gamma)$.

PROPOSITION 9.3. The fundamental group of the complement of the bifurcation diagram of a germ of type A_k, D_k, or E_k is the braid group of the corresponding Weyl group: $\pi_1(C^k - \Sigma) = B(\Gamma)$; $C^k - \Sigma$ is homeomorphic to X/Γ.

The complement $C^k - \Sigma$ of the bifurcation surface serves as the base of a series of important fibrations. One of them has fiber over the point $\lambda \in C^k - \Sigma$ the hypersurface in C^n with equation $F(\cdot, \lambda) = 0$. Since λ does not belong to the bifurcation diagram Σ, the fiber is a nonsingular hypersurface. We denote it by $V = V_\lambda$.

PROPOSITION 9.4. The nonsingular fiber V_λ of a versal deformation of a singularity A_k, D_k, or E_k is homotopically equivalent to a bouquet of spheres of dimension $n - 1$.

As these spheres one can take the vanishing cycles of Picard-Lefschetz. Moreover, if n is odd (so that the intersection index in $H_{n-1}(V_\lambda)$ is symmetric), then these k cycles can be chosen so that their intersection indices will give the Dynkin diagram:

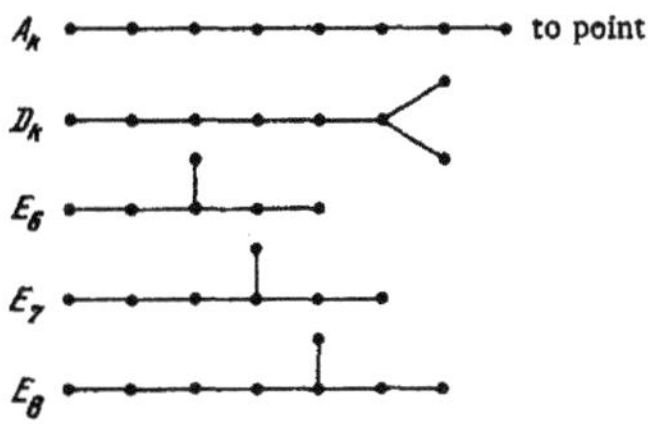

Here each point represents a sphere (a generator in $H_{n-1}(V_\lambda)$). The self-intersection indices of the generators are equal to $(-1)^{n-1/2}2$, if the points α and β are joined by a segment, and $(\alpha, \beta) = (-1)^{n+1/2}$, if the points α and β are joined by a segment, and $(\alpha, \beta) = 0$, if there is no segment.

PROPOSITION 9.5. The action of the fundamental group $\pi_1(C^k - \Sigma)$ of the base of the fibration considered on the homology of the fiber V_λ (for odd n) is the standard representation of the generalized braid group as the Weyl group; the Picard-Lefschetz transformation is realized as a map into $H_{n-1}(V_\lambda)$ (for A_k, see the work of A. N. Varchenko [11]).

Remark 9.6. An interesting special case is the case $n = 3$. In this case the hypersurface V_λ is an ordinary surface. The singularities of surfaces of types A_k, D_k, E_k have been thoroughly studied under the name of "double rational points" (see [12], [13], [14]). In the theory of double rational points, in particular, it is proved that a singular point of a surface which satisfies some rigidity condition, can be reduced to one of the types A_k, D_k, E_6, E_7, E_8. Moreover, it turns out that the set of lines joined in minimal resolutions of double rational points is described by a Dynkin diagram (points – lines, segments – intersections).

Surfaces with double rational points, their versal deformations, and their resolutions can be obtained from the corresponding Lie groups (A_k, D_k, E_k): they appear in the description of the singularities of a map, associating with a matrix (an element of the Lie algebra) its characteristic polynomial. In this connection, see [15], [16], [17]. With the indicated map is also connected a family of algebraic manifolds of other (but always even) dimensions – it would not be surprising to meet among them $\{V_\lambda\}$ not only for $n - 1 = 2$ but also for $n - 1 = 2l > 2$.

Despite the facts mentioned above, the connection of singularities with Weyl groups does not seem to be well understood. For example, it is not clear whether connected with singularities are only coxeter groups generated by reflections, or real Lie groups with their Weyl groups – in the latter case one could hope to obtain some information about nonsimple germs.

§10. Lagrangian Singularities

The investigation above of the classification of simple germs of functions has applications in the theory of singularities of projections of so-called lagrangian manifolds.

Actually, this classification was found for the solution of the problems of asymptotic integrals of rapidly oscillating functions [18], which is closely connected with lagrangian singularities. In the present paragraph is communicated (without detailed proof) preliminary information about lagrangian singularities; the classification results obtained with its help is contained in §11.

Definition 10.1. A symplectic manifold is a pair (M^{2n}, ω^2), where M^{2n} is a smooth even-dimensional manifold, and ω^2 is a closed nondegenerate differential 2-form on it. A diffeomorphism of symplectic manifolds $f: M_1^{2n} \to M_2^{2n}$ is called symplectic if $f^*\omega_2^2 = \omega_1^2$.

Definition 10.2. A submanifold L^n of a symplectic manifold (M^{2n}, ω^2) is called a lagrangian manifold if the form $i^*\omega^2$ induced by the inclusion $i: L \to M$ is equal to zero.

Definition 10.3. A fibration $p: M \to B$ is called a lagrangian fibration if its fiber is lagrangian.

Definition 10.4. A lagrangian map $\pi: L \to B$ is the map induced by the projection $\pi = pi$ of a lagrangian manifold onto the base of a lagrangian fibration.

Examples of lagrangian fibrations are the cotangent fibrations T^*B of smooth manifolds B. More particular examples are coordinate 2n-dimensional space R^{2n} with coordinates x_i, y_i $(i = 1, \ldots, n)$, with form $\omega^2 = \sum dx_i \wedge dy_i$ and projections $\pi(x, y) = y$.

PROPOSITION 10.5. Any lagrangian fibration is locally isomorphic to the standard one just described, and the affine structure on the fiber in a neighborhood of each point is defined invariantly.

The proof is based on an application of Darboux's theorem.

An example of a lagrangian submanifold is a submanifold $L = \{x, y : y = \partial S/\partial x\}$, where S is a function on $R^n = \{x\}$, of the standard 2n-dimensional space R^{2n}. The corresponding lagrangian map $\pi: L \to B$ is called the gradient. If x_i are taken as coordinates on L, and y_i on B, then π is given by the formula

$$\pi(x_1, \ldots, x_n) = \left(\frac{\partial S}{\partial x_1}, \ldots, \frac{\partial S}{\partial x_n}\right). \qquad (10.1)$$

Every lagrangian submanifold L^n in R^{2n} is locally given by its "derivative function" S(x) in a neighborhood of each of its points, in which the tangent space to L^n is transversal to the y-space.

PROPOSITION 10.6. In a neighborhood of each point without exception a lagrangian manifold is given by at least one of the 2^n formulas of the following form:

$$y_i = \frac{\partial F}{\partial x_i}, \quad x_j = -\frac{\partial F}{\partial y_j} \quad (i \in I,\ j \in J), \qquad (10.2)$$

where $F = F(x_I, y_J)$ is the "derivative function" and $I = (i_1, \ldots, i_k)$ is one of the 2^n subsets of the set $(1, \ldots, n)$, and J is its complement. Here, if the kernel of the projection of the tangent plane onto the y-space is k-dimensional, then as I one can take a set of k elements.

The proof is found, for example, in [19].

Parallel to the general theory of singularities of smooth maps there is a theory of singularities of lagrangian maps, which are about as frequently met in applications (caustics, envelopes, Huygens principle, Hamilton-Jacobi equations, etc.).

Definition 10.7. A lagrangian equivalence of lagrangian maps $\pi_{1,2} : L_{1,2} \to B_{1,2}$ is a symplectic diffeomorphism $s: M_1 \to M_2$, $d: B_1 \to B_2$ of the lagrangian fibrations $p_{1,2}: M_{1,2} \to B_{1,2}$, such that L_1 is carried into L_2. If such an equivalence exists, then the maps π_1, π_2 are said to be lagrangian equivalent.

Proposition 10.8. From lagrangian equivalence follows equivalence in the sense of the ordinary theory of singularities of smooth maps ($f_{1,2}: U \to V$ are equivalent if there exist diffeomorphisms $h: U \to U$ and $k: V \to V$ such that $kf_1 = f_2h$).

The proof is obvious. The converse proposition is not true, as is shown by the example of smooth equivalent but lagrangian inequivalent germs at zero $f_1 = x^3$ and $f_2 = x^3 + x^4$.

Definition 10.9. A lagrangian map is said to be lagrange stable if every close lagrangian map is lagrangian equivalent to it.

Remark. In the definition of lagrange stability one must choose among several possible concepts. Here is one of them.

Definition 10.10. A lagrangian map is said to be weakly lagrange stable if every close lagrangian map is equivalent to it in the sense of the ordinary theory of singularities.

Lagrange stable maps are weakly lagrange stable according to proposition 10.8. I do not know examples of weakly lagrange stable germs of maps which are not lagrange stable.*

*There exist germs of smooth maps $R^2 \to R^2$ which are not equivalent to gradients in the sense of the ordinary theory of singularities (and also local rings which are not realized by gradients). The author thanks V. P. Palamodova, for showing the example: $y_1 = x_1^3$, $y_2 = x_1^2 x_2 + x_2^3$.

Remark. The concepts of lagrangian equivalence and stability in the large are not very reasonable. For example, the lagrangian curves $(x^2 + y^2 - 4)((x - \lambda)^2 + y^2 - 1) = 0$ in the plane (x, y) for different λ_1, $\lambda_2 (0 \leq \lambda_{1,2} < 1)$ are lagrange inequivalent because of the invariance of the affine structure on the fiber of a lagrangian fibration.

Local variants of definitions of lagrangian stability and weak lagrangian stability of germs are constructed on the model of the ordinary theory of singularities (see [1]). Below "stability" always means "lagrangian stability," if the contrary is not asserted.

PROPOSITION 10.11. Every germ of a lagrangian map is lagrange equivalent to the germ of a gradient map.

For the proof, a slight bending of the coordinate system suffices. In fact, every germ of a lagrangian map can be written in the form (10.2). The symplectic diffeomorphism given by $x'_i = x_i$, $y'_i = y_i$ $(i \in I)$, $x'_j = x_j + \lambda y_j$, $y'_j = y_j$ $(j \notin I)$, preserves fibers and defines a lagrangian equivalence of the initial map with some new lagrangian map, depending on the number λ. It is easy to verify that this new germ is a gradient [i.e., can be written in the form (10.1)] for almost all λ.

In fact, a derivative function of the form (10.2) for the new germ will be $F - \frac{1}{2}\lambda \sum y_j^2$ $(j \in J)$. The condition of local solvability of the new equation (10.2) with respect to y_J will be $|\partial^2 F/\partial y_J^2 - \lambda E| \neq 0$. If λ is not an eigenvalue of the hessian of F with respect to y_J, then the new germ is a gradient; thus, the exceptional values of λ are finite in number (not more than n).

Proposition 10.11 formally reduces the study of all germs of lagrangian maps to the study of gradient germs. However actually in working with lagrangian singularities it is more convenient to use the charts (10.2).

We proceed now to the infinitesimal analogs of the concepts introduced.

Definition 10.12. An infinitesimal lagrangian equivalence of a lagrangian fibration is a vector field on the total space which preserves both the symplectic structure and the fiber structure.

Just like any vector field which preserves the symplectic structure on M, the field of an infinitesimal lagrangian equivalence X is locally given by a real function (the Hamiltonian function) H by the formula $\omega^2(X(x), \xi) = dH(\xi)$ $(\forall \xi \in T_x M)$.

LEMMA 10.13. The Hamiltonian function of an infinitesimal lagrangian equivalence is linear (inhomogeneous) along each fiber. Conversely, any function which is linear (inhomogeneous) along each fiber gives an infinitesimal lagrangian equivalence.

Proof. By virtue of proposition 10.5, it suffices to consider the coordinate fibration $(x, y) \mapsto (y)$. Then the field X has components $\partial H/\partial y, -\partial H/\partial x$. For the field to be a lagrangian equivalence, the second component should not depend on x. Consequently, $H = a(y)x + b(y)$, which is what was needed.

Remark. Lagrangian equivalences are classically called "extended point transformations."

Definition 10.14. An infinitesimal lagrangian deformation of a lagrangian manifold is an element of the tangent space to the manifold of a lagrangian manifold.

Definition 10.15. A lagrangian map is called infinitesimally stable if every infinitesimal lagrangian deformation of its lagrangian manifold is induced by some infinitesimal lagrangian equivalence.

The definition of the corresponding local concepts is analogous: it is only necessary to replace manifolds and maps by their germs everywhere.

Remark. In order to make definition 10.14 for germs completely precise, we note that if the germ of a lagrangian manifold is given by formula (10.1) or (10.2), then the tangent space to the manifold germ is identified in the germs considered with the space of germs of smooth functions of x (or of x_I, y^J). From Lemma 10.13 it is easy to deduce

PROPOSITION 10.16. For infinitesimal stability of a gradient germ (10.1) at zero it is necessary and sufficient that the local ring of the gradient map be generated by linear functions.

Here, the local ring of a gradient map is the quotient ring of the germs of smooth functions of x at 0 by the ideal spanned by the n germs of the partial derivatives $\partial S/\partial x_i$.

Thus, the condition of infinitesimal stability consists in the fact that for every germ α of a smooth function at 0 there exists a decomposition $\alpha(x) = \sum \frac{\partial S}{\partial x_i} h_i(x) + c_0 + c_1 x_1 + \ldots + c_n x_n$, where h_i are germs of smooth functions at zero, and c_i are numbers.

Now we shall formulate the condition of infinitesimal stability for germs of the form (10.2) at zero. We assume that at zero $F = 0$ and $\partial F/\partial x_I = 0$ (the latter can be achieved by translation of the origin of coordinates y_I). We introduce the functions $f(x) = F(x, 0)$ and $\varphi_j(x) = (\partial F/\partial y_j)(x, 0)$ ($j \in J$, $x = x_I$).

From Lemma 10.13 follows

THEOREM 10.17. Infinitesimal stability of the germ (10.2) at zero is equivalent to versality of the n + 1-parameter deformation G of the germ of the function f at 0 which is obtained from F if the y_j are considered as parameters and a general linear inhomogeneous function of x is added:

$$G(x, \lambda) = f(x) + \sum \lambda_j \varphi_j(x) + \lambda_0 + \sum \lambda_i x_i \quad (i \in I,\ j \in J).$$

The proof is based on the fact that the condition of infinitesimal stability (existence of a decomposition of any germ $\alpha(x, y)$ into a sum $\alpha(x, y) = A_0 + \sum A_i x_i + \sum A_j (\partial F/\partial y_j)$, $i \in I$, $j \in J$, where A_m are germs of functions of $\partial F/\partial x_i$ and of y_J at zero) coincides with the condition of infinitesimal versality from Par. 8.2 (existence of a decomposition of any germ $\alpha(x)$ into a sum $(x) = \sum h_i \partial f/\partial x_i + \sum c_j \varphi_j(x_I) + c_0 + \sum c_i x_i$ $(i \in I,\ j \in J)$ (where h are germs of functions of x_I at zero, and c_m are numbers) by the preparation theorem of Weierstrass-Malgrange.

The classification of lagrangian germs is obtained by combining the theorems on functions from §§ 2-8, Theorem 10.17 and the following proposition.

THEOREM 10.18. Every infinitesimally stable germ of a lagrangian map is stable.

This follows from general theorems on actions of infinite-dimensional groups, which are not formulated in [1], and can be proved on the model of the proofs of the stability theorems in [1] or on the model of the proofs of J. Mather in [7].

§11. Classification of Simple Lagrangian Germs

Now we shall define and classify simple stable germs of lagrangian maps. Germs of lagrangian maps of an n-dimensional lagrangian manifold in general position are stable and simple for $n < 6$. Hence our classification reduces to normal form the germs of lagrangian maps in general position for $n < 6$.

Definition 11.1. A germ of a lagrangian map at the point x_0 is called simple, if there exists a finite set of germs of lagrangian maps such that every finite-parameter family of lagrangian maps containing a map with the given germ for the value zero of the parameters has for all close values of the parameters at all points sufficiently close to x_0, only germs which are lagrangian equivalent to germs of the given set.

Remark 11.12. Stable germs are not necessarily simple. In fact, the manifold of singularities of a stable germ can contain curves along which the lagrangian type of the germ varies continuously. Simple germs can be unstable (an example is the gradient map of the line with $S(x) = x^4$). The basic result of this paragraph is the classification of simple stable germs.

THEOREM 11.3. Every stable simple germ of a lagrangian map of an n-dimensional lagrangian manifold is lagrangian equivalent to one of the germs of the following list:

$$\begin{aligned}
A_k&:\ F = \pm x_1^{k+1} + y_{k-1}x_1^{k-1} + \ldots + y_2 x_1^2 && (k \leqslant n+1,\ I = \{1\}),\\
D_k&:\ F = \pm x_1^2 x_2 \pm x_2^{k-1} + y_{k-1}x_2^{k-2} + \ldots + y_3 x_2^2 && (k \leqslant n+1,\ I = \{1, 2\}),\\
E_6&:\ F = \pm x_1^3 \pm x_2^4 + y_5 x_1 x_2^2 + y_4 x_1 x_2 + y_3 x_2^2 && (5 \leqslant n,\ I = \{1, 2\}),\\
E_7&:\ F = \pm x_1^3 \pm x_1 x_2^3 + y_6 x_1 x_2 + y_5 x_2^4 + y_4 x_2^3 + y_3 x_2^2 && (6 \leqslant n,\ I = \{1, 2\}),\\
E_8&:\ F = \pm x_1^3 \pm x_2^5 + y_7 x_1 x_2^3 + y_6 x_1 x_2^2 + y_5 x_1 x_2 + y_4 x_2^3 + y_3 x_2^2 && (7 \leqslant n,\ I = \{1, 2\}).
\end{aligned}$$

269

Here F is a function which gives the equations of a lagrangian submanifold of the space $R^{2n} = \{(x, y)\}$ by the formulas $y_i = \partial F/\partial x_i$, $x_j = -\partial F/\partial y_j$ $(i \in I, j \notin I)$. The lagrangian map is given by the projection $(x, y) \mapsto y$.

Remark 11.4. The normal forms 11.3 are written by throwing out summands of degree 0 and 1 in x from the formulas for versal deformations in Par. 8.4. The stability of the germs follows from Theorems 10.17 and 10.18; the completeness of the list is proved (with the help of these theorems) by the same arguments which in §§3-6 proved the completeness of the list of simple germs of functions.

COROLLARY 11.5. Every stable simple germ of a lagrangian map of an n-dimensional lagrangian manifold is lagrangian equivalent to one of the germs of gradient maps given by the following functions:

$$A_k:\ S = \pm x_1^{k+1} + (x_{k-1} + x_1^{k-1})^2 + \ldots + (x_2 + x_1^2)^2 + Q,$$
$$D_k:\ S = \pm x_1^2 x_2 \pm x_2^{k-1} + (x_{k-1} + x_2^{k-2})^2 + \ldots + (x_3 + x_2^2)^2 + Q,$$
$$E_6:\ S = \pm x_1^3 \pm x_2^4 + (x_5 + x_1 x_2^2)^2 + (x_4 + x_1 x_2)^2 + (x_3 + x_2^2)^2 + Q,$$
$$E_7:\ S = \pm x_1^3 \pm x_1 x_2^3 + (x_6 + x_1 x_2)^2 + (x_5 + x_2^4)^2 + (x_4 + x_2^3)^2 + (x_3 + x_2^2)^2 + Q,$$
$$E_8:\ S = \pm x_1^3 \pm x_2^5 + (x_7 + x_1 x_2^3)^2 + (x_6 + x_1 x_2^2)^2 + (x_5 + x_1 x_2)^2 + (x_4 + x_2^3)^2 + (x_3 + x_2^2)^2 + Q,$$

where $Q = x_k^2 + \ldots + x_n^2$.

The corollary is proved with the help of proposition 10.11 and the normal forms given here are obtained from the normal forms of Theorem 11.3 by the transformations indicated in the proof of proposition 10.11.

THEOREM 11.6. For $n < 6$ a lagrangian map of an n-dimensional lagrangian manifold in general position has at each point a simple stable germ.

COROLLARY 11.7. A lagrangian map of a lagrangian manifold of dimension $n < 6$ can, by a small perturbation (in the class of lagrangian maps), be transformed into one such that in a neighborhood of each of its points it will reduce by a lagrangian equivalence to one of the following normal forms:

for $n = 1$

$$A_1:\ F = x_1^2, \quad A_2:\ F = \pm x_1^3;$$

for $n = 2$, in addition,

$$A_3:\ F = \pm x_1^4 + y_2 x_1^2;$$

for $n = 3$, in addition,

$$A_4:\ F = \pm x_1^5 + y_3 x_1^3 + y_2 x_1^2,$$
$$D_4:\ F = \pm x_1^2 x_2 \pm x_2^3 + y_3 x_2^2;$$

for $n = 4$, in addition,

$$A_5:\ F = \pm x_1^6 + y_4 x_1^4 + y_3 x_1^3 + y_2 x_1^2,$$
$$D_5:\ F = \pm x_1^2 x_2 \pm x_2^4 + y_4 x_2^3 + y_3 x_2^2;$$

for $n = 5$, in addition,

$$A_6:\ F = \pm x_1^7 + y_5 x_1^5 + \ldots + y_2 x_1^2,$$
$$D_6:\ F = \pm x_1^2 x_2 \pm x_2^5 + y_5 x_2^4 + y_4 x_2^3 + y_3 x_2^2,$$
$$E_6:\ F = \pm x_1^3 \pm x_2^4 + y_5 x_1 x_2^2 + y_4 x_1 x_2 + y_3 x_2^2.$$

Here the lagrangian manifold is given in the space $R^{2n} = \{(x, y)\}$ by the equations $y_i = \partial F/\partial x_i$, $x_j = -\partial F/\partial y_j$, where $i = 1$, $j \neq 1$ for the cases A_k and $i = 1$ or 2, $j \neq 1$ and 2 for D_k and E_k. The lagrangian map is the projection onto the y-space.

For example, a typical lagrangian map of a three-dimensional manifold has at isolated points singularities of type A_4 and D_4 (three tangents come together, two of them can be minimal), on curves joining these points there can be singularities of type A_3 (cusps) and on surfaces passing through these curves there can be singularities of type A_2 (folds).

270

We remark here further that the codimension $c(I) = c(i_1, \ldots, i_s)$ of the Boardman class Σ^I (see [1] or [20], [21]) in the space of jets of lagrangian maps of an n-dimensional manifold can be computed by the following formula: $c(I) = \nu_{n,1}(n, I) - n$, where ν is the codimension of the Boardman class $\Sigma^{n, i_1, \ldots, i_s}$ in the space of jets of functions $R^n \to R^1$. To compute ν, in [1] or [20] there are formulas. For example, $A_k = \Sigma^{1_{k-1},0}$ and has codimension $k - 1$, Σ^2 has codimension 3 and consists (with a residual part of codimension 4) of jets of type D_4, $\Sigma^{2,1}$ has codimension 5 and consists (with a residual part of codimension 6) of jets of type E_6, and $\Sigma^{2,1,1}$ has codimension 7 and contains E_8.

COROLLARY 11.8. A lagrangian map of a typical lagrangian manifold of dimension $n < 6$ can, by a small perturbation (in the class of lagrangian maps), be transformed into one such that in a neighborhood of any point it can be reduced by a lagrangian equivalence to one of the following normal forms $x \mapsto \partial S/\partial x$:

for n = 1

$$A_1:\ S = x_1^2,\quad A_2:\ S = \pm x_1^3,$$

for n = 2

$$A_1:\ S = x_1^2 + x_2^2,\quad A_2:\ S = \pm x_1^3 + x_2^2,\quad A_3:\ S = \pm x_1^4 + (x_2 + x_1^2)^2.$$

for n = 3

$$\begin{aligned}
A_1&:\ S = x_1^2 + x_2^2 + x_3^2,\\
A_2&:\ S = \pm x_1^3 + x_2^2 + x_3^2,\\
A_3&:\ S = \pm x_1^4 + (x_2 + x_1^2)^2 + x_3^2,\\
A_4&:\ S = \pm x_1^5 + (x_3 + x_1^3)^2 + (x_2 + x_1^2)^2,\\
D_4&:\ S = \pm x_1^2 x_2 \pm x_2^3 + (x_3 + x_2^2)^2,
\end{aligned}$$

for n = 4

$$\begin{aligned}
A_1&:\ S = x_1^2 + x_2^2 + x_3^2 + x_4^2,\\
A_2&:\ S = \pm x_1^3 + x_2^2 + x_3^2 + x_4^2,\\
A_3&:\ S = \pm x_1^4 + (x_2 + x_1^2)^2 + x_3^2 + x_4^2,\\
A_4&:\ S = \pm x_1^5 + (x_3 + x_1^3)^2 + (x_2 + x_1^2)^2 + x_4^2,\\
A_5&:\ S = \pm x_1^6 + (x_4 + x_1^4)^2 + (x_3 + x_1^3)^2 + (x_2 + x_1^2)^2,\\
D_4&:\ S = \pm x_1^2 x_2 \pm x_2^3 + (x_3 + x_2^2)^2 + x_4^2,\\
D_5&:\ S = \pm x_1^2 x_2 \pm x_2^4 + (x_4 + x_2^3)^2 + (x_3 + x_2^2)^2,
\end{aligned}$$

for n = 5

$$\begin{aligned}
A_1&:\ S = x_1^2 + x_2^2 + x_3^2 + x_4^2 + x_5^2,\\
A_2&:\ S = \pm x_1^3 + x_2^2 + x_3^2 + x_4^2 + x_5^2,\\
A_3&:\ S = \pm x_1^4 + (x_2 + x_1^2)^2 + x_3^2 + x_4^2 + x_5^2,\\
A_4&:\ S = \pm x_1^5 + (x_3 + x_1^3)^2 + (x_2 + x_1^2)^2 + x_4^2 + x_5^2,\\
A_5&:\ S = \pm x_1^6 + (x_4 + x_1^4)^2 + (x_3 + x_1^3)^2 + (x_2 + x_1^2)^2 + x_5^2,\\
A_6&:\ S = \pm x_1^7 + (x_5 + x_1^5)^2 + (x_4 + x_1^4)^2 + (x_3 + x_1^3)^2 + (x_2 + x_1^2)^2,\\
D_4&:\ S = \pm x_1^2 x_2 \pm x_2^3 + (x_3 + x_2^2)^2 + x_4^2 + x_5^2,\\
D_5&:\ S = \pm x_1^2 x_2 \pm x_2^4 + (x_4 + x_2^3)^2 + (x_3 + x_2^2)^2 + x_5^2,\\
D_6&:\ S = \pm x_1^2 x_2 \pm x_2^5 + (x_5 + x_2^4)^2 + (x_4 + x_2^3)^2 + (x_3 + x_2^2)^2,\\
E_6&:\ S = \pm x_1^3 \pm x_2^4 + (x_5 + x_1 x_2^2)^2 + (x_4 + x_1 x_2)^2 + (x_3 + x_2^2)^2.
\end{aligned}$$

Remark 11.9. For $n \geq 6$ there exist lagrangian maps which cannot be approximated by lagrangian maps whose germs at each point are lagrange stable (or simple, or even weakly lagrange stable). All this follows from the existence of projectively invariant cubical curves on the projective plane and from the fact that the codimension of Σ^3 in the lagrangian case is equal to 6.

Note Added in Proof. At the time of publication of this paper the author received the preprint of J. Guckenheim on "Catastrophes and partial differential equations," pp. 1-29 (Princeton, 1972); among the results announced by J. Guckenheim are some of the propositions of § 10 of the present paper; he also indicated that some of them were found by L. Hörmander and A. Weinstein.

LITERATURE CITED

1. V. I. Arnol'd, "Singularities of smooth maps," UMN, 23, No. 1, 3-44 (1968).

2. J. C. Tougeron, "Ideaux de fonctions differentiables. I," Ann. Inst. Fourier, 18, No. 1, 177-240 (1968).
3. A. M. Samoilenko, "On equivalences of Taylor's polynomials of smooth functions in a neighborhood of critical points of finite type," Funktsional'. Analiz i Ego Prilozhen., 2, No. 4, 63-69 (1968).
4. J. Milnor, Morse Theory, Princeton Univ. Press (1963).
5. G. N. Tyurina, "Local semiuniversal flat deformations of isolated singularities of complex spaces," Izv. AN SSSR, Ser. Matem., 33, 1026-1058 (1969).
6. V. I. Arnol'd, "Lectures on bifurcation and versal families," UMN, 27, No. 5, 120-184 (1972).
7. J. Mather, "Structural stability of maps," in: Singularities of Differentiated Reflections [Russian translation], Mir, Moscow (1968), pp. 216-267.
8. V. I. Arnol'd, "Rings of algebraic functions and the cohomology of swallow's tails," UMN, 23, No. 4, 247-248 (1968).
9. E. Brieskorn, "Die Fundamentalgruppe des Raumes der regulären Orbits einer endlichen komplexen Spiegelungsgruppe," Invent. Math., 12, No. 1, 57-61 (1971).
10. E. Brieskorn, "Braids," Seminaire N. Bourbaki, 24, No. 401, November, 1971.
11. A. N. Varchenko, "On the branching of multiple integrals depending on a parameter," Functional'. Analiz i Ego Prilozhen., 3, No. 4, 79-80 (1969).
12. M. Artin, "Some numerical criteria for contractibility of curves on algebraic surfaces," Amer. J. Math., 84, 485-496 (1962).
13. E. Brieskorn, "Die Auflösung der rationalen Singularitäten holomorphen Abbildungen," Math. Ann., 178, 255-270 (1968).
14. G. N. Tyurina, "Resolution of singularities of flat deformations of double rational points," Funktional'. Analiz i Ego Prilozhen., 4, No. 1, 77-83 (1970).
15. V. I. Arnol'd, "On matrices depending on parameters," UMN, 26, No. 2, 101-114 (1971).
16. E. Brieskorn, Nice Mathematical Congress, 1970.
17. T. Springer, Budapest Symposium on Group Representations, 1971.
18. V. I. Arnol'd, "Integrals of rapidly oscillating functions and singularities of projections of lagrangian manifolds," Funktional'. Analiz i Ego Prilozhen., 6, No. 3, 61-62 (1972).
19. V. I. Arnol'd, "On characteristic classes appearing in quantization conditions," Funktional'. Analiz i Ego Prilozhen., 1, No. 1, 1-14 (1967).
20. J. Boardman, in: Singularities of Differentiated Reflections [Russian translation], Mir, Moscow (1968), pp. 102-153.
21. B. Morin, Calcul Jacobien, These, Paris-Sud (1972).

NORMAL FORMS OF FUNCTIONS IN NEIGHBOURHOODS OF DEGENERATE CRITICAL POINTS

V. I. Arnol'd

An analysis of the normal forms to which functions can be reduced in neighbourhoods of degenerate critical points shows that many of them are quasihomogeneous or semiquasihomogeneous.

A semiquasihomogeneous function is a sum of a quasihomogeneous (or weighted homogeneous) polynomial with an isolated critical point and summands of a higher degree of quasihomogeneity. The normal form to which a semiquasihomogeneous function can be reduced is described in terms of the local ring of the gradient mapping given by the quasihomogeneous part of the function. The number of parameters in this normal form is called the inner modality of the quasihomogeneous part.

A classification is given of all quasihomogeneous critical points of inner modality 1: up to stable equivalence they are exhausted by three one-parameter families of parabolic singularities and 14 exceptional polynomials, 8 of which are functions of two variables, and 6 functions of three variables.

Contents

§1. Introduction

1.1. Degenerate critical points. A critical point of a function is said to be *non-degenerate* if the second differential of the function at the point is

Originally publ. in: Usp. Mat. Nauk 29(2), 11-49, © Russ. Acad. Sciences 1974
English transl.: Russ. Math. Surv. 29(2), 10-50, © London Mathematical Society 1974
Reprint by Springer-Verlag Berlin Heidelberg 2016 with kind permission.

a non-degenerate quadratic form. The behaviour of a function in the neighbourhood of a non-degenerate critical point is described by Morse's lemma: there is a system of coordinates in the neighbourhood of such a point in which the function takes a simple "normal form", namely

$$f(x_1, \ldots, x_n) = \pm x_1^2 \pm \ldots \pm x_n^2 + \text{const.}$$

A function "in general position" has only non-degenerate critical points. Every function can be brought to general position by arbitrarily small displacements, so that all its critical points become non-degenerate.

Degenerate critical points appear naturally in cases when the functions depend on parameters. For instance, in the one-parameter family

$$f(x, t) = x^3 + tx$$

of functions of a single variable x, a degenerate critical point arises for an exceptional value of the parameter ($t = 0$). This point cannot be removed by small displacement of the family: whatever the displacement, a degenerate singularity always occurs for a nearby value of the parameter.

Thus, in the investigation of critical points of functions depending on parameters we have to consider degenerate points as well as non-degenerate. The larger the number of parameters, the more complicated the critical points that can occur.

1.2. **Terminology**. This article is the fifth in a series ([1]–[4]) on the classification of critical points of functions. To make this paper independent of its predecessors, we give here certain definitions and results from [1]–[4].

At first glance, the problem of classifying critical points of l-parameter families of functions of n variables that are irremovable under small displacements seems hopeless for large l and n. Nonetheless, when the first piece of the classification was executed, it turned out to be not as complicated as all that, and the singularities arising were closely connected with objects that seem very remote from them at first sight: namely with Lie groups and the Weyl series A_k, D_k, E_k, Coxeter groups generated by reflections, the braid groups of Artin and Brieskorn, and finally, with the classification of regular polygons in ordinary three-space.

At first the classification of critical points is discrete: if the number l of parameters is less than six, then we can find a finite number of normal forms such that every l-parameter family of functions can be brought to "general position" by a small displacement under which every function in the family reduces to one of the given normal forms in the neighbourhood of every critical point, after a smooth change of variables.

Beginning with $l = 6$ (and the number of variables $n \geqslant 3$), "moduli" emerge, and the normal forms inevitably involve parameters.

The number of moduli (that is, the number of parameters in the normal forms) remains finite for finite l. The fact is that a smooth function is equivalent in a neighbourhood of a critical point to a fairly long segment of its Taylor series, except in the case of critical points of infinite multiplicity (which can be removed under a small displacement of an l-parameter system, for any finite l) (see [5]–[7]).

The class of functions whose Taylor series at 0 coincide up to the terms of degree k with the Taylor series of a given function is called the *k-jet* of the function at 0. A k-jet is said to be *sufficient* if all functions with this k-jet are mutually equivalent (that is, can be transformed into one another under a smooth change of variables).

In these terms we can state that *every function with a critical point of finite multiplicity has a sufficient jet.*

Finiteness of the multiplicity of a critical point is defined easily in the complex case: a function $f\colon (\mathbf{C}^n, 0) \to (\mathbf{C}, 0)$ has 0 as a critical point of finite multiplicity if 0 is an isolated critical point. The *multiplicity* μ of an isolated critical point of a complex function can be defined as the number of non-degenerate critical points into which it splits under a small displacement, or as the degree of the gradient function mapping a small sphere S^{2n-1} around 0 in $\mathbf{C}^n$ onto itself.

In the real case the multiplicity μ of a smooth mapping $F\colon (\mathbf{R}^n, 0) \to (\mathbf{R}^n, 0)$ at 0 is defined as the $\mathbf{R}$-dimension of the local ring

$$Q(F) = \mathbf{R}[[x_1, \ldots, x_n]]/(F_1, \ldots, F_n),$$

where $(F_1, \ldots, F_n)$ is the ideal spanned by the Taylor series of the functions specifying F. Here $\mathbf{R}[[\ldots]]$ is the ring of formal power series in the coordinates $(x_1, \ldots, x_n)$ in the spatial inverse image (instead of formal power series we can use convergent series, polynomials, or smooth functions, provided that $\mu < \infty$).

The multiplicity μ of the critical point 0 of a function $f\colon (\mathbf{R}^n, 0) \to (\mathbf{R}^n, 0)$ is defined as the multiplicity of the gradient mapping: $\mu = \dim_{\mathbf{R}} Q(F)$, where $F = \operatorname{grad} f$. We denote the local ring of the gradient mapping by Q_f, and call it the *local ring of the function* at the critical point. Of course, an analogous ring is defined in the complex case, and then $\mu = \dim_{\mathbf{C}} Q_f$ [8].

In investigating critical points of finite multiplicity, the function may be replaced by a sufficient jet or a polynomial.

Thus, the question of classifying critical points reduces to an algebraic question about the orbits of the action of a finite-dimensional Lie group on a finite-dimensional manifold.

The number of moduli (or the modality) of a function in the neighbourhood of a critical point is defined as follows.

Let G be a Lie group acting on a (finite-dimensional) manifold X. The orbits of G can generate discrete stratifications or continuous families in the neighbourhood of a given point $x \in X$. We say that a point x *has modality* m (under the given action) if a sufficiently small neighbourhood of x in X can be covered by finitely many families of orbits, depending on not more than m parameters (and an arbitrarily small neighbourhood of x intersects some m-parameter family of orbits).

By the modality of a function at a critical point we understand the modality of the action of the group of changes of independent variables on the space of functionals of the function.

More accurately, we consider the finite-dimensional manifold of k-jets at 0 of the function f: $(\mathbf{C}^n, 0) \to (\mathbf{C}, 0)$ with the critical point 0 and critical value 0. The group of k-jets at 0 of diffeomorphisms $(\mathbf{C}^n, 0) \to (\mathbf{C}^n, 0)$ fixing 0 acts on this manifold. By the *modality* of f at the critical point 0 we mean the modality of its k-jet under the action of the group of k-jets of diffeomorphisms, where k is sufficiently large.

1.3. 0-modal critical points. The simplest degenerate critical points are the 0-modal ones. The 0-modal critical points can be classified as follows.

Two functions with critical point 0 and critical value 0 are said to be *equivalent* if one is transformed into the other under a diffeomorphism fixing 0 of a sufficiently small neighbourhood of 0. Two functions are *stably equivalent* if they become equivalent under direct addition of non-degenerate quadratic forms. For example, the functions $f(x) = x^3$ and $g(x, y) = x^3 + y^2$ are stably equivalent.

In a neighbourhood of a 0-modal critical point 0, every function f: $(\mathbf{C}^n, 0) \to (\mathbf{C}^n, 0)$ is equivalent to one of the following functions [2]:

$$\begin{aligned} A_k &: f(x) = x^{k+1}, \\ D_k &: f(x, y) = x^2 y + y^{k-1}, \\ E_6 &: f(x, y) = x^3 + y^4, \\ E_7 &: f(x, y) = x^3 + xy^3, \\ E_8 &: f(x, y) = x^3 + y^5. \end{aligned}$$

The classification of 0-modal critical points has given rise to the hope that the natural classes of critical points allowing a simple description are those of small modality (and not the classes with small μ, or the classes with small codimension l, which are needed in analysis and topology).

1.4. Unimodal critical points. The classification of critical points of modality 1 turned out to be not too complicated [3]. Indeed, there is one three-index series of one-parameter families of unimodal singularities,

$$T_{p,q,r}: f(x, y, z) = axyz + x^p + y^q + z^r,$$

where $\frac{1}{p} + \frac{1}{q} + \frac{1}{r} \leqslant 1$, and 14 "exceptional" one-parameter families. Every function f: $(\mathbf{C}^n, 0) \to (\mathbf{C}^n, 0)$ is equivalent in the neighbourhood of a critical point of modality 1 to one of those listed.

There are three special cases of the series T for which $\frac{1}{p} + \frac{1}{q} + \frac{1}{r} = 1$, namely (3, 3, 3), (2, 4, 4) and (2, 3, 6). These singularities are said to be *parabolic*, and the remaining singularities in the series T are *hyperbolic* (corresponding to the signature of the quadratic form given by the intersection index in the homology of average dimension of the local manifold of non-singular level of the function – see [4]; *elliptic* singularities turn up only as the 0-modal singularities A, D, E).

The 14 exceptional families of unimodal singularities are also given by simple formulae:

$$K_{12} = x^3 + y^7 + axy^5, \qquad K_{13} = x^3 + xy^5 + ay^8,$$
$$K_{14} = x^3 + y^8 + axy^6, \qquad Z_{11} = x^3y + y^5 + axy^4,$$
$$Z_{12} = x^3y + xy^4 + ax^2y^3, \qquad Z_{13} = x^3y + y^6 + axy^5,$$
$$W_{12} = x^4 + y^5 + ax^2y^3, \qquad W_{13} = x^4 + xy^4 + ay^6,$$
$$Q_{10} = x^3 + y^4 + yz^2 + axy^3, \qquad Q_{11} = x^3 + y^2z + xz^3 + az^5,$$
$$Q_{12} = x^3 + y^5 + yz^2 + axy^4, \qquad S_{11} = x^4 + y^2z + xz^2 + ax^3z,$$
$$S_{12} = x^2y + y^2z + xz^3 + az^5, \qquad U_{12} = x^3 + y^3 + z^4 + axyz^2.$$

An analysis of these formulae shows that in each such family there is exactly one quasihomogeneous (weighted homogeneous) polynomial, and that the whole family can be considered as a one-parameter deformation of this polynomial having the special property of "semiquasihomogeneity".

The quasihomogeneous polynomials of these 14 types are connected with the 14 triangles in the Lobachevskii plane, as was shown by I. V. Dolgachev (and for K_{12} already by Klein).

1.5. Semiquasihomogeneous critical points. The aim of this paper is the investigation of quasihomogeneous functions and their semiquasihomogeneous deformations. We indicate the normal forms to which functions having non-degenerate quasihomogeneous "principal part" can be reduced. We call the number of moduli in these normal forms the *inner modality* of the quasi-homogeneous function (the exact definition of the inner modality μ_0 is in §8.6; μ_0 does not depend on the choice of the normal form).

Quasihomogeneous functions of inner modality 1 can be listed independently of the classification of all unimodal functions. The main result of our article is an algebraic classification of quasihomogeneous functions of inner modality 1.

It turns out that (up to stable equivalence) there are just three one-parameter families of parabolic singularities and 14 polynomials generating exceptional families.

The technique developed here for working with semiquasihomogeneous singularities enables us to shorten significantly the calculations necessary for classifying unimodal singularities. The technique is based on a discussion of various filtrations in the ring of functions (or power series). With every such filtration we can associate its space of jets of functions, its filtered group of jets of diffeomorphisms, and the Lie algebra of vector fields. In the case of a "quasihomogeneous" filtration there arises also a Lie group of quasihomogeneous diffeomorphisms, which plays a rôle in this theory like that of the full linear group in the case of ordinary jets (which corresponds to the filtration given by the powers of the maximal ideal).

The author is grateful to A. G. Kushirenko for numerous discussions that have put on a solid foundation calculations that had previously seemed a matter of pure luck. The author is also grateful to A. M. Gabrielov and D. B. Fuks for helpful discussions.

§2. Quasihomogeneous functions and filtrations

In this section we give the definitions of the principal objects that are studied in the sequel: quasihomogeneous and semiquasihomogeneous functions.

2.1. DEFINITION. We consider the arithmetical space $\mathbf{C}^n$ with fixed coordinates $x_1, \ldots, x_n$. *A function* f: $(\mathbf{C}^n, 0) \to (\mathbf{C}, 0)$ is said to be *quasihomogeneous of degree d with exponents* $\alpha_1, \ldots, \alpha_n$ if $f(\lambda^{\alpha_1} x_1, \ldots, \lambda^{\alpha_n} x_n) = \lambda^d f(x_1, \ldots, x_n)$ *for all* λ.

In terms of the Taylor series $f = \Sigma f_{\mathbf{k}} \mathbf{x}^{\mathbf{k}}$, quasihomogeneity of degree 1 means that all the exponents of non-zero terms lie on the hyperplane

$$\Gamma = \{\mathbf{k} : \alpha_1 k_1 + \ldots + \alpha_n k_n = 1\}.$$

In what follows, we consider quasihomogeneous functions of degree 1 with rational exponents, $0 < \alpha_s \leqslant 1/2$. Such functions are automatically polynomials. We call the hyperplane Γ the *diagonal*. The diagonal cuts the coordinate axes in segments of length $a_s = 1/\alpha_s$.

2.2. DEFINITION. A quasihomogeneous function f is said to be *non-degenerate* if 0 is an isolated critical point (that is, if the multiplicity μ of the critical point 0 is finite). The degenerate quasihomogeneous functions form an algebraic hypersurface in the linear space of all quasihomogeneous polynomials with fixed quasihomogeneous exponents.

With every type of quasihomogeneity (that is, with every set $\boldsymbol{\alpha}$ of quasihomogeneous exponents) there is associated a filtration in the ring of power series (functions, germs etc), defined as follows:

2.3. DEFINITION. We say that a monomial $\mathbf{x}^{\mathbf{k}} = x_1^{k_1} \ldots x_n^{k_n}$ has (generalized) *degree d* if $\langle \alpha, \mathbf{k} \rangle = \alpha_1 k_1 + \ldots + \alpha_n k_n = d$.

The degree of a monomial is a rational number. The exponents of all monomials of degree d (of given type) lie on a hyperplane parallel to the diagonal Γ. We fix the quasihomogeneous type, that is, we fix the set $\boldsymbol{\alpha}$ of exponents.

2.4. DEFINITION. A polynomial (power series, germ, function) *has filtration d* if all its monomials are of degree d or higher: when the (generalized) degree of all monomials is d, we call d the (generalized) *degree* of the polynomial; the degree of 0 is $+\infty$.

The polynomials (series, germs) of filtration d form a linear space E_d; $E_{d'} \subseteq E_d$ if $d < d'$; the filtration of a product is the sum of the filtrations of the factors, so that E_d is an ideal in the ring of polynomials (series, germs). Denoting this ring by A, we call the factor-ring A/E_d the *ring of d-jets,* and its elements *d-jets.*

By the filtration $\varphi(f)$ of a polynomial (series, germ) f we understand, as a rule, the largest d such that $f \in E_d$. The filtrations of all polynomials (series, germs) lie in a rational arithmetical progression: $\varphi(f) \in \mathbf{Z}_+ d_0$, where d_0 is the greatest common divisor of the numbers α_s (an initial segment of the progression may not be completely filled out by the values of φ).

2.5. DEFINITION. A polynomial (power series, germ) is said to be *semiquasihomogeneous of degree d with exponents* $\alpha_1, \ldots, \alpha_s$ if it is of the form $f = f_0 + f'$, where f_0 is a non-degenerate quasihomogeneous polynomial of degree d with exponents α, and f' is a polynomial (series, germ) of filtration strictly greater than d.

In other words, a semiquasihomogeneous function is obtained from a non-degenerate quasihomogeneous function by adding monomials whose exponents lie *above the diagonal.* Note that a quasihomogeneous function is not semiquasihomogeneous if it is degenerate.

§3. Multiplicity and generators of the local ring of a semiquasihomogeneous function

In this section we show that a monomial basis for the local ring of a quasihomogeneous non-degenerate function is a basis for all semiquasihomogeneous functions with the given quasihomogeneous part.

3.1. THEOREM. *The multiplicity of the critical point* 0 *of a semiquasihomogeneous function f is that of its quasihomogeneous part*: $\mu(f) = \mu(f_0)$.

PROOF. We consider the family of topological spheres

$$S_t = \{\mathbf{x} \in \mathbf{C}^n : |x_1|^{a_1} + \ldots + |x_n|^{a_n} = t\}, \qquad a_s = 1/\alpha_s, \quad d = 1.$$

The number $\mu(f)$ is the degree of the mapping $\mathbf{x} \to (\partial f/\partial \mathbf{x})/\|\partial f/\partial \mathbf{x}\|$, $\mathbf{x} \in S_t$, for small values of t. For every point $\mathbf{x} \in S_1$, at least one of the derivatives $\partial f_0/\partial x_s$ is different from 0 (since f_0 is non-degenerate). Thus, there is a constant c such that $\max_s |\partial f_0/\partial x_x| \geqslant c > 0$ on S_1.

Note that $S_t = T_t S_1$, where $T_t(x_1, \ldots, x_n) = (t^{\alpha_1} x_1, \ldots, t^{\alpha_n} x_n)$. Further, the partial derivative $\partial f_0/\partial x_s$ is quasihomogeneous of degree $1 - \alpha_s$ and of type α. Therefore, at every point of S_t, $|\partial f_0/\partial x_s| \geqslant c t^{1-\alpha_s}$ for at least one s.

On the other hand, the function f' has filtration not less than $1 + d_0$. Therefore there exists a constant C such that $|\partial f'/\partial x_s| \leqslant C t^{1+d_0-\alpha_s}$ on S_t for all s.

Comparing with the preceding inequality, we see that for small enough t there are no critical points of the functions $f_0 + \theta f'$, $0 \leqslant \theta \leqslant 1$, on S_t. Thus, the degree of the mappings of the sphere onto itself given by the gradients of f_0 and $f_0 + f'$ coincide, as required.

REMARK. It can be shown in a similar way that *all quasihomogeneous functions sufficiently near f_0 and having the same degree of quasihomogeneity* *have the same multiplicity* μ. Further, since the set of non-degenerate quasihomogeneous functions of given degree is connected, *the multiplicity* μ *is the same for all non-degenerate quasihomogeneous functions of given degree (and therefore for all semiquasihomogeneous functions of given degree and given type).*

The multiplicity μ of the critical point 0 of a function f may also be defined as the dimension of the local ring

$$Q_f = \mathbf{C}[[x_1, \ldots, x_n]]/(\partial f/\partial x_1, \ldots, \partial f/\partial x_n).$$

Allowing freedom of speech, we call a set of μ series (polynomials, germs) *a basis of the local ring of* f if they become a basis for Q_f over $\mathbf{C}$ after factorization by this ideal.

From the theorem just proved we deduce:

3.3. COROLLARY. *Assume that the system of monomials* $e_1, \ldots, e_\mu$ *is a basis of the local ring of the quasihomogeneous part* f_0 *of a semiquasihomogeneous function* f. *Then this same system of monomials is a basis of the local ring of* f.

The proof is based on the following general lemma.

3.4. LEMMA. *Assume that a system of smooth functions* f *depending continuously on a finite number of parameters has* 0 *as a critical point of constant multiplicity* μ *for all values of the parameters. Then every basis of the local ring of the function corresponding to the value* 0 *of the parameter remains a basis for nearby values of the parameter.*

PROOF OF THE LEMMA. The lemma follows from this fact: given a subspace of a finite-dimensional vector space depending smoothly on some parameters, and a system of vectors forming a basis of the transversal space, then the system remains a basis of the transversal space for nearby values of the parameters.

To make the space finite-dimensional, it is enough to factor the ring $\mathbf{C}[[x_1, \ldots, x_n]]$ by a sufficiently high power of the maximal ideal (the power depending only on μ, see [5]).

PROOF OF COROLLARY 3.3. We consider a semiquasihomogeneous function $f = f_0 + f'$. We claim that the transition from f_0 to f may be regarded as a *small* deformation. We construct a one-parameter family of functions, $f_t(x) = t^{-1}f(T_t x)$, where $T_t x = (t^{\alpha_1}x_1, \ldots, t^{\alpha_n}x_n)$, $d = 1$. We have $f_t(x) = f_0 + t^{-1}f'(T_t x)$, where all the coefficients in the second summand depend continuously on t, since the filtration of f' is greater than 1. By the lemma, a basis of the local ring for f_0 is a basis for f_t for sufficiently small t. A basis of the local ring for f_t goes over to a basis of that for f under the action of the diffeomorphism T_t connecting the functions f and f_t. But under T_t, every monomial goes over to a monomial proportional to it. Thus, a *monomial* basis of Q_{f_0} is not only a basis of Q_{f_t} for small t, but also of Q_f; this is what we wanted to prove.

3.5. REMARK.*The number of basis monomials of the local ring of a* *quasihomogeneous or semiquasihomogeneous function f having given generalized degree d does not depend on the choice of the basis in the local ring.*

PROOF. We consider the factor-space

$$E_\delta/(E_{>\delta} + E_\delta \cap I), \text{ where } I = (\partial f/\partial x_1, \ldots, \partial f/\partial x_n),$$

E_δ is the space of series of filtration δ in $\mathbf{C}[[x_1, \ldots, x_n]]$,

$E_{>\delta}$ is the space of series of filtration greater than δ.

The number of basis monomials of degree δ is equal to the dimension of this factor-space, therefore, does not depend on the basis.

3.6. COROLLARY. *The number of basis monomials of the local ring of a function having given generalized degree δ (for given type $\boldsymbol{\alpha}$) is the same for all semiquasihomogeneous functions f of type $\boldsymbol{\alpha}$ and degree d.*

PROOF. It is enough to consider non-degenerate quasihomogeneous functions (for a semiquasihomogeneous function the basis is the same, by Corollary 3.3). The manifold of non-degenerate quasihomogeneous functions of given degree d (and type $\boldsymbol{\alpha}$) is arcwise connected (it is the complement of a hypersurface in a linear space). The number of basis monomials of degree δ of the local ring is locally constant along a curve joining two points of this manifold, because the same basis works for neighbouring functions (Lemma 3.4). Thus, it is constant, as required.

§4. Quasihomogeneous mappings

In this section we compute various numerical invariants of quasihomogeneous mappings, in particular, the multiplicity μ and the generating polynomial χ.

We fix a quasihomogeneous type $\boldsymbol{\alpha} = (\alpha_1, \ldots, \alpha_n)$ in $\mathbf{C}^n$ with a fixed coordinate system. We consider a mapping F: $(\mathbf{C}^n, 0) \to (\mathbf{C}^n, 0)$ and use the older notation $F(x_1, \ldots, x_n) = (F_1(x_1), \ldots, F_n(x_n))$. Let $\mathbf{d} = (d_1, \ldots, d_n)$ be a vector with non-negative components.

4.1. DEFINITION. The mapping F is said to be *quasihomogeneous of degree* $\mathbf{d}$ (and type $\boldsymbol{\alpha}$) if every component F_s is a quasihomogeneous function of degree d_s of one and the same type $\boldsymbol{\alpha}$.

The local ring of F is the factor ring

$$Q(F) = \mathbf{C}[[x_1, \ldots, x_n]]/(F_1, \ldots, F_n).$$

F is said to be *non-degenerate* if its multiplicity at 0 is finite, that is, if the local ring $Q(F)$ has finite dimension over $\mathbf{C}$; this dimension $\mu = \dim_{\mathbf{C}} Q(F)$ is called the *multiplicity of F at* 0.

F is said to be *semiquasihomogeneous* if $F = F_0 + F'$, where F_0 is a non-degenerate quasihomogeneous mapping, and each component F'_s has filtration greater than the degree of the corresponding component F_{0s}.

If a function f is semiquasihomogeneous of degree d (and type $\boldsymbol{\alpha}$), then the mapping $x \mapsto \operatorname{grad}(f(x))$ is semiquasihomogeneous of degree $d_s = d - \alpha_s$.

4.2. PROPOSITION. *The assertions of Theorem* 3.1, *Remark* 3.2, *Corollary* 3.3, *Lemma* 3.4, *Remark* 3.5 *and Corollary* 3.6 *hold not only for the gradient mappings* $x \mapsto \operatorname{grad}(f(x))$ *as formulated above, but also for any quasihomogeneous or semiquasihomogeneous mappings.*

For example, Remark 3.2 runs as follows: *all semiquasihomogeneous mappings of the same degree* $\mathbf{d}$ *(and type* $\boldsymbol{\alpha}$*) have the same multiplicity* μ.

Kushirenko has proved analogues of these assertions over an arbitrary algebraically closed field. For $\mathbf{C}$ the proof proceeds as in §3.

The value of the class of quasihomogeneous mappings for our problem lies in the facts that in them homotopies can be carried out more freely, and changes of variables are easier than for gradients. In particular, the following proposition (see [10]) makes it possible to go by a simple substitution from a quasihomogeneous to a genuine homogeneous mapping.

4.3. PROPOSITION. *Suppose that* $F\colon \mathbf{C}^n \to \mathbf{C}^n$ *is a quasihomogeneous mapping whose type and degree have the common denominator* N: $\alpha_s = A_s/N$, $d_s = D_s/N$, *where now* A_s, D_s, N *are integers. Consider the mapping* $T\colon \mathbf{C}^n \to \mathbf{C}^n$ *given by the formula* $T(y_1, \ldots, y_n) = (y_1^{A_1}, \ldots, y_n^{A_n})$. *Then*

1) *the mapping* $F \circ T\colon \mathbf{C}^n \to \mathbf{C}^n$ *has as its components homogeneous functions in the usual sense, of degrees* $D_1, \ldots, D_n$.

2) $$\mu(F \circ T) = \mu(F) \prod A_s.$$

3) *If* $e_1, \ldots, e_\mu$ *is a monomial basis of the local ring of* F, *then the following functions form a monomial basis of the local ring of* $F \circ T$:

$$e'_{i,u} = (T^* e_i) y_1^{u_1} \ldots y_n^{u_n}, \text{ where } 1 \leqslant i \leqslant \mu,\ 0 \leqslant u_s < A_s.$$

PROOF. 1) The monomial $\mathbf{x}^{\mathbf{k}}$ determines in the s-th component of $F \circ T$ the monomial $\prod y_s^{k_s A_s}$ of degree $\sum k_s A_s = N(\mathbf{k}, \boldsymbol{\alpha}) = N d_s = D_s$. 2) The formula for the multiplicity is obtained from considering the system of equations $(F \circ T)(y) = \varepsilon$ in y. 3) The functions $e'_{i,u}$ generate the whole local ring. For every function of y can be written in the form $\varphi = \sum_u y^u T^* \varphi_u$, and every function of x in the form $\varphi_u = \sum c_{i,u} e_i + \sum F_s h_{s,u}$. Thus,

$$\varphi = \sum_{u,i} c_{i,u} y^u T^* e_i + \sum_{u,s} y^u h_{s,u} T^* F_s,$$

that is, the $e'_{i,u}$ generate $Q(F \circ T)$. Since the number of functions $e'_{i,u}$ is $\mu(F \circ T)$, they form a basis. This proves Proposition 4.3.

4.4. DEFINITION. The *generating function* of type $\boldsymbol{\alpha}$ of a semiquasihomogeneous mapping F (where $\alpha_s = A_s/N$, A_s and N integers) is the polynomial $\chi_F(z) = \sum \mu_i z^i$, where μ_i is the number of basis monomials in the local ring of F having generalized degree i/N.

We remark that even for fixed quasihomogeneous type χ depends on the integer N. However, the possible values of N are multiples of one of them (the least common denominator of the fractions α_s), and $\chi_{F;\,kN,\,\alpha}(z) = \chi_{F;\,N,\,\alpha}(z^k)$.

The dimension of the local ring is given by the formula $\mu = \chi_F(1)$. Further, the degree of χ_F is the greatest of the (generalized) degrees of the monomial generators in a basis of the local ring.

4.5. THEOREM. (see [9], [14]). *The generating function of a (semi)-quasihomogeneous mapping F of degree d and type α for which $\alpha_s = A_s/N$, $d_s = D_s/N$, where A_s, D_s, N are integers, is given by the formula*

$$\chi_F(z) = \prod_{s=1}^{n} \frac{z^{D_s}-1}{z^{A_s}-1}.$$

EXAMPLE. If $F = \operatorname{grad} f$, where f is a (semi)quasihomogeneous function of type α (and degree 1), then

$$\chi_F(z) = \prod_{s=1}^{n} \frac{z^{N-A_s}-1}{z^{A_s}-1}.$$

Several useful formulae follow immediately from this theorem.

4.6. COROLLARY (see [10], [14]). *The dimension of the local ring of a semiquasihomogeneous mapping is given by the "generalized Bezout formula":*

$$\mu = \prod_{s=1}^{n} \frac{d_s}{\alpha_s}.$$

4.7. COROLLARY (see [11]). *The local ring of a semiquasihomogeneous mapping F has exactly one basis monomial of degree*

$$d_{\max} = \sum_{s=1}^{n} (d_s - \alpha_s);$$

all monomials of higher filtration lie in the ideal generated by the components $(F_1, \ldots, F_s)$.

Let us look, in particular, at the local ring of a semiquasihomogeneous function f of type $(\alpha_1, \ldots, \alpha_n)$ of degree 1. In this case $\mathbf{d}_s = 1 - \alpha_s$, and we obtain the next result.

4.8. COROLLARY (see [10], [14]). *The dimension of the local ring of a semiquasihomogeneous function f of type $(\alpha_1, \ldots, \alpha_n)$ and degree 1 is given by the formula*

$$\mu = \prod \left(\frac{1}{\alpha_s} - 1\right)$$

4.9. COROLLARY ([11]). *A monomial basis of the local ring of a semi-quasihomogeneous function f of type $(\alpha_1, \ldots, \alpha_n)$ and degree 1 has exactly* *one generator of (generalized) degree* $d_{\max} = \Sigma(1 - 2\alpha_s)$; *all monomials of higher degree lie in the ideal* $(\partial f/\partial x_1, \ldots, \partial f/\partial x_s)$.

Here are some more immediate corollaries of Theorem 4.5.

4.10. COROLLARY. *The generating polynomial of a semiquasihomogeneous mapping is always recurrent*:

$$\mu_i = \mu_{k-i} \text{ where } k = \Sigma D_s - \Sigma A_s.$$

4.11. COROLLARY. *The generalized degree of the penultimate monomial (according to filtration) in a monomial basis of the local ring of a quasihomogeneous function of type* $(\alpha_1, \ldots, \alpha_n)$ *and degree* 1 *is* $d_{\max} - \alpha_{\min}$, *where* $\alpha_{\min} = \min(\alpha_1, \ldots, \alpha_n)$.

4.12. COROLLARY. *A non-degenerate quasihomogeneous mapping of type* $\boldsymbol{\alpha} = \mathbf{A}/N$ *and degree* $\mathbf{d} = \mathbf{D}/N$ *can exist only when the polynomial* $\prod(z^{D_s} - 1)$ *is divisible by* $\prod(z^{A_s} - 1)$.

4.13. COROLLARY. *A non-degenerate quasihomogeneous mapping of type* $\boldsymbol{\alpha}$ $(\alpha_s = A_s/N)$ *can exist only when* $\prod(z^{N-A_s} - 1)/\prod(z^{A_s} - 1)$ *is a polynomial.*

REMARK. In the case of functions of two or three variables, the fact that the fraction $\Pi(z^{N-A_s} - 1)/\Pi(z^{A_s} - 1)$ can be cancelled is sufficient as well as necessary for the existence of a non-degenerate quasihomogeneous function with exponents A_s/N (see §§ 10, 11). This is not true for four variables, as can be seen from the following example, which V. M. Izlev has shown to me:

$$N = 265, A_1 = 1, A_2 = 24, A_3 = 33, A_4 = 58.$$

In this example the quotient is a polynomial with non-negative coefficients, while all quasihomogeneous functions with exponents A_s/N are degenerate.

The results of this section have been rediscovered over and over again. At the time when this paper is going to press, Gabrielov has informed me that Theorem 4.5 and Corollaries 4.6 and 4.8 can be found in papers of Milnor, Orlik and Wagreich. Saito and Hironaka have given other proofs of Corollaries 4.7 and 4.9. Kushirenko has told me that all the results, are, in fact, in Bourbaki [14] (see Proposition 2 in Ch. 5, § 5.1, in the section "Poincaré series of a graded algebra").

PROOF OF THEOREM 4.5. It is enough to consider the case of a non-degenerate quasihomogeneous mapping F (see Corollary 3.6 and Proposition 4.2). We change T as in Proposition 4.3. It follows from the form of the generators of the local ring of $F \circ T$ that

$$\chi_{F \circ T;\, 1, \mathbf{1}}(z) = \chi_{F;\, N, \boldsymbol{\alpha}}(z)\chi_{T;\, 1, \mathbf{1}}(z), \text{ where } \mathbf{1} = (1, \ldots, 1).$$

The generating polynomials of T and $F \circ T$ occurring in this formula are homogeneous in the usual sense, and can be calculated explicitly. Indeed, for the mapping $x = y^A$ we have $\chi(z) = \dfrac{z^A - 1}{z - 1}$. It follows that

$$\chi_T(z) = \prod_{s=1}^{n} \frac{z^{A_s} - 1}{z - 1}$$

(here and later the pair $(N, \alpha) = (1, 1)$ is omitted from the notation for χ).

On the other hand, $F \circ T$ is a non-degenerate mapping whose components are homogeneous functions of degree D_s. Thus (by Proposition 4.2 and Corollary 3.6), it has the same generating polynomial as every other non-degenerate homogeneous mapping with these degrees. We can take, for example, the mapping T' given by the formula

$$T'(y_1, \ldots, y_n) = (y_1^{D_1}, \ldots, y_n^{D_n}).$$

Thus,

$$\chi_{F \circ T}(z) = \chi_{T'}(z) = \prod_{s=1}^{n} \frac{z^{D_s} - 1}{z - 1}.$$

The formula for χ_F is obtained by dividing the formulae for $\chi_{F \circ T}$ and for χ_T. This proves Theorem 4.5.

§5. Quasihomogeneous diffeomorphisms and quasijets

Several Lie groups and Lie algebras are connected with the filtrations defined by a quasihomogeneous type $\boldsymbol{\alpha}$. In the case of ordinary homogeneity, these are the full linear group, the group of k-jets of diffeomorphisms, the subgroup of k-jets with identity $(k-1)$-jets, and their factor groups. The analogues for quasihomogeneous filtrations are defined as follows.

We consider the space $\mathbf{C}^n$, with a fixed coordinate system $(x_1, \ldots, x_n)$. The ring of formal power series[1] in these coordinates is denoted by $A = \mathbf{C}[[x_1, \ldots, x_n]]$. We assume that a type of quasihomogeneity $\alpha = (\alpha_1, \ldots, \alpha_n)$ is given. We denote by E_d the ideal of A generated by series of filtration d. Further, let $E_{>d}$ stand for the ideal of E_d consisting of series of filtration strictly greater than d.

A *formal diffeomorphism* $g: (\mathbf{C}^n, 0) \to (\mathbf{C}^n, 0)$ is given by a collection of n power series without free terms and gives a ring isomorphism $g^*: A \to A$ by the formula $g^*f = f \circ g$, where $\circ$ denotes the substitution of a series in a series.

5.1. DEFINITION. A diffeomorphism g has *filtration* d if, for all λ,

$$(g^* - 1)E_\lambda \subset E_{\lambda+d}.$$

5.2. PROPOSITION. *Let $d \geqslant 0$. Then the set $G_d = G_d(\boldsymbol{\alpha})$ of all diffeomorphisms of filtration d is a group under* $\circ$.

PROOF. We remark, first of all, that $g^*E_\lambda = E_\lambda$ for all λ when $d \geqslant 0$ ($g^*E_\lambda \supset E_\lambda$, since $d \geqslant 0$, and $g^{*-1}E_\lambda \supset E_\lambda$, since the factor space A/E_λ is finite-dimensional). Thus, for a, b in G_d we have

[1] The greater part of what follows carries over at once to the case when A is the ring of convergent series over $\mathbf{C}$ or $\mathbf{R}$, or the ring of germs of smooth functions.

$$[(a \circ b)^* - 1]E_\lambda = [b^*(a^* - 1) + (b^* - 1)]\, E_\lambda \subset E_{\lambda+d},$$
$$(a^{-1*} - 1)\, E_\lambda = a^{-1*}\,(1 - a^*)\, E_\lambda \subset E_{\lambda+d},$$

as required.

5.3. PROPOSITION. *For $q > p \geqslant 0$, G_q is a normal subgroup of G_p.*

PROOF. The definition of G_q uses only the filtration $\{E_\lambda\}$. This filtration is invariant under G_0 and a fortiori under G_p. Thus, a subgroup defined in terms of this filtration is normal, as required.

The group G_0 is especially important, because it plays for the quasihomogeneous case the rôle played by the full group of jets of diffeomorphisms in the homogeneous case. It must be stressed that in the quasihomogeneous case certain diffeomorphisms have negative filtrations and do not lie in G_0.

5.4. DEFINITION. *The group of d-jets of type α* is the factor group of the group of diffeomorphisms by the subgroup consisting of the diffeomorphisms of filtration greater than d,

$$J_d = J_d(\alpha) = G_0/G_{>d}.$$

It is clear that J_d is a finite-dimensional Lie group. There are natural factorizations $\pi_{p,q}: J_p \to J_q$ $(p > q \geqslant 0)$.

Attention should be drawn to the fact that in the ordinary homogeneous case our numbering differs by 1 from the standard one: our J_0 is called the group of 1-jets, etc.

5.5. PROPOSITION. *The group J_p is obtained from J_0 by a chain of extensions with commutative factors. More accurately, let E_p be the term of the filtration immediately following E_q. Then the kernel K of the homomorphism $\pi_{p,q}$ is commutative.*

PROOF. Let $A, B \in K$; we consider any representatives $a, b \in G_0$. Then

$$(ab)^* - 1 = (a^* - 1) + (b^* - 1) + (b^* - 1)(a^* - 1).$$

Further, for every λ,

$$(a^* - 1)\, E_\lambda \subset E_{\lambda+p}, \quad (b^* - 1)\, E_\lambda \subset E_{\lambda+p},$$

since the q-jets a and b are trivial. Thus, $[(ab)^* - (ba)^*]E_\lambda \subset E_{\lambda+2p}$. Therefore, ab and ba determine the same element of J_p, which is what we wanted to prove.

The group J_0 has especial value, because it is the quasihomogeneous generalization of the full linear group.

5.6. DEFINITION. A diffeomorphism $g \in G_0$ is said to be *quasihomogeneous of type α* if every space of quasihomogeneous functions of degree d (and type α) is mapped into itself by g.

The set of all quasihomogeneous diffeomorphisms (of fixed type) forms a group. We denote it by $H(= H(\alpha))$ and call it the *group of quasihomogeneous diffeomorphisms.*

We consider the natural embedding i: $H \to G_0$ and factorization π: $G_0 \to J_0$.

5.7. PROPOSITION. *The group J_0 is naturally isomorphic to the group H of quasihomogeneous diffeomorphisms; in fact, the compound map πi: $H \to J_0$ is an isomorphism of Lie groups.*

PROOF. a) $\operatorname{Ker} \pi i = e$. For $\operatorname{Ker} \pi i = H \cap G_{>0}$. Hence for $h \in \operatorname{Ker} \pi i$ and for every monomial f of degree d, $(h^* - 1)f$ lies in the space of homogeneous functions of degree d and also has filtration greater than d. Thus, $(h^* - 1)f = 0$ for every monomial f, and so $h = e$.

b) $\operatorname{Im} \pi i = J_0$. For this proof we construct the inverse mapping $J_0 \to H$ explicitly. Let $x_1, \ldots, x_n$ be coordinates in $\mathbf{C}^n$, and let a diffeomorphism $g \in G_0$ be a representative of the jet $j \in J_0$. We consider the series $g^*x_i \in E_{\alpha_i}$. We select in it the homogeneous component y_i of degree α_i, so that $g^*x_i = y_i + z_i$, $z_i \in E_{>\alpha_i}$. We define a polynomial mapping h^{-1}: $\mathbf{C}^n \to \mathbf{C}^n$ by the relation $h^*x_i = y_i$. To check that h is a diffeomorphism, we calculate the Jacobian:

$$\det\left|\frac{\partial (y_i+z_i)}{\partial x_j}\right| = \det\left|\frac{\partial y_i}{\partial x_j}\right| + R.$$

The term R containing the derivatives with respect to z is 0 at the origin. For every summand of the determinant y is homogeneous of degree 0 in x. All other summands containing z have positive filtration since $z_i \in E_{>d_i}$. Therefore, $R \in E_{>0}$ and $R(0) = 0$. Thus, the Jacobians of g and h at 0 are the same, so that the Jacobian of h at zero is different from 0, which means that h is a diffeomorphism. The ring automorphism h^* preserves degrees of all monomials, because it preserves the degrees of the coordinates x_i. Thus, $h \in H$. It is clear that $\pi i h = j$, and this proves the assertion.

5.8. PROPOSITION. *Suppose that $d \geqslant 0$. Then the group J_d of d-jets of diffeomorphisms acts as a group of linear transformations on the space $A/E_{>d}$ of d-jets of functions.*

PROOF. Let $g \in G_{>d}$. Then application of g does not change the α-jet of a function f, since $f \circ g - f \in E_{>d}$. Thus, $(h, f) \mapsto f \circ h$ gives a mapping $J_d \times (A/E_{>d}) \to A/E_{>d}$, as required.

5.9. REMARK. In the case of ordinary homogeneity, even the group of $(d-1)$-jets of diffeomorphisms acts on the space of d-jets (all this in our notation). This fact has an analogue in the quasihomogeneous case in the action of the group of $(d - \min \alpha_s)$-jets.

§6. Quasihomogeneous vector fields

The infinitesimal analogues of the concepts we have introduced run as follows.

6.1. DEFINITION. A formal vector field $\mathrm{v} = \Sigma v_i \partial/\partial x_i$ *has filtration d* if the directional derivative of v raises the filtration by not less than

d: $L_{\mathbf{v}}E_\lambda \subset E_{\lambda+d}$. We denote the set of all vector fields of filtration d by $\mathfrak{g}_d$. Our filtration in the module of vector fields (that is, derivation of A) is compatible with the filtration of the ring:

$$a \in E_d,\ \mathbf{v} \in \mathfrak{g}_\delta \Rightarrow a\mathbf{v} \in \mathfrak{g}_{d+\delta},\ L_{\mathbf{v}}a \in E_{d+\delta}.$$

6.2. PROPOSITION. *Suppose that $d \geqslant 0$. Then* 1) *the Poisson bracket of vector fields defines on $\mathfrak{g}_d$ a Lie algebra structure*; 2) *the Poisson bracket of elements $\mathfrak{g}_{d_1}$ and $\mathfrak{g}_{d_2}$ lies in $\mathfrak{g}_{d_1+d_2}$ so that each $\mathfrak{g}_d$ is an ideal in the Lie algebra $\mathfrak{g}_0$.*

PROOF. If $f \in E_\lambda$, $\mathbf{v}_1 \in \mathfrak{g}_{d_1}$, $\mathbf{v}_2 \in \mathfrak{g}_{d_2}$ then $(L_{\mathbf{v}_1}L_{\mathbf{v}_2} - L_{\mathbf{v}_2}L_{\mathbf{v}_1})f \in E_{\lambda+d_1+d_2}$, as required.

The filtration of a vector field is connected with the filtrations of its components in the following manner.

6.3. PROPOSITION. *The field $\mathbf{v} = \Sigma v_i \partial/\partial x_i$ has filtration d (and type α) if and only if each component v_i is a function (series) of filtration $d + \alpha_i$.*

For the proof we introduce the following notation. A vector field of the form $\mathbf{x}^{\mathbf{k}}\partial/\partial x_i$ is called a *vector monomial.* The *degree* of such a vector monomial (for given quasihomogeneous type) is the (possibly negative) rational number $\langle \mathbf{k}, \alpha\rangle - \alpha_i = \langle \mathbf{k} - \mathbf{1}_i, \alpha\rangle$ in the arithmetical progression containing the degrees of the ordinary monomials. A vector field is said to be *homogeneous of degree d* if all vector monomials occurring in it with non-zero coefficients are of degree d.

PROOF OF PROPOSITION 6.3. If $\mathbf{v} \in \mathfrak{g}_d$, then $v_i = L_{\mathbf{v}}x_i \in E_{d+\alpha_i}$, since $x_i \in E_{\alpha_i}$. Let $v_i = \Sigma\, v_{i,\mathbf{k}}\mathbf{x}^{\mathbf{k}}$. For every monomial $f = \mathbf{x}^{\mathbf{l}}$ we have

$$L_{\mathbf{v}}f = \sum v_i \frac{\partial f}{\partial x_i} = \sum l_i v_{i,\mathbf{k}}\mathbf{x}^{\mathbf{l}+\mathbf{k}-\mathbf{1}_i}.$$

Here, $\langle \mathbf{k}, \alpha\rangle \geqslant d + \alpha_i$ if $v_i \in E_{d+\alpha_i}$. Thus, $\langle \mathbf{l} + \mathbf{k} - \mathbf{1}_i, \alpha\rangle \geqslant \langle \mathbf{l}, \alpha\rangle + d$, that is, $L_{\mathbf{v}}E_\lambda \subset E_{\lambda+d}$, as required.

We obtain immediately from Proposition 5.5:

6.4. COROLLARY. *The Lie algebra $\mathfrak{j}_d$ of the group J_d of d-jets of diffeomorphisms is the factor algebra $\mathfrak{j}_d = \mathfrak{g}_0/\mathfrak{g}_{-d}$. The mapping $\pi_{p,q}$: $J_p \to J_q$ induces a homomorphism $\pi_{p,q*}$: $\mathfrak{j}_p \to \mathfrak{j}_q$ of Lie algebras. The kernel of the mapping of $\mathfrak{j}_p$ into the algebra of jets that immediately precedes it in filtration is commutative.* .

Finally, from Proposition 5.7:

6.5. COROLLARY. *The quasihomogeneous vector fields of degree* 0 *form a finite-dimensional Lie subalgebra $\mathfrak{h}$ of the Lie algebra of all vector fields. The Lie algebra $\mathfrak{h}$ is naturally isomorphic to the Lie algebra of the group j_0 of* 0*-jets of diffeomorphisms.*

In what follows we sometimes identify the vector field $\mathbf{v}$ with the set of n functions (or series) v_i. The next two propositions are used later in the reduction of semiquasihomogeneous functions to normal form.

6.6. LEMMA. *Let F be a power series of filtration d, and let $\mathbf{v}$ be a formal vector field of positive filtration δ. Then the Taylor series*

$$F(\mathbf{x}+\mathbf{v}(\mathbf{x}))=F(\mathbf{x})+\frac{\partial F}{\partial \mathbf{x}}\mathbf{v}+R$$

has remainder term R of filtration strictly greater than $d+\delta$.

PROOF. In view of the linearity of R relative to F, it is enough to show this for the case when F is a monomial. Let $F=\mathbf{x}^{\mathbf{k}}$, $\mathbf{v}=\Sigma v_i\partial/\partial x_i$. We consider the term of the Taylor series containing $\partial^{|m|}F/\partial\mathbf{x}^{\mathbf{m}}$ ($\mathbf{m}=(m_1, \dots, m_n)$). The monomials occurring in this term have exponents $\mathbf{p}=\mathbf{k}-\mathbf{m}+\sum_{i=1}^{n}\mathbf{l}_i$, where $\mathbf{l}_i$ is the exponent of one of the monomial functions $v_i^{m_i}$. Thus, $\mathbf{l}_i=\sum_{j=1}^{m_i}\mathbf{l}_{i,j}$, where $\mathbf{l}_{i,j}$ is one of the exponents $\mathbf{l}$ in the decomposition $v_i=\Sigma v_{i,\mathbf{l}}\mathbf{x}^{\mathbf{l}}$. So

$$\mathbf{p}=\mathbf{k}+\sum_{i=1}^{n}\sum_{j=1}^{m_i}(\mathbf{l}_{i,j}-\mathbf{1}_i).$$

But $\langle\mathbf{k},\alpha\rangle\geqslant d$, $\langle\mathbf{l}_{ij}-\mathbf{1}_i,\alpha\rangle\geqslant\delta>0$, by assumption. Therefore, $\langle\mathbf{p},\alpha\rangle\geqslant d+|\mathbf{m}|\delta$, where $|\mathbf{m}|=m_1+\ldots+m_n$. Hence all monomials occurring in terms of degree higher than 1 in the Taylor series relative to v have filtration not less than $d+2\delta$, as required.

6.7. COROLLARY. *Suppose that* $F=F_0+F_1+F_2$, *where* $F_0\in F_d$, $F_1\in E_{>d}$, $F_2\in E_{>d+\delta}$: $\mathbf{v}=\mathbf{v}_0+\mathbf{v}_1$, *where* $v_0\in\mathfrak{g}_\delta$, $v_1\in\mathfrak{g}_{>\delta}$, $\delta>0$. *Then*

$$F(\mathbf{x}+\mathbf{v}(\mathbf{x}))=F_0(\mathbf{x})+\left[F_1(\mathbf{x})+\frac{\partial F_0}{\partial\mathbf{x}}\mathbf{v}_0\right]+R',\quad R'\in E_{>d+\delta}.$$

PROOF. We set $F_0+F_1=F'$. We have $R'=R_1+R_2+R_3+R_4$, where

$$R_1=F'(\mathbf{x}+\mathbf{v}(\mathbf{x}))-F'(\mathbf{x})-\frac{\partial F'}{\partial\mathbf{x}}\mathbf{v}\in E_{>d+\delta}$$

(Lemma 6.6),

$$R_2=\frac{\partial F_0}{\partial\mathbf{x}}\mathbf{v}_1=L_{\mathbf{v}_1}F_0\in E_{>d+\delta},$$

$$R_3=\frac{\partial F_1}{\partial\mathbf{x}}\mathbf{v}=L_{\mathbf{v}}F_1\in E_{>d+\delta},$$

$$R_4=F_2(\mathbf{x}+\mathbf{v}(\mathbf{x}))\in E_{>d+\delta},$$

as required.

§7. The normal form of a semiquasihomogeneous function

We consider the local ring of a quasihomogeneous or semiquasihomogeneous function f of degree d. We fix a system of monomials forming a basis for this ring.

7.1. DEFINITION. A monomial is said to be *upper* or *to lie above the diagonal* (or *lower*, or *diagonal*) *if it has degree greater than* d (or less than d, or equal to d) for given quasihomogeneous exponents.

Note that the *number of upper, diagonal, and lower basis monomials does not depend on the basis of the local ring* (see Remark 3.5).

Let $e_1, \ldots, e_s$ be the system of all upper basis monomials in a fixed basis of the local ring of the function f_0.

7.2. THEOREM. *Every semiquasihomogeneous function with quasihomogeneous part* f_0 *is equivalent to a function of the form* $f_0 + \Sigma c_k e_k$, *where the* c_k *are constants.*

The proof of Theorem 7.2 is obtained on application of the following lemma.

7.3. LEMMA. *Let* f_0 *be a quasihomogeneous function of degree* d *and* $e_1, \ldots, e_r$ *the set of all basis monomials of fixed degree* $d' > d$ *in the local ring of* f_0. *Then every series of the form* $f_0 + f_1$, *where the filtration of* f_1 *is greater than* d, *can be brought by a formal diffeomorphism to the form* $f_0 + f_1'$, *where the terms in* f_1' *of degree less than* d' *are the same as in* f_1, *and the terms of degree* d' *reduce to* $c_1 e_1 + \ldots + c_r e_r$.

PROOF. Let g denote the sum of the terms of degree d' in f_1. There exists a decompostion (if convenient, as far as the terms of filtration higher than d', but certainly without them)

$$g = \sum \frac{\partial f_0}{\partial x_i} v_i(\mathbf{x}) + c_1 e_1 + \ldots + c_r e_r,$$

since $e_1, \ldots, e_r$ are basis monomials. The vector field $\mathbf{v}$ occurring in this formula can be replaced by a homogeneous one of degree $\delta = d' - d > 0$ without invalidating the formula (to prove this it is enough to decompose $\mathbf{v}$ into its homogeneous parts).

We consider now the formal substitution $\mathbf{x} = \mathbf{y} - \mathbf{v}(\mathbf{y})$, where $\mathbf{v}$ is the vector field with components v_i as defined above. We claim that this is a formal diffeomorphism. For the field $\mathbf{v}$ has positive degree δ, so that if the coordinates are numbered according to decreasing exponents α_i, the Jacobian matrix of the substitution at 0 is unitriangular. Applying Corollary 6.7, we find that

$$f(\mathbf{y} - \mathbf{v}(\mathbf{y})) = f_0(\mathbf{y}) + [f_1(\mathbf{y}) + (c_1 e_1(\mathbf{y}) + \ldots + c_r e_r(\mathbf{y})) - g(\mathbf{y})] + R'(\mathbf{y})$$

(in the old notation). Since the filtration of R' is greater than d', this proves Lemma 7.3.

PROOF OF THEOREM 7.2. Applying Lemma 7.3 to the function f_0 and the monomials next to those of highest degree d', we come to the desired form of the term of degree d'. Applying the same lemma to the series $f_0 + f_1'$ so obtained and the monomials following those of degree d', we come to the desired form of the term of degree d' without changing the terms of degree d and d'. Continuing in this way, we obtain the desired normal form up to terms of degree as high as required (and even, if convenient, *we can reduce the formal series completely to formal normal form by a formal diffeomorphism*; this follows from the fact that the degrees of fields $\mathbf{v}$ that arise at various stages grow).

Up to this moment we have not used anywhere the finiteness of the multiplicity μ, so that the formal assertion has been proved without this assumption. If μ is finite, then a fairly long section of the Taylor series (of length bounded in terms of μ; see, for example, [5]) of the function is equivalent to the function itself, so that reduction to normal form is realized by a genuine diffeomorphism.

§8. The normal form of a quasihomogeneous function

Let f_0 be a non-degenerate quasihomogeneous function. We consider the linear space E of all quasihomogeneous functions of the same type and degree of quasihomogeneity as f_0.

8.1. THEOREM. *Every quasihomogeneous function of the type and degree of f_0 and sufficiently near to f_0 is equivalent to a function of the form $f_0 + c_1 e_1 + \ldots + c_r e_r$, where $e_1, \ldots, e_r$ is the collection of all diagonal basis monomials of the local ring of f_0.*

PROOF. We consider a diffeomorphism of the form $\mathbf{x} \to \mathbf{x} + \varphi(\mathbf{x})$, where φ is a quasihomogeneous vector field of degree 0. All such diffeomorphisms form a Lie group. This group acts (linearly) on E. An orbit through f_0 and one through any nearby point have one and the same dimension (see Remark 3.2 and Lemma 3.4). They intersect the plane $f_0 + Ce_1 + \ldots + Ce_r$ transversally at the point f_0 and neighbouring points. Thus, the union of the orbits intersecting this plane near f_0 contains a neighbourhood of f_0 in the whole space E, as we wanted to show.

8.2. EXAMPLE. We show that *every non-degenerate binary form of degree $n \geqslant 4$ reduces to a normal form with $n - 3$ parameters*:

$$x^{n-1}y + c_1x^{n-2}y^2 + \ldots + c_{n-3}x^2y^{n-2} + xy^{n-1}.$$

For the non-degeneracy of this form yields the existence of two simple linear factors, which can be taken as the coordinates x and y. Then the terms x^n and y^n are absent from the expression of the form; after this we make the coefficients of $x^{n-1}y$ and xy^{n-1} equal to 1. In this way the form reduces to the one above.

It is easy to check that the monomials $x^{n-2}y^2, \ldots, x^2y^{n-2}$ form a diagonal basis for every form of the type indicated (even in the degenerate case).

8.3. PROPOSITION. *A non-degenerate binary form of degree 4 can be reduced by a linear transformation to the (Legendre) normal form X_9: $x^4 + ax^2y^2 + y^4$, $a^2 \neq 4$.*

PROOF. By Proposition 8.2, the form reduces to $xy(x^2 + 2cxy + y^2)$. This is the product of the quadratic forms $a = xy$, $b = x^2 + 2cxy + y^2$. There are two *independent* degenerate forms among the linear combinations $pa + qb$ of a and b ($p:q$ is defined from the characteristic equation $p^2 + 2cpq + (c^2 - 1)q^2 = 0$, whose discriminant is not 0). Taking the square roots of these degenerate forms as coordinates, we reduce the original form to $c_1x^4 + c_2x^2y^2 + c_3y^4$. The condition for such a form to be non-degenerate is $c_1c_3(4c_1c_3 - c_2^2) \neq 0$. By a magnification of coordinates we reduce the form to that indicated in Proposition 8.3, which is thereby proved.

8.4. PROPOSITION. *A non-degenerate quasihomogeneous function of degree 1 with quasihomogeneous exponents 1/3, 1/6 can be reduced by a quasihomogeneous diffeomorphism to the normal form J_{10}: $x^3 + ax^2y^2 + y^6$, where $4a^3 + 27 \neq 0$.*

PROOF. By definition, the function has the form $c_1x^3 + c_2x^2y^2 + c_3xy^4 + c_4y^6$. From the non-degeneracy it follows that $c_1 \neq 0$. By changing x to $x + \lambda y^2$ we can achieve $c_3 = 0$. Then $c_4 \neq 0$ as a consequence of non-degeneracy, and after a magnification of coordinates the form reduces to that in Proposition 8.4, which is thereby proved.

8.5. PROPOSITION. *Every non-degenerate quasihomogeneous function of degree 1 with exponents* $\left(\frac{1}{3}, \frac{1}{3k}\right)$, $\left(\frac{k}{3k+1}, \frac{1}{3k+1}\right)$, $\left(\frac{1}{4}, \frac{1}{2k}\right)$, $\left(\frac{k}{4k+2}, \frac{1}{2k+1}\right)$ *can be reduced by a quasihomogeneous diffeomorphism to one of the respective normal forms:*

$$x^3 + ax^2y^k + y^{3k}, \quad x^3y + ax^2y^{k+1} + y^{3k+1} \quad (4a^3 + 27 \neq 0);$$
$$x^4 + ax^2y^k + y^{2k}, \quad x^4y + ax^2y^{k+1} + y^{2k+1} \quad (a^2 - 4 \neq 0).$$

The proof repeats the calculations in Propositions 8.3 and 8.4.

8.6. DEFINITION. The *inner modality* of a quasihomogeneous function is the total number of diagonal and superdiagonal monomials in some (and thus in any) monomial basis of the local ring.

8.7. THEOREM. *The modality of any semiquasihomogeneous function (in particular, of a non-degenerate quasihomogeneous function) is not less than the inner modality of its quasihomogeneous part.*

For suppose that $f = f_0 + f_1$ is the function in question and that $e_1, \ldots, e_r$ is the set of all diagonal and superdiagonal monomials in any basis of the local ring. Then the same monomials form the diagonal and superdiagonal part of a basis of the local ring of each of the functions $f + c_1e_1 + \ldots + c_re_r$, at least for sufficiently small c (see Corollary 3.3 and Lemma 3.4). Thus, there exists an s-dimensional plane through f in the space of functions with critical point and critical value 0 having the property that all non-zero tangent vectors to the orbits of points of the plane near f do not lie in the plane. Thus, the modality of f is not less than s.

8.8. REMARK. Gabrielov has shown that the modality is equal to the dimension of the submanifold of a base of versal deformations along which μ does not change. In all the examples of semiquasihomogeneous functions I know this dimension is the total number of diagonal and superdiagonal basis monomials.

§9. Piecewise filtrations

It often turns out to be useful to consider filtrations in which the rôle of the diagonal is played by a Newton open polygon (or, in the case of several dimensions, by a polyhedron convex towards 0). The formal definition is as follows.

Let $\boldsymbol{\alpha}_1, \ldots, \boldsymbol{\alpha}_p$ be a fixed collection of p quasihomogeneous types. We recall that the monomial $\mathbf{x^k}$ is of degree $\langle \boldsymbol{\alpha}_i, \mathbf{k} \rangle = \varphi_i(\mathbf{k})$ in the i-th filtration. We define the *piecewise degree* of $\mathbf{x^k}$ to be $\varphi(\mathbf{k}) = \min[\varphi_1(\mathbf{k}), \ldots, \varphi_p(\mathbf{k})]$.

9.1. DEFINITION. A power series has *piecewise filtration d* if all its monomials have piecewise degree d or higher.

Note that the equation $\varphi(\mathbf{k}) = 1$ defines a hypersurface Γ in the space of exponents $\mathbf{k}$ that is convex towards 0. We call Γ an *open polyhedron.* In these terms we can say that a monomial has (piecewise) degree d if and only if its exponent vector lies on the open polyhedron $d\Gamma$ obtained from Γ by a homothety with the coefficient d. In exactly the same way a series has (piecewise) filtration d if the exponent vectors of all its monomials lie on or outside $d\Gamma$.

The sum of the terms of lowest (piecewise) degree in a given power series is called the *principal part* of the series. A (piecewise) *homogeneous function of degree d* is a polynomial whose monomials all have (piecewise) degree d.

Analogous concepts are defined for vector fields; the degree of the monomial $\mathbf{x^l}\partial/\partial x_i$ is defined to be

$$\varphi(\mathbf{l}-\mathbf{1}_i) = \min_{1\leqslant j\leqslant p} \langle \boldsymbol{\alpha}_j, \mathbf{l}-\mathbf{1}_j \rangle.$$

Note that for all functions f, g and every vector field $\mathbf{v}$ we have[1]

filtration of $fg \geqslant$ filtration of f + filtration of g,

filtration of $\sum \frac{\partial f}{\partial x_i} v_i \geqslant$ filtration of f + filtration of $\mathbf{v}$.

---

[1] The filtration φ is naturally connected with the filtration arising naturally from the Koszul complex constructed from the derivative of a piecewise-homogeneous function; this was pointed out to me by Kushirenko.

The *group of diffeomorphisms of filtration d, the group of d-jets* of diffeomorphisms and the corresponding Lie algebras are defined just as in the case of quasihomogeneous filtrations. There is no analogue for piecewise filtrations, except for the group of quasihomogeneous diffeomorphisms.

9.2. DEFINITION. A piecewise-homogeneous function f_0 of degree d *satisfies condition A* if for every function g of filtration $d + \delta > d$ in the ideal spanned by the derivatives of f_0 there is a decomposition

$$g = \sum \frac{\partial f_0}{\partial x_i} v_i + g',$$

where the vector field $\mathbf{v}$ has filtration δ, and the function g' has filtration greater than $d + \delta$.

Note that a quasihomogeneous function always satisfies condition A.

We consider a basis of the local ring of a piecewise-homogeneous function f_0 of finite multiplicity μ.

9.3. DEFINITION. A basis $e_1, \ldots, e_\mu$ of homogeneous elements is said to be *regular* if, for each D, the elements of the basis of degree D are independent modulo the sum of the ideal $I = (\partial f/\partial x_0)$ and the space $E_{>D}$ of functions of filtrations greater than D.

9.4. PROPOSITION. *There always exists a regular basis, in fact, one consisting entirely of monomials.*

PROOF. The monomials whose exponent vectors lie on $D\Gamma$ generate $E_D \bmod E_{>D}$. Thus, their images in $E_D/(E_D \cap I) + E_{>D}$ generate this linear space, hence a basis for the factor space can be extracted from their images. The inverse images of the basis vectors so chosen are monomials in E_D, and we include these monomials in a basis of the local ring.

For large enough D we have $E_D \subset I$ (since $\mu < \infty$). Therefore, the system of monomials constructed is finite. It is clear from the construction that every vector in A is representable as a linear combination of the chosen monomials and of elements of this ideal. Finally, if the least degree of a monomial occurring in a relation $c_1 e_1 + \ldots \in I$ with non-zero coefficient is $D < \infty$, then the images of monomials e_i of degree D in the factor space $E_D/(E_D \cap I) + E_{>D}$ would be dependent, against the choice of the e_i. Thus, $\{e_i\}$ is a basis of the local ring, as required.

The number of elements in a regular monomial basis having given (piecewise) homogeneous degree does not depend on the choice of a basis of the local ring. A monomial in a regular basis is said to be *diagonal* (*superdiagonal*) if its degree is equal to (greater than) the degree of the function f_0 under discussion.

9.5. THEOREM. *If the principal part f_0 of a function f satisfies condition A and has finite multiplicity μ, then f can be reduced by a diffeomorphism to the form $f_0 + c_1 e_1 + \ldots + c_s e_s$, where $e_1, \ldots, e_s$ are the superdiagonal monomials in a regular basis.*

The proof of Theorem 9.5 repeats that of Theorem 7.2.

9.6. EXAMPLE. *We consider the function* $f_0 = x^a + \lambda x^2 y^2 + y^b$, *where*

$a \geqslant 4$, $b \geqslant 5$, $\lambda \neq 0$. *We claim that* 1) $\mu = a + b + 1$; 2) *the system* $1, x, \ldots, x^{a-1}, y, \ldots, y^b, xy$ *of monomials is a regular basis*; 3) *condition A is satisfied by the filtration given by the open polygon* Γ *with vertices* $(a, 0), (2, 2), (0, b)$.

For the proof of all these assertions it is useful to carry out certain geometrical constructions in the plane of exponents. These constructions reduce the analysis of the local ring to a sequence of geometrical operations reminiscent of the solution of crossword puzzles. This "crossword solution" technique is applicable not only in this example, and we explain it to a somewhat wider extent than is necessary for the analysis of our example.

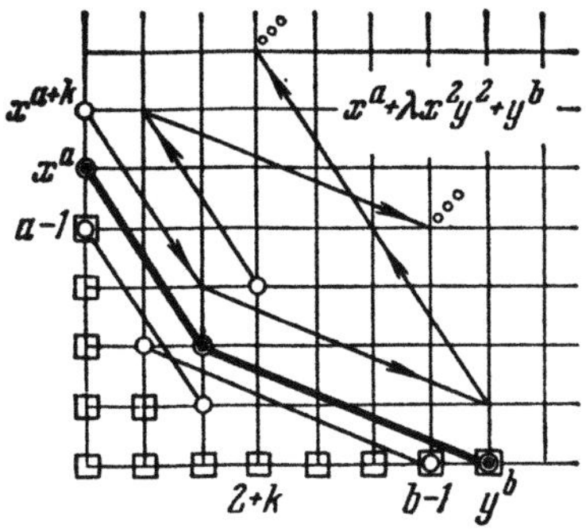

Fig. 1.

So let us assume that each partial derivative of the given function contains not more than two monomials (which is obviously satisfied by f_0). We join the monomials of the partial derivative with respect to x by a segment, which we call a *fundamental x-segment.* Similar fundamental segments are defined for the other variables. In our example there are two fundamental segments parallel to the open polygon Γ (Fig. 1).

The fundamental segments depict relationships between the images of their monomials in the local ring. We consider some consequences of these relationships. We call every translate of a fundamental segment by an integral vector with non-negative components a *permissible x-segment.* We note that two permissible segments may lie one on the other and even coincide geometrically (if one is an x-segment and the other a y-segment). But this does not happen in our example.

Two permissible segments are said to be *joined* if they have a common end. A *permissible chain* is a collection of permissible segments such that every one of the segments is joined to every other by a sequence of consecutive joined segments in the collection. A chain is said to be *maximal* if it is not part of any larger permissible chain.

A *cycle* is a finite sequence of consecutive joined permissible segments in which the last segment is joined to the first. A cycle is said to be *regular* if in its journey along the x-(y-)segments each direction is traversed the same number of times as every other. (In our example cycles are trivial, but if we consider the case $a = b = 4$, we get a non-trivial cycle $(1, 3) \to (3, 1) \to (1, 3)$ with one x-segment and one y-segment.)

We can now formulate the "crossword solution" rule for a function f with a critical point of finite multiplicity.

9.7. PROPOSITION. 1. *If the exponent of a monomial lies in an infinite permissible chain, then it lies in the ideal generated by the partial* *derivatives of the function.*

2. *If all cycles are regular, then the dimension μ of the local ring is equal to the number of maximal cycles. In this case we can find a basis of the local ring by taking an (arbitrary) monomial from each of the maximal chains.*

3. *Moreover, given a filtration, we obtain a regular basis by choosing a monomial of highest filtration from each of the maximal chains.*

The proof of Proposition 9.7 follows immediately from the definitions.

9.8. PROPOSITION. *The maximal permissible chains in Example 9.6 are as follows:*

1) *each of the points* $1, x, \ldots, x^{a-2}, y, \ldots, y^{b-2}, xy$ *has an empty permissible chain;*

2) *there are three finite maximal permissible chains* $x^{a-1} \to xy^2$, $x^2y \to y^{b-1}$, $x^a \to x^2y^2 \to y^b$;

3) *each of the remaining points is the beginning of an infinite permissible chain.*

It is clear from Fig. 1 that the permissible segment $x^py^q \to y^{p-2}y^{q-1+b}$ raises the filtration for $p > q \geqslant 1$, while the segment $x^py^q \to x^{p-1+a}y^{q-2}$ raises the filtration for $q > p \geqslant 1$. The filtrations of monomials with $p = 0$ or $q = 0$ or $p = q$ rises in two steps:

$$x^{1+k} \to x^{2+k}y^2 \to x^ky^{b+1}, \quad x^{2+k}y^{2+k} \to x^{k+a-1}y^{k+b-1}.$$

Consequently, apart from those described in 1) and 2), every maximal permissible chain of every monomial contains monomials of arbitrarily high degree and so is infinite.

Thus, the monomials figuring in 9.6 form a regular basis, and $\mu = a + b + 1$. To check condition A it is enough to compute the filtrations of the coefficients of the relations constructed above that describe the permissible chains. This calculation is not complicated and we omit it.

Therefore, condition A is satisfied, and the number of basis monomials of a regular basis above Γ turns out to be 0. Thus, from Theorem 9.5 we derive:

9.9. COROLLARY. *Every function f with principal part $f_0 = x^a + \lambda x^2y^2 + y^b$, where $\lambda \neq 0$, $a \geqslant 4$, $b \geqslant 5$, is equivalent to its principal part.*

REMARK. The following rule for calculating the modality of functions of two variables with a non-degenerate Newton diagram is corroborated by a large collection of diverse examples, though it has not been proved in general as yet.

From the point (2, 2) of the plane of exponents, we draw horizontal and vertical rays in the direction of increasing exponents and consider the polygon bounded by the segments of these rays in the Newton open polygon. *The modality of the function is the number of integral points inside and on the boundary of the polygon.* The non-degeneracy condition can be stated here as follows. The coefficients of the monomials corresponding to the links in the Newton open polygon must be such that the polynomial obtained on adding them is the product of a monomial and a non-degenerate quasihomogeneous function.

Kushirenko has proved that the inner modality (that is, the number of monomials in a regular basis on the Newton open polygon and above) of a function of two variables with a non-degenerate Newton diagram is, in fact, equal to the number of integral points in the polygon described above.

9.10. CONDITION FOR SUFFICIENCY OF A JET. *Suppose that, in addition to the conditions of Theorem* 9.5, *all monomials of degree* $s + 1$ *lie in the ideal* $(\partial f_0 / \partial x_i)$ *and are superdiagonal. Then the s-jet of f is sufficient.*

This is because the algorithm for reducing to the normal form as indicated in the proof of Theorem 7.2 does not alter the coefficients of superdiagonal monomials, no matter how we alter the terms of degree $s + 1$ and above in the series for f.

In particular, the b-jets in Example 8.6 are sufficient for $b \geqslant a$.

§ 10. Semiquasihomogeneous functions of two variables

In this section and the next we give tables in which we describe generators for local rings, normal forms of semiquasihomogeneous singularities, and sufficient jets for the simplest quasihomogeneous singularities of degree 1. The tables include all inner unimodal singularities and the first non-unimodal ones.

The non-degenerate quasihomogeneous functions of two variables fall into three (intersecting) classes. We recall that a function of two variables of corank 1 is a function whose second differential at zero is identically 0.

10.1. PROPOSITION. *Every non-degenerate quasihomogeneous function of corank* 2 *of two variables* x, y *can contain only the following monomials with non-zero coefficients:* x^a *and* y^b, *or* x^a *and* xy^b, *or* $x^a y$ *and* y^b *or* $x^a y$ *and* $y^b x$.

PROOF. Otherwise the function would be divisible by x^2 or y^2, and 0 would not be an isolated critical point.

We recall that all the exponents of the monomials of a quasihomogeneous function of degree 1 lie on a line Γ, the so-called diagonal (given by the equation $\langle \mathbf{k}, \alpha \rangle = 1$).

10.2. THEOREM. *We assume that exactly two monomials of a non-degenerate function of two variables lie on the diagonal* Γ. *Then the following system of monomials forms a basis for the local ring:*

f	α_1, α_2	μ	Basis monomials $x^k y^l$
$x^a + y^b$	$\frac{1}{a}, \frac{1}{b}$	$(a-1)(b-1)$	$0 \leqslant k \leqslant a-2,\ 0 \leqslant l \leqslant b-2$
$x^a y + y^b$	$\frac{b-1}{ab}, \frac{1}{b}$	$(a-1)b+1$	$0 \leqslant k \leqslant a-2,\ 0 \leqslant l \leqslant b-1;\ x^{a-1}$
$x^a y + y^b x$	$\frac{b-1}{ab-1}, \frac{a-1}{ab-1}$	ab	$0 \leqslant k \leqslant a-1,\ 0 \leqslant l \leqslant b-1$

(a function f can be reduced to the form shown in the table by magnification and renumbering of coordinates).

PROOF. Consider Fig. 2. In this figure the domain of monomials containined in the ideal (f_x, f_y) is hatched-in. The thin slanting lines

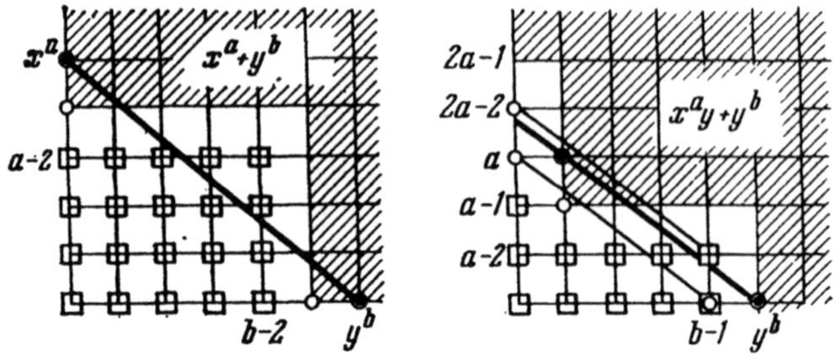

Fig. 2.

are permissible chains (see 9.6). It is clear from the figure that every monomial not contained in the ideal is joined to a monomial in the table by a permissible path. This means that the monomials in the table generate the local ring. But they are equal in number to the dimension of the local ring as calculated in 4.8. Thus they form a basis, as asserted.

By Theorem 7.2, we derive from Theorem 10.2 the normal forms for semiquasihomogeneous singularities with the quasihomogeneous parts shown in the table, and also a formula for their versal deformations. These formulae allow us to extract explicitly the characteristic values of the classical monodromy operator (see [10]).

We recall that we have called the total number μ_0 of diagonal and superdiagonal basis monomials of the local ring the *inner modality.*

10.3. THEOREM. 1) *The quasihomogeneous functions of two variables with $\mu_0 = 0$ are exhausted (up to equivalence) by the following list:*

Type	Normal form	v_1, v_2; N	Basis monomials and their weights
A_k	$x^{k+1}+y^2$	2, $k+1$; $2k+2$	$1, x, \ldots, x^{k-1}$ $0, 2, \ldots, 2k-2$
D_k	x^2y+y^{k-1}	$k-2$, 2; $2k-2$	$1, y, \ldots, y^{k-2}, x$ $0, 2, \ldots, 2k-4, k-2$
E_6	x^3+y^4	4, 3; 12	$1, y, x, y^2, xy, xy^2$ 0, 3, 4, 6, 7, 10
E_7	x^3+xy^3	6, 4; 18	$1, y, x, y^2, xy, y^3, y^4$ 0, 4, 6, 8, 10, 12, 16
E_8	x^3+y^5	10, 6; 30	$1, y, x, y^2, xy, y^3, xy^2, xy^3$ 0, 6, 10, 12, 16, 18, 22, 28

All non-degenerate functions with quasihomogeneous exponents U_i/N as in the table can be reduced to the normal forms of the table.

2) *The quasihomogeneous functions of two variables with $\mu_0 = 1$ are exhausted (up to equivalence) by two one-parameter families and 8 individual functions as in the following table:*

Type	Normal form	$U_1, U_2; N$	Basis monomials and their weights
X_9	$x^4+ax^2y^2+y^4$, $a^2-4\neq 0$	1, 1; 4	$1, x, y, x^2, xy, y^2, x^2y, xy^2, x^2y^2$ 0, 1, 1, 2, 2, 2, 3, 3, 4
Z_{11}	x^3y+y^5	4, 3; 15	$1, y, x, y^2, xy, x^2, y^3, xy^2, y^4, xy^3, xy^4$ 0, 3, 4, 6, 7, 8, 9, 10, 12, 13, 16
Z_{12}	x^3y+xy^4	3, 2; 11	$1, y, x, y^2, xy, x^2, y^3, xy^2, x^2y, xy^3, x^2y^2, x^2y^3$ 0, 2, 3, 4, 5, 6, 6, 7, 8, 9, 10, 12
Z_{13}	x^3y+y^6	5, 3; 18	$1, y, x, y^2, xy, y^3, x^2, xy^2, y^4, xy^3, y^5, xy^4, xy^5$ 0, 3, 5, 6, 8, 9, 10, 11, 12, 14, 15, 17, 20
W_{12}	x^4+y^5	5, 4; 20	$1, y, x, y^2, xy, x^2, y^3, xy^2, x^2y, xy^3, x^2y^2, x^2y^3$ 0, 4, 5, 8, 9, 10, 12, 13, 14, 17, 18, 22
W_{13}	x^4+xy^4	4, 3; 16	$1, y, x, y^2, xy, x^2, y^3, xy^2, x^2y, y^4, x^2y^2, y^5, y^6$ 0, 3, 4, 6, 7, 8, 9, 10, 11, 12, 14, 15, 18
J_{10}	$x^3+ax^2y^2+y^6$, $4a^3+27\neq 0$	2, 1; 6	$1, y, x, y^2, xy, y^3, xy^2, y^4, xy^3, xy^4$ 0, 1, 2, 2, 3, 3, 4, 4, 5, 6
K_{12}	x^3+y^7	7, 3; 21	$1, y, y^2, x, y^3, xy, y^4, xy^2, y^5, xy^3, xy^4, xy^5$ 0, 3, 6, 7, 9, 10, 12, 13, 15, 16, 19, 22
K_{13}	x^3+xy^5	5, 2; 15	$1, y, \ldots, y^8; x, xy, xy^2, xy^3$ 0, 2, ..., 16; 5, 7, 9, 11
K_{14}	x^3+y^8	8, 3; 24	$1, y, \ldots, y^6; x, xy, \ldots, xy^6$ 0, 3, ..., 18; 8, 11, ..., 26

All non-degenerate functions with quasihomogeneous exponents U_i/N as in the table can be reduced to the normal forms of the table.

In these tables, the lower index in the notation for functions is the value of μ. The numbers below the monomials are their weights, that is, the numerators of the generalized degrees (the monomial $x^p y^q$ is of degree $(U_1 p + U_2 q)/N$).

PROOF OF THE THEOREM. 1) If the second differential is not identically 0, then the function is equivalent to A_k (see [2], for example). If the corank of the function is 2, then the homogeneous exponents are given by the table in 10.2. By 4.9, the generalized degree of a basis monomial of highest degree is $d_{\max} = n - 2\Sigma\alpha_i = 2 - 2(\alpha_1 + \alpha_2)$. The condition

$d_{max} < 1$ is equivalent to $\mu_0 = 0$, and in each of the three cases of Proposition 10.1 it defines the domain under a hyperbola in the (a, b)-plane. Calculation of the integral points in these domains gives the series A, D, E.

What has been said becomes a little more understandable, perhaps, if we observe that the *classification of singularities with* $\mu_0 = 0$ *reduces to an enumeration of the lines in the plane passing below the point* (2, 2), *intersecting the coordinate axes at distances not less than* 2 *from* 0, *and having integral points in the positive octant with abscissae and ordinates not exceeding* 1.

This is because the condition $d_{max} < 1$ means that the diagonal Γ lies below (2, 2).

It is easy to check that such lines are exhausted by our list (apart from re-naming the axes), and that non-degenerate quasihomogeneous functions with quasihomogeneous exponents A, D or E reduce to the forms shown in the table.

The basis monomials are easy to find from the crosswords described in 9.6; the formula for μ_i from 4.5 is also useful.

2) Suppose that $\mu_0 = 1$. Then a basis monomial of highest generalized degree lies on the diagonal or above, and the last but one strictly below it. But by 4.11, the last and last but one of the (generalized) degrees

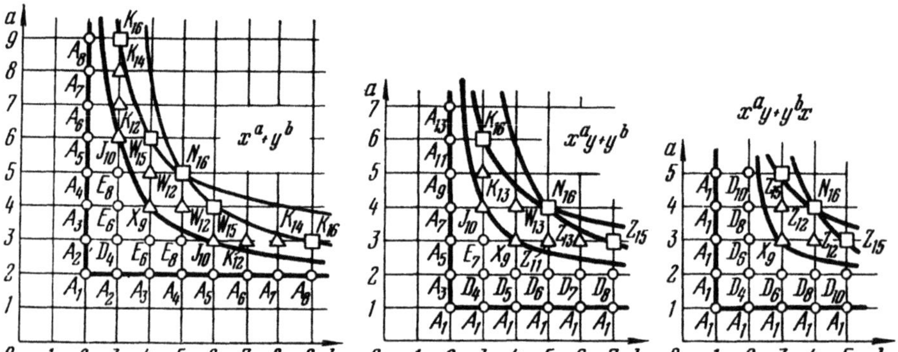

Fig. 3.

of the basis monomials are d_{max} and $d_{max} - \alpha_{min}$ (because the second degree is $\alpha_{min} = \min(\alpha_1, \alpha_2)$). Thus, the condition $\mu_0 = 1$ assumes the form

$$2\alpha_1 + 2\alpha_2 \leqslant 2, \quad 3\alpha_1 + 2\alpha_2 > 2, \quad 2\alpha_1 + 3\alpha_2 > 2.$$

Substituting the values of α_1 and α_2 from the table 10.2, we obtain in each of the three cases a domain between hyperbolae in the (a, b)-plane. The enumeration of the integral points in these domains (Fig. 3) gives the exponents of homogeneity shown in Table 10.3 ($X_9 - K_{14}$). For all these exponents there are exactly two integral points on the diagonal, apart from the cases X_9 and J_{10}. Magnification of coordinates reduces the function to a sum of monomials; the normal forms for X_9 and J_{10} are given in Propositions 8.3 and 8.4. The basis monomials are computed using 9.6, 4.5. This proves the theorem.

10.4. REMARK. The classification of singularities with $\mu_0 = 1$ comes to an enumeration of the lines Γ in the plane of exponents passing through the point (2, 2) or above it, but below (2, 3) and (3, 2).

We can also consider the boundary cases when Γ passes through one of the points (2, 3), (3, 2). The quasihomogeneous singularities corresponding to these boundary points are:

Type	Normal form	Non-degeneracy condition	$U_1,\ U_2$	N
Z_{15}	$x^3y+ax^2y^3+y^7$	$4a^3+27\neq 0$	2, 1	7
W_{15}	$x^4+ax^2y^3+y^6$	$a^2-4\neq 0$	3, 2	12
K_{16}	$x^3+ax^2y^3+y^9$	$4a^3+27\neq 0$	3, 1	9
N_{16}	$x^4y+ax^3y^2+bx^2y^3+y^4x$	$12(a^3+b^3)-36ab-5a^2b^2+81\neq 0$	1, 1	5

The following systems of monomials are examples of basis of the local rings:

Type	Basis monomials below the diagonal	On	Above
Z_{15}	$1, y, x, y^2, xy, y^3, x^2, xy^2, y^4, xy^3, y^5, xy^4, y^6$ 0, 1, 2, 2, 3, 3, 4, 4, 4, 5, 5, 6, 6	xy^5 7	xy^6 8
W_{15}	$1, y, x, y^2, xy, x^2, y^3, xy^2, x^2y, y^4, xy^3, x^2y^2, xy^4$ 0, 2, 3, 4, 5, 6, 6, 7, 8, 8, 9, 10, 11	x^2y^3 12	x^2y^4 14
K_{16}	$1, y, y^2, y^3, x, y^4, xy, y^5, xy^2, y^6, xy^3, y^7, xy^4, xy^5$ 0, 1, 2, 3, 3, 4, 4, 5, 5, 6, 6, 7, 7, 8	xy^6 9	xy^7 10
N_{16}	$1, x, y, x^2, xy, y^2, x^3, x^2y, xy^2, y^3, x^3y, x^2y^2, xy^3$ 0, 1, 1, 2, 2, 2, 3, 3, 3, 3, 4, 4, 4	$x^2y^3,\ x^3y^2$ 5, 5	x^3y^3 6

In the case of N_{16}, the system of monomials written down forms a basis only when $ab \neq 9$. If we replace the last monomial (x^3y^3) by x^4y^2, the resulting system determines a basis if $35a + 9b^2 \neq 0$. In the cases Z_{15}, W_{15}, K_{16} one and the same system works for all values of the parameters for which the function remains non-degenerate.

10.5. COROLLARY. *Every semiquasihomogeneous function of inner modality $\mu_0 = 1$ (that is, having as principal part one of the types $X_9 - K_{14}$ or with quasihomogeneous exponents as in Table* 10.3 *can be reduced by a diffeomorphism to a normal form of one of the following types:*

Type	Normal form	Non-degeneracy condition	s	Type	Normal form	Non-degeneracy condition	s
X_9	$x^4+ax^2y^2+y^4$	$a^2-4\neq 0$	4	W_{13}	$x^4+xy^4+ay^6$	$a\in C$	6
Z_{11}	$x^3y+y^5+axy^4$	$a\in C$	5	J_{10}	$x^3+ax^2y^2+y^6$	$4a^3+27\neq 0$	6
Z_{12}	$x^3y+xy^4+ax^2y^3$	$a\in C$	6	K_{12}	$x^3+y^7+axy^5$	$a\in C$	7
Z_{13}	$x^3y+y^6+axy^5$	$a\in C$	6	K_{13}	$x^3+xy^5+ay^8$	$a\in C$	8
W_{12}	$x^4+y^5+ax^2y^3$	$a\in C$	5	K_{14}	$x^3+y^8+axy^6$	$a\in C$	8

Bases for the local rings of these singularities are the same as for the quasihomogeneous singularities of Theorem 10.3. *The number s in the table denotes the (ordinary) degree of a sufficient jet.*

10.6. COROLLARY. *Semiquasihomogeneous functions with homogeneous exponents Z_{15}, W_{15}, K_{16}, N_{16} (see* 10.4) *can be reduced to the following forms:*

Type	Normal form	Non-degeneracy condition	s
Z_{15}	$x^3y+ax^2y^3+y^7+bxy^6$	$4a^3+27\neq 0$	8
W_{15}	$x^4+ax^2y^3+y^6+bx^2y^4$	$a^2-4\neq 0$	7
K_{16}	$x^3+ax^2y^3+y^9+bxy^7$	$4a^3+27\neq 0$	10
N_{16}	$x^4y+ax^3y^2+bx^2y^3+xy^4+cx^3y^3$	$12(a^3+b^3)+81\neq 36ab+5a^2b^2$	6

In the case N_{16} it is assumed that $ab\neq 9$. If $ab = 9$, but $35a+9b^2\neq 0$, then cx^3y^3 must be replaced by cx^4y^2.

10.7. COROLLARY. *Each of the singularities in the list* 10.5 *has modality not less than* 1. *The singularities Z_{15}, W_{15}, K_{16} have modality not less than* 2, *and N_{16} not less than* 3.

For the modalities of all these singularities are exactly these values, but we do not prove this here.

§11. Quasihomogeneous functions of three variables

The non-degenerate quasihomogeneous functions of three variables fall into seven (intersecting) classes. The classification has been considered by Orlik and Wagreich [12], but they missed two classes (III and VI).

11.1. PROPOSITION. *Every non-degenerate quasihomogeneous function of three-variables of degree* 1 *and corank* 3 *contains at least one of the seven systems of monomials in the following tables with non-zero coefficients (with a suitable numbering of the variables)*:

Class	Monomials	α_1,	α_2,	α_3	μ
I	$x^a,\ y^b,\ z^c$	$\frac{1}{a}$,	$\frac{1}{b}$,	$\frac{1}{c}$	$(a-1)(b-1)(c-1)$
II	$x^a,\ y^b,\ z^cy$	$\frac{1}{a}$,	$\frac{1}{b}$,	$\frac{b-1}{bc}$	$(a-1)(bc-b+1)$
III	$x^a,\ y^bx,\ z^cx$	$\frac{1}{a}$,	$\frac{a-1}{ab}$	$\frac{a-1}{ac}$	$\frac{(ab-a+1)(ac-a+1)}{a-1}$
IV	$x^a,\ y^bz,\ z^cy$	$\frac{1}{a}$,	$\frac{c-1}{bc-1}$,	$\frac{b-1}{bc-1}$	$(a-1)bc$
V	$x^a,\ y^bz,\ z^cx$	$\frac{1}{a}$,	$\frac{ac-a+1}{abc}$,	$\frac{a-1}{ac}$	$ac(b-1)+a-1$
VI	$x^ay,\ y^bx,\ z^cx$	$\frac{b-1}{ab-1}$,	$\frac{a-1}{ab-1}$,	$\frac{(a-1)b}{(ab-1)c}$	$\frac{a(abc-c-ab+b)}{a-1}$
VII	$x^ay,\ y^bz,\ z^cx$	$\frac{bc-c+1}{abc+1}$,	$\frac{ac-a+1}{abc+1}$,	$\frac{ab-b+1}{abc+1}$	abc

PROOF. We begin a classification for arbitrarily many variables $x_1, \ldots, x_n$. We fix the suffix i of the coordinate x_i. When all monomials of the form $x_i^a x_j$ are absent, the x_i-axis consists entirely of critical points.

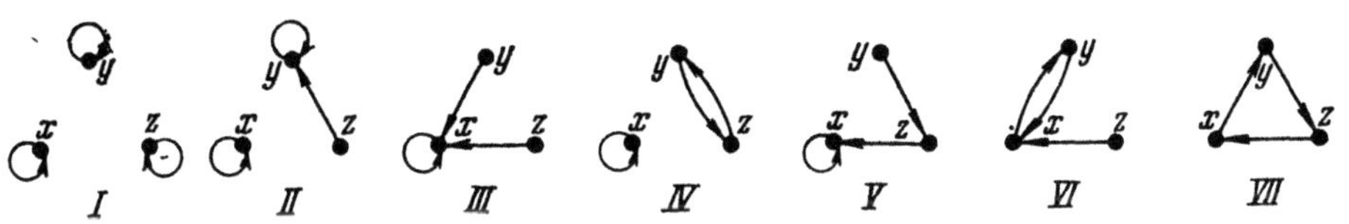

Fig. 4.

Therefore, at distance not more than 1 from each coordinate axis in the space of exponents there is the exponent of an existing monomial. Taking a monomial near to each of the axes (which is possible since the second differential is $\equiv 0$), we obtain a mapping $i \to j$ of the set of coordinate axes to itself. Thus, we have to classify mappings of a finite set to itself. For $n = 3$ this is not hard to do. A set with three elements has seven endomorphisms (apart from a re-naming of points), and this is what gives the seven classes of the table.

The values of the homogeneous exponents α_i and the multiplicity μ are calculated according to the formulae in §4.

11.2. PROPOSITION. 1) *A non-degenerate quasihomogeneous function of class* III *exists if and only if the least common multiple* $[b, c]$ *of* b *and* c *is divisible by* $a - 1$.

2) *A non-degenerate quasihomogeneous function of class* VI *exists if and only if* $(b - 1)c$ *is divisible by the product of* $a - 1$ *and the greatest common divisor* (b, c) *of* b *and* c.

3) *Non-degenerate quasihomogeneous functions of the remaining five classes exist for arbitrary* a, b, c.

PROOF. For the proof of 3) it is enough to add up the monomials shown in the table. To prove 1) and 2), we note that a quasihomogeneous function of class III or VI is degenerate if it contains none of the monomials $y^p z^q$ $(p \geqslant 0,\ q \geqslant 0)$. For the zero level set consists in that case of two components (one of which is the plane $x = 0$). This means that the critical point is not isolated (the set of critical points contains the line of intersection of the components), and the function is degenerate.

Conversely, it is easy to check that the quasihomogeneous functions

$$\text{III: } x^a + xy^b + xz^c + \varepsilon y^p z^q \quad \text{and VI: } x^a y + y^b x + z^c x + \varepsilon y^p z^q$$

are non-degenerate for almost all ε.

It remains to prove that a diagonal monomial $y^p z^q$ exists under exactly the above divisibility conditions.

In case III the generalized degree of $y^p z^q$ is $(pc + qb)(a - 1)/abc$. The monomial $y^p z^q$ is diagonal if and only if the degree is 1, that is, $(pc + qb)(a - 1) = (a - 1)bc + bc$. Thus, bc is divisible by $a - 1$, and the quotient (which is $pc + qb + bc$) is divisible by (b, c). In other words, bc is divisible by the product of $a - 1$ and (b, c), that is, $[b, c]$ is divisible by $a - 1$.

Conversely, suppose that $[b, c]$ is a multiple of $a - 1$. Then the number $\frac{abc}{a-1} = bc + \frac{bc}{a-1}$ is an integer and divisible by bc. But every number greater than bc and a multiple of (b, c) can be written in the form[1] $pc + qb$ $(p \geqslant 0,\ q \geqslant 0)$.

[1] For there are not less than 2 integral points on the line $\{p, q: pc + qb = bc\}$ in the quadrant $p \geqslant 0$, $q \geqslant 0$. The distance between consecutive integral points on every parallel line $pc + qb = m > bc$ is the same, so that there is an integral point on the segment of the line within the square (even when $m \geqslant (b-1)(c-1)$).

Thus, $abc/(a-1) = pc + qb$, and the monomial $y^p z^q$ is the diagonal.

In case VI the diagonality condition takes the form

$$(a-1)(pc+qb) = (a-1)bc + (b-1)c.$$

Thus, $(b-1)c$ is divisible by $(a-1)(b, c)$. Conversely, if $(b-1)c$ is divisible by $(a-1)(b, c)$, then $bc + \frac{(b-1)c}{a-1}$ can be written in the form $pc + qb$, where $p \geqslant 0$, $q \geqslant 0$, as required.

An example of a non-degenerate function of class III is $x^7 + xy^3 + xz^4 + \varepsilon x^2y^2$; and one of class VI is $x^5y + xy^3 + xz^4 + \varepsilon x^2y^2$.

We recall that the inner modality μ_0 of a quasihomogeneous function is the number of monomials in the local ring lying on the diagonal or above.

11.3. THEOREM. *The inner unimodal quasihomogeneous functions (that is, those with $\mu_0 = 1$) of corank 3 in three variables are exhausted (up to equivalence) by the following list:*

Type	Normal form	α_1, α_2, α_3
P_8	$x^2z+y^3+ay^2z+z^3$, $4a^3+27 \neq 0$	$\frac{1}{3}, \frac{1}{3}, \frac{1}{3}$
Q_{10}	$x^3+y^4+yz^2$	$\frac{1}{3}, \frac{1}{4}, \frac{3}{8}$
Q_{11}	$x^3+y^2z+xz^3$	$\frac{1}{3}, \frac{7}{18}, \frac{2}{9}$
Q_{12}	$x^3+y^5+yz^2$	$\frac{1}{3}, \frac{1}{5}, \frac{2}{5}$
S_{11}	$x^4+y^2z+xz^2$	$\frac{1}{4}, \frac{5}{16}, \frac{3}{8}$
S_{12}	$x^2y+y^2z+xz^3$	$\frac{4}{13}, \frac{5}{13}, \frac{3}{13}$
U_{12}	$x^3+y^3+z^4$	$\frac{1}{3}, \frac{1}{3}, \frac{1}{4}$

Every non-degenerate quasihomogeneous function with quasihomogeneous exponents shown in the table can be reduced to the indicated normal form.

PROOF. By §4, the last and last but one of the generalized degrees of the basis monomials are $d_{\max}$ and $d_{\max} - \alpha_{\min}$. The condition $\mu_0 = 1$ can therefore be rewritten in the form $d_{\max} \geqslant 1$, $d_{\max} - \alpha_{\min} < 1$. By §4, $d_{\max} = n - 2\Sigma\alpha_s$. Thus, for $n = 3$ the condition $\mu_0 = 1$ takes the form

$$\alpha_1+\alpha_2+\alpha_3\leqslant 1,\quad 2\alpha_1+2\alpha_2+2\alpha_3>2-\alpha_s,\quad s=1,2,3.$$

Now we must sort out the seven classes I–VII of Proposition 11.1, and choose an α in each case for which the given inequalities hold. We discard the case when monomials of degree 2 occur, since we are interested in functions of corank 3. Substituting the values of α, from Table 11.1, we get the following integral points corresponding to singularities of corank 3:

Class	(a, b, c) and notation for the function
I	$(3, 3, 3)=P_8$; $(3, 3, 4)$, $(3, 4, 3)$, $(4, 3, 3)=U_{12}$
II	$(3, 3, 2)=P_8$; $(4, 3, 2)=U_{12}$; $(3, 4, 2)=Q_{10}$; $(3, 5, 2)=Q_{12}$
III	$(3, 2, 2)$, $(2, 3, 2)$, $(2, 2, 3)=P_8$
IV	$(3, 2, 2)=P_8$; $(4, 2, 2)=U_{12}$; $(3, 3, 2)$, $(3, 2, 3)=Q_{12}$
V	$(3, 2, 2)=P_8$; $(3, 2, 3)=Q_{11}$; $(4, 2, 2)=S_{11}$
VI	$(2, 2, 2)=P_8$
VII	$(2, 2, 3)$, $(2, 3, 2)$, $(3, 2, 2)=S_{12}$; $(2, 2, 2)=P_8$

On the boundary of the domain of permissible α (that is, for $2\alpha_1+2\alpha_2+2\alpha_3\geqslant 2-\alpha_s$), there are some further integral points with integral a, b, c, namely:

Class	(a, b, c) and notation for the function
II	$(3, 3, 3)=U_{14}$; $(4, 4, 2)=V_{15}$; $(3, 6, 2)=Q_{14}$
III	$(3, 2, 3)$, $(3, 3, 2)=U_{14}$; $(4, 2, 3)$, $(4, 3, 2)=V_{15}$
V	$(3, 2, 4)=Q_{14}$; $(4, 2, 3)=V_{15}$; $(5, 2, 2)=S_{14}$; $(3, 3, 2)=U_{14}$
VI	$(2, 3, 2)=S_{14}$; $(2, 2, 3)=U_{14}$; $(3, 3, 2)=V_{15}$
I, IV, VII	—

For classes III and VI we have shown here only the integral points satisfying the divisibility relations of Proposition 11.2.

We show now that non-degenerate quasihomogeneous functions of the types listed really exist, and exhibit normal forms for them.

By Table 11.1, the homogeneous exponents α_s correspond to integral points (a, b, c). Sometimes triples (a, b, c) in different classes give rise to the same exponents. Such triples have been denoted in the preceding tables by the same letters (for instance, (4,3,3) in class I and (4, 2, 2) in class IV are denoted by U_{12}, and both give $\alpha_1 = 1/4$, $\alpha_2 = \alpha_3 = 1/3$).

When we know the homogeneous exponents, we can find all diagonal monomials. For example, for the values of α_s of type U_{12} just indicated, the equation of the diagonal Γ has the form $3p + 4q + 4r = 12$. The diagonal monomial $x^p y^q z^r$ corresponds to a non-negative solution of this equation in integers. There are five such solutions:

$$(4,0,0),\ (0,3,0),\ (0,2,1),\ (0,1,2),\ (0,0,3).$$

When we carry out the analogous computations for all the cases listed above, we come to the following table (in which the cases reducing to a permutation of the coordinate axes are omitted):

Type	$A_1, A_2, A_3; N$	Exponents of diagonal monomial $x^p y^q z^r$
$I_{333} = P_8$	1, 1, 1; 3	(300), (210), (201), (120), (102), (111), (021), (012), (030), (003)
$II_{342} = Q_{10}$	8, 6, 9; 24	(300), (040), (012)
$V_{323} = Q_{11}$	6, 7, 4; 18	(300), (021), (103)
$II_{352} = Q_{12}$	5, 3, 6; 15	(300), (050), (031), (012)
$V_{422} = S_{11}$	4, 5, 6; 16	(400), (021), (102)
$VII_{223} = S_{12}$	4, 5, 3; 13	(210), (021), (103)
$I_{433} = U_{12}$	3, 4, 4; 12	(400), (030), (021), (012), (003)

Here $\alpha_s = A_s/N$, so that $A_1 p + A_2 q + A_3 r = N$.

The boundary points of the domain of permissible α give four more types of quasihomogeneity:

Type	A_1, A_2, A_3; N	Exponents of diagonal monomials $x^p y^q z^r$
$II_{362} = Q_{14}$	4, 2, 5; 12	(300), (220), (140), (060), (012)
$VI_{232} = S_{14}$	4, 2, 3; 10	(050), (130), (210), (102), (022)
$II_{333} = U_{14}$	3, 3, 2; 9	(300), (210), (120), (030), (103), (013)
$II_{442} = V_{15}$	2, 2, 3; 8	(400), (310), (220), (130), (040), (102), (012)

We now choose normal forms for the functions with these 11 types of quasihomogeneity.

CASES Q_{10}, Q_{11}, S_{11} and S_{12}. All of these have three diagonal monomials each. The coefficients must be different from 0 (otherwise the function is degenerate). After magnification of coordinates the function reduces to a sum of diagonal monomials, that is,

$$Q_{10} = x^3 + y^4 + yz^2, \quad Q_{11} = x^3 + y^2z + xz^3,$$
$$S_{11} = x^4 + y^2z + xz^2, \quad S_{12} = x^2y + y^2z + xz^3.$$

CASE Q_{12}. The general linear combination of the diagonal monomials in the table is

$$Ax^3 + By^5 + Cy^3z + Dyz^2.$$

For a non-degenerate function, $AD \neq 0$. Replacing z by $z + \lambda y^2$ we reduce the function to a form in which $C = 0$. After this, magnification of the coordinates x, y, z converts the coefficients A, B, C to 1, and we obtain

$$Q_{12} = x^3 + y^5 + yz^2.$$

CASE U_{12}. According to the table, the general form of a quasihomogeneous function of type U_{12} is $Ax^4 + By^3 + Cy^2z + Dyz^2 + Ez^3$. For a non-degenerate function $A \neq 0$, and the cubic form in y and z is non-degenerate. A non-degenerate binary cubic form can be reduced by a linear change of variables to $y^3 + z^3$. The coefficient A can be made unity by magnifying the x-axis. So we obtain

$$U_{12} = x^4 + y^3 + z^3.$$

CASE P_8. It is well known that a non-degenerate ternary cubic form can be reduced to the Weierstrass normal form

$$x^2z - 4y^3 + g_2yz^2 + g_3z^3, \quad g_2^3 - 27g_3^2 \neq 0.$$

Replacing y by $y + \lambda z$, we reduce the form to $x^2z - 4y^3 + Ay^2z + Bz^3$. By non-degeneracy, $B \neq 0$. By a magnification of the x- and z-coordinates we can reduce the form to

$$P_8 = x^2z + y^3 + ay^2z + z^3.$$

This form is non-degenerate if $4a^3 + 27 \neq 0$.

This completes the proof of Theorem 11.3.

11.4. PROPOSITION. *Non-degenerate quasihomogeneous functions with quasihomogeneous exponents Q_{14}, S_{14}, U_{14} and V_{15} (see above) can be reduced by a quasihomogeneous diffeomorphism to the following normal forms:*

Type	Normal form	Non-degeneracy condition
Q_{14}	$yz^2+x^3+ax^2y^2+y^6$	$4a^3+27\neq 0$
S_{14}	$xz^2+x^2y+axy^3+y^5$	$a^2-4\neq 0$
U_{14}	$x^3+y^3+xz^3+ayz^3$	$a^3-1\neq 0$
V_{14}	$xz^2+ayz^2+x^4+bx^2y^2+y^4$	$(b^2-4)(a^4+a^2b+1)\neq 0$

PROOF. CASE Q_{14}. From the table of diagonal monomials we find

$$Q_{14} = A_1x^3 + A_2x^2y^2 + A_3xy^4 + A_4y^6 + A_5yz^2.$$

The non-degeneracy condition gives $A_1 \neq 0$. Replacement of x by $x + \lambda y^2$ removes A_3. In the expression so obtained, $A_4A_5 \neq 0$ (non-degeneracy). By magnifying coordinates we make $A_1 = A_4 = A_5 = 1$. When this is done, the non-degeneracy conditions takes the form $4A_2^3 + 27 \neq 0$.

CASE S_{14}. From the table of diagonal monomials,

$$S_{14} = A_1xz^2 + A_2x^2y + A_3xy^3 + A_4y^5 + A_5y^2z^2.$$

It follows from non-degeneracy that $A_1 \neq 0$. Replacement of x by $x + \lambda y^2$ removes A_5. After this, non-degeneracy gives $A_1A_2A_4 \neq 0$. We make $A_1 = A_2 = A_4 = 1$ by magnifiying coordinates. The non-degeneracy condition then takes the form $A_3^2 \neq 4$.

CASE U_{14}. From the table of diagonal monomials,

$$U_{14} = A_1x^3 + A_2x^2y + A_3xy^2 + A_4y^3 + A_5xz^3 + A_6yz^3.$$

The cubic form $A_1x^3 + \ldots + A_4y^3$ is non-degenerate, and by a linear change of the variables x, y it can be reduced to the form $x^3 + y^3$. It follows from non-degeneracy that one of the coefficients A_5, A_6 is not zero, say A_5. We make $A_5 = 1$ by magnifying the z-coordinate. The non-degeneracy condition then takes the form $A_6^3 \neq 1$.

CASE V_{15}. From the table of diagonal monomials,

$$V_{15} = A_1z^2x + A_2z^2y + A_3x^4 + A_4x^3y + A_5x^2y^2 + A_6xy^3 + A_7y^4.$$

By non-degeneracy, the binary 4-form $A_3x^4 + \ldots + A_7y^4$ is itself non-degenerate. By a linear change of x and y we can annihilate A_4 and A_6 (see 8.3). By non-degeneracy, one of A_1, A_2 is not 0, say A_1. By a magnification of coordinates we achieve that $A_1 = A_3 = A_7 = 1$. The non-degeneracy condition then assumes the form $A_5^2 \neq 4$, $A_2^4 + A_5A_2^2 + 1 \neq 0$. This proves Proposition 11.4.

11.5. PROPOSITION. *The following sets of monomials are bases for the local rings of the functions in Theorem* 11.3:

Type	Normal form	Basis monomials and their weights
P_8	$x^2z+y^3+ay^2z+z^3$ $4a^3+27\neq 0$ $N=3$	$1, x, y, z, xy, yz, z^2, yz^2$ $0, 1, 1, 1, 2, 2, 2, 3$
Q_{10}	$x^3+y^4+yz^2$ $N=24$	$1, y, x, z, y^2, xy, yz, y^3, xy^2, xy^3$ $0, 6, 8, 9, 12, 14, 17, 18, 20, 26$
Q_{11}	$x^3+y^2z+xz^3$ $N=18$	$1, z, x, y, z^2, xz, z^3, xy, xz^2, z^4, z^5$ $0, 4, 6, 7, 8, 10, 12, 13, 14, 16, 20$
Q_{12}	$x^3+y^5+yz^2$ $N=15$	$1, y, x, z, y^2, xy, y^3, xz, xy^2, y^4, xy^3, xy^4$ $0, 3, 5, 6, 6, 8, 9, 11, 11, 12, 14, 17$
S_{11}	$x^4+y^2z+xz^2$ $N=16$	$1, x, y, z, x^2, xy, xz, x^3, x^2y, x^2z, x^3z$ $0, 4, 5, 6, 8, 9, 10, 12, 13, 14, 18$
S_{12}	$x^2y+y^2z+xz^3$ $N=13$	$1, z, x, y, z^2, xz, yz, xy, xz^2, yz^2, z^4, z^5$ $0, 3, 4, 5, 6, 7, 8, 9, 10, 11, 12, 15$
U_{12}	$x^3+y^3+z^4$ $N=12$	$1, z, x, y, z^2, xz, yz, xy, xz^2, yz^2, xyz, xyz^2$ $0, 3, 4, 4, 6, 7, 7, 8, 10, 10, 11, 14$

Here the number below the monomial denotes its weight: $x^p y^q z^r$ has weight $i = A_1p + A_2q + A_3r$, where the $\alpha_s = A_s/N$ are the homogeneous exponents (the values of A_s and N are shown in pages 44/5).

The proof is based on the formula in §4 for the number μ_i of monomials of weight i. The choice of μ_i monomials out of all solutions of the equation $i = A_1p + A_2q + A_3r$ proceeds by means of the "crossword solution" as described in 9.6.

Similar calculations for the functions $Q_{14} - V_{15}$ lead to the following result:

11.6. PROPOSITION. *The following sets of monomials are bases for the local rings of the functions in Proposition* 11.4:

Type	Monomials below the diagonal	On	Above
Q_{14}	$1,\ y,\ x,\ y^2,\ z,\ y^3,\ xy,\ y^4,\ xy^2,\ xz,\ y^5,\ xy^3$ $0,\ 2,\ 4,\ 4,\ 5,\ 6,\ 6,\ 8,\ 8,\ 9,\ 10,\ 10$	y^6 12	y^7 14
S_{14}	$1,\ y,\ z,\ x,\ y^2,\ yz,\ y^3,\ xy,\ y^2z,\ xy^2,\ y^4,\ y^3z$ $0,\ 2,\ 3,\ 4,\ 4,\ 5,\ 6,\ 6,\ 7,\ 8,\ 8,\ 9$	xy^3 10	xy^4 12
U_{14}	$1,\ z,\ x,\ y,\ z^2,\ xz,\ yz,\ xy,\ z^3,\ yz^2,\ z^4,\ xyz$ $0,\ 2,\ 3,\ 3,\ 4,\ 5,\ 5,\ 6,\ 6,\ 7,\ 8,\ 8$	yz^3 9	yz^4 11
V_{15}	$1,\ x,\ y,\ z,\ x^2,\ xy,\ y^2,\ yz,\ x^2y,\ xy^2,\ z^2,\ y^2z$ $0,\ 2,\ 2,\ 3,\ 4,\ 4,\ 4,\ 5,\ 6,\ 6,\ 6,\ 7$	$yz^2,\ x^2y^2$ 8, 8	y^2z^2 10

Thus, in all the cases listed we have succeeded in giving a single system of monomials furnishing a basis of the local ring for all values of the parameters for which the function is non-degenerate. The possibility of such a choice of normal forms and bases in the cases P_8, Q_{14}, S_{14}, U_{14}, V_{15} has not been clear up to now.

11.7. COROLLARY. *Every semiquasihomogeneous function with quasihomogeneous part of any of the types $P_8 - U_{12}$ (or with quasihomogeneous exponents as in Table 11.3) reduces under diffeomorphism to one of the following normal forms:*

Type	Normal form	s
P_8	$x^3z+y^3+ay^2z+z^3$, $4a^3+27\neq 0$	3
Q_{10}	$x^3+y^4+yz^2+axy^3$	4
Q_{11}	$x^3+y^2z+xz^3+az^5$	5
Q_{12}	$x^3+y^5+yz^2+axy^4$	5
S_{11}	$x^4+y^2z+xz^2+ax^3z$	4
S_{12}	$x^2y+y^2z+xz^3+az^5$	5
U_{12}	$x^3+y^3+z^4+axyz^2$	4

Bases for the local rings are the same as for the quasihomogeneous singularities shown in Table 11.5. The number s is the order of a sufficient s-jet.

11.8. COROLLARY. *Every semiquasihomogeneous function with quasihomogeneous exponents of the singularities Q_{14}, S_{14}, U_{14}, V_{15} can be* *reduced by a diffeomorphism to one of the following normal forms:*

Type	Normal form	Non-degeneracy condition	s
Q_{14}	$yz^2+x^3+ax^2y^2+y^6+by^7$	$4a^3+27\neq 0$	7
S_{14}	$xz^2+x^2y+axy^3+y^5+bxy^4$	$a^2-4\neq 0$	6
U_{14}	$x^3+y^3+xz^3+ayz^3+byz^4$	$a^3-1\neq 0$	5
V_{15}	$xz^2+ayz^2+x^4+bx^2y^2+y^4+cy^2z^2$	$b^2-4\neq 0$ $a^4+a^2b+1\neq 0$	5

Proofs of these corollaries are obtained from Theorem 7.2 and Tables 11.3, 11.4, 11.5, 11.6. The number s is computed on the basis of 9.10 (the proof of the equation $s(S_{11}) = 4$ uses the fact that x^5 lies in the ideal for arbitrary a).

11.9. COROLLARY. *Each of the singularities in the list* 11.7 *has inner modality* $\mu_0 = 1$ *and modality not less than* 1. *Singularities of the types* Q_{14}, S_{14}, U_{14} *have* $\mu_0 = 2$ *and modality not less than* 2; *the type* V_{15} *has* $\mu_0 = 3$ *and modality not less than* 3.

11.10. REMARK. In fact *the modalities of the singularities listed in* §§10 *and* 11 *are the same as the corresponding inner modalities* μ_0 (see [3]); however, we do not prove this here.

11.11. THEOREM. *Every quasihomogeneous function of inner modality* $\mu_0 = 1$ *is stably equivalent to a function of one of the* 17 *types in Theorems* 10.3 (part 2) *and* 11.3.

PROOF. The finiteness of μ follows from that of μ_0. By a theorem of Saito (1.3 of [13]), every quasihomogeneous function with finite μ is stably equivalent to a quasihomogeneous function (possibly in fewer variables) whose second differential is identically 0. The inner modality of the new function is the same as that of the original one. The new function is not a function of a single variable, or else $\mu_0 = 0$. If it is a function of two variables, then by Theorem 10.3(2), the function can be reduced to one of the 10 forms shown in this theorem. If it is a function of three variables, Theorem 11.3 says that it can be reduced to one of the seven forms shown in this theorem.

For corank greater than 3 there are no functions of inner modality 1. For Saito proved that the (generalized) degree of a basis monomial of highest degree in the local ring of any quasihomogeneous function of generalized degree 1 with zero second differential is not less than that of a non-degenerate cubic form in the same number of variables (Theorem 2.11 of [11]): $d_{\max} = n - 2\sum\alpha_s \geqslant n - \frac{2}{3}n$. On the other hand, it follows from this inequality that $\Sigma\alpha_s \leqslant n/3$, which means that $\min\alpha_s$ is not more than

1/3. On the other hand, for $n \geqslant 4$ we have $d_{\max} = n - 2\sum \alpha_s \geqslant 4/3$. Thus, for $n \geqslant 4$ we have not only $d_{\max} \geqslant 1$, but $d_{\max} - \alpha_{\min} \geqslant 1$. Therefore, $\mu_0 \geqslant 2$ by Corollary 7.11. This proves the theorem.

References

[1] V. I. Arnol'd, Integrals of quickly oscillating functions and singularities of projections of Langrange manifolds, Funktsional. Anal. i Prilozhen. **6** : 3 (1972), 61–62. = Functional. Anal. Appl. **6** (1972). 222–224.

[2] V. I. Arnol'd, Normal forms of functions near degenerate critical points, the Weyl groups A_k, D_k, E_k and Lagrange singularities, Funktsional. Anal. i Prilozhen. **6** : 4 (1972), 3–25. = Functional. Anal. Appl. **6** (1972), 254–272.

[3] V. I. Arnol'd, Classification of unimodal critical points of functions, Funktsional. Anal. i Prilozhen. **7** : 3 (1973), 75–76. = Functional. Anal. Appl. **7** (1973), 230–231.

[4] V. I. Arnol'd, Remarks on the stationary phase method and Coxeter numbers, Uspekhi Mat. Nauk **28** : 5((1973), 17–44. = Russian Math. Surveys **28** : 5 (1973), 19–48.

[5] V. I. Arnol'd, Singularities of smooth mappings, Uspekhi Mat. Nauk. **23** : 1 (1968), 3–44. MR **37** #2243. = Russian Math. Surveys **23** : 1 (1968), 1–43.

[6] A. M. Samoilenko, The equivalence of a smooth function to a Taylor polynomial in the vicinity of a critical point of finite type, Funktsional. Anal. i Prilozhen, **2** : 4 (1968), 63–69. MR **42** #5132. = Functional Anal. Appl. **2** (1968), 318–323.

[7] J. C. Tougeron, Idéaux de fonctions differentiables, Ann. Inst. Fourier (Grenoble) **18** (1968), 177–240.

[8] V. P. Palamodov, On the multiplicity of a holomorphic transformation, Funktsional. Anal. i Prilozhen. **1** : 3 (1967), 54–65. MR **38** #4720. = Functional. Anal. Appl. **1** (1967), 218–226.

[9] P. Orlik and P. Wagreich, Singularities of algebraic surfaces with C^* action, Math. Ann. **93** (1971), 121–135.

[10] J. Milnor and P. Orlik, Isolated singularities defined by weighted homogeneous polynomials, Topology **9** (1970), 385–393.

[11] K. Saito, Einfach elliptische Singularitäten, Göttingen 1973, 1–46.

[12] P. Orlik and P. Wagreich, Isolated singularities of algebraic surfaces with C^* action, Ann. of Math. (2) **93** (1971), 205–228.

[13] K. Saito, Quasihomogene isolierte Singularitäten von Hyperflächen, Inventiones Math. **14** (1971), 123–142.

[14] N. Bourbaki, Groupes et algèbres de Lie, Hermann & Cie, Paris 1960. Translation: *Gruppy i algebry Li,* Mir, Moscow 1972.

Received by the Editors 28 September 1973

Translated by J. Wiegold

Critical Points of Functions and Classification of Caustics

V.I. Arnold

(The Petrovskiĭ seminar on Differential equations and mathematical problems of physics; meetings on September 19 and 26, 1973.[1])

1. The talk was largely an overview of the results of papers [1, 2, 3, 4, 5]. Two theorems not included in those papers are formulated below.

2. **Theorem 1** (O.V. Lyashko). *The space of complex Morse functions in a neighborhood of A_μ, D_μ, and E_μ functions (see* [2]*) is of local homotopy type of a $K(\pi,1)$ space. This group π is a subgroup of finite index ν in the Artin braid group on μ strands. Here $\nu = \mu! h^\mu / |W|$, where h is the Coxeter number and $|W|$ is the order of the Weyl group. Thus, ν attains the following values*:

Type of singularity	A_μ	D_μ	E_6	E_7	E_8
ν	$(\mu+1)^{\mu-1}$	$2(\mu-1)^\mu$	$2^9 \cdot 3^4$	$2 \cdot 3^{12}$	$2^2 \cdot 3^5 \cdot 5^7$

The $K(\pi,1)$ space mentioned in Theorem 1 can be described as follows. Let $f(x)$, $x \in \mathbb{C}^n$ be a quasi-homogeneous polynomial with the critical point $x=0$ of one of the types A_μ, D_μ, or E_μ. Let $F(x,\lambda) = f(x) + \lambda_1 e_1 + \ldots + \lambda_\mu e_\mu$ be a μ-parameter versal deformation of the function f, where $\lambda = (\lambda_1, \ldots, \lambda_\mu) \in \mathbb{C}^\mu$ is the parameter and $e_s(x)$ are monomials.

Denote by G the set of all values $\lambda \in \mathbb{C}^\mu$ for which the function $F(\cdot,\lambda) : \mathbb{C}^n \to \mathbb{C}$ has neither multiple critical points nor multiple critical values. Then G is a $K(\pi,1)$ space.

3. **Theorem 2.** *An arbitrary holomorphic function of two variables with an isolated critical point at* 0 *and the* 3*-jet x^3 by means of a biholomorphic change of*

[1] Translation of Uspekhi Mat. Nauk, **29** (1974), no. 3, 243–244.

Originally publ. in: Usp. Mat. Nauk 29(3), 243-244, © Russ. Acad. Sciences 1974
Translated by Boris Khesin and Andrey Khesin.
Springer-Verlag Berlin Heidelberg 2016 with kind permission.

coordinates in a vicinity of 0 *can be put into one of the normal forms from the following list*:

$$\begin{array}{ccccccccccccc} & & J_{k,2} \leftarrow J_{k,3} \leftarrow J_{k,4} \leftarrow \dots & & & & & & J_{k+1,2} \leftarrow J_{k+1,3} \leftarrow \dots \\ & & \downarrow & & & & & & \downarrow \\ \dots \leftarrow & & J_{k,1} & \leftarrow E_{6k} & \leftarrow E_{6k+1} & \leftarrow E_{6k+2} & \leftarrow J_{k+1,0} & \leftarrow & J_{k+1,1} & \leftarrow E_{6(k+1)} \leftarrow \dots \end{array}$$

Here, the letters stand for the following functions of the variables x and y:

$$\begin{aligned} E_{6k} &= x^3 + y^{3k+1} + a(y)xy^{2k+1}, \quad k \geqslant 1; \\ E_{6k+1} &= x^3 + xy^{2k+1} + a(y)y^{3k+2}, \quad k \geqslant 1; \\ E_{6k+2} &= x^3 + y^{3k+2} + a(y)xy^{2k+2}, \quad k \geqslant 1; \\ J_{k,l} &= x^3 + x^2y^k + a(y)y^{3k+l}, \quad k > 1,\ l > 0,\ a_1 \neq 0; \\ J_{k,0} &= x^3 + bx^2y^k + y^{3k} + c(y)xy^{2k+1}, \quad k > 1,\ 4b^3 + 27 \neq 0, \end{aligned}$$

where $a(y) = a_1 + a_2y + \ldots + a_{k-1}y^{k-2}$, $c(y) = c_1 + c_2y + \ldots + c_{k-2}y^{k-3}$.

The modality of the germs at 0 *of all the above functions is equal to* $k-1$, *while the multiplicity* (*i.e. the Milnor number*) *is given by the formulas* $\mu(J_{k,l}) = 6k + l - 2$ *and* $\mu(E_s) = s$.

4. I am seizing the opportunity to point out several typos in [1, 2, 3, 4].[2] In [2] on page 13, lines 24 and 26, and on page 14, line 27, instead of $x_1x_2^4$, it should read $ax_1x_2^4$. In [4] on page 32, line 5 from bottom, it should read -2 instead of $+2$. On the same page (page 32 in [4]), the diagrams for X_9 and J_{10} are incorrect. The correct diagrams for $X_9 = T_{2,4,4}$ and $J_{10} = T_{2,3,6}$ can be found on the next page (page 33 in [4]).

References

[1] V.I. Arnold, *Integrals of rapidly oscillating functions and singularities of projections of Lagrangian manifolds.* Funkt. Anal. i Prilozh., 1972, 6(3), 61–62; English transl.: Funct. Analysis and Its Appl., 1972, 6(3), 222–224.

[2] V.I. Arnold, *Normal forms for functions near degenerate critical points, the Weyl groups of A_k, D_k, E_k, and Lagrangian singularities.* Funkt. Anal. i Prilozh., 1972, 6(4), 3–25; English transl.: Funct. Analysis and Its Appl., 1972, 6(4), 254–272.

[3] V.I. Arnold, *Classification of unimodal critical points of functions.* Funkt. Anal. i Prilozh., 1973, 7(3), 75–76; English transl.: Funct. Analysis and Its Appl., 1973, 7(3), 230–231.

[2] Pages correspond to the Russian originals.

[4] V.I. Arnold, *Remarks on the stationary phase method and Coxeter numbers.* Uspekhi Mat. Nauk, 1973, 28(5), 17–44; English transl.: Russian Math. Surveys, 1973, 28(5), 19–48.

[5] V.I. Arnold, *Normal forms of functions in neighborhoods of degenerate critical points.* Uspekhi Mat. Nauk, 1974, 29(2), 11–49; English transl.: Russian Math. Surveys, 1974, 29(2), 10–50.

CLASSIFICATION OF UNIMODAL CRITICAL POINTS OF FUNCTIONS

V. I. Arnol'd

1. Definitions. Two germs of smooth functions f: $(\mathbf{C}^n, 0) \to (\mathbf{C}^n, 0)$ are equivalent if they belong to the same orbit of the group of germs of diffeomorphisms $(\mathbf{C}^n, 0) \to (\mathbf{C}^n, 0)$. Two germs are stably equivalent if they become equivalent after taking their direct sums with nondegenerate quadratic forms. The modality of a point $x \in X$ under the action $G \times X \to X$ of a Lie group G on a manifold X is the least number m such that a sufficiently small neighborhood of the point x in X meets not more than a finite number $\leq$ m-parametric families of orbits of the group G. A germ of a function f is m-modal if its sufficient k-jet is m-modal under the action of the group of k-jets of diffeomorphisms on the manifold of k-jets of functions at critical points with critical value 0. By the codimension of a set of germs given by conditions on k-jets, one means the codimension in the indicated manifold of k-jets. The multiplicity μ of a germ is the dimension over C of the local ring of gradient maps at 0. We have $\mu = \mu_+ + \mu_0 + \mu_-$, where $\mu_{\pm}$ are the positive and negative indices of inertia of the intersection index in homology of the middle dimensions of local level sets of functions of $n \equiv 3$ (4) variables stably equivalent to the given one. A germ is elliptic if $\mu = \mu_-$, and parabolic if $\mu_+ = 0$.

2. THEOREM. Every finite-multiplicity germ of a function at a critical point is either 0-modal and can be reduced to one of the normal forms A, D, E (see [1]), or belongs to one of the following classes:

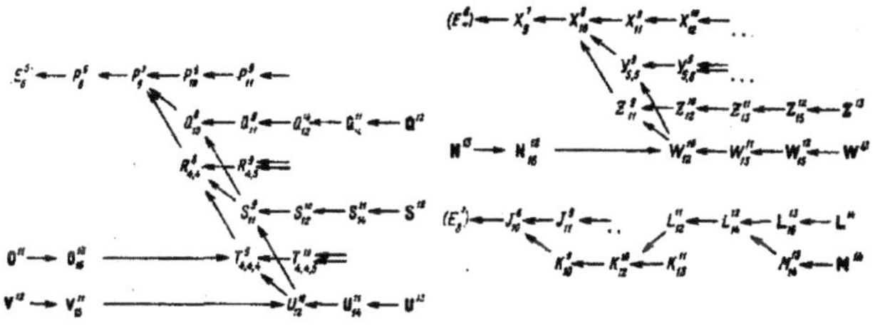

Here the letters denote classes of germs stably equivalent to the following:

$P_{p+5}^{p+3}: x^2z + y^3 + \varepsilon y^2 z + az^p,\ a \neq 0,\ \varepsilon^2 = 1,\ p > 3;\ a(4\varepsilon + 27a) \neq 0,\ \varepsilon^3 = \varepsilon,\ p = 3;$

$Q_{10}^{8}: x^2z + y^3 + ayz^3 \pm z^4,\ a \in \mathbf{C};\quad Q_{11}^{9}: x^2z + y^3 \pm yz^3 + az^5,\ a \in \mathbf{C};$

$Q_{12}^{10}: x^2z + y^3 + ayz^4 \pm z^5,\ a \in \mathbf{C};\quad Q_{14}^{11}: x^2z + y^3 \pm yz^4 + az^6 + bz^7,\ 27a^2 \pm 4 \neq 0;$

$R_{p,q}^{p+q}: x^3 + 3xyz \pm y^p + az^q,\ 4 \leqslant p \leqslant q,\ \mu = p + q + 2,\ a \neq 0;$

$S_{11}^{9}: x^2z + yz^2 \pm y^4 + ay^3z,\ a \in \mathbf{C};\quad S_{12}^{10}: x^2z + yz^2 + xy^3 + ay^5,\ a \in \mathbf{C};$

$S_{14}^{11}: x^2z + yz^2 \pm y^3z + ay^5 + by^6,\ a \neq 0,\ a \neq 1/4;$

$T_{p,q,r}^{p+q+r-3}: axyz \pm x^p \pm y^q \pm z^r,\ 4 \leqslant p \leqslant q \leqslant r,\ \mu = p + q + r - 1,\ a \neq 0;$

$U_{12}^{10}: x^3 + y^3 \pm z^4 + 3axyz^2,\ a \in \mathbf{C};\quad U_{14}^{11}: x^3 + y^3 + 3xz^3 + 3ayz^3 + 9byz^4,\ a^3 \neq 1;$

$V_{15}^{11}: x^2y \pm z^4 \pm y^2z^2 + ay^3z + by^4 + cy^4z,\ a(12b + 1)\Delta(a, b) \neq 0;$

Moscow State University. Translated from Funktsional'nyi Analiz i Ego Prilozheniya, Vol. 7, No. 3, pp. 75-76, July-September, 1973. Original article submitted March 12, 1973.

Originally publ. in: Funkts. Anal. Prilozh. 7(3), 75-76, © Russ. Acad. Sciences 1973
English transl.: Funct. Anal. Appl. 7(3), 230-231, © Consultants Bureau, New York 1973
Reprint by Springer-Verlag Berlin Heidelberg 2016 with kind permission.

O_{16}^{10}: $x^3 + y^3 + z^3 + u^3 + (ax + by + cz + du)^3 + exyzu$, $\Delta(a, b, c, d) \neq 0$;

X_{p+5}^{p+3}: $x^4 + \varepsilon x^2y^2 + ay^p$, $a \neq 0$, $\varepsilon^2 = 1$, $p > 4$; $a(4a - \varepsilon^2) \neq 0$, $\varepsilon^3 = \varepsilon$, $p = 4$;

$Y_{p,q}^{p+q-1}$: $\pm x^2y^2 \pm x^p + ay^q$, $5 \leqslant p \leqslant q$, $\mu = p + q + 1$, $a \neq 0$;

Z_{11}^{9}: $x^3y + 3axy^4 + y^5$, $a \in C$; Z_{12}^{10}: $x^3y + 3xy^4 + ay^6$, $a \in C$;

Z_{13}^{11}: $x^3y + 3axy^5 \pm y^6$, $a \in C$; $\mathbf{Z}_{15}^{12}$: $x^3y + 3xy^5 + ay^7 + by^8$, $a^2 + 4 \neq 0$;

W_{12}^{10}: $x^4 + y^5 + 2ax^2y^3$, $a \in C$; W_{13}^{11}: $x^4 + 4xy^4 + ay^6$, $a \in C$;

$\mathbf{W}_{15}^{12}$: $x^4 + 2x^2y^3 + ay^6 + by^7$, $a^2 \neq a$;

$\mathbf{N}_{16}^{12}$: $x^4y + ax^3y^2 + bx^2y^3 + xy^4 + cx^3y^3$, $ab \neq 9$;

J_{p+4}^{p+2}: $x^3 + \varepsilon x^2y^2 + ay^p$, $a \neq 0$, $\varepsilon^2 = 1$, $p > 6$; $a(4\varepsilon + 27a) \neq 0$, $\varepsilon^3 = \varepsilon$, $p = 6$;

K_{10}^{9}: $x^3 \pm y^6$; K_{12}^{10}: $x^3 + xy^5 + ay^7$, $a \neq 0$; K_{13}^{11}: $x^3 + xy^5 + ay^8$, $a \in C$;

L_{12}^{11}: $x^3 + y^7$; L_{14}^{12}: $x^3 + xy^6 + ay^8$, $a \neq 0$;

$\mathbf{L}_{16}^{13}$: $x^3 + xy^6 + ay^9 + by^{10}$, $a^2 + 1 \neq 0$, M_{14}^{13}: $x^3 \pm y^8$.

The indicated polynomials of degree k define sufficient k-jets. The upper index is the codimension of the class, the lower (if there is one) is the multiplicity μ. Arrows indicate abutment: in any neighborhood of each point of the set of k-jets of the given class one has points of all classes to which arrows go from the given one (and also some others: not all abutments are indicated).

3. Some Corollaries. A) All 1-modal germs are stably equivalent to germs of the list given. All germs of the list (excluding bold-faced ones) are 1-modal. B) The set of germs of modality greater than 1 has codimension 10 for $n \geq 4$, 11 for $n = 3$, and 12 for $n = 2$. C) The germs Q_{14}^{11}, S_{14}^{11}, U_{14}^{11}, Z_{15}^{12}, W_{15}^{12}, L_{16}^{13} are bimodal, V_{15}^{11} and N_{16}^{12} are trimodal, O_{16}^{10} is 5-modal. D) The set of germs stably inequivalent to A, D, E and germs of the given list has codimension 11 for $n \geq 4$, 12 for $n = 3$, 13 for $n = 2$. E) The given list (together with A, D, E) contains all singularities with $\mu < 14$. F) The parabolic germs are precisely P_8, X_9, J_{10}, and K_{10}. They are also defined by the fact that a sufficiently small neighborhood of a k-jet intersects only a finite number of orbits of lower codimension, and by considering infinitely many of the same. G) Germs with $\mu_+ = 1$ are precisely the nonparabolic P, R, T, X, Y, J. For them $\mu_0 = 1$. They are naturally joined in one series T:

$$P_{p+5} = T_{3,3,p};\quad R_{p,q} = T_{3,p,q};\quad X_{p+5} = T_{2,4,p};\quad Y_{p,q} = T_{2,p,q};\quad J_{p+1} = T_{2,3,p}.$$

4. Remarks. 1) An analogous classification is obtained in the C^∞-R-case. 2) The list of singularities with $\mu \leq 9$ was given by D. Siersma (1972). 3) All elliptic singularities were found by G. N. Tyurina [2]: these are A, D, E. 4) All 0-modal singularities in another setting were given by Wagreich [3]. 6) Corollary F was formulated as a conjecture by J. Milnor. 7) Probably the singularities T are determined by the property $\beta = 1/2$ (see [4]). 8) The proof of Corollary G is based on the calculation of the Dynkin diagram of "weak change of bases of vanishing cycles" (see [5]), given by A. M. Gabriélov. 9) For all singularities known to me, $\mu_+ + \mu_0$ is even.

The author thanks A. M. Gabriélov, S. M. Gusein-Zada, and J. Milnor for helpful discussions.

LITERATURE CITED

1. V. I. Arnol'd, Funktsional'. Analiz i Ego Prilozhen., 6, No. 4, 3-25 (1972).
2. G. N. Tyurina, Izv. Akad. Nauk SSSR, Ser. Matem., 32, 605-620 (1968).
3. P. Wagreich, Topology, 11, No. 1, 51-72 (1972).
4. V. I. Arnol'd, Funktsional'. Analiz i Ego Prilozhen., 6, No. 3, 61-62 (1972).
5. A. M. Gabriélov, Funktsional'. Analiz i Ego Prilozhen., 7, No. 3, 18-32 (1973).

CLASSIFICATION OF BIMODAL CRITICAL POINTS OF FUNCTIONS

V. I. Arnol'd

1. Definitions. Two germs of smooth functions $f : (\mathbf{C}^n, 0) \to (\mathbf{C}, 0)$ are equivalent, if they belong to the same orbit of the group of germs of diffeomorphisms $(\mathbf{C}^n, 0) \to (\mathbf{C}^n, 0)$. Two germs are stably equivalent, if they become equivalent after taking their direct sum with nondegenerate quadratic forms. The modality of the point $x \in X$ under the action of the Lie group G on the manifold X is the least number m such that a sufficiently small neighborhood of the point x in X is covered by no more than a finite number of not more than m-parameter families of orbits of the group G. A germ of a function is m-modal, if its sufficient k-jet is m-modal under the action of the group of k-jets of diffeomorphisms on the manifold of k-jets of functions with critical point 0 and critical value 0.

Germs of modality 0 and 1 are classified in [1] and [2] (see also [3-5]).

2. THEOREM. Up to stable equivalence, germs of modality 2 are completely exhausted by the infinite series of two-parameter families and the fourteen exceptional two-parameter families of the following three lists.

4 series of functions of two variables $J_{3,i}$, $Z^1_{1,i}$, $W_{1,i}$, $W^\#_{1,i}$:

$$
\begin{array}{llll}
J_{3,0} = x^3 + bx^2y^3 + y^9 + cxy^7, & & 4b^3 + 27 \neq 0, & \mu = 16; \\
J_{3,p} = x^3 + x^2y^3 + ay^{9+p}, & p > 0, & a_0 \neq 0, & \mu = 16 + p; \\
Z^1_{1,0} = x^3y + bx^2y^3 + cxy^6 + y^7, & & 4b^3 + 27 \neq 0, & \mu = 15; \\
Z^1_{1,p} = y(x^3 + x^2y^2 + ay^{6+p}), & p > 0, & a_0 \neq 0, & \mu = 15 + p; \\
W_{1,0} = x^4 + ax^2y^3 + y^6, & & a_0^2 \neq 4, & \mu = 15; \\
W_{1,p} = x^4 + x^2y^3 + ay^{6+p}, & p > 0, & a_0 \neq 0, & \mu = 15 + p; \\
W^\#_{1,2q-1} = (x^2 + y^3)^2 + axy^{4+q}, & q > 0, & a_0 \neq 0, & \mu(W^\#_{1,p}) = 15 + p; \\
W^\#_{1,2q} = (x^2 + y^3)^2 + ax^2y^{3+q}, & q > 0, & a_0 \neq 0, & \mu(W^\#_{1,p}) = 15 + p;
\end{array}
$$

4 series of functions of 3 variables $Q_{2,i}$; $S_{1,i}$; $S^\#_{1,i}$; $U_{1,i}$:

$$
\begin{array}{llll}
Q_{2,0} = x^3 + yz^2 + ax^2y^2 + xy^4, & & a_0^2 \neq 4, & \mu = 14; \\
Q_{2,p} = x^3 + yz^2 + x^2y^2 + ay^{6+p}, & p > 0, & a_0 \neq 0, & \mu = 14 + p; \\
S_{1,0} = x^2z + yz^2 + y^5 + azy^3, & & a_0^2 \neq 4, & \mu = 14; \\
S_{1,p} = x^2z + yz^2 + x^2y^2 + ay^{5+p}, & p > 0, & a_0 \neq 0, & \mu = 14 + p; \\
S^\#_{1,2q-1} = x^2z + yz^2 + zy^3 + axy^{3+q}, & q > 0, & a_0 \neq 0, & \mu(S^\#_{1,p}) = 14 + p; \\
S^\#_{1,2q} = x^2z + yz^2 + zy^3 + ax^2y^{2+q}, & q > 0, & a_0 \neq 0, & \mu(S^\#_{1,p}) = 14 + p; \\
U_{1,0} = x^3 + xz^2 + xy^3 + ay^3z, & a_0(a_0^2 + 1) \neq 0, & & \mu = 14; \\
U_{1,2q-1} = x^3 + xz^2 + xy^3 + ay^{1+q}z^2, & q > 0, & a_0 \neq 0, & \mu(U_{1,p}) = 14 + p; \\
U_{1,2q} = x^3 + xz^2 + xy^3 + ay^{3+q}z, & q > 0, & a_0 \neq 0, & \mu(U_{1,p}) = 14 + p;
\end{array}
$$

14 exceptional families:

Moscow State University. Translated from Funktsional'nyi Analiz i Ego Prilozheniya, Vol. 9, No. 1, pp. 49-50, January-March, 1975. Original article submitted September 30, 1974.

Originally publ. in: Funkts. Anal. Prilozh. 9(1), 49-50, © Russ. Acad. Sciences 1975
English transl.: Funct. Anal. Appl. 9(1), 43-44, © Consultants Bureau, New York 1975
Reprint by Springer-Verlag Berlin Heidelberg 2016 with kind permission.

$$E_{18}=x^3+y^{10}+axy^7,\quad E_{19}=x^3+xy^7+ay^{11},\quad E_{20}=x^3+y^{11}+axy^8;$$
$$Z^1_{17}=x^3y+y^8+axy^6,\quad Z^1_{18}=x^3y+xy^6+ay^9,\quad Z^1_{19}=x^3y+y^9+axy^7;$$
$$W_{17}=x^4+xy^5+ay^7,\quad W_{18}=x^4+y^7+ax^2y^4;$$
$$Q_{16}=x^3+yz^2+y^7+axy^5,\quad Q_{17}=x^3+yz^2+xy^5+ay^8,\quad Q_{18}=x^3+yz^2+y^8+axy^6,$$
$$S_{16}=x^2z+yz^2+xy^4+ay^6,\quad S_{17}=x^2z+yz^2+y^6+azy^4,\quad U_{16}=x^3+xz^2+y^5+ax^2y^2.$$

In all these formulas, $a = a_0 + a_1y$, and all functions of the indicated families are bimodal.

3. Hierarchy of singularities in the series:

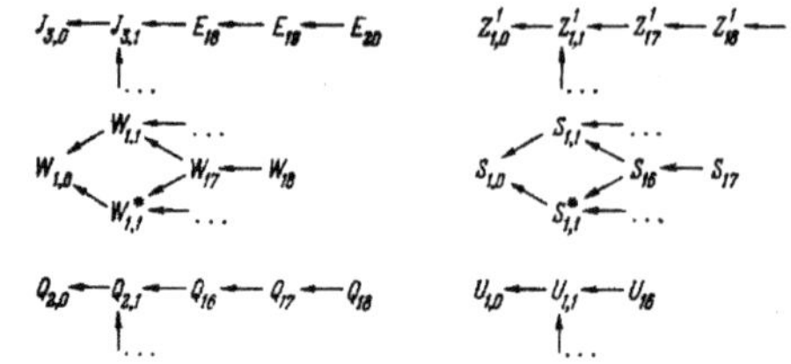

The 14 exceptional families of bimodal singularities form a quadrangular pyramid in three-dimensional space, similar to the fourteen exceptional unimodal families (see [5]):

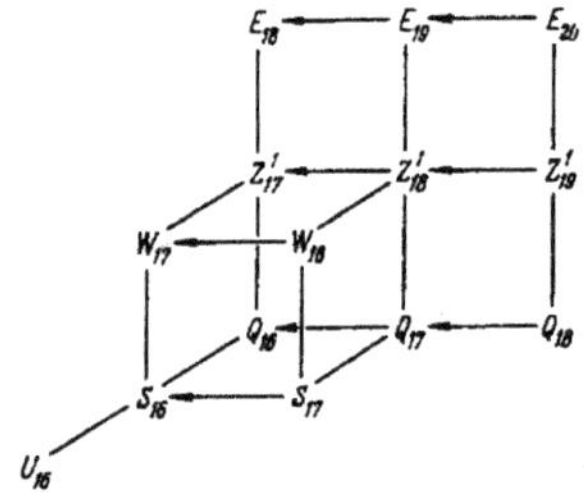

The vertex of the pyramid is U_{16}. Horizontal arrows indicate abutment. The multiplicity μ depends linearly on the coordinates. The vertical lines, which were not explained in [5], join (in the unimodal as well as the bimodal case) singularities obtained from one Kodaira class in the construction of V. S. Kulikov [7].

4. Remarks. 1. The cumbersome notation of the series is explained by the fact that the classification of bimodal singularities obtained is a fragment of the classification one now has of all singularities of corank 2 with nonzero 4-jet and of corank 3 with reduced 3-jet [6]. 2. From the theorem cited follows the classification of bimodal quasihomogeneous singularities: they are all 6 one-parameter families ($J_{3,0}$, $Z^1_{1,0}$, $W_{1,0}$, $Q_{2,0}$, $S_{1,0}$, $U_{1,0}$; $c = a_1 = 0$) and 14 isolated singularities $E_{18}-U_{16}$ (where $a = 0$).

LITERATURE CITED

1. V. I. Arnol'd, Funktsional. Anal. i Ego Prilozhen., 6, No. 4, 3-25 (1972).
2. V. I. Arnol'd, Funktsional. Anal. i Ego Prilozhen., 7, No. 3, 75-76 (1973).
3. V. I. Arnol'd, Uspekhi Matem. Nauk, 28, No. 5, 17-44 (1973).
4. V. I. Arnol'd, Uspekhi Matem. Nauk, 29, No. 2, 11-49 (1974).
5. V. I. Arnol'd, "Critical points of smooth functions," Vancouver International Congress (1974).
6. V. I. Arnol'd, Uspekhi Matem. Nauk, 29, No. 3, 243-244 (1974).
7. V. S. Kulikov, Funktsional. Anal. i Ego Prilozhen., 9, No. 1, 72-73 (1975).

SPECTRAL SEQUENCE FOR REDUCTION OF FUNCTIONS TO NORMAL FORM

V. I. Arnol'd

Reduction of functions to normal forms in neighborhoods of critical points is connected with many computations (see [1, 2, 3]). These computations are successfully carried out in a few complicated cases only with the help of a spectral sequence defined by a suitable filtration of some subcomplex of the Koshule complex constructed with respect to the derivative of the given function. Below we give statements and examples of application of the corresponding theorems.

1. Definitions and Notation

A quasihomogeneous filtration $A_0 \supset A_1 \supset \ldots$ of type $w \in N^n$ in the ring $A = C[[x_1, \ldots, x_n]]$ is given by the calibration $\deg x_k = w_k$. An induced filtration $\ldots \mathfrak{A}_{-1} \supset \mathfrak{A}_0 \supset \mathfrak{A}_1 \supset \ldots$ in the Lie algebra $\mathfrak{A}$ of formal vector fields of the form $\sum a_s \partial/\partial x_s$ $(a_s \in A)$ is given by the condition $a \in \mathfrak{A}_p \Leftrightarrow aA_q \subset A_{p+q}$ $\forall q$.

Let $f \in A_N$. We consider the complex $0 \to \mathfrak{A}_0 \to A_N \to 0$ given by the differential $\partial a = af$ $\forall a \in \mathfrak{A}_0$. The differential ∂ and filtrations in $\mathfrak{A}_0$ and in A_N define a spectral sequence. We write the r-th term of this sequence in the form $d^r: S_p^r \to A_{p+r}^r$ $(r \geqslant 0,\ p \geqslant 0)$, $S_p^0 = \mathfrak{A}_p/\mathfrak{A}_{p+1}$, $A_p^0 = A_{N+p}/A_{N+p+1}$ (the standard notation [4] is: $S_p^r = E_{-p,p+1}^r$, $A_p^r = E_{-p,p}^r$). The decomposition of f into quasihomogeneous components of degrees N, $N+1$, ... is expressed below as $f = f_0 + f_1 + \ldots$.

2. Statements of Results

<u>PROPOSITION.</u> For every $p \geq 0$ the spaces S_p^r and A_p^r are stabilized when $r \to \infty$, where the A-modules $\oplus S_p^\infty$ and $\oplus A_p^\infty$ are isomorphic to the calibrated A-modules adjoint to the kernel and cokernel of ∂; these isomorphisms are defined by the natural mappings $A_{N+p} \to A_r^p$, $(\operatorname{Ker} \partial) \cap \mathfrak{A}_p \to S_p^r$.

<u>THEOREM $T_{r,p}$.</u> Let $p > r \geq 0$. Let $e_1, \ldots, e_u$ be the quasihomogeneous polynomials of degree $N + p$, C-generating A_p^{r+1} under the natural mapping $A_{N+p} \to A_p^{r+1}$. Then there exists a formal diffeomorphism

$$y_1 = x_1 + g_1,\ \ldots,\ y_n = x_n + g_n, \text{ where } \sum g_k \partial/\partial x_k \in \mathfrak{A}_{p-r},$$

such that the series $f = f_0 + f_1 + \ldots$ takes the following form after a permutation of y:

$$f(y_1, \ldots, y_n) = f_0(x) + \ldots + f_{p-1}(x) + \sum c_i e_i(x) + R,$$

where $R \in A_{N+p+1}$ and $c_i \in C$.

<u>THEOREM T_r.</u> Let $r \geq 0$. Let $e_1, e_2, \ldots$ be the quasihomogeneous polynomials of all possible degrees $N + p$, $p > 0$, generating over C all the spaces A_p^{r+1} under the natural mappings $A_{N+p} \to A_p^{r+1}$. Then there exists a formal diffeomorphism $y_1 = x_1 + g_1, \ldots, y_n = x_n + g_n$ where $\sum g_k \partial/\partial x_k \in \mathfrak{A}_0$, such that the series $f = f_0 + f_1 + \ldots$ takes the following form after a permutation of y:

$$f(y_1, \ldots, y_n) = f_0(x) + \ldots + f_r(x) + \sum c_i e_i,$$

where $\deg e_i > N + r$ and $c_i \in C$.

Moscow State University. Translated from Funktsional'nyi Analiz i Ego Prilozheniya, Vol. 9, No. 3, pp. 81–82, July–September, 1975. Original article submitted April 7, 1975.

Originally publ. in: Funkts. Anal. Prilozh. 9(3), 81-82, © Russ. Acad. Sciences 1975
English transl.: Funct. Anal. Appl. 9(3), 251-253, © Consultants Bureau, New York 1975
Reprint by Springer-Verlag Berlin Heidelberg 2016 with kind permission.

Definition. A negative Lie algebra $\mathfrak{A}$ (respectively, a zero algebra $\mathfrak{A}^0$) of type w is defined to be an algebra of fields of the form $\sum a_s\partial/\partial x_s$, where a_s are polynomials and deg $a_s \leq w_s$ (respectively, deg $a_s = w_s$).

Definition. A series $f = f_0 + f_1 + \ldots \in A_N$ is said to satisfy condition C (respectively, condition B) if the stationary algebra of the point f_0 under the action of a negative (respectively, zero) Lie algebra on the space of polynomials of quasihomogeneous degree less than (respectively, equal to) N is trivial, i.e., equals 0.

Condition B can be written also in the form $S_0^1 = 0$.

THEOREM BT. If f satisfies condition B, then Theorem $T_{r,p}$ holds even for $p = r \geq 1$, while in Theorem T_r the condition $p > r$ can be replaced by $p \geq r \geq 1$, if at the same time we replace in the conclusion f_r by f_{r-1} and deg $e_i > N + r$ by deg $e_i \geq N + r$.

THEOREM CT. If f satisfies condition C, then $\mathfrak{A}_0 f = A_N \cap \mathfrak{A}f$. If $e_1, e_2, \ldots$ are quasihomogeneous polynomials of all possible degrees $N + p$ $(p \geq 0)$ defining C-bases in the spaces A_p^∞ under the natural mappings $A_{N+p} \to A_p^\infty$ then the images of e_i in the local ring $A/\mathfrak{A}f$ are C-linearly independent.

3. Example

Let

$$f_0 = x^4 + x^2y^{2k+1}, \quad k \geqslant 1.$$

This function is quasihomogeneous with weights deg x = 2k + 1, deg y = 2, and of degree N = deg f_0 = 8k + 4. It satisfies conditions B and C.

THEOREM W. Every formal series $f = f_0 + f_1 + \ldots$, where deg $f_p = N + p$, is reduced by a formal diffeomorphism either to the normal form

$$W_{k,i} = x^4 + x^2y^{2k+1} + ax^3y^{k+1} + by^{4k+2+i},$$

where $a = a_0 + \ldots + a_{k-2}y^{k-2}$, $b = b_0 + \ldots + b_{2k-1}y^{2k-1}$ $(i > 0, b_0 \neq 0; a = 0$ for k = 1), or to an analogous form with $b \equiv 0$ (the latter holding only in the case when the multiplicity μ of the critical point O is infinite). The number of modules (see [2]) of the series f is not less than the number of parameters in the normal form (i.e., less than 3k − 1).

Proof. To abbreviate the notation we identify the factor spaces A_p/A_{p+1} and $\mathfrak{A}_p/\mathfrak{A}_{p+1}$ with the spaces of quasihomogeneous polynomials and vector fields of degree p, respectively. Let s_p be the quasihomogeneous field of degree p. Then $d^0 s_p = s_p f_0$. From this we easily deduce the following lemma.

LEMMA 1. The homogeneous ideal $\mathfrak{A}_0 f_0$ contains the monomials $x^4, x^3y^{2k}, x^2y^{2k+1}, xy^{4k+1}$. The spaces A_p^1 $(p \geq 0)$ are generated over C by the cosets of the monomials x^3y^α $(k+1 \leqslant \alpha \leqslant 2k-1)$ and y^β $(\beta \geqslant 4k+2)$.

By Theorem T_0 and Lemma 1 we can reduce f to the form $F = f_0 + ax^3y^{k+1} + b$, where $a = a_0 + a_1y + \ldots + a_{k-2}y^{k-2}$, and b is a series in powers of y beginning with terms of degree higher than 4k + 2 in y. We denote by 4k + 2 + i the exponent of y in the first nonzero term b_0y^{4k+2+i} of the series b.

LEMMA 2. In the spectral sequence of the series F we have $F\ d^1 = \ldots = d^{2i-1} = 0$ and $d^{2i}[s_p] = [s_p b_0 y^{4k+2+i}]$. (Here and in the sequel the square brackets denote cosets.)

For the proof of Lemma 2 we note that the stationary algebra of the function f_0 in $\mathfrak{A}$ is generated over A by the field of degree 4k:

$$v = xy^2(2k+1)\partial/\partial x - (4x^2 + 2y^{2k+1})\partial/\partial y.$$

Therefore, $S_p^1 = 0$ for $p < 4k$. Hence, for every field $s = s_p + s_{p+1} + \ldots$, for which $0 \neq s_p \in [s_p] \in S_p^r$ $(r > 0)$, we have $s \in \mathfrak{A}_{4k}$.

But $\mathfrak{A}_{4k}(ax^3y^{k+1}) \subset \mathfrak{A}_0 f_0$. Indeed, all elements of the ideal on the left-hand side are divided by x^2 and have degrees not less than 10k + 4, while elements $x^4, x^3y^{2k}, x^2y^{2k+1}$ of the ideal on the right-hand side have degrees less than 10k + 4. Therefore, every monomial of each element of the ideal on the left-hand side is divided by one of the three monomials we have just mentioned, which proves the adduced inclusion. Thus, the summand ax^3y^{k+1} in the normal form does not affect the spectral sequence and the assertion of the lemma follows.

252

LEMMA 3. For $r = 2i$, $p \geq 4k$ the equality $d^r S^r_p = A^r_{p+r}$ holds and, consequently, we have $d^q = 0$ for $q > 2i$.

Indeed, by Lemma 2, $d^{2i} S^{2i}_{4k} \supset [Cvy^{4k+2+i}]$. But $[vy^{4k+2+i}] = [(-4x^2 + 2y^{2k+1}) cy^{4k+1+i}]$ where $c \neq 0$. Moreover, $x^2 y^{4k+1} \in \mathfrak{A}_0 f_0$. Hence, $[Cvy^{4k+2+i}] = [Cy^{6k+2+i}] = A^{2i}_{4k+2i}$. The lemma is proved.

THEOREM W follows from Theorems T_{2i} and CT.

LITERATURE CITED

1. V. I. Arnol'd, Funktsional'. Analiz i Ego Prilozhen., 6, No. 4, 3-25 (1973).
2. V. I. Arnol'd, Usp. Matem. Nauk, 28, No. 5, 17-44 (1973).
3. V. I. Arnol'd, Funktsional'. Analiz i Ego Prilozhen., 9, No. 1, 49-50 (1975).
4. S. MacLane, Homology, Springer-Verlag (1967).

Sel Math Sov
Vol. 1, No. 1, 1981

0272-9903/81/01/0003-16$4.70

Spectral Sequences for Reducing Functions to Normal Forms*

V. I. Arnold

Reducing functions to normal forms in a neighborhood of critical points involves many computations [1–4]. The computations can be considerably simplified with the aid of a spectral sequence constructed on the basis of an appropriate filtration of the Koszul complex (among others) arising from partial derivatives. In this paper we formulate and prove theorems on normal forms obtained using this technique.

Also, we give examples illustrating the computation of spectral sequences and their application to the problem of classification of critical points of the function.

1. Definitions and notations

1. Quasihomogeneous filtration. We denote by $A = C[[x_1, \ldots, x_n]]$ the ring of formal power series with complex coefficients in n variables $x_1, \ldots, x_n$. By a *quasihomogeneity type* in A we mean a set of integers $(w_1, \ldots, w_n; N)$. Next, we assume that a quasihomogeneity type (w, N) is fixed in A. By the *degree* of a monomial x^{m} we mean the integer $\mathrm{m} \cdot \mathrm{w}$. The series f is said to be a *quasihomogeneous function* of the type (w, N), if all the terms have degree N. The type of quasihomogeneity determines a ring filtration in A: $A = A_0 \supset A_1 \supset \cdots$, where A_d is the ideal consisting of series whose terms have degree d or higher.

By $\mathfrak{A}$ we denote the Lie algebra of formal vector fields of the form $a = \sum a_s \partial/\partial x_s$, $a_s \in A$. The Lie algebra $\mathfrak{A}$ acts on A and has the structure of a free A-module with n generators $\partial/\partial x_s$.

The filtration in A induces the filtration $\mathfrak{A} \supset \ldots \supset \mathfrak{A}_{-1} \supset \mathfrak{A}_0 \supset \mathfrak{A}_1 \ldots$, where $\mathfrak{A}_d$ consists of all fields increasing filtration by d:

$$\mathfrak{A}_d = \{ a \in \mathfrak{A} : aA_p \subset A_{p+d} \forall p \}.$$

*Originally published in Zadachy mekhaniky i matematicheskoi fiziky, Moscow, Nauka (1976). Translated by A. B. Aries.

Originally publ. in: Problems in Mechanics and Mathematical Physics, 7-20, © Nauka, Moscow 1976
English transl.: Selecta Math. Sov. 1(1), 3-18, © Birkhäuser Boston, Inc. 1981
Reprint by Springer-Verlag Berlin Heidelberg 2016 with kind permission.

Therefore, the field a has the filtration d if and only if each component has the appropriate filtration, namely:

$$a=\sum a_s\partial/\partial x_s\in\mathfrak{A}_d\Leftrightarrow a_s\in A_{d+w_s}\forall s.$$

In other words, the field a has the filtration d if for each vector-monomial $x^m\partial/\partial x_s$ in a with non-zero coefficients, the inequality $(m,w)-w_s\geqslant d$ is satisfied.

We note that $\mathfrak{A}_d$ for $d\geqslant 0$ is a subalgebra of the Lie algebra $\mathfrak{A}$. In particular, the algebra $\mathfrak{A}_0$ of fields of nonnegative degree is a subalgebra.

2. *Koszul complex and short complex.* Let f be a series in A. By a *Koszul complex* constructed from the derivatives of f we mean an exterior algebra over A with n one-dimensional generators e_s and the differential defined by the formula $de_s=\partial f/\partial x_s$. By a short complex constructed from the derivatives of f we mean the complex

$$0\to\mathfrak{A}\xrightarrow{\partial}A\to 0$$

composed of homogeneous terms of degrees 0 and 1 of the Koszul complex. Therefore, for any field $a\in\mathfrak{A}$ we have

$$\partial a=af=\sum a_s\partial f/\partial x_s.$$

The image of the differential ∂ is said to be the *gradient ideal* of f and is denoted by I_f. The group of zero-dimensional homology classes of the short complex (relative to A) is said to be the *local ring* for f and is denoted by $Q_f=A/I_f$. The local ring has the structure of a C-algebra and of an A-module.

The group of one-dimensional homology classes of the short complex (relative to $\mathfrak{A}$) is said to be the *stationary algebra* for f and is denoted by $S_f=\operatorname{Ker}\partial$. The stationary algebra has the structure of a Lie subalgebra and of an A-submodule of the Lie algebra as well as of the A-module $\mathfrak{A}$.

The series f has a *finite-order* critical point O, if the dimension Q_f over C is finite. This dimension Q_f is said to be the *multiplicity* of the critical point or the *Milnor number* and is denoted by $\mu=\dim_C Q_f$. In order that the multiplicity of the critical point be finite, it is necessary and sufficient that the Koszul complex have no homology except in dimension zero (the zero-dimensional homology is the local ring Q_f).

3. *Positive complex.* We fix the type of quasihomogeneity (w,N) as well as a series $f\in A_N$. Then we can define the *(short) positive complex* of A-modules

$$0\to\mathfrak{A}^+\xrightarrow{\partial^+}A^+\to 0,$$

where $\mathfrak{A}^+=\mathfrak{A}_0\subset\mathfrak{A}$ is the Lie algebra of all fields of nonnegative filtrations, ∂^+ is the restriction of ∂ to $\mathfrak{A}^+$ (that is, $\partial^+a=af$), $A^+=A_N$ is the ideal consisting of series whose terms have degree N and higher.

The image of the differential ∂^+ is said to be the *upper gradient ideal* for f and is denoted by I_f^+. The group of zero-dimensional homology classes of the short positive complex (derived from A^+) is said to be the *upper local ring* and is denoted by $Q_f^+ = A^+/I_f^+$.

The group of one-dimensional homology classes of the short positive complex is said to be the *upper stationary algebra* for f and is denoted by $S_f^+ = \operatorname{Ker} \partial^+$.

We note that Q_f^+ and S_f^+ have structures of A-modules and, in addition, $S_f^+ = \mathfrak{A}^+ \cap S_f$. One can define the *positive Koszul complex* in a similar way.

In contrast to the Koszul complex and the short complex defined by the function f, the short positive complex and the positive Koszul complex depend also on the initial filtration $A_0 \supset A_1 \supset \cdots$.

4. Spectral sequence of short positive complex. A-modules of a positive complex are equipped with quasihomogeneous filtrations (see 1.1). Let $\mathfrak{a}_p = \mathfrak{A}_p$ for $p \geq 0$. Then

$$\mathfrak{A}^+ = \mathfrak{a}_0 \supset \mathfrak{a}_1 \supset \cdots, \qquad A^+ = A_N \supset A_{N+1} \cdots.$$

For our future discussion it is convenient to enumerate the terms of the filtration A^+, beginning with the zero term:

$$A^+ = \mathcal{A}_0 \supset \mathcal{A}_1 \supset \cdots, \qquad \mathcal{A}_p = A_{N+p}.$$

Furthermore, we define terms of negative filtrations by the equalities $\mathcal{A}_p = \mathcal{A}_0$, $\mathfrak{A}_p = \mathfrak{A}_0$ for $p < 0$. The differential ∂^+ respects the filtration: $\partial^+ \mathfrak{A}_p \subset \mathcal{A}_p$. We consider the spectral sequence generated by the differential ∂^+ and the filtrations $\{\mathfrak{A}_p\}$, $\{\mathcal{A}_p\}$. It is convenient to write the rth term of this sequence as

$$d^r : S_p^r \to A_p^r \qquad (r \geq 0, p \geq 0),$$

where $S_p^r = \mathfrak{a}_p / \mathfrak{a}_{p+1}$, $A_p^r = \mathcal{A}_p / \mathcal{A}_{p+1}$ (standard notations [5]:

$$S_p^r = E_{-p,1+p}^r; \qquad A_p^r = E_{-p,p}^r).$$

2. Formulations of the results

The theorems formulated below will be proved in Section 4.

1. Convergence of a spectral sequence.

Theorem. *For each $p \geq 0$ there exists an $r_0(p)$ such that for $r \geq r_0(p)$*

$$S_p^r \cong S_f^+ \cap \mathfrak{a}_p / S_f^+ \cap \mathfrak{a}_{p+1}, A_p^r \cong \mathcal{A}_p / (\mathcal{A}_p \cap I_f^+) \cup \mathcal{A}_{p+1}.$$

In other words, S_p^r and A_p^r stabilize as $r \to \infty$, the stable A-modules $\oplus S_p^\infty$, $\oplus A_p^\infty$ being isomorphic to graded A-modules connected with homologies. The isomorphisms mentioned above are given by the mappings $\mathcal{A}_p \to A_p^r$, $S_f^+ \cap \mathfrak{a}_p \to S_p^r$.

2. *The normal form of terms of degree p.* We fix $p > r \geq 0$.

Theorem $T_{r,p}$. *Let $e_1, \ldots, e_u$ be quasihomogeneous polynomials of degree $N+p$, C-generating A_p^{r+1} under the natural mapping $\mathfrak{C}_p \to A_p^{r+1}$. Then there exists a formal diffeomorphism*

$$y_1 = x_1 + g_1, \ldots, \qquad y_n = x_n + g_n, \qquad \Sigma g_k \partial/\partial x_k \in \mathfrak{a}_{p-r},$$

such that the series $f = f_0 + f_1 + \cdots$ assumes the following form after the substitution of y:

$$f(y_1, \ldots, y_n) = f_0(x) + \cdots + f_{p-1}(x) + \Sigma c_i e_i(x) + R, \qquad R \in \mathfrak{C}_{p+1},$$

where c_i are constants.

Here and below, $f = f_0 + f_1 + \cdots$ denotes the expansion of the series f as quasihomogeneous components of degrees $N, N+1, \ldots$.

3. *The normal form of the rth approximation.* We fix $r \geq 0$.

Theorem T_r. *Let $e_1, e_2, \ldots$ be quasihomogeneous polynomials of all possible degrees $N+p$, where $p > r$, C-generating all spaces A_p^{r+1} under the natural mappings $\mathfrak{C}_p \to A_p^{r+1}$. Then there exists a formal diffeomorphism*

$$y_1 = x_1 + g_1, \ldots, \qquad y_n = x_n + g_n, \quad \text{where } \Sigma g_k \partial/\partial x_k \in \mathfrak{A}^+$$

such that the series $f = f_0 + f_1 + \cdots$ assumes the following form after the substitution of y:

$$f(y_1, \ldots, y_n) = f_0(x) + \cdots + f_r(x) + \Sigma c_i e_i \qquad \deg e_i > N + r,$$

where c_i are constants.

4. *Condition B.* We now consider the principal quasihomogeneous part f_0 of the series $f = f_0 + f_1 + \cdots$.

Definition. The series f *satisfies condition B* if the stationary Lie algebra of the point f_0 is trivial (equal to zero) under the action of the Lie algebra of quasihomogeneous diffeomorphisms on the space of quasihomogeneous polynomials of degree $N = \deg f_0$.

In other words, f satisfies condition B, if $S_0^1 = 0$. Therefore, condition B is imposed only on f_0.

Theorem BT. *If f satisfies condition B, Theorem $T_{r,p}$ holds for $r = p \geq 1$.*

5. *Condition C.*

Definition. By the *negative Lie algebra* $\mathfrak{A}^-$ of the type w we mean the Lie algebra of vector fields of the form $\Sigma a_s' \partial/\partial x_s$, where all monomials of each polynomial a_s' have a strictly lower degree than the degree of the monomial $x_s (\deg x_s = w_s)$.

We note that $\mathfrak{A}^-$ is a finite-dimensional Lie algebra.

Definition. The series $f=f_0+f_1+\cdots$ *satisfies condition* C if the stationary algebra of the point f_0 is trivial (equal to zero) under the action of the negative Lie algebra $\mathfrak{A}^-$ on the space of polynomials of (quasi) degree not higher than $N=\deg f_0$.

We note that condition C is imposed only on f_0.

Theorem CT. *If f satisfies condition C, then $I_f^+=I_f^+=A^r\cap I_f$.*

Corollary. *Let f satisfy condition C. Furthermore, let $e_1, e_2, e_3, \cdots$ be quasihomogeneous polynomials of all possible degrees $N+p$, $p>0$, defining bases in spaces A_p^∞ of the spectral sequence under the natural mappings $\mathcal{Q}_p\to A_p^\infty$. Then the images of the polynomials $e_1, e_2, \ldots$ in the local ring $Q_f=A/I_f$ are C-linearly independent.*

In other words, the tangent space of the deformation $f+\sum\lambda_i e_i$ intersects the tangent space of the orbit of f at one point.

3. Description of a spectral sequence

Let $r>0$, $p>0$. We identify the factor-spaces

$$S_p^0=\mathfrak{a}_p/\mathfrak{a}_{p+1},\qquad A_p^0=\mathcal{Q}_p/\mathcal{Q}_{p+1}$$

with spaces of quasihomogeneous vector fields of degree p and polynomials of degree $N+p$, respectively. The spaces S_p^r entering the spectral sequence can be described as subspaces of the space of quasihomogeneous fields S_p^0, and also, the spaces A_p^r can be described as factor-spaces $A_p^r=A_p^0/I_p^r$ of the space of quasihomogeneous polynomials for some subspace I_p^r (where $I_p^0=0$); finally, the differential $d^r: S_p^r\to A_{p+r}^r$ associates to a quasihomogeneous field $s_p\in s_p^r$ a quasihomogeneous polynomial of degree $N+p+r$ defined up to a polynomial from I_{p+r}^r.

We shall give below definitions of the spaces S_p^r, I_p^r, A_p^r and of the mappings d^r. Let $f=f_0+f_1+\cdots$ be the expansion of the series f as quasihomogeneous components of degrees $N, N+1, \ldots$.

1. Description of the spaces S_p^r, $r>0$.

Definition. S_p^r is the space of quasihomogeneous vector fields s_p of degree p, permitting "extensions" to polynomials $s_p+\cdots+s_{p+r-1}$ satisfying the system of r equations

$$\begin{aligned}
&s_pf_0=0,\\
&s_pf_1+s_{p+1}f_0=0,\\
&\cdots\cdots\cdots\cdots\\
&s_pf_{r-1}+\cdots+s_{p+r-1}f_0=0.
\end{aligned}$$

2. Description of the differential d^r, $r>0$.

Definition. The differential d^r acts on the quasihomogeneous field $s_p\in S_p^r$ in accordance with the formula

$$d^rs_p=s_pf_r+\cdots+s_{p+r}f_0 \pmod{I_{p+r}^r}.$$

where s_q satisfies the equations given in 3.1, I^r_{p+r} is defined in 3.3, and where, finally, we may take $s_{p+r}=0$ for $r>0$.

3. *Description of the spaces* I_p^{r+1}, $r>0$.

Definition. I_p^{r+1} is defined as the set of all quasihomogeneous polynomials of degree $N+p$, representable as $s_{p-r}f_r+\cdots+s_pf_0$, where the quasihomogeneous fields s_q of degrees q satisfy the r conditions

$$\begin{gathered} s_{p-r}f_0=0,\\ s_{p-r}f_1+s_{p-r+1}f_0=0,\\ \cdots\cdots\cdots\cdots\cdots\\ s_{p-r}f_{r-1}+\cdots+s_{p-1}f_0=0 \end{gathered}$$

and belong to $\mathfrak{A}^+$ (that is, all s_q with $q<0$ are equal to zero).

Definition. For any $r>0$, $A^r_p=A^0_p/I^r_p$.

4. *Algebraic properties.* It is easy to verify

Proposition. *The following equalities hold*:

$$S_p^{r+1}=\operatorname{Ker}(d^r: S_p^r\to A^r_{p+r}),\qquad A^{r+1}_{p+r}=A^r_{p+r}/d^rS^r_p.$$

These equalities will be proved in §4.

5. *Example.* d^0: $S_p^0\to A_p^0$ is given by the equality $d^0s_p=s_pf_0$. Therefore, I_p^1 is a homogeneous $N+p$-component of the gradient ideal of the quasihomogeneous part f_0 of degree N of the function $f=f_0+\cdots$.

Hence, we can identify A_p^1 with the $N+p$-component of the local ring Q_{f_0} of the quasihomogeneous function f_0. Further, S_p^1 is the p-component of the stationary algebra of the function f_0.

6. *Example.* d^1: $S_p^1\to A^1_{p+1}$ is given by the equality $d^1s_p=s_pf_1\operatorname{mod}\{s_{p+1}f_0\}$, where $s_pf_0=0$. Therefore,

$$\begin{gathered} s_p^2=\{s_p: s_pf_0=0, \exists s_{p+1}: s_pf_1=s_{p+1}f_0\},\\ I_p^2=\{s_pf_0\}+\{s_{p-1}f_1: s_{p-1}f_0=0\} \end{gathered}$$

(for $p=0$ the second term disappears).

Therefore, for $r=0$ and $r=1$ the value d^rs_p of the differential d^r is defined as the coset of the polynomial s_pf_r. This simple formula for a differential does not hold for large r.

7. *Example.* d^2: $S_p^2\to A^2_{p+2}$ is given by the equality

$$d^2s_p=s_pf_2+s_{p+1}f_1\operatorname{mod}\{s_{p+2}f_0\}+\{s_{p+1}f_1: s_{p+1}f_0=0\},$$

where $s_pf_0=0$, $s_pf_1+s_{p+1}f_0=0$.

4. Proof

We shall first prove that the spaces and differentials described in §3 in fact form the spectral sequence of the short positive complex.

By the general definition (see [5]),

$$S_p^r = \mathrm{S}_p^r / \mathrm{S}_p^r \cap \mathfrak{a}_{p+1}, \qquad A_p^r = \mathcal{Q}_p / \mathrm{I}_p^r \cup \mathcal{Q}_{p+1},$$

where

$$\mathrm{S}_p^r = \{ s \in \mathfrak{a}_p : \partial s \in \mathcal{Q}_{p+r} \}, \qquad \mathrm{I}_p^r = \partial \mathrm{S}_{p-r+1}^{r-1};$$

here $-\infty < p < +\infty$, $r > -1$ for S, $r > 0$ for I; everywhere $\partial s = sf$.

1. Properties of approximations to the stationary algebra. The definitions directly imply the following:

Proposition. *S_p^r is a subalgebra and a sub-A-module in $\mathfrak{a}_p$. We have the inclusions $\mathrm{S}_p^r \supset \mathrm{S}_p^{r+1}$, $\mathrm{S}_p^r \supset S_{p+1}^{r-1}$, $A_q \mathrm{S}_p^r \subset \mathrm{S}_{p+q}^r$. Also, we have the equalities*

$$\mathrm{S}_p^{-1} = \mathrm{S}_p^0 = \mathfrak{a}_p \quad \textit{for all } p,$$

$$\mathrm{S}_p^r = \mathfrak{a}_0 \quad \textit{for } p < 0, p+r < 0,$$

$$\mathrm{S}_p^r = \mathrm{S}_0^{r+p} \quad \textit{for } p < 0, p+r > 0.$$

For $p > 0$, $r > 1$, the condition $s \in S_p^r$ imposes a restriction only on the first r terms $s_p + \cdots + s_{p+r-1}$ of the expansion of s into quasihomogeneous components. Furthermore, these restrictions depend only on r terms $f_0 + \cdots + f_{r-1}$ of the expansion f.

These conditions are given in explicit form in 3.1.

2. Properties of approximations to the gradient ideal. The definitions and the proposition in 4.1 directly imply the following:

Proposition. *I_p^r is an ideal in A. We have the inclusions $\mathrm{I}_p^r \subset \mathrm{I}_p^{r+1}$, $\mathrm{I}_p^r \supset \mathrm{I}_{p+1}^{r+1}$, $A_q \mathrm{I}_p^r \subset \mathrm{I}_{p+q}^r$. Also, we have the equalities*

$$\mathrm{I}_p^0 = \partial \mathfrak{a}_{p+1}, \; I_p^1 = \partial \mathfrak{a}_p \quad \textit{for all } p;$$

$$\mathrm{I}_p^r = \partial \mathfrak{a}_0 \quad \textit{for } p < 0, r > 1;$$

$$\mathrm{I}_p^{p+q} = \mathrm{I}_p^{p+1} \quad \textit{for all } p, q > 1;$$

$$\mathrm{I}_p^r = (\partial \mathfrak{a}_{p-r+1}) \cap \mathcal{Q}_p \quad \textit{for all } r, p.$$

3. Properties of spaces and differentials of a spectral sequence. The definitions and the propositions in 4.1 and 4.2 directly imply the following:

Proposition. *The spaces S_p^r have the following properties:*

$$S_p^r \supset S_p^{r+1}, \qquad (A_q / A_{q+1}) S_p^r \subset S_{p+q}^r, \qquad S_p^r \subset \mathfrak{a}_p / \mathfrak{a}_{p+1};$$

$$S_p^{-1} = S_p^0 = \mathfrak{a}_p / \mathfrak{a}_{p+1}; \qquad S_p^r = 0 \quad \textit{for } p < 0;$$

$$S_p^r \{ [s_p] : \exists s = s_p + \cdots + s_{p+r-1}, sf \in \mathcal{Q}_{p+r} \} \quad \textit{for } p > 0, r > 1,$$

where the brackets denote the coset of the quasihomogeneous field s_p of degree p modulo $\mathfrak{a}_{p+1}$.

The spaces $I_p^r = \mathrm{I}_p^r / \mathrm{I}_p^r \cap A_{p+1}$ have the following properties:

$$I_p^r \subset I_p^{r+1},\ (A_q/A_{q+1}) I_p^r \subset I_{p+q}^r, \qquad I_p^r \subset \mathfrak{A}_p/\mathfrak{A}_{p+1};$$

$$I_p^0 = 0 \quad \text{for all } p; \qquad I_p^r = 0 \quad \text{for } p<0 \text{ and any } r;$$

$$I_p^{p+q} = I_p^{p+1} \quad \text{for all } p \text{ and for } q \geq 1;$$

$$I_p^r = \{[a_p] : \exists s = s_{p-r+1} + \cdots + s_p : [sf] = [a_p]\} \quad \text{for } r \geq 1, p \geq 0$$

(*where the brackets denote the coset modulo $\mathfrak{A}_{p+1}$ and a_p and s_q are quasihomogeneous polynomials and fields of degrees $N+p$ and q; $s_q = 0$ for $q<0$*).

The spaces A_p^r have the following properties:

$$A_p^r = (\mathfrak{A}_p/\mathfrak{A}_{p+1})/I_p^r = \mathfrak{A}_p/(\mathrm{I}_p^r \cup \mathfrak{A}_{p+1});$$

$$A_p^r = 0 \quad \text{for } p<0, \qquad A_p^0 = \mathfrak{A}_p/\mathfrak{A}_{p+1} \quad \text{for } p \geq 0;$$

$$A_p^{p+q} = A_p^{p+1} \quad \text{for } q \geq 0.$$

The mapping ∂ induces mappings d^r: $S_p^r \to A_{p+r}^r$ ($r \geq 0$) having the following properties:

$d^r S_p^r = 0$ for $p<0$, d^0: $\mathfrak{a}_p/\mathfrak{a}_{p+1} \to \mathfrak{A}_p/\mathfrak{A}_{p+1}$ is given by the condition $d^0[s_p] = [s_p f_0]$ (where s_p is a quasihomogeneous field of degree p and the brackets denote the coset modulo $\mathfrak{a}_{p+1}$ and modulo $\mathfrak{A}_{p+1}$); for $r \geq 1$, $p \geq 0$ the value d^r on the class $[s_p] \in S_p^r \subset \mathfrak{A}_p/\mathfrak{A}_{p+1}$ is defined as the coset of the series sf modulo I_{p+r}^r, where $s = s_p + \cdots + s_{p+r} \in \mathrm{S}_p^r$.

4. *Proofs of the assertions given in §3.* After the factor-spaces $\mathfrak{a}_p/\mathfrak{a}_{p+1}$ and $\mathfrak{A}_p/\mathfrak{A}_{p+1}$ have been identified with the spaces of quasihomogeneous vector fields and polynomials in $\mathfrak{a}_p$ and in $\mathfrak{A}_p$, the proposition in 4.3 turns into the formulas given in §3. We have thereby proved that the spectral sequence described in §3 is the spectral sequence of the short positive complex ∂^+. In particular, the proposition in 3.4 follows from the usual properties of spectral sequences [5].

5. *Proof of the Theorem in 2.1.* The Theorem in 2.1 follows from the standard results on convergence of spectral sequences and from the following assertion which is easy to prove.

Lemma. *If $s_p \in S_p^r$ for all r, there exists a formal series $s = s_p + \cdots$, such that $sf = 0$.*

Proof. By definition (see §4)

$$S_p^r = ((\mathfrak{a}_p \cap \partial^{-1}\mathfrak{A}_{p+r}) \cup \mathfrak{a}_{p+1})/\mathfrak{a}_{p+1}.$$

Therefore,

$$\bigcap_r S_p^r = \left(\left(\mathfrak{a}_p \cap \left(\bigcap_r \partial^{-1}\mathfrak{A}_{p+r}\right)\right) \cup \mathfrak{a}_{p+1}\right) / \mathfrak{a}_{p+1}.$$

But

$$\bigcap_r \partial^{-1}\mathfrak{A}_{p+r} = \partial^{-1} \bigcap_r \mathfrak{A}_{p+r} = \partial^{-1} 0.$$

Hence

$$\bigcap_r S_p^r = ((\mathfrak{a}_p \cap S_f^+) \cup \mathfrak{a}_{p+1}) / \mathfrak{a}_{p+1},$$

as required.

6. *Proof of Theorem* $T_{r,p}$, $p > r \geq 0$. Using the notations from §3, prove that $N+p$-quasijets $[f] \bmod \mathfrak{A}_{p+1}$, $[f+\varphi_p] \bmod \mathfrak{A}_{p+1}$ map into each other by means of formal substitution of the variables of nonnegative filtrations, $y = x + g(x)$, $\Sigma g_s \partial/\partial x_s \in \mathfrak{a}_{p-r}$ if $[\varphi_p] \in I_p^{r+1}$. The latter condition implies that there exists an expansion $\varphi_p = s_{p-r} f_r + \cdots + s_p f_0$, where the quasihomogeneous fields s_q of nonnegative filtrations satisfy the r conditions

$$s_{p-r} f_0 = 0, \ldots, s_{p-r} f_{r-1} + \cdots + s_{p-1} f_0 = 0.$$

It is crucial to note that the expansion φ_p and the restrictions on s_q depend only on the terms $f_0, \ldots, f_r$ of the expansion f and do not depend on the a posteriori terms.

We now consider the one-parameter family of $N+p$-quasijets $F(t) = [f + t\varphi_p] \bmod \mathfrak{A}_{p+1}$, $0 \leq t \leq 1$.

Since $p > r$, the expansion $F(t)$ for any t into quasihomogeneous components with respect to x begins with the terms $f_0 + \cdots + f_r$ not dependent on t. Therefore, the space of quasihomogeneous polynomials I_p^{r+1} constructed from $F(t)$ for any t, is the same.

We define the vector field s in a neighborhood of the point $O \in C^n$ by the formula $s = s_{p-r} + \cdots + s_p$. Then, for any t the derivative of the $N+p$-quasijet of $F(t)$ in the direction of the field s is seen to be

$$\begin{aligned} sF(t) &= \left[(s_{p-r} + \cdots + s_p)(f_0 + \cdots + f_p + t\varphi_p)\right] \bmod \mathfrak{A}_{p+1} \\ &= \left[s_{p-r} f_0 + \cdots + (s_{p-r} f_r + \cdots + s_p f_0)\right] \bmod \mathfrak{A}_{p+1} = [\varphi_p] \bmod \mathfrak{A}_{p+1}. \end{aligned}$$

(it is crucial that $p > r$; for $p = r$ the term $t\varphi_p$ should influence $sF(t)$).

The field s at the point $O \in C^n$ goes to zero since $s \subset \mathfrak{A}_{p-r}$. Hence the phase flow of the field s is defined in a sufficiently small neighborhood of the point $O \in C^n$ for all t in the interval $0 \leq t \leq 1$. The transformation of this flow at $t = 1$ defines the p-jet of the local diffeomorphism lifting $F(0)$ into $F(1)$ and having the form $y = x + g(x)$, $\Sigma g_k \partial/\partial x_k \in \mathfrak{a}_{p-r}$. This diffeomorphism is the one we have sought.

7. Proof of Theorem T_r, $r>0$. We shall subsequently use Theorems $T_{r,\,r+1}$, $T_{r,\,r+2}, \ldots$ in order to reduce the terms of (quasi) degree $N+p$, where $p=r+1, r+2, \ldots$, to the normal form. Under these operations, the first $r+1$ terms of the expansion $f=f_0+\cdots+f_r+\cdots$ do not change. Therefore, when the terms of the (quasi) degrees indicated are normalized, the spaces I_p^{r+1} and A_p^{r+1} do not change. Therefore, the second theorem, etc. can be applied to the series obtained upon the first substitution (the substitution of Theorem $T_{r,\,r+1}$).

The resultant sequence of formal substitutions of variables is convergent since the terms of each fixed degree stabilize (because $y-x\in \mathfrak{a}_{p-r}$ in Theorem $T_{r,\,p}$). The limiting formal substitution is the one we have sought.

8. Proof of Theorem BT. Let $p=r>1$ and let f satisfy condition B. We shall prove Theorem $T_{r,\,r}$. Assuming that $p=r$ in the proof of Theorem $T_{r,\,p}$ from 4.6, we obtain

$$\varphi_r=s_0f_r+\cdots+s_rf_0,$$

where s_q satisfies the r conditions $s_0f_0=0$, $s_0f_1+s_1f_0=0, \ldots$, $s_0f_{r-1}+\cdots+s_{r-1}f_0=0$. But $s_0=0$ by condition B. Therefore, neither the expansion φ_r nor the restrictions on s_q depend on f_r (they depend only on $f_0, \ldots, f_{r-1}$). Hence, for $p=r$ the space I_p^{r+1} constructed from $F(t)$ does not depend on t. The remaining proof of Theorem $T_{r,\,r}$ is the same as the final part of the proof of Theorem $T_{r,\,p}$ for $p>r$ (4.6).

9. Proof of Theorem CT. By definition, $I_f^+=\mathfrak{A}^+f$, $I_f=\mathfrak{A}f$. Hence $I_f^+\subset I_f$ and $I_f^+\subset \mathfrak{C}_0$. Therefore, $I_f^+\subset I_f\cap \mathfrak{C}_0$.

We shall prove the inverse inclusion. Let $u\in I_f\cap\mathfrak{C}_0$. Then $u=af$, $a\in\mathfrak{A}$. We consider a nonzero quasihomogeneous component a_q of the field a of the lowest quasidegree q. Quasidegrees of all quasihomogeneous components of the series af, with the exception of a_qf_0, are higher than the quasidegrees of the polynomial a_qf_0. If $q<0$, $a_qf_0\neq 0$ by condition C. Therefore, $q<0$ contradicts $u\in\mathfrak{C}_0$. Therefore, $q>0$, that is, $a\in\mathfrak{A}^+$, $u\in I_f^+$, which was to be proved.

5. Examples

1. The series W. Let $f_0=x^4+x^2y^{2k+1}$, $k>1$. This function is quasihomogeneous, with weights $\deg x=2k+1$, $\deg y=2$, of degree $N=\deg f_0=8k+4$. This function satisfies conditions B and C.

Theorem W. *Each formal series $f=f_0+f_1+\cdots$, where $\deg f_p=N+p$, is reduced, via a formal diffeomorphism, either to the normal form*

$$W_{k,\,i}=f_0+ax^3y^{k+1}+by^{4k+2+i},$$

where $a = a_0 + \cdots + a_{k-2}y^{k-2}$, $b = b_0 + \cdots + b_{2k-1}y^{2k-1}$, $i > 0$, $b_0 \neq 0$ *(and where* $a = 0$ *for* $k = 1$*), or to a similar normal form with* $b \equiv 0$ *(the latter only in the case when the multiplicity* μ *of the critical point O is infinite). The number of moduli (see* [2]*) of the series f is not smaller than the number of parameters in the normal form (that is, not smaller than* $3k-1$*).*

Proof. We shall use the identifications of the factor-spaces A_p/A_{p+1} and $\mathfrak{A}_p/\mathfrak{A}_{p+1}$ with the spaces of quasihomogeneous polynomials and vector fields, which were introduced in §3.

Let s_p be a quasihomogeneous field of degree p. Then, according to §3, $d^0 s_p = s_p f_0$. It is easy to prove (using, for example, [3]) the following:

Lemma 1. *The homogeneous ideal* $\mathfrak{A}_0 f_0$ *contains monomials* $x^4, x^3y^{2k}, x^2y^{2k+1}, xy^{4k+1}$ *and a binomial* $2x^3y^{k+1} + xy^{3k+2}$*. The spaces* A^1_p $(p > 0)$ *of the first term of a spectral sequence are generated over C by the images of the monomials* x^3y^α*, where* $k+1 \leqslant \alpha \leqslant 2k-1$*, and of* y^β*, where* $\beta > 4k+2$.

By Lemma 1 and Theorem T_0, it is possible to reduce f to the form

$$F = f_0 + ax^3y^{k+1} + b,$$

where $a = a_0 + \cdots + a_{k-2}y^{k-2}$ and b is a series in y, beginning with the terms of degree higher than $4k+2$ in y. We denote by $4k+2+i$ the index of degree y in the first nonzero term $b_0\varphi$, $\varphi = y^{4k+2+i}$ of the series b. Let $r = \deg\varphi - N = 2i$.

Lemma 2. *In the spectral sequence of the series F*

$$d^1 = \cdots = d^{r-1} = 0,\ d^r[s_p] = b_0[s_p\varphi]$$

(here and below the brackets denote cosets).

Proof. We note that the stationary algebra of the function f_0 is generated over A by the field $v = xy^{2k}(2k+1)\partial/\partial x - (4x^2 + 2y^{2k+1})\partial/\partial y$ of degree $4k$. Hence $s^1_p = 0$ for $p < 4k$. Therefore, for any field $s = s_p + s_{p+1} + \cdots$, for which $0 \neq s_p \in [s_p] \in s^r_p$, $r \geqslant 1$, we have $s \in \mathfrak{A}_{4k}$. But $\mathfrak{A}_{4k}(as^3y^{k+1}) \subset \mathfrak{A}_0 f_0$. In fact, the elements of the ideal on the left side are divisible by x^2 and have degree not lower than $10k+4$, and the elements x^4, x^3y^{2k} and x^2y^{2k+1} of the ideal on the right side have degree lower than $10k+4$. Hence, each monomial of each element of the ideal on the left side is divisible by one of the three monomials indicated, which proves the inclusion given above.

Therefore, the term ax^3y^{k+1} in the normal form has no effect on the spectral sequence, thus proving the assertion of the lemma.

Lemma 3. *For* $r = 2i$, $p > 4k$, *we have the equality* $d^r S^r_p = A^r_{p+r}$, *and therefore,* $d^q = 0$ *for* $q > 2i$.

Proof. By Lemma 2, $d^r S^r_{4k} \supset C[v\varphi]$. But $v\varphi = (-4x^2 + 2y^{2k+1})c_0 y^{4k+1+i}$, where $c_0 \neq 0$. Furthermore, $x^2y^{4k+1} \in \mathfrak{A}_0 f_0$. Hence, $C[v\varphi] = C[y^{6k+2+i}]$

$=A^r_{4k+r}$. Thus, $d^r S^r_{4k}=A^r_{4k+r}$. Multiplying this by the appropriate power of y, we obtain what was required for $p>4k$.

Theorem W follows now from Theorems BT_{2i} and CT.

2. *The table of normal forms.* In Table 1 the normal forms $f_0+f_1+\cdots$ are listed, to which the functions with the indicated principal quasihomogeneous parts f_0 are reduced via the method described above. In Table 1 everywhere

$$a=a_0+\cdots+a_{k-2}y^{k-2}\quad \textit{for } k>1,\qquad a=0\quad \textit{for } k=1;$$

$$b=b_0+\cdots+b_{2k-1}y^{2k-1},\ c=c_0+\cdots+c_{k-1}y^{k-1}$$

$$a'=a_1y+\cdots+a_{k-2}y^{k-2},\ b'=b_0+\cdots+b_{2k-2}y^{2k-2}.$$

The number b_0 is nonzero in all cases.

3. *Computation of a spectral sequence of functions of 2 variables.* The theorem whose formulation is given in Table 1 is proved on the model of Theorem W in 5.1. In Table 2 the results of intermediate computations (corresponding to Lemmas 1, 2 and 3 in 5.1) for the functions of two variables from Table 1 are given. We denote by $\{e_j\}$ the union of quasihomogeneous functions generating C-bases of all the spaces A^1_p, $p>0$. After Theorem T_0 has been applied, f is reduced to the form $F=f_0+\Sigma c_j e_j$. In Table 2 the following information about the spectral sequence constructed from F is given: v is the generator of the stationary algebra of f_0 over A:

$$\rho=\deg v=\min\{\sigma\colon S^1_\sigma\neq 0\};$$

φ is the quasihomogeneous function of the lowest quasihomogeneous degree from among the number e_j entering with the nonzero coefficient c_j into the expansion $F=f_0+\Sigma c_j e_j$, for which $v\varphi$ does not belong to the ideal $\mathfrak{A}_0 f_0$:

$$r=\deg\varphi-\deg f_0=\deg\varphi-N=\min\{\sigma>0\colon d^\sigma\neq 0\};$$

τ is the monomial for the coset $[\tau]$ for which the relation $C[v\varphi]=C[\tau]$ is satisfied in $A^r_{r+\rho}$.

It is important to choose the generators $\{e_j\}$ as indicated in Table 2. Then the terms x^3y^α have no effect on the spectral sequence by the same reasoning as in the example in 5.1.

4. *Computation of a spectral sequence for the function of 3 variables given in Table 1.* In Table 3, v denotes an element of the stationary algebra of a function generating the homolog y of the complete Koszul complex of f_0 as an A-module.

TABLE 1

Type	f_0	Weight; N	$f_1 + \cdots$	Boundaries
$J_{k,i}$	$x^3 + x^2y^k$	$k, 1; 3k$	$ay^{3k+i}, a_0 \neq 0$	$k>1, i>0$
$Z_{k-2,i}$	$x^3y + x^2y^k$	$k-1, 1; 3k-2$	$ay^{3k+i-2}, a_0 \neq 0$	$k>2, i>0$
$W_{k,i}$	$x^4 + x^2y^{2k+1}$	$2k+1, 2; 8k+4$	$ax^3y^{k+1} + by^{4k+2+i}$	$k>0, i>0$
$W^*_{k,i}$	$(x^2 + y^{2k+1})^2$	$2k+1, 2; 8k+4$	$ay^{4k+2+q} + bxy^{3k+1+q}, i=2q-1;$ $axy^{3k+2+q} + bx^2y^{2k+1+q}, i=2q.$	$k>0, i>0$
$X_{k,i}$	$x^4 + a_0x^3y^k + x^2y^{2k}, a_0^2 \neq 4$	$k, 1; 4k$	$a'x^3y^k + b'y^{4k+i}$	$k>1, i>0$
$Q_{k,i}$	$x^3 + yz^2 + x^2y^k$	$2k, 2, 3k-1; 6k$	$cy^{3k+i}, c_0 \neq 0$	$k>1, i>0$
$S_{k,i}$	$x^2z + yz^2 + x^2y^{2k}$	$2k+1, 2, 4k; 8k+2$	$ax^3y^k + by^{4k+1+i}$	$k>0, i>0$
$S^*_{k,i}$	$x^2z + yz^2 + zy^{2k+1}$	$2k+1, 2, 4k; 8k+2$	$ay^{4k+1+q} + bxy^{3k+q}, i=2q-1;$ $axy^{3k+1+q} + bx^2y^{2k+q}, i=2q$	$k>0, i>0$
$U_{k,i}$	$x^3 + xz^2 + xy^{2k+1}$	$2k+1, 2, 2k+1; 6k+3$	$ax^2y^{k+1} + czy^{2k+1+q} + bz^2y^{k+q}, i=2q-1;$ $ax^2y^{k+1} + cy^{3k+2+q} + bzy^{2k+1+q}, i=2q$	$k>0, i>0$
$V_{1,i}$	$x^2y + z^4 + bz^2y + z^2y^2, b^2 \neq 4$	$3, 2, 2; 8$	$(a_0 + a_1y)y^{4+i}, a_0 \neq 0$	$i>0$
$V^*_{1,i}$	$x^2y + z^3y + ay^2z^2 + y^4, 4a^3 + 27 \neq 0$	$3, 2, 2; 8$	$(b_0 + b_1z)xz^{2+q}, i=2q-1;$ $(b_0 + b_1z)z^{4+q}, i=2q$	$i>0$

TABLE 2

Type	Elements of $\mathfrak{A}_0 f_0$	$\{e_j\}$	r	ρ	φ	ν	τ
$J_{k,i}$	x^3, x^2y^k, xy^{2k}	$y^\beta, \beta>3k$	i	$k-1$	y^{3k+i}	$-kxy^{k-1}\partial/\partial x+$ $+(3x+2y^k)\partial/\partial y$	y^{4k-1+i}
$Z_{k-2,i}$	$x^4, x^3y, x^2y^k, xy^{2k-1}$	$y^\beta, \beta>3k-2$	i	$k-1$	y^{3k-2+i}	$(x^2+kxy^{k-1})\partial/\partial x-$ $-(3xy+2y^k)\partial/\partial y$	y^{4k-3+i}
$W_{k,i}$	$x^4, x^3y^{2k}, x^2y^{2k+1}, xy^{4k+1}$, $2x^3y^{k+1}+xy^{3k+z}$	$x^3y^\alpha, k+1\leqslant\alpha\leqslant 2k-1$ $y^\beta, \beta>4k+2$	$2i$	$4k$	y^{4k+2+i}	$(2k+1)xy^{2k}\partial/\partial x-$ $-(4x^2+2y^{2k+1})\partial/\partial y$	y^{6k+2+i}
$kW^*_{k,2q-1}$	$x^4+x^2y^{2k+1}$, $x^3y^{k+1}+xy^{3k+2}$, $x^2y^{2k+1}+y^4$	$xy^\alpha, \alpha\geqslant 3k+2$ $y^\beta, \beta>4k+2$	$2q-1$	$2k-1$	xy^{3k+1+q}	$(2k+1)y^{2k}\partial/\partial x-$ $-2x\partial/\partial y$	y^{5k+1+q}
$W^*_{k,2q}$	$x^4+x^2y^{2k+1}$, $x^3y^{k+1}+xy^{3k+2}$, $x^2y^{2k+1}+y^4$	$xy^\alpha, \alpha\geqslant 3k+2$ $x^2y^\beta, \beta>2k+1$	$2q$	$2k-1$	x^2y^{2k+1+q}	$(2k+1)y^{2k}\partial/\partial x-$ $-2x\partial/\partial y$	xy^{4k+1+q}
$X_{k,i}$	$x^5, x^4y^{k-1}, x^3y^{2k-1}$, x^2y^{3k-1}, xy^{4k-1}, $2x^4+a_0x^3y^k$, $a_0x^3y^k+2x^2y^{2k}$, $(8-3a_0^2)x^3y^k+4xy^{3k}$	$x^3y^\alpha, k<\alpha\leqslant 2k-2$; $y^\beta, \beta>4k$	i	$2k-1$	y^{4k+i}	$k(a_0x^2y^{k-1}+$ $+2xy^{2k-1})\dfrac{\partial}{\partial x}-(4x^2+$ $+3a_0xy^k+2y^{2k})\partial/\partial y$	y^{6k-1+i}

TABLE 3

Type	Elements of $\mathfrak{A} f_0$	$\{e_j\}$	r	ρ	φ	ν	τ
$Q_{k,i}$	$x^3, yz, x^2y^k, xz^2, x^2z, z^3, xy^{2k}$	$y^\beta, \beta > 3k$	$2i$	$2k$	y^{3k+i}	$2kxy^k\partial/\partial x+(6xy+4y^{k+1})\times \times\partial/\partial y-(3xz+y^kz)\partial/\partial z$	y^{4k+i}
$S_{k,i}$	$x^2z, yz^2, x^2y^{2k}, x^4, z^3, zy^{2k+1}, xz^2, xzy^{2k}, x^3y^{2k-1}, xy^{4k}$	$x^3y^\alpha, k \leqslant \alpha \leqslant 2k-2, y^\beta, \beta > 4k+1$	$2i$	$4k$	y^{4k+1+i}	$(4kxy^{2k}-xz)\partial/\partial x- -4(y^{2k+1}+yz)\partial/\partial y+ +2(y^{2k}z+z^2)\partial/\partial z$	y^{6k+i+1}
$S^*_{k,i}$	$xz, yz^2, z^3, x^2y^{2k}+y^{4k+1}, x^4+x^2y^{2k+1}, x^3y^k+xy^{3k+1}$	$y^\alpha, \alpha > 4k+1, xy^\beta, \beta \geqslant 3k+1,$ $i=2q-1$;	$2q-1$	$2k-1$	xy^{3k+q}	$(z+(2k+1)y^{2k})\partial/\partial x- -2x\partial/\partial y$	y^{3k+q}, $i=2q-1$;
		$xy^\alpha, \alpha \geqslant 3k+1, x^2y^\beta, \beta > 2k,$ $i=2q$	$2q$	$2k-1$	x^2y^{2k+q}	$(z+(2k+1)y^{2k})\partial/\partial x- -2x\partial/\partial y$	xy^{4k+q}. $i=2q$
$U_{k,i}$	$xz, xy^{2k}, x^3, z^3+zy^{2k+1}, z^2y^k+y^{4k+1}$	$x^2y^\alpha, k \leqslant \alpha \leqslant 2k, z^2y^\beta, \beta \geqslant k+1, zy^\gamma, \gamma > 2k+1,$ $i=2q-1$;	$2q-1$	$2k-1$	z^2y^{k+q}	$-2z\partial/\partial y+y^{2k}(2k+i)\partial/\partial z$	zy^{3k+q}. $i=2q-1$;
		$x^2y^\alpha, k \leqslant \alpha < 2k, zy^\beta, \beta > 2k+1, y^\gamma, \gamma \geqslant 3k+2,$ $i=2q$	$2q$	$2k-1$	zy^{2k+1+q}	$-2z\partial/\partial y+y^{2k}(2k+i)\partial/\partial z$	y^{4k+1+q}. $i=2q$
$V_{1,i}$	$xy, z^3, z^4y, z^3y^2, z^2y^3, zy^4, x^3, x^2z^2, xz^3$	$y^\beta, \beta > 4$	$2i$	4	y^{4+i}	$(4xz^2+3bxyz+2xy^2)\partial/\partial x- -(8z^2y+6by^2z+4y^3)\partial/\partial y+ +(2bz^2y+4y^2z)\partial/\partial z$	y^{6+i}
$V^*_{1,i}$	$xy, z^3, z^4y, z^3y^2, z^2y^3, zy^4, x^3+xz^2, x^2z^2+z^5$	$xz^\alpha, \alpha \geqslant 3$ $z^\beta, \beta > 4$	i	1	xz^{2+q}, $i=2q-1$; $z^{4+q}, i=2q$	$(3z^2+2ayz)\partial/\partial x-2x\partial/\partial z$ $(3z^2+2ayz)\partial/\partial x-2x\partial/\partial z$	z^{4+q}, $i=2q-1$ xz^{3+q}. $i=2q$

References

1. V. I. Arnold, *Normal forms of functions near degenerate critical points, the Weyl groups A_K, D_K, E_K and Lagrange singularities*, Functional Analysis and its Applications 6 (1972), 254–272. Originally published in Funkts. Analyz 6 (1972).
2. ______, *Remarks on the stationary phase method and Coxeter numbers*, Russian Math. Surveys 28 (1973), 19–48. Originally published in Uspekhi Mat. Nauk. 28 (1973).
3. ______, *Normal forms of functions in neighborhoods of degenerate critical points*, Russian Math. Surveys 29 (1974), 10–50. Originally published in Uspekhi Mat. Nauk 29 (1974).
4. ______, *Classification of bimodal critical points of functions*, Functional Analysis and its Applications 9 (1975), 230–231. Originally published in Funkts. Analyz 9 (1975).
5. S. MacLane, Homology, New York, Springer-Verlag, 1967.

Russian Math. Surveys 30:5 (1975), 1–75
From Uspekhi Mat. Nauk 30:5 (1975), 3–65

CRITICAL POINTS OF SMOOTH FUNCTIONS AND THEIR NORMAL FORMS

V. I. Arnol'd

This paper contains a survey of research on critical points of smooth functions and their bifurcations. We indicate applications to the theory of Lagrangian singularities (caustics), Legendre singularities (wave fronts) and the asymptotic behaviour of oscillatory integrals (the stationary phase method). We describe the connections with the theories of groups generated by reflections, automorphic forms, and degenerations of elliptic curves. We give proofs of the theorems on the classification of critical points with at most one modulus, and also a list of all singularities with at most two moduli. The proofs of the classification theorems are based on a geometric technique associated with Newton polygons, on the study of the roots of certain Lie algebras resembling the Enriques-Demazure technique of fans, and on spectral sequences that are constructed with respect to quasihomogeneous filtrations of the Koszul complex defined by the partial derivatives of a function.

Contents

CHAPTER I [1]

General Survey

A *critical point* of a smooth function is a point at which the differential of the function is equal to 0. A critical point is said to be *non-degenerate* if the second differential is a non-degenerate quadratic form. Every degenerate critical point splits into several non-degenerate critical points under a small perturbation (a so-called *morsification*).

Functions in general position have only non-degenerate critical points. Degenerate critical points show up naturally only when the functions depend on parameters. For example, the family of functions $f(x, t) = x^3 - tx$ has a degenerate critical point for the parameter value $t = 0$, and every nearby family also has a degenerate critical point for some small value of the parameter. For a large number of parameters more complicated critical points appear.

The study of families of functions depending on parameters arises everywhere in analysis (or mathematical physics). Below we shall discuss three applications: to the study of *Lagrangian singularities* (or *caustics*), *Legendre singularities* (or *wave fronts*) and *oscillatory integrals* (or the *stationary phase method*).

The classification of the simplest singularities in all these problems turns out to be related to the Lie, Coxeter, and Weyl groups of the series A, D and E, to the theory of braids of Artin and Brieskorn [22], and to the classification of regular polyhedra in Euclidean three-space.

The appearance of the diagrams A, D and E and of the Coxeter groups in such diverse situations as the theory of simple Lie groups, the classification of simple categories of linear spaces (in papers of Gabriel [*133]–[*135], Gel'fand and Ponomarev [*138]–[*140], Roiter and Nazarova [*136] [*137], and of Dlab and Ringel [*141]–[*142], Kodaira's classification of the degenerations of elliptic curves, the theory of regular polyhedra and the

[1] * This chapter is a slightly modified version of the author's one hour address to the International Congress of Mathematicians, Vancouver 1974.

theory of simple singularities seems an astonishing series of coincidences in the results of independent classifications. As we now see, the classification of more complicated singularities also leads to astonishing coincidences, in which Lobachevskii triangles and automorphic functions are involved.

§ 1. Classification of critical points

Let f be the germ of a smooth function at an isolated critical point $O \in \mathbf{C}^n$. The *multiplicity* (or *Milnor number*) μ of O is the number of non-degenerate critical points into which O splits under a morsification.

Two germs of functions at O are said to be *equivalent* if one of them is taken to the other under a diffeomorphism of the domain space that leaves O fixed.

The jet (Taylor polynomial) of a function at O is *sufficient* if it determines the germ of that function at O up to equivalence.

A smooth function with an isolated critical point is always equivalent to a polynomial, namely, to a segment of its Taylor series, and it has a sufficient jet (Tougeron [79], M. Artin [14], Arnol'd [3], Samoilenko [67])

Thus, the problem of classifying isolated critical points reduces to algebraic problems concerning the action of finite-dimensional Lie groups on finite-dimensional spaces of jets (Thom [74], Mather [54], Siersma [69]).

The classification of the simplest singularities is discrete, but strongly degenerate singularities depend on parameters (moduli).

The *modality* m of a point $x \in X$ under the action of a Lie group G on a variety X is the smallest number such that a sufficiently small neighbourhood of x is covered by a finitely many m-parameter families of orbits. The point x is called *simple* if its modality is 0, that is, if a neighborhood of it is intersected by finitely many orbits.

The modality of the germ of a function at a critical point with critical value 0 is defined as the modality of its sufficient jet in the space of jets of functions with critical point O and critical value 0.

Two germs are said to be *stably equivalent* if they become equivalent after direct addition of non-degenerate quadratic forms of a suitable number of variables.

THEOREM I (see [6]). *Up to stable equivalence, the simple germs (germs with m = 0) are exhausted by the following list*:

$$A_k\colon f(x) = x^{k+1},\quad k \geqslant 1;\qquad D_k\colon f(x, y) = x^2y + y^{k-1},\quad k \geqslant 4;$$
$$E_6 : f(x, y) = x^3 + y^4;\quad E_7 : f(x, y) = x^3 + xy^3;\quad E_8 : f(x, y) = x^3 + y^5$$

THEOREM II (see [7]). *Up to stable equivalence, the unimodal germs (germs with m = 1) are exhausted by the three-index series of one-parameter families*

$$T_{p,q,r}: f(x, y, z) = axyz + x^p + y^q + z^r, \quad \frac{1}{p}+\frac{1}{q}+\frac{1}{r} < 1, \quad a \neq 0;$$

$$T_{3,3,3}: f(x, y, z) = x^3 + y^3 + z^3 + axyz, \quad a^3 + 27 \neq 0;$$
$$T_{2,4,4}: f(x, y, z) = x^4 + y^4 + z^2 + ax^2y^2, \quad a^2 \neq 4;$$
$$T_{2,3,6}: f(x, y, z) = x^3 + y^6 + z^2 + ax^2y^2, \quad 4a^3 + 27 \neq 0,$$

and 14 *more exceptional one-parameter families, listed in Table* 1 *below (the meaning of its columns will be explained later).*

Table 1

Notation	Normal form	Indices of homogeneity	Coxeter numbers	Dolgachev numbers	Gabrielov numbers	Dual class
Q_{10}	$x^2z+y^3+z^4+ayz^3$	8 9 6	−24	2 3 9	3 3 4	K_{14}
Q_{11}	$x^2z+y^3+yz^3+az^5$	7 6 4	−18	2 4 7	3 3 5	Z_{13}
Q_{12}	$x^2z+y^3+z^5+ayz^4$	6 5 3	−15	3 3 6	3 3 6	Q_{12}
S_{11}	$x^2z+yz^2+y^4+ay^3z$	6 5 4	−16	2 5 6	3 4 4	W_{13}
S_{12}	$x^2z+yz^2+xy^3+ay^5$	5 4 3	−13	3 4 5	3 4 5	S_{12}
U_{12}	$x^3+y^3+z^4+axyz^2$	4 4 3	−12	4 4 4	4 4 4	U_{12}
Z_{11}	$x^3y+y^5+z^2+axy^4$	15 8 6	−30	2 3 8	2 4 5	K_{13}
Z_{12}	$x^3y+xy^4+z^2+ay^6$	11 6 4	−22	2 4 6	2 4 6	Z_{12}
Z_{13}	$x^3y+y^6+z^2+axy^5$	9 5 3	−18	3 3 5	2 4 7	Q_{11}
W_{12}	$x^4+y^5+z^2+ax^2y^3$	10 5 4	−20	2 5 5	2 5 5	W_{12}
W_{13}	$x^4+xy^4+z^2+ay^6$	8 4 3	−16	3 4 4	2 5 6	S_{11}
K_{12}	$x^3+y^7+z^2+axy^5$	21 14 6	−42	2 3 7	2 3 7	K_{12}
K_{13}	$x^3+xy^5+z^2+ay^8$	15 10 4	−30	2 4 5	2 3 8	Z_{11}
K_{14}	$x^3+y^8+z^2+axy^6$	12 8 3	−24	3 3 4	2 3 9	Q_{10}

There are also classifications of all functions of two variables with non-zero 3-jet [9] or with non-zero 4-jet (§13). Similar tables exist in the real case. Furthermore, the following theorem holds.

THEOREM III (see [6], [7]). *The set of non-simple germs of functions of $n \geqslant 3$ variables has codimension* 6, *and the set of germs of modality greater than* 1 *has codimension* 10 *in the space of germs of functions with critical value* 0.

Thus, every s-parameter family of functions, where $s < 6$ (respectively, $s < 10$), can by a sufficiently small perturbation be brought into general position, so that the germs of the functions of the family at all the critical points become stably equivalent to the germs of Theorem I (respectively, to those of Theorems I and II), up to additive constants.

§2. Quotient singularities

The group $SU(2)$ acts linearly on $\mathbf{C}^2$. The discrete subgroups of $SU(2)$ are called *binary groups*: the binary polygonal group, the binary dihedral group, binary tetrahedral group, binary group of the cube (or quaternion group), and the binary icosahedral group. (After factoring $SU(2)$ by its centre $\{\pm E\}$ they go to the corresponding subgroups of the group of rotations of the sphere.)

The quotient variety $M = \mathbf{C}^2/\Gamma$ of $\mathbf{C}^2$ by the action of a binary group Γ is an algebraic surface with a singular point. The ring of Γ-invariant polynomials of two variables has 3 generators. The relation (syzygy) connecting these 3 generators is also the equation of the quotient variety M as a surface in $\mathbf{C}^3$. At the time of H. A. Schwarz the following result was already known.

THEOREM IV (see [43], [45], [57]). *The surface* $\mathbf{C}^2/\Gamma$ *has a simple singularity of one of the types* A_k *(for a polygon),* D_k *(for a dihedron),* E_6 *(for the tetrahedron),* E_7 *(for the cube),* E_8 *(for the icosahedron).*

The group $SU(1, 1)$ of (2×2)-unimodular matrices that preserve the quadratic form $|z_1|^2 - |z_2|^2$ acts on the set P of vectors in $\mathbf{C}^2$ on which the form is positive. Each discrete group Γ_0 of motions of the Lobachevskii plane (with compact fundamental domain) determines a "binary group" $\Gamma \subset SU(1, 1)$ and an algebraic surface $M = (P/\Gamma) \cup O$ with singular point O. It can be shown that the coordinate ring of M is isomorphic to the algebra of integral automorphic forms with respect to Γ_0 (I. V. Dolgachev).

Let Δ be a Lobachevskii triangle with the angles π/p, π/q, π/r. The reflections in the sides of Δ generate a discrete group. The motions form a subgroup of it of index 2; we call it the group of Δ.

In $SU(1, 1)$ the *binary triangular group* $\Gamma(\Delta)$ corresponds to it.

An analysis of the 14 exceptional singularities led Dolgachev to the following result.

THEOREM V (see [28]). *There exist exactly* 14 *Lobachevskii triangles for which the surface M is imbedded in* $\mathbf{C}^3$ *(in other words, precisely* 14 *triangles for which the algebra of integral automorphic forms has* 3 *generators. These* 14 *surfaces M have exceptional unimodal singularities at O (see Theorem* II).

The corresponding numbers (p, q, r) are indicated under the heading "Dolgachev numbers" in Table 1 on p.4.

§3. Quadratic forms of singularities

With each isolated critical point of a holomorphic function there is associated a manifold V with boundary ∂V, called the *non-singular level manifold*. More precisely, we consider a sufficiently small ball with centre at the critical point. Then the part of the level set inside the ball and sufficiently close to the critical set is a smooth manifold with boundary (Fig. 1, p. 6).

Fig. 1

(The boundary ∂V supplies standard examples in differential topology. For example, for the simple singularity E_8 in $\mathbf{C}^5$ the manifold ∂V is an exotic 7-sphere of Milnor, which is homeomorphic, but not diffeomorphic, to S^7. Contracting the boundary of V (E_8) in $\mathbf{C}^7$ to a point we obtain a non-smoothable topological 12-dimensional manifold. See [44], [19], [57], [46].)

Milnor [57] has proved that the manifold V^{2n-2} is homotopy equivalent to a bouquet of μ spheres S^{n-1} of middle dimension, so that $H_{n-1}(V, \mathbf{Z}) = \mathbf{Z}^\mu$. The intersection index defines an integral bilinear form on H_{n-1}.

The *quadratic form of a singularity* is the index of self-intersection on homology of the non-singular level manifold of a function of $n \equiv 3$ (mod 4) variables that is stably equivalent to the given function (the behaviour of the intersection index on writing down the squares of the new variables is described by a theorem of Thom and Sebastiani [73]; by adding two squares the intersection index "changes sign" (that is, becomes symmetric)).

A singularity is said to be *elliptic* (respectively, *parabolic, hyperbolic*) if its quadratic form is negative definite (respectively, semidefinite, has positive index of inertia 1).

THEOREM VI (see [76], [6], [7]). *The elliptic singularities are precisely the simple singularities A, D and E of Theorem* I; *the parabolic (respectively, hyperbolic) singularities are the* $T_{p,q,r}$ *of Theorem* II *with* $1/p+1/q+1/r=1$ (*respectively, with* $1/p+1/q+1/r<1$).

The assertion about parabolic singularities was conjectured by Milnor, based on work of Wagreich [83].

It is convenient to describe the quadratic forms of singularities by *Dynkin diagrams.* Such a diagram is a graph whose vertices correspond to the vanishing cycles (basis vectors in H_{n-1} with square -2). Two vertices are joined by k simple (respectively, dotted) edges if the scalar product of the corresponding vectors is equal to k (respectively, $-k$). For example, the diagram •—• corresponds to the form $-2x_1^2+2x_1x_2-2x_2^2$.

Very effective methods for computing Dynkin diagrams have been worked out by Gabrielov ([34]–[36]) and Gusein-Zade ([39]–[40]). The latter's method allows us to draw the diagram of any function of two variables immediately from the picture of the level curves of a real morsification. (At the congress in Vancouver, N. A'Campo ([*131]) announced that he had independently discovered Gusein-Zade's method.)

The quadratic forms of the simple singularities A, D, and E are given by the usual Dynkin diagrams (see Hirzebruch [43]):

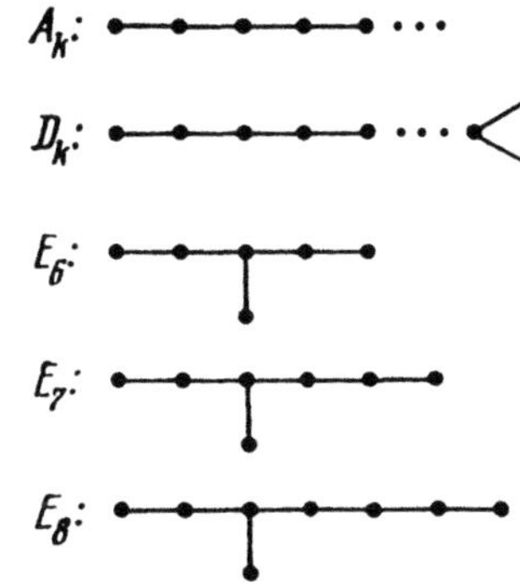

The quadratic forms of all unimodal singularities have been found by Gabrielov ([34]–[35]). We denote by $\tau_{p,q,r}$ the quadratic form with a diagram in the form of the letter T, on the three segments of which there are, respectively, p, q, r vertices (for example $E_7 = \tau_{2,3,4}$).

THEOREM VII (see [7]). *The quadratic form of each hyperbolic or parabolic singularity* $T_{p,q,r}$ *is a direct sum* $\tau_{p,q,r} \oplus 0$, *where* 0 *is the zero form of one variable. The quadratic form of each of the* 14 *exceptional singularities is the direct sum* $\tau_{p,q,r} \oplus \begin{pmatrix} 0 & 1 \\ 1 & 0 \end{pmatrix}$, *where* p, q, r *are indicated under the heading "Gabrielov number" in Table* 1.

§4. A strange duality

A comparison of the Dolgachev numbers and the Gabrielov numbers of the 14 exceptional singularities leads to the following conclusions:

THEOREM VIII. *The Gabrielov numbers of each exceptional singularity coincide with the Dolgachev numbers of some (generally speaking, another) singularity, and the Gabrielov numbers of the latter are the Dolgachev numbers of the former.*

Thus, the involution

$$Q_{10} \leftrightarrow K_{14}, \quad Q_{11} \leftrightarrow Z_{13}, \quad Z_{11} \leftrightarrow K_{13}, \quad S_{11} \leftrightarrow W_{13},$$

acts on the set of the 14 singularities, leaving fixed all six singularities with $\mu = 12$. We note that a priori no connection whatever is apparent between the singularities that are dual to each other (or their Lobachevskii triangles or their quadratic forms), nor between the Gabrielov and Dolgachev numbers.

Upon analysing the table of Dolgachev numbers, D. B. Fuks remarked that their sum with μ is always equal to 24 (this observation permitted him to correct an error in the original computations).

Thus, the sum of all six Gabrielov and Dolgachev numbers of each of the fourteen singularities is equal to 24. We also note that the singularities that are dual to each other are precisely those for which the Coxeter numbers (defined below) coincide. At the present stage the theory of singularities is an experimental science.

§5. Versal deformation and the level bifurcation set

A *deformation* of a function f is the germ at O of a smooth mapping of a finite-dimensional space (the *base* of the deformation) into a function space, taking O into f. A deformation is said to be *versal* if this mapping is transversal to the orbit of f under the action of the (pseudo-) group of diffeomorphisms of the argument space; if the dimension of the base is minimal (equal to the codimension of the orbit), then the deformation is said to be *miniversal.*

The germ of f at O with finite multiplicity μ always has a μ-parameter miniversal deformation ([77], [54], [47], [87]).

The *local ring* of the germ of f at the critical point $O \in \mathbf{C}^n$ is the factor ring of the ring of (formal or convergent) power series of the coordinates at O by the ideal spanned by the series of the partial derivatives of f. The dimension of this ring over $\mathbf{C}$ is equal to the multiplicity μ (Palamodov [61]).

As a miniversal deformation of the function f one can take the deformation $\lambda \mapsto f + \lambda_1 e_1 + \ldots + \lambda_\mu e_\mu$, where the functions e_s generate the local ring of f at O as a $\mathbf{C}$-module. Every deformation of f is equivalent to one induced from a versal deformation of the germ of f at O.

The *level bifurcation set* for f is the germ at O of the hypersurface in the base that is formed by those λ for which 0 is the critical value of the corresponding function close to O.

The complement to the level bifurcation set is the base of a fibration whose fibre is a non-singular level manifold of f. The action of the fundamental group of the base of the fibration on the homology of a fibre is called the *monodromy* of the singularity and its image is the *monodromy group*.

THEOREM IX (see [6]). *The complement to the level bifurcation set of a simple singularity has the homotopy type of a $K(\pi, 1)$-space, where π is the Artin-Brieskorn braid group (see [21]–[23]). This complement is the variety of regular orbits of the action of the corresponding Coxeter group on the complexification of the Euclidean space $\mathbf{R}^\mu$. The monodromy group of a simple singularity is the natural representation of the braid group on the*

Coxeter group.

The proof of Theorem IX in the cases E_6, E_7, E_8 is not simple: it is based on theorems of Deligne [26] and Brieskorn [20]. The topology of the complements to bifurcation sets has been little studied (see [4], [22]), but the recently discovered connections with loop spaces (Segal [68], Fuks [32]–[33]), with pseudo-isotopies (Cerf [25], Thom [75]) with algebraic *K*-theory (Volodin [82], Wagoner [84], Hatcher [41], [*41a]) seem very promising.

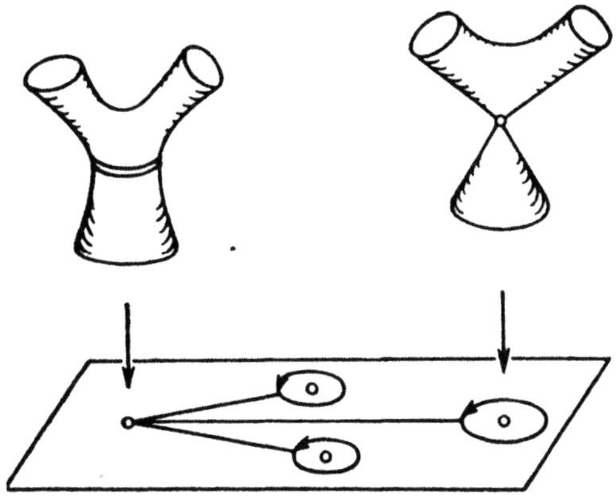

Fig. 2

We return to the level bifurcation set Σ for an arbitrary singularity. A line $\mathbf{C}^1$ in general position in the base intersects Σ at μ points close to O. We call these points *distinguished.*

On the line $\mathbf{C}^1$ we fix a base point (different from the distinguished points); this point corresponds to a non-singular fibre V. We choose μ non-intersecting *distinguished paths* in $\mathbf{C}^1$, going from the base point to the distinguished points (Fig. 2). In the fibre over a point of a distinguished path, close to the end of the path, a Picard-Lefschetz *vanishing cycle* (a middle-dimensional sphere) is defined. Returning it along the path to the fibre over the base point, we define there a *distinguished vanishing cycle.* These μ cycles form a *distinguished basis* of the homology of the fibre (Lamotke [129], Gabrielov [34]).

The fundamental group of the complement to the bifurcation set is generated by μ *distinguished loops* on $\mathbf{C}^1$; they are obtained from the distinguished paths if, without having reached a distinguished point, we go around it and return to the start (theorem of Zariski [88]; see also [80], [49]).

For simplicity we assume that the number of variables is $n \equiv 3 \pmod 4$. Then the action of each distinguished generator of the fundamental group on the homology of the fibre reduces to reflection in the orthogonal complement to the distinguished vanishing cycle (Picard-Lefschetz theorem).

Thus, to compute the monodromy group of a singularity it suffices to find the Dynkin diagram of some distinguished basis. The first examples were found by Pham (see Pham [62], Brieskorn [19]), using the function $f = \Sigma x_i^{a_i}$; Pham's basis is in fact distinguished. The papers of Gabrielov [34]–[35] and Gusein-Zade [39]–[40] contain many other examples. Lazzeri proved that the Dynkin diagram of a distinguished basis is always connected, and Gusein-Zade [40] proved that for any function of two variables this diagram can be obtained from the picture of the level lines of a real morsification.

The *monodromy operator* of a singularity is the action on $H_{n-1}(V)$ of the product of the distinguished generators of the fundamental group. This operator is associated with the asymptotic behaviour of integrals in which f is contained, and many articles are devoted to computing it (Milnor and Orlik [58], Brieskorn [24], A'Campo [2]).

A'Campo has proved that the monodromy operator of any degenerate singularity is non-trivial. The calculation of the monodromy operator with respect to the Dynkin diagram of a distinguished basis reduces to multiplication of matrices. The monodromy operator of a simple singularity is the Coxeter element of the corresponding Coxeter group. Its order is called the *Coxeter number* of the simple singularity.

§6. Orbit transversal and the function bifurcation set

Consider the set $\mathfrak{m}^2$ of germs at $O \in \mathbf{C}^n$ of functions with critical point O and critical value 0. The group of O-fixing germs of diffeomorphisms of the argument space acts on $\mathfrak{m}^2$. A minimal transversal to the orbit of a function f in $\mathfrak{m}^2$ has dimension $\mu - 1$. This transversal T gives a $(\mu - 1)$-parameter deformation of f. This deformation, like any other, is induced from a miniversal deformation of f under some mapping of the base $\tau: T^{\mu-1} \to \mathbf{C}^\mu$.

The level bifurcation set Σ is precisely the image of the transversal under the mapping τ. The set Σ is irreducible and has a non-singular normalization T (see [36], [71]).

Next, we consider the set $\mathfrak{m}$ of functions equal to 0 at O. Deformations in the class of such functions are called *restricted*. A restricted miniversal deformation has $\mu - 1$ parameters, and if we add an arbitrary constant to functions of a family, then we obtain an *extended* (μ-parameter) miniversal deformation in the usual sense of the term miniversal.

We fix a restricted miniversal deformation of f. A point of the base $\mathbf{C}^{\mu-1}$ of this deformation is called a *function bifurcation point* if it corresponds to a function with fewer than μ critical values close to O. The set of all such points forms a hypersurface Δ in a neighbourhood of $O \in \mathbf{C}^{\mu-1}$ (Fig. 3). This hypersurface (more precisely, its germ at O) is called the *function bifurcation set* for f.

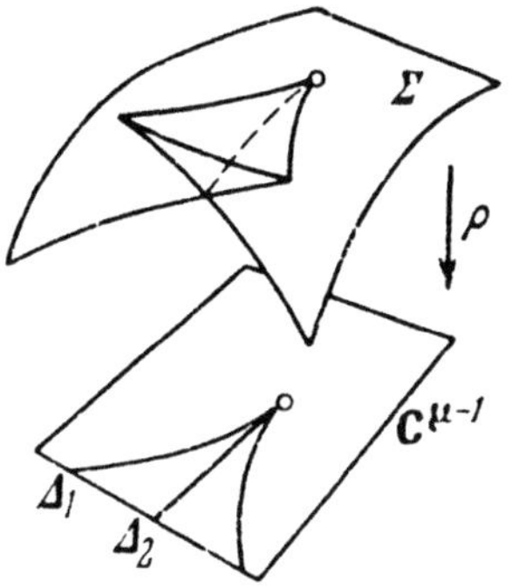

Fig. 3

THEOREM X (see [48], [36]). *The projection ρ of the base of an extended miniversal deformation onto the base of a restricted miniversal deformation, when restricted to the level bifurcation set Σ, gives a μ-sheeted covering of the complement to Δ. The group of this covering is the symmetric group S(μ).*

THEOREM XI (Lyashko [9], Looijenga [51]). *For simple functions the complement to the function bifurcation set,* $\mathbf{C}^{\mu-1} - \Delta$, *is a* $K(\pi, 1)$*-space, where π is a subgroup of finite index* $\nu = \mu! N^{\mu} W^{-1}$ *(where N is the Coxeter number, W the order of the Weyl group) in the Artin braid group on μ strings.*

The bifurcation set Δ splits in a natural way into two hypersurfaces Δ_1 and Δ_2: Δ_1 corresponds to functions with multiple critical points, Δ_2 to functions with coincident critical values.

The smooth mapping $\rho \circ \tau$: $T^{\mu-1} \to \mathbf{C}^{\mu-1}$ from the orbit transversal in $\mathfrak{m}^2$ to the base of a restricted deformation has Δ_1 as its set of critical values and gives a μ-sheeted cover of the complement to Δ_1 near O.

The hypersurface Δ_1 is called the *caustic*, and Δ_2 the *cut locus* or the *Maxwell stratum.*

§7. Lagrangian singularities and the classification of caustics

One can see caustics clearly on a wall, lit up by sun-rays reflected off a curved surface (for example, the inside of a cup). By moving the cup one can see that caustics in general position have only standard singularities, which do not disappear under a small perturbation (see, for example, [*144]).

The analysis of caustics is a part of the theory of *Lagrangian singularities* (see [6] and the articles of Guckenheimer [38] and Weinstein [85]), which is analogous to the usual theory of singularities of Whitney [86], Thom [72], and Mather [54].

We recall that a *symplectic manifold* is a smooth manifold M^{2n} with a closed non-degenerate 2-form ω. A *Lagrangian submanifold* of M^{2n} is one of dimension n on which $\omega = 0$. A fibration $p: M^{2n} \to B^n$ is said to be *Lagrangian* if its fibres are Lagrangian submanifolds. The standard example is the cotangent bundle $T^*B \to B$ (the "phase space" of classical mechanics).

Let $i: L \to M$ be the embedding of a Lagrangian submanifold into the total space of a Lagrange fibration $p: M \to B$. The projection $p \circ i: L \to B$ is called a *Lagrange mapping*, and the set of its critical values a *caustic*.

Two Lagrange mappings are *equivalent* if the corresponding Lagrangian manifolds are transformed into each other by a diffeomorphism of the fibration that preserves the 2-form; in this case the caustics are obviously diffeomorphic. A Lagrange mapping is *stable* at a point O if every nearby Lagrange mapping has, at some point near O, a germ equivalent to the germ of the given mapping at O.

The germ of a Lagrange mapping is *simple* if all nearby germs belong to finitely many equivalence classes. (A simple germ can be non-stable, and a stable germ need not be simple.)

THEOREM XII (see [6]). *Simple stable germs of Lagrange mappings are classified by the series A, D, and E. Every Lagrange mapping with $n < 6$ can be approximated by a mapping whose germ at each point is simple and stable.*

A description of the Lagrangian singularities A, D, and E is given below (§9) in coordinate form. From this description it follows, in particular, that the stable singularities of caustics in three-space are only the cuspidal edges (A_3), swallow tails (A_4), and the points of contact of three cuspidal edges, two of which may be imaginary ($D_4^{\pm}$) (Fig. 4). All other singularities split into these under a small perturbation.

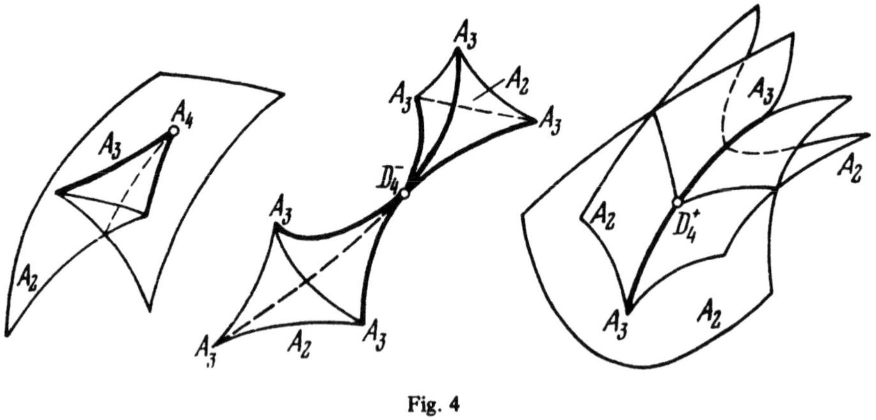

Fig. 4

§8. Legendre singularities and the classification of wave fronts

An example of a wave front is the curve that is obtained if we select a segment of length t on each interior normal to an ellipse. It is easy to check that a wave front may have singularities and that these singularities cannot be removed by a small deformation of the original ellipse.

The analysis of wave front singularities is a part of the theory of Legendre singularities (see [10], [11]).

The theory of Legendre singularities parallels that of Lagrangian singularities with the following differences: the role of the symplectic structure is taken by a contact structure, that of the affine structure by a projective structure, gradients by Legendre transformations, functions by hypersurfaces, etc.

The analogy between the two theories is none other than Hamilton's "optics-mechanics" analogy.

We recall that a *contact manifold* is an odd-dimensional manifold M^{2n+1} equipped with a maximally non-integrable "contact field" of hyperplanes (if the contact field is locally given by a form α, then the form $\alpha \wedge (d\alpha)^n$ is non-degenerate). The standard examples of contact manifolds are the projective cotangent bundle PT^*B^{n+1} and the manifold of 1-jets of functions $J^1(W^n, \mathbf{R})$ with their natural contact fields (given by the "integrability conditions").

The integral manifold of a contact field on M^{2n+1} is called a *Legendre submanifold* if it has the highest possible dimension (namely n). A fibration $p\colon M^{2n+1} \to B^{n+1}$ is called a *Legendre fibration* if its fibres are Legendre submanifolds (an example is the projective cotangent bundle $PT^*B \to B$). All the Legendre fibrations of the same dimension are locally Legendre equivalent.

If $i\colon L^n \to M^{2n+1}$ is an embedding of a Legendre submanifold in the total space of the Legendre fibration $p\colon M^{2n+1} \to B^{n+1}$, then $p \circ i\colon L^n \to B^{n+1}$ is called a *Legendre mapping* and the image is a *wave front*.

Stability and simplicity are defined as in the Lagrange case.

THEOREM XIII. *The simple stable germs of Legendre mappings are classified by the series A, D, and E. Every Legendre mapping with $n < 6$ can be approximated by a mapping with simple stable germs.*

In three-space the wave fronts in general position may have only self-intersections, cuspidal edges (A_2) and swallow tails (A_3); in the propagation of a wave front its singularity moves along a caustic, and at certain moments of time rearrangements of the three types A_4, $D_4^{\pm}$ occur (Fig. 5, p. 14).

We note that there exists a *symplectization* functor, which associates a symplectic manifold E^{2n+2} with a contact manifold M^{2n+1}, Lagrangian submanifolds with Legendre submanifolds, etc. However, symplectification takes Legendre singularities in general position to very special "conical" Lagrangian singularities. The correct way of obtaining Legendre singularities is by means of the *contactization* functor $M^{2n} \mapsto E^{2n+1}$ (the contactification functor is only defined for germs or for manifolds with integral class ω defining the symplectic structure).

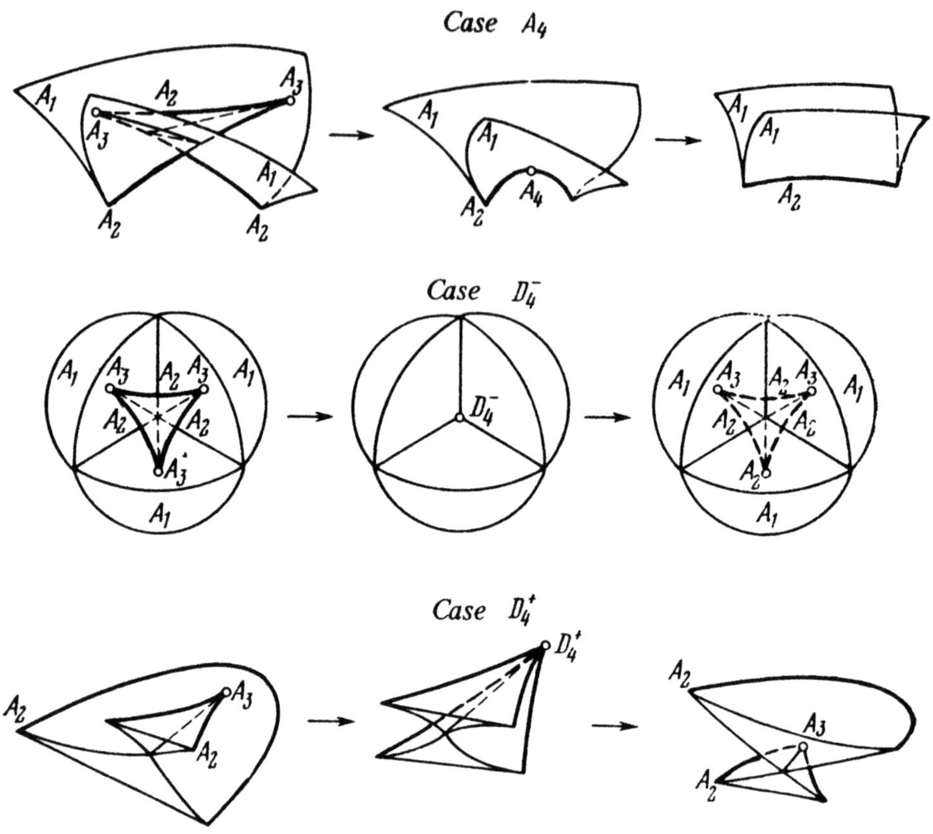

Fig. 5

§9. Normal forms for caustics and wave fronts

We use the old-fashioned coordinate notation. We consider a deformation $F(x, \lambda)$ of a function $f(x)$ of a variable $x \in \mathbf{R}^k$ with parameter $\lambda \in \mathbf{R}^l$. Suppose that $n = k + l$. We consider a symplectic space $\mathbf{R}^{2n}$ with coordinates $x \in \mathbf{R}^k$, $y \in \mathbf{R}^{k*}$, $\lambda \in \mathbf{R}^l$, $\varkappa \in \mathbf{R}^{l*}$ and with the 2-form

$$\omega = dx \wedge dy + d\varkappa \wedge d\lambda$$

(in intelligible shortened notation). The equations

$$(1) \qquad y = \frac{\partial F}{\partial x}, \quad \varkappa = -\frac{\partial F}{\partial \lambda}$$

define a Lagrangian submanifold, and together with the Lagrange fibration $(x, y, \lambda, \varkappa) \to (y, \lambda)$ they define a Lagrange mapping L.

With respect to the deformation F we construct two more families of functions of x:

$$\Phi(x; \lambda, y, z) = F(x, \lambda) - z - xy$$

with parameters $\lambda \in \mathbf{R}^l$, $y \in \mathbf{R}^{k^*}$, $z \in \mathbf{R}$;

$$G(x;\ a,\ \lambda) = F(a + x,\ \lambda) - F(a,\ \lambda) - xF'_a(a,\ \lambda)$$

with parameters $a \in \mathbf{R}^k$, $\lambda \in \mathbf{R}^l$.

Let $G(x;\ O,\ O) = g(x)$.

THEOREM XIV. *The following conditions are equivalent*:

1) *The germ of L at the point $x = O$, $\lambda = O$ is Lagrange stable.*

2) *The deformation G is transversal to the orbit of g in m^2.*

If $f \in \mathfrak{m}^2$, then each of the conditions 1) *and* 2) *is equivalent to*:

3) *The deformation Φ is versal.*

THEOREM XV. *The simple stable germs of Lagrange mappings are equivalent to germs of the form* (1), *where F is a deformation of a simple germ of f such that Φ is versal.*

(For example, if $f = x^4$ (type A_3), we can take $F = x^4 + \lambda x^2$; a complete table of the functions F for all the cases A, D, and E can be found in [6].)

We now extend $\mathbf{R}^{2n}$ to $\mathbf{R}^{2n+1} = \mathbf{R}^{2n} \times \mathbf{R}^1$ and denote by z the coordinate in $\mathbf{R}^1$. In $\mathbf{R}^{2n+1}$ we define a contact structure by the form $\alpha = x dy + \varkappa\, d\lambda + dz$. The formula

$$(2) \qquad y = \frac{\partial F}{\partial x}, \qquad \varkappa = -\frac{\partial F}{\partial \lambda}, \qquad z = F - x\frac{\partial F}{\partial x}$$

gives a Legendre manifold, and together with the Legendre fibration $(x,\ y,\ \lambda,\ \varkappa\ ;\ z) - (y,\ \lambda;\ z)$ a Legendre mapping L'.

THEOREM XVI (see [11]). *The simple stable germs of Legendre mappings are equivalent to germs of the form* (2), *where F is the same as in the previous theorem.*

In addition to the group of diffeomorphisms of the argument space, the group of multiplication by non-vanishing functions acts on the space of functions of x and hence, the direct product of these two groups acts on that space. A deformation of the germ of a function is said to be *versal for levels* if it is transversal to the orbit of this (pseudo-) group.

As an example of such a deformation of a germ of f at O we can take

$$\lambda \mapsto f + \lambda_1 e_1 + \ldots + \lambda_r e_r,$$

where the e_s generate the quotient ring of the ring of power series at O modulo the ideal $(f,\ \partial f/\partial x)$ as an $\mathbf{R}$-module.

The direct product of the group of multiplications by non-vanishing germs at O and the group of diffeomorphisms leaving O fixed acts on $\mathfrak{m}^2$.

THEOREM XVII. *The following assertions are equivalent*:

1) *The germ of L' at the point $x = O$, $\lambda = O$ is Legendre stable.*

2) *The deformation G is transversal to the orbit of g under the action of the product group on $\mathfrak{m}^2$.*

If $f \in \mathfrak{m}^2$, *then each of the conditions* 1) *and* 2) *is equivalent to*:

3) *The deformation* Φ *is versal for levels.*

Comparing (1) and (2) with the results of §§5 and 6, we are led to the following conclusion:

THEOREM XVIII. *The mapping* τ *from the orbit transversal space to the base of the miniversal deformation gives a stable Legendre germ*; *the level bifurcation set is its wave front. The mapping* $\rho \circ \tau$ *from the orbit transversal space to the base of the restricted miniversal deformation defines a stable Lagrangian germ*; *the function bifurcation set* Δ_1 *is its caustic.*

Thus, the study of the topology of the bifurcation sets Σ and Δ_1 in the bases of versal deformations and that of the topology of caustics and wave fronts of stable Lagrangian and Legendre mappings are the same problem.

The above theorems may become clearer if we consider the germ of the *restricted critical set of the deformation* $F(x, \lambda)$, defined as

$$C = \{x, \lambda \colon \partial F/\partial x = 0, \quad F(x, \lambda) = 0\}.$$

If F is miniversal, then C is a germ of a smooth manifold of dimension $\mu - 1$. The natural projection $(x, \lambda) \mapsto \lambda$ defines a mapping $\pi\colon C \to \Sigma$. The coordinate system x defines a diffeomorphism $j\colon C \to T$ onto the orbit transversal (j is defined by the translation of the critical point to the origin). The diagram

$$\begin{array}{ccc} C & \xrightarrow{\;j\;} & T \\ & {}_{\tau}\searrow \quad \swarrow_{\pi} & \\ & \Sigma & \end{array}$$

is commutative. Therefore π, like τ, normalizes Σ, the mapping $\rho \circ \pi$ has the properties of $\rho \circ \tau$ and so on.

§10. Integrals of oscillatory functions

The determination of the intensity of light in a neighbourhood of a caustic leads to the problem of the asymptotic behaviour of integrals of rapidly oscillating functions of the type of Fresnel or Airy integrals

$$I(h, \lambda) = \int_{\mathbf{R}^h} e^{\frac{i}{h} F(x, \lambda)} \varphi \, dx, \qquad x \in \mathbf{R}^k, \qquad \lambda \in \mathbf{R}^l,$$

depending on a parameter λ, as the wave length tends to 0 (that is, as $h \to 0$). Here the parameter λ denotes the point of observation, φ a function with compact support, F a smooth real "phase function". Of course, similar integrals also turn up in other domains of physics and mathematics, for example, in number theory and the theory of partial differential equations (see [81], [55], [42]).

If the light is intense enough to destroy the medium, the destruction will begin at the points of caustics with the largest I. Thus, the problem arises of determining the asymptotic behaviour, as $h \to 0$, of the largest values of I in λ that occurs intrinsically for generic phase functions F. The classification of simple singularities was found as a by-product during the solution of this problem, which was communicated to the author by V. P. Maslov (see [56], [5]).

According to the stationary phase principle the fundamental contribution of the asymptotics is given by the critical points of F in x (for constant λ). In the absence of parameters ($l = 0$) for a function in general position the critical points are Morse (that is, non-degenerate), and I decreases like $h^{n/2}$ as $h \to 0$ (Fresnel [31]).

If there is a parameter, then degenerate critical points become irremovable, and for isolated "caustic" values of the parameter, the integral decreases more slowly than $h^{\frac{k}{2}-\beta}$. The so defined number β is called the *index of singularity* of the corresponding critical point.

Strictly speaking, for singularities with finite multiplicity the integral $I(h)$ has an asymptotic expansion

$$I \sim \sum_{\alpha, \varkappa} c_{\alpha, \varkappa} h^{\frac{k}{2}-\alpha} \ln^{\varkappa} h,$$

for fixed λ, where α runs through finitely many rational arithmetic progressions, $0 \leqslant \varkappa \leqslant k-1$ (see articles by I. N. Bernstein and Gel'fand [16], Atiyah [15], I. N. Bernstein [17], Malgrange [52], [53] and [*53a]).[1] The index of singularity β is the minimum of the α for which $c_{\alpha,\varkappa} \neq 0$ for some φ concentrated in an arbitrarily small neighbourhood of the critical point x in question.

THEOREM XIX (see [5], [7]). *For simple singularities* $\beta = \frac{1}{2} - \frac{1}{N}$, *where N is the Coxeter number*:

$N(A_{k+1}) = k+1$, $N(D_k) = 2k-2$, $N(E_6) = 12$, $N(E_7) = 18$, $N(E_8) = 30$. *For parabolic and hyperbolic singularities* $\beta = \frac{1}{2}$.

For all other singularities it appears that $\beta > \frac{1}{2}$.

We define the Coxeter number N of any singularity by the formula $\beta = \frac{1}{2} - \frac{1}{N}$, where β is the index of singularity.

THEOREM XX (see [7]). *For $l \leqslant 10$ and $k \geqslant 3$, the maximum index of singularity that inherently occurs in generic families has the form*

$$\beta_l = \frac{1}{2} - \frac{1}{N},$$

where the number N is given by the table

l	0	1	2	3	4	5	6	7	8	9	10, $k=3$	11, $k=3$	10, $k>3$
N	+2	+3	+4	+6	+8	+12	∞	∞	−24	−16	−12	−8	−6

[1] (Translator's note.) These functions have also been studied by M. Sato, M. Kashiwara and their co-workers under the name of *b*-functions. For a summary of their work, see M. Kashiwara, Sur la *b*-fonction, Séminaire Goulaouic-Lions-Schwartz 1974/75, Exp. 25. For more detailed expositions (in Japanese), see the collection "*b*-functions and singular points of hypersurfaces", Proc. Conf. Research Inst. Math. Sciences, Kyôto, Japan, 27 September–10 October, 1973, Sûrikaisekikenkyuo Kôkyûroku 225 (1975).

All the numbers $\beta_l = \beta_l(k)$ are rational and for sufficiently large k do not depend on k.

The calculation of the β_l seems to be very difficult; it appears that $\beta_l \sim \sqrt{(2l/6)}$. It seems probable that β is semicontinuous in λ and that for λ near λ_0 we have the uniform estimate

$$|I(h, \lambda)| \leqslant C(\varepsilon, \varphi)\, h^{\frac{h}{2} - \beta(\lambda_0) - \varepsilon}$$

for any $\varepsilon > 0$.

This estimate has been proved by I. M. Vinogradov [81] for the simple singularities of the series A, and by Duistermaat [30] for all simple singularities.

§ 11. Semiquasihomogeneous functions and the Newton diagram

The original proofs of all the classification theorems were based on long computations, which can be replaced by a distinctive geometric technique connected with the Newton diagram.

A function $f(x_1, \ldots, x_n)$ is said to be *quasihomogeneous* of degree d with weights $(\alpha_1, \ldots, \alpha_n)$ if $f(t^{\alpha_1} x_1, \ldots, t^{\alpha_n} x_n) = t^d f(x_1, \ldots, x_n)$ identically in $t \in \mathbf{C}^*$. Here $0 < \alpha_s \leqslant \frac{1}{2}$ are rational numbers. (For more about quasihomogeneous functions, see [8] and also papers by Saito [65], [66] Milnor and Orlik [57], Orlik and Wagreich [60].)

A function f is *semiquasihomogeneous* if $f = f_0 + f'$, where f_0 is quasihomogeneous of degree 1 and has a singularity of finite multiplicity at O, and all the monomials of f' are of degree greater than 1.

THEOREM XXI (see [8]). *A semiquasihomogeneous function can be reduced by a diffeomorphism to the normal form $f_0 + c_1 e_1 + \ldots + c_r e_r$, where the c_s are numbers and the e_s are basis monomials of the local ring of the function f_0 of degree greater than* 1.

The Newton diagram of an arbitrary germ f is a convex polyhedron in $\mathbf{R}^n$ constructed from the exponents of the monomials occurring in the Taylor expansion of f; it contains much useful information. Here I only state one result of Kushnirenko, which allows us to compute the multiplicity.

We assume that the Newton diagram of the function $f(x_1, \ldots, x_n)$ has points on each of the coordinate axes (this is not a restriction, since f has a sufficient jet).

THEOREM XXII (Kushnirenko). *Let V denote the volume of the subdomain of the positive orthant in $\mathbf{R}^n$ lying below the Newton diagram, V_i the $(n-1)$-dimensional volume under the Newton diagram on the i-th coordinate hyperplane, V_{ij} the $(n-2)$-dimensional volume on the coordinate plane orthogonal to the i-th and j-th basis vectors, etc. Then for all functions f with a given Newton diagram*

$$\mu(f) \geqslant n!V - (n-1)! \sum_i V_i + (n-2)! \sum_{i<j}' V_{i,j} - \ldots \pm 1,$$

with equality for almost all f.

For example, for almost all functions of two variables with a given Newton diagram $\mu = 2S - a - b + 1$, where S is the area under the diagram, and a and b are the coordinates of the points of the diagram on the axes (Fig. 6).

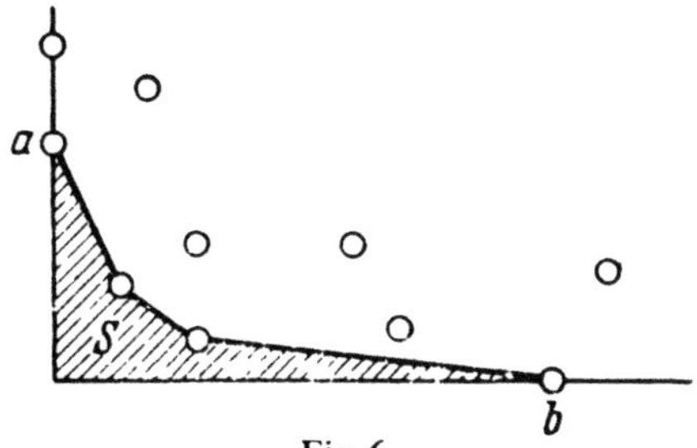

Fig. 6

Very recently D. N. Bernstein, Kushnirenko, and Khovanskii have obtained far-reaching generalizations of this theorem. In particular, the number of solutions of a system of polynomial equations $P_1(x_1, \ldots, x_n) = 0, \ldots, P_n(x_1, \ldots, x_n) = 0$, for which none of the coordinates x_s is zero, is equal to $n!$ times the mixed Minkowski volume of the convex hulls of the supports of the polynomials P_s (for almost all sets of polynomials P_s whose supports have given convex hulls). They have also obtained analogous formulae, expressing other numerical invariants of affine complete intersections in terms of the geometry of polyhedra. For example, the number of holomorphic forms on a hypersurface is equal to the number of lattice points inside the convex hull of the support (for almost all hypersurfaces whose supports have a given convex hull). (This formula seems to have been known to Hodge [*143], but according to Khovanskii, Hodge's arguments are not satisfactory.) D. N. Bernstein and Khovanskii have proved a formula for the Euler characteristics of complete intersections (spring 1975); another proof of this formula was given by Kushnirenko for the case of hypersurfaces. At the Bonner Arbeitstagung (June 1975) Varchenko discussed these problems with A'Campo and other Western mathematicians. In September 1975 Varchenko proved a formula for the ζ-function of the monodromy in terms of the Newton polyhedron. He uses the Bernstein-Khovanskii-Kushnirenko formula for Euler characteristics and A'Campo's results on the ζ-function of a monodromy [*145]. According to a letter from Brieskorn to Varchenko dated 30 December 1975, this formula was also found by A'Campo in October 1975 for functions of two variables, and by F. Ehlers in November 1975 for the general case. Varchenko has also found a formula for the index of singularity of oscillatory integrals in terms of Newton polyhedra. Among other things, he has proved that:

1) for functions of two variables with isolated singularities, the index of singularity can be calculated from the point of intersection of the Newton polygon with the diagonal in some system of coordinates, that is, for those systems of coordinates for which this point of intersection is as far from the origin as possible;

2) the same is true for functions of more variables if the coordinates of the point of intersection are greater than 1;

3) the index of singularity is not upper semicontinuous for some functions of three variables and for some Γ-non-degenerate functions of five variables for which the above formula fails (that is, the coordinates of the point of intersection are less than 1);

4) the index of singularity depends on the *real* form: it is different for some singularities of functions of three variables at isolated critical points even though the singularities are equivalent over **C** (this is the same example as in 3)).

Varchenko's paper on oscillatory integrals contains many other, more precise results, with fewer restrictions, but these are too long to be stated here [105].

Among the numerous unsolved problems in the theory of singularities there is the important problem of the smoothness of the stratum μ = const in the base of a miniversal deformation.

Lê and Ramanujam [50] have proved that for $n \neq 3$ neither the topological type of the singular level set nor the topological type of the "Milnor fibration" formed by the non-singular fibres varies. It also seems likely that neither the topological type of a function nor the index of singularity β varies. Pham [63] has noted the change of the topology of the bifurcation set along the stratum μ = const. Gabrielov [36] has proved that the dimension of the stratum μ = const is equal to the modality of the singularity. Using results of Teissier [71], Kushnirenko and Gabrielov have proved that for (semi)homogeneous singularities this dimension is equal to the number of generators of a monomial basis of the local ring on the Newton diagram and above it [37]. This result is probably true for (semi)quasihomogeneous functions.

In [8] I have expressed the conjecture that the modality of a function of two variables is equal to the number of (lattice) points between the Newton diagram and the coordinate rays passing through the point (2,2), counting the boundary points (for almost all functions with a given diagram); this was proved by Kushnirenko, using results of Varchenko and Lê Dũng Tráng.

In this survey I have left to one side many important questions of the theory of critical points of functions, particularly, algebraic ones (see the ponderous tome "Singularités à Cargèse", [71]). From the results listed above we see the power of "transcendental" and topological methods, based on an analysis of the hierarchy of singularities, on a detailed study of the classes of low codimension, on considerations of semicontinuity and genericity, going back to Poincaré's theory of bifurcations [12], and formalized in Thom's transversality theorems. These ideas were first applied to the study of singular points of hypersurfaces by Galina N. Tyurina (see [76]–[78], [13]).

CHAPTER II

The hierarchy of singularities

§ 12. Normal forms

A large part of the results discussed in Chapter I is based on the classification theorems or is found by means of these theorems. However, to obtain these theorems there is no other approach so far than the experimental one. Below we state these experimental results (including a complete classification of the singularities with at most two moduli) and prove many of them (including the theorem on the classification of unimodal singularities).

Before proceeding to the technical statements of the results (see §§13 and 14), I indicate some conclusions of a general nature to which our list of singularities leads (§13). I do not state these conclusions in the form of a theorem, because I do not know the extent of the class of singularities to which they apply.

1. **Normal forms.** *For many classes of singularities there exist simple normal forms.*

The following definitions give a precise meaning to these words.

The group of germs (or jets) of diffeomorphisms of $\mathbf{C}^n$ leaving the critical point O fixed acts on the space of germs (or jets) of functions at O. A *class of singularities* is a subset of the space of germs (or jets) of functions that is invariant under this action. Orbits are examples of classes of singularities. Two germs (or jets) are said to be equivalent if they belong to the same orbit.

Another example of a class is the so-called μ = const *stratum.* The *multiplicity* (Milnor number) μ of a critical point $O \in \mathbf{C}^n$ of a function f is the index of the singular point O of the vector field grad f. The μ = const *stratum for* f is defined as the connected component containing f of the space of germs with fixed multiplicity μ at O.

To define a *normal form,* we regard the space of polynomials $M = \mathbf{C}[x_1, \ldots, x_n]$ as a subset of the space of germs of functions $f(x_1, \ldots, x_n)$ at O.

A *normal form for a class K of functions* is given by a smooth mapping $\Phi: B \to M$ of a finite-dimensional linear *space of parameters B* into the space of polynomials for which the following three conditions hold:

1) $\Phi(B)$ intersects all the orbits of K;

2) the inverse image in B of each orbit is finite;

3) the inverse image of the whole complement to K is contained in some proper hypersurface in B.

A normal form is said to be *polynomial* (respectively, *affine*) if the mapping Φ is polynomial (respectively, linear and inhomogeneous). An affine normal form is called *simple* if Φ has the form

$$\Phi(b_1, \ldots, b_r) = \varphi_0 + b_1\mathbf{x}^{\mathbf{m}_1} + \ldots + b_r\mathbf{x}^{\mathbf{m}_r},$$

where φ_0 is a fixed polynomial, the b_i are numbers and the $\mathbf{x}^{\mathbf{m}_i}$ are monomials. (In the applications the polynomial φ_0 is usually "simple" itself, that is, a sum of a few monomials.)

The existence of a unique normal form (at least a polynomial one) for the whole stratum μ = const is by no means obvious a priori. An astonishing conclusion from our calculations is the existence of such normal forms for all the singularities of the list in § 13 (therefore, in particular, for all singularities with one and two moduli). The majority of our forms are simple; it is probable that all the singularities in § 13 have simple normal forms. I do not know how extensive the class of functions is for which the stratum μ = const admits a simple (or at least polynomial) normal form (this question is naturally related to stable equivalence classes; see § 12.5).

2. **Series of singularities.** In the list of § 13 all the singularities are split into *series,* denoted by capital letters (we use light-face letters $A, \ldots, Z$ with various suffixes to denote the strata μ = const and bold-face letters $\boldsymbol{A}, \ldots, \boldsymbol{Z}$ with or without suffixes to denote classes of singularities that are unions of μ = const strata). Although the series undoubtedly exist, it is not at all clear what a series of singularities is.

For example, consider the series $\boldsymbol{A}$ and $\boldsymbol{D}$, which are formed by the orbits of the germs A_k: $f(x, y) = x^{k+1} + y^2$ and D_k: $f(x, y) = x^2y + y^{k-1}$. The classes A_k and D_k degenerate to each other as follows:

$$\begin{array}{ccccccccc} A_1 & \leftarrow & A_2 & \leftarrow & A_3 & \leftarrow & A_4 & \leftarrow & \ldots \\ & & & & \uparrow & & \uparrow & & \\ & & & & D_5 & \leftarrow & D_6 & \leftarrow & \ldots \end{array}$$

(A class of singularities L *degenerates* (or *is adjacent*) to a class K (notation: $K \leftarrow L$) if every function $f \in L$ can be deformed into a function of K by a suitably small perturbation.)

It is clear that there are two series in this example, $\boldsymbol{A}$ and $\boldsymbol{D}$. However, what is the formal meaning of this assertion, and does it exceed the limits of the arbitrary names?

Thus, to define the series $\boldsymbol{A}$ is to learn how to invert the arrows of the adherences, so that we go from A_k to A_{k+1}, without deviating through

D_{k+1}. In the given case this is not hard to do (the singularities of A have a second differential of corank $\leqslant 1$). In more complicated cases (see §§13, 14) we can also state a rule for inverting the arrows (in each case separately). As a result there arise series with one or several suffixes (for example, the three-suffix series $T_{k,l,m} = axyz + x^k + y^l + z^m$), and the functions of the series may depend on parameters.

As in the example of the series A and D above, so in all cases, after a series is found, we can define it. However, a general definition of a series of singularities is not known. It is only clear that the series are associated with singularities of infinite multiplicity (for example, $D \sim x^2y$, $T \sim xyz$), so that the hierarchy of series reflects the hierarchy of non-isolated singularities.

3. Periodicity. The partition of many classes of singularities into μ = const strata reveals a peculiar periodicity, which can be described as follows. The whole stratification (partition) is a chain of identical fragments (beasts). Each beast consists of points (strata), two of which (the head and the tail) are distinguished. Apart from the head and tail, the beast can contain the strata joining them by arrows of adjacency, and extremities (series of infinite length). The head of each beast adheres to the tail of the preceding beast.

For example, the stratification of singularities of corank 2 with 3-jet x^3 is given by a chain of beasts, each of which consists of 5 points and one infinite leg:

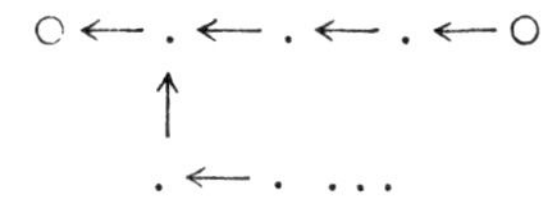

(J_k in the list of §13).

The form of the beasts corresponding to the other classes of singularities is also indicated in §13.

The reason for periodicity in the general case is not clear. A partial explanation has only been obtained for quasihomogeneous singularities by means of a technique resembling the fans of Enriques and Demazure [27] (see §16).

However, the periodicity becomes apparent not only under reduction of quasihomogeneous functions to normal forms, but also in all the calculations associated with the classifications (so that for all the calculations it suffices, in fact, to consider only one beast from a chain). This periodicity is an unexpected and, in the general case, unexplained experimental fact. Like the existence of series, periodicity suggests that there is some algebraic structure in the set of strata.

4. Classes of small modality. From the point of view of applications the most important characteristic of a class of singularities is its *codimension c* in the space of germs of functions with critical point O and critical value 0.

A generic function has only singularities of codimension $c = 0$ (non-degenerate). Degenerate singularities are irremovable only in the case of a *family* of functions depending on parameters. In this case the class of codimension c is irremovable by a small perturbation only if the number of parameters is $l \geqslant c$.

Thus, *in applications it is always necessary to study all classes up to codimension* l (that is, classes for which the complement to their union has codimension greater than l). This problem should not be confused with that of classifying singularities with orbital codimension $\leqslant l$ (that is, with $\mu \leqslant l + 1$). The latter problem occurs in applications only as a tool for solving the former.

From the topological point of view the most important characteristic of a singularity is the multiplicity μ of a critical point (equal to the number of simple critical points into which a composite point decomposes under a small perturbation).

The unexpected conclusion from these calculations is that *the algebraically most natural results are obtained not for the classification of classes of singularities up to a definite codimension* c *or multiplicity* μ, *but for the classification of classes of singularities of small modality* m.

I recall that the *modality* (or *number of moduli*) of an isolated critical point O of a function f: $(\mathbf{C}^n, O) \to (\mathbf{C}, 0)$ is the smallest number M such that a sufficiently small neighbourhood of any k-jet of f at O is covered by at most finitely many of at most M-parameter families of orbits of the action of the group of diffeomorphisms $(\mathbf{C}^n, O) \to (\mathbf{C}^n, O)$ on the space of jets of functions with critical point O and critical value 0 (see §1).

The modality m is equal to the dimension of the stratum $\mu = \text{const}$ in the base of a versal deformation minus 1 (Gabrielov [36]). Therefore, the codimension c of the stratum $\mu = \text{const}$ in the space of germs of functions with critical point O and critical value 0, the multiplicity μ and the modality m are connected by the relation

$$\mu = c + m + 1.$$

At the present time the following are completely classified:

(1) all the singularities for which $c \leqslant 10$;

(2) all the singularities for which $\mu \leqslant 16$;

(3) all the singularities for which $m \leqslant 2$;

Singularities with $m = 0$, 1 and 2 are called, respectively, *simple, unimodal*, and *bimodal*. Lists of them are given in §13. An analysis of these lists shows that

1) simple singularities are classified precisely by the Coxeter groups A_k, D_k. E_6, E_7, E_8 (that is, by the regular polyhedra in 3-space) [6];

2) unimodal singularities form a single infinite three-suffix series and 14 "exceptional" one-parameter families generated by quasihomogeneous singularities [7].

The quasihomogeneous unimodal singularities are obtained from automorphic functions connected with 14 distinguished triangles on the Lobachevskii plane and three distinguished triangles on the Euclidean plane in precisely the same way as simple singularities are connected with regular polyhedra (Dolgachev [28]).

3) bimodal singularities form 8 infinite series and 14 exceptional two-parameter families generated by quasihomogeneous singularities.

The quasihomogeneous bimodal singularities are associated with the 6 quadrilaterals and the 14 triangles on the Lobachevskii plane (in the latter case one must consider automorphic functions with automorphy factors corresponding to 2-, 3- or 5-sheeted coverings). (Dolgachev [29]).

All singularities with one and two moduli are classified precisely by the degenerations of elliptic curves, which were classified by Kodaira (Kulikov [89]). Kulikov's construction is that 1, 2 or 3 points are blown up in a minimal resolution of the degenerate fibre, after which the original fibre is blown down.

Unfortunately, all the results listed here are obtained by a comparison of independently proved classification theorems, none of which can (so far) be derived from another. Therefore all these results depend on our calculations.

5. Stable equivalence. I recall that two germs f: $(\mathbf{C}^n, O) \to (\mathbf{C}, 0)$ and g: $(\mathbf{C}^m, O) \to (\mathbf{C}, 0)$ are said to be *stably equivalent* if they become equivalent after adding non-degenerate quadratic forms (for example, $f(x) = x^3$ is stably equivalent to $g(x, y) = x^3 + y^2$). If two functions of the same number of variables are stably equivalent, then they are equivalent. Thus, passage to stable equivalence classes is not concerned with the number of variables of functions,[1] because nothing changes in the classification of functions of a fixed number of variables.

The number of stable μ-equivalence clasess for small μ that are powers of 2 is itself a power of 2:

μ	1	2	4	8	16
Number of classes	1	1	2	4	32

The *corank* of the germ of a function at a critical point is the dimension of the zero space of its second differential. Every function of corank r is stably equivalent to a function of r variables ("generalized Morse lemma"). In classification up to stable equivalence one usually takes normal forms with the number of variables equal to the corank.

---

[1] In § 13 we do not indicate the number of arguments of a function when all the arguments are explicitly included in the formula. For example, the stable equivalence class of the function x^3 includes the functions $f(x) = x^3$ and $g(x, y) = x^3 + y^2$, but not the function $f(x, y) = x^3$.

Calculations show, however, that in many cases more natural normal forms are obtained by passing to functions of a larger number of variables that are stably equivalent to the given function. The simplest example is the inclusion of the singularities $x^p + ax^2y^2 + y^q$ into the series T; a more interesting example is given by the series $W^{\#} \sim (x^2 + y^3)^2$.

I recall the definition of the *Newton polyhedron* of the series $f \in \mathbf{C}[[x_1, \ldots, x_n]]$. The *support* supp f of a series f is the set of all exponents $\boldsymbol{m} \in \mathbf{Z}^n$ of the monomials $\boldsymbol{x}^{\boldsymbol{m}}$ that occur in f with non-zero coefficients. We assume the lattice $\mathbf{Z}^n$ to be embedded in $\mathbf{R}^n$. For each point $\boldsymbol{m} \in$ supp f we consider the positive orthant $\boldsymbol{m} + \mathbf{R}^n_+$. The Newton polyhedron Γ for f is the boundary of the convex hull of these orthants.

A function f is called Γ-non-degenerate if the multiplicity μ of its critical point O has the least possible value $\nu < \infty$ for functions with the Newton polyhedron Γ. An explicit formula for the Newton number $\nu(\Gamma)$ and a sufficient condition for Γ-non-degeneracy were found by Kushnirenko (Theorem XXII in § 11, and [90]).

A large part of our normal forms consists of Γ-non-degenerate functions.

Is every function in a neighbourhood of a critical point of finite multiplicity stably equivalent to a Γ-non-degenerate function? If the answer to this question is positive, then the classification of singularities and the reduction to normal forms can be carried out in a rather explicit fashion: the study of Γ-non-degenerate functions leads to constructions basic to the stereometric constructions (see [8]).

The arrangement of material in this chapter is as follows. In § 13 we give lists of normal forms and some adherency diagrams. Here we list all the singularities with at most 2 moduli, all the singularities of corank 2 with non-zero 4-jet, all the singularities of corank 3 with 3-jet giving an irreducible cubic; we also indicate which singularities remain unclassified.

In § 14 we state 105 theorems, which form the main results of this chapter. These theorems, together with the list of singularities of § 13, constitute a determination of singularities, similar to botanical determination. Actually, in § 14 we describe an algorithm to find the place of an arbitrary singularity in the lists of § 13.

In §§ 15-18 we prove the majority of the theorems of § 14. All these proofs are based on the geometric technique developed in [8]. The formulae of §§ 13-14 can be interpreted as an abbreviated notation for the supports and the Newton polyhedra, and the proofs of the theorems consist not in calculations, but in the geometric manipulation of these supports and polyhedra.

The classification of quasihomogeneous singularities in § 16 is based on the general technique of the roots of a quasihomogeneous Lie algebra, developed here in greater detail than is necessary for proving the theorems of § 14.

The theorems in §§15–18 have as a consequence, in particular, the classification theorem for unimodal singularities (Theorem II of §1), which was announced in [7]. The basic part of the proof is contained in §18.

§13. Lists of singularities

(In this section the letters $A, \ldots, Z$ indicate *stable* equivalence classes of functions or families of functions.)

I. SINGULARITIES WITH $m = 0$, 1, and 2.

I.0. SIMPLE SINGULARITIES ($m = 0$). There are two infinite series A, D, and three exceptional singularities E_6, E_7, E_8:

$$A_k \ni x^{k+1} \quad (k \geqslant 1),$$
$$D_k \ni x^2x + y^{k-1} \quad (k \geqslant 4);$$
$$E_6 \ni x^3 + y^4, \qquad E_7 \ni x^3 + xy^3, \qquad E_8 \ni x^3 + y^5.$$

The diagram of adjacencies:

$$\begin{array}{ccccccccc} \boldsymbol{A} \leftarrow \boldsymbol{D} \leftarrow & E_6 & \leftarrow & E_7 & \leftarrow & E_8, & \quad & \boldsymbol{A} = A_1 \leftarrow A_2 \leftarrow A_3 \leftarrow A_4 \leftarrow \ldots \\ & \uparrow & & \uparrow & & \uparrow & & \uparrow \\ & & & & & & & (\boldsymbol{D}) \\ & (\boldsymbol{P}) & & (\boldsymbol{X}) & & (\boldsymbol{J}) & & \boldsymbol{D} = D_4 \leftarrow D_5 \leftarrow D_6 \leftarrow \ldots \\ & & & & & & & \uparrow \\ & & & & & & & (\boldsymbol{E}) \end{array}$$

The classes $\boldsymbol{P}$, $\boldsymbol{X}$ and $\boldsymbol{J}$ are defined below: see II.

I.1. UNIMODAL SINGULARITIES ($m = 1$). There are 3 families of parabolic singularities, a three-suffix series of hyperbolic singularities and 14 families of exceptional singularities.

Parabolic:

P_8	$x^3 + y^3 + z^3 + axyz$	$a^3 + 27 \neq 0$
X_9	$x^4 + y^4 + ax^2y^2$	$a^2 \neq 4$
J_{10}	$x^3 + y^6 + ax^2y^2$	$4a^3 + 27 \neq 0$

Hyperbolic:

$$T_{p,q,r} \ni x^p + y^q + z^r + axyz, \quad a \neq 0, \quad \frac{1}{p} + \frac{1}{q} + \frac{1}{r} < 1.$$

14 *exceptional families*:

E_{12}	$x^3 + y^7 + axy^5$	E_{13}	$x^3 + xy^5 + ay^8$
E_{14}	$x^3 + y^8 + axy^6$	Z_{11}	$x^3y + y^5 + axy^4$
Z_{12}	$x^3y + xy^4 + ax^2y^3$	Z_{13}	$x^3y + y^6 + axy^5$
W_{12}	$x^4 + y^5 + ax^2y^3$	W_{13}	$x^4 + xy^4 + ay^6$
Q_{10}	$x^3 + y^4 + yz^2 + axy^3$	Q_{11}	$x^3 + y^2z + xz^3 + az^5$
Q_{12}	$x^3 + y^5 + yz^2 + axy^4$	S_{11}	$x^4 + y^2z + xz^2 + ax^3z$
S_{12}	$x^2y + y^2z + xz^3 + az^5$	U_{12}	$x^3 + y^3 + z^4 + axyz^2$

Some adjacencies of unimodal singularities:

$$\begin{array}{ccccccccccc}
 & & & & J_{10}=T_{2,3,6} & \to & (E_8) & & & & \\
 & & & & \uparrow & & & & & & \\
\ldots & \to & T_{2,3,8} & \to & T_{2,3,7} & \leftarrow & E_{12} & \leftarrow E_{13} & \leftarrow E_{14} & \leftarrow & (J_3)
\end{array}$$

The parentheses contain classes of singularities whose modality is not equal to 1.

$$\begin{array}{ccccccccccc}
 & & & & X_9=T_{2,4,4} & \to & (E_7) & & & & \\
 & & & & \uparrow & & & & & & \\
\ldots & \to & T_{2,4,6} & \to & T_{2,4,5} & \leftarrow & Z_{11} & \leftarrow & Z_{12} & \leftarrow Z_{13} & \leftarrow (Z_1^1) \\
 & & \uparrow & & \uparrow & & \uparrow & & \uparrow & & \\
\ldots & \rightrightarrows & T_{2,5,6} & \to & T_{2,5,5} & \leftarrow & W_{12} & \leftarrow & W_{13} & \leftarrow & (W_{1,0};\ N)
\end{array}$$

$$\begin{array}{ccccccccccc}
 & & & & P_8=T_{3,3,3} & \to & (E_6) & & & & \\
 & & & & \uparrow & & & & & & \\
\ldots & \to & T_{3,3,5} & \to & T_{3,3,4} & \leftarrow & Q_{10} & \leftarrow & Q_{11} & \leftarrow Q_{12} & \leftarrow (Q_2) \\
 & & \uparrow & & \uparrow & & \uparrow & & \uparrow & & \\
\vdots & \rightrightarrows & T_{3,4,5} & \to & T_{3,4,4} & \leftarrow & S_{11} & \leftarrow & S_{12} & \leftarrow & (S_{1,0}) \\
 & & \uparrow & & \uparrow & & \uparrow & & & & \\
\vdots & \substack{\to\\ \to\\ \to} & T_{4,4,5} & \to & T_{4,4,4} & \leftarrow & U_{12} & \leftarrow & (U_{1,0};\ V) & & \\
 & & & & \uparrow & & & & & & \\
 & & & & (O) & & & & & &
\end{array}$$

I.2. BIMODAL SINGULARITIES (m = 2). There are 8 infinite series and 14 exceptional families. Let $\boldsymbol{a} = a_0 + a_1 y$.

4 *infinite series of bimodal singularities of corank* 2:

Notation	Normal form	Restrictions	Multiplicity μ
$J_{3,0}$	$x^3+bx^2y^3+y^9+cxy^7$	$4b^3+27 \neq 0$	16
$J_{3,p}$	$x^3+x^2y^3+\boldsymbol{a}y^{9+p}$	$p>0,\ a_0 \neq 0$	$16+p$
$Z_{1,0}$	$x^3y+dx^2y^3+cxy^6+y^7$	$4d^3+27 \neq 0$	15
$Z_{1,p}$	$x^3y+x^2y^3+\boldsymbol{a}y^{7+p}$	$p>0,\ a_0 \neq 0$	$15+p$
$W_{1,0}$	$x^4+\boldsymbol{a}x^2y^3+y^6$	$a_0^2 \neq 4$	15
$W_{1,p}$	$x^4+x^2y^3+\boldsymbol{a}y^{6+p}$	$p>0,\ a_0 \neq 0$	$15+p$
$W^{\#}_{1,2q-1}$	$(x^2+y^3)^2+\boldsymbol{a}xy^{4+q}$	$q>0,\ a_0 \neq 0$	$15+2q-1$
$W^{\#}_{1,2q}$	$(x^2+y^3)^2+\boldsymbol{a}x^2y^{3+q}$	$q>0,\ a_0 \neq 0$	$15+2q$

4 *infinite series of bimodal singularities of corank* 3:

Notation	Normal form	Restrictions	Multiplicity μ
$Q_{2,0}$	$x^3+yz^2+ax^2y^2+xy^4$	$a_0^2\neq 4$	14
$Q_{2,p}$	$x^3+yz^2+x^2y^2+ay^{6+p}$	$p>0,\ a_0\neq 0$	$14+p$
$S_{1,0}$	$x^2z+yz^2+y^5+azy^3$	$a_0^2\neq 4$	14
$S_{1,p}$	$x^2z+yz^2+x^2y^2+ay^{5+p}$	$p>0,\ a_0\neq 0$	$14+p$
$S^{\#}_{1,2q-1}$	$x^2z+yz^2+zy^3+axy^{3+q}$	$q>0,\ a_0\neq 0$	$14+2q-1$
$S^{\#}_{1,2q}$	$x^2z+yz^2+zy^3+ax^2y^{2+q}$	$q>0,\ a_0\neq 0$	$14+2q$
$U_{1,0}$	$x^3+xz^2+xy^3+ay^3z$	$a_0(a_0^2+1)\neq 0$	14
$U_{1,2q-1}$	$x^3+xz^2+xy^3+ay^{1+q}z^2$	$q>0,\ a_0\neq 0$	$14+2q-1$
$U_{1,2q}$	$x^3+xz^2+xy^3+ay^{3+q}z$	$q>0,\ a_0\neq 0$	$14+2q$

14 *exceptional families*

E_{18}	$x^3+y^{10}+axy^7$	E_{19}	$x^3+xy^7+ay^8$
E_{20}	$x^3+y^8+axy^8$	Z_{17}	$x^3y+y^8+axy^6$
Z_{18}	$x^3y+xy^6+ay^9$	Z_{19}	$x^3y+y^9+axy^7$
W_{17}	$x^4+xy^5+ay^7$	W_{18}	$x^4+y^7+ax^2y^4$
Q_{16}	$x^3+yz^2+y^7+axy^5$	Q_{17}	$x^3+yz^2+xy^5+ay^8$
Q_{18}	$x^3+yz^2+y^8+axy^6$	S_{16}	$x^2z+yz^2+xy^4+ay^6$
S_{17}	$x^2z+yz^2+y^6+azy^4$	U_{16}	$x^3+xz^2+y^5+ax^2y^2$

All the functions of these families (under the restrictions indicated in the tables) are bimodal.

Some adjacencies of bimodal singularities:

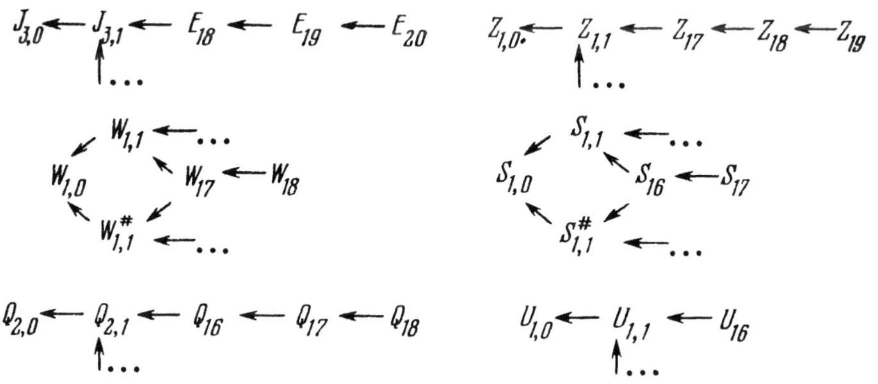

Pyramids of the exceptional singularities of modality 1 *and* 2:

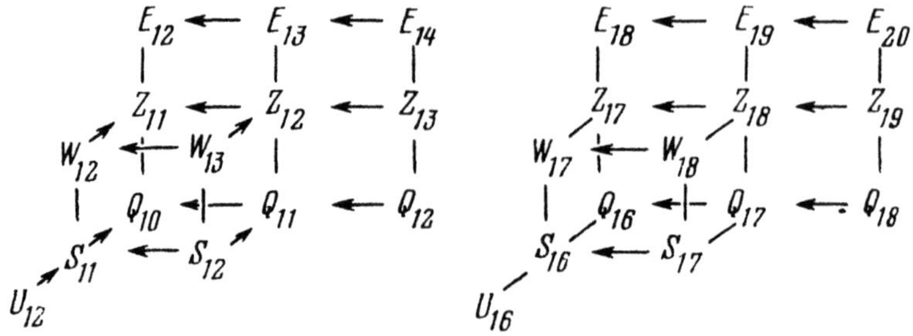

The vertical lines join singularities obtained from the same Kodaira class by means of the construction of Kulikov [89].

II. SINGULARITIES OF CORANK 2 WITH NON-ZERO 4-JET.

Throughout this section $a = a_0 + \ldots + a_{k-2}y^{k-2}$ $(a = 0$ for $k = 1)$.

II.1. SINGULARITIES OF CORANK 2 WITH NON-ZERO 3-JET. Besides the simple singularities A, D, E_6, E_7, E_8 there is yet another infinite series of classes

$$J = \leftarrow J_2 \leftarrow J_3 \leftarrow \ldots,$$

where

$$\begin{array}{l} J_k = J_{k,0} \leftarrow J_{k,1} \leftarrow E_{6k} \leftarrow E_{6k+1} \leftarrow E_{6k+2} \leftarrow (J_{k+1}) \\ \qquad\qquad\quad \uparrow \\ \qquad\qquad J_{k,2} \leftarrow J_{k,3} \leftarrow \ldots \end{array}$$

Notation	Normal form	Restrictions	Multiplicity μ	Modality m
$J_{k,0}$	$x^3+bx^2y^k+y^{3k}+cxy^{2k+1}$	$k>1, 4b^3+27\neq 0$	$6k-2$	$k-1$
$J_{k,i}$	$x^3+x^2y^k+ax^{3k+i}$	$k>1,\ i>0,\ a_0\neq 0$	$6k-2+i$	$k-1$
E_{6k}	$x^3+y^{3k+1}+axy^{2k+1}$	$k\geqslant 1$	$6k$	$k-1$
E_{6k+1}	$x^3+xy^{2k+1}+ay^{3k+2}$	$k\geqslant 1$	$6k+1$	$k-1$
E_{6k+2}	$x^3+y^{3k+2}+axy^{2k+2}$	$k\geqslant 1$	$6k+2$	$k-1$

Here $c = c_0 + \ldots + c_{k-3}y^{k-3}$ for $k > 2$; for $k = 2$ we set $c = 0$.

II.2. SINGULARITIES OF CORANK 2 WITH ZERO 3-JET AND NON-ZERO 4-JET. All such singularities form a single infinite series of classes

$$X = \leftarrow X_1 \leftarrow X_2 \leftarrow \ldots,$$

where

$$X_k = X_{k,0} \leftarrow X_k^* \begin{array}{c} \swarrow\ Y_k\ \nwarrow \\ \nwarrow\ Z_k\ \swarrow \end{array} W_k \leftarrow (X_{k+1}),$$

and

$$\begin{array}{ll}
\quad\quad\downarrow & \quad\quad\downarrow \\
X_k^* = \leftarrow X_{k,1} \leftarrow X_{k,2} \leftarrow \dots, & Y_k = \leftarrow Y^k_{1,1} \leftarrow Y^k_{2,1} \leftleftarrows \dots, \\
\quad\quad\uparrow & \\
\quad\quad\downarrow & \quad\quad\downarrow \\
Z_k = \leftarrow Z_0^k \leftarrow Z_1^k \leftarrow \dots, & Z_0^k = \leftarrow Z^k_{12k-1} \leftarrow Z^k_{12k} \leftarrow Z^k_{12k+1} \leftarrow,
\end{array}$$

$$Z_i^k = \leftarrow Z^k_{i,0} \leftarrow Z^k_{i,1} \leftarrow Z^k_{12k+6i-1} \leftarrow Z^k_{12k+6i} \leftarrow Z^k_{12k+6i+1} \leftarrow (Z^k_{i+1})$$

$$\begin{array}{c}
\uparrow \\
Z^k_{i,2} \leftarrow Z^k_{i,3} \leftarrow \dots \qquad (i>0),
\end{array}$$

$$\begin{array}{ccccc}
 & & W_{k,1} \leftarrow W_{k,2} \leftarrow \dots & & \\
 & \swarrow & & \nwarrow & \\
W_k = \leftarrow W_{12k} \leftarrow W_{12k+1} \leftarrow W_{k,0} & & & & W_{12k+5} \leftarrow W_{12k+6} \leftarrow (X_{k+1}). \\
 & \nwarrow & & \swarrow & \\
 & & W^{\#}_{k,1} \leftarrow W^{\#}_{k,2} \leftarrow \dots & &
\end{array}$$

Singularities of the classes X and Y:

Notation	Normal form ($k>1$)	Restrictions	Multiplicity μ	Modality m
$X_{k,0}$	$x^4+bx^3y^k+ax^2y^{2k}+xy^{3k}$	$\Delta\neq 0,\ a_0b_0\neq 9$	$12k-3$	$3k-2$
$X_{k,p}$	$x^4+ax^3y^k+x^2y^{2k}+by^{4k+p}$	$a_0^2\neq 4,\ b_0\neq 0,\ p>0$	$12k-3+p$	$3k-2$
$Y^k_{r,s}$	$[(x+ay^k)^2+by^{2k+s}]\,(x^2+y^{2k+r})$	$1\leqslant s\leqslant r,\ a_0\neq 0\neq b_0$	$12k-3+r+s$	$3k-2$

In the case $k = 1$ these formulae need some modification:

$X_{1,0}$	$x^4+a_0x^2y^2+y^4$	$a_0^2\neq 4$	9	1
$X_{1,p}$	$x^4+x^2y^2+a_0y^{4+p}$	$a_0\neq 0$	$9+p$	1
$Y^1_{r,s}$	$x^{4+r}+a_0x^2y^2+y^{4+s}$	$a_0\neq 0$	$9+r+s$	1

Finally, $X_{1,0}= X_9$, $X_{1,p}= T_{2,4,4+p}$, $Y^1_{r,s}= T_{2,4+r,4+s}$ (see I). Here

$$\Delta = 4(a_0^3 + b_0^3)- a_0^2b_0^2 - 18a_0b_0 + 27,\ b = b_0 + \dots + b_{2k-2}y^{2k-2}.$$

Singularities of the class Z.

The singularities $Z^k_{i,0}$ and Z^k_μ $(k > 1)$ have a normal form of the kind $f = (x + ay^k)f_2$, where $a_0 \neq 0$ and f_2 is given by the following table:

Notation	f_2	Restrictions ($k>1$)	Multiplicity μ	Modality m
$Z^k_{i,0}$	$x^3+dx^2y^{k+i}+cxy^{2k+2i+1}+y^{3k+3i}$	$4d^3+27\neq 0,$ $i>0$	$12k+6i-3$	$3k+i-2$
$Z^k_{12k+6i-1}$	$x^3+bxy^{2k+2i+1}+y^{3k+3i+1}$	$i\geqslant 0$	$12k+6i-1$	$3k+i-2$
Z^k_{12k+6i}	$x^3+xy^{2k+2i+1}+by^{3k+3i+2}$	$i\geqslant 0$	$12k+6i$	$3k+i-2$
$Z^k_{12k+6i+1}$	$x^3+bxy^{2k+2i+2}+y^{3k+3i+2}$	$i\geqslant 0$	$12k+6i+1$	$3k+i-2$

The singularity $Z^k_{i,p}$ $(k > 1,\ i > 0,\ p > 0)$ has the normal form $Z^k_{i,p} \ni (x^2 + axy^k + by^{2k+i})(x^2 + y^{2k+2i+p})$, $a_0 \neq 0$, $b_0 \neq 0$; its multiplicity is $\mu = 12k + 6i + p - 3$, its modality is $m = 3k + i - 2$.

For $k = 1$ the preceding formulae are modified in the following way:

1) the upper index k is dropped;

2) the singularities $Z_{i,0}$, Z_{6i+11}, Z_{6i+12}, Z_{6i+13} $(i > 0)$ have normal forms of the kind $f = yf_2$, where f_2 is given by the preceding table;

3) $Z_{i,p} \ni y(x^3 + x^2y^{i+1} + by^{3i+p+3})$, $b_0 \neq 0, i > 0, p > 0$; $\mu = 9 + 6i + p$, $m = i + 1$.

Throughout the formulae above we have used

$$b = b_0 + \ldots + b_{2k+i-2}y^{2k+i-2}, \quad c = c_0 + \ldots + c_{2k+i-3}y^{2k+i-3}.$$

Singularities of the class W:

Notation	Normal form	Restrictions	Multiplicity μ	Modality m
W_{12k}	$x^4 + y^{4k+1} + axy^{3k+1} + cx^2y^{2k+1}$	$k \geqslant 1$	$12k$	$3k-2$
W_{12k+1}	$x^4 + xy^{3k+1} + ax^2y^{2k+1} + cy^{4k+2}$	$k \geqslant 1$	$12k+1$	$3k-2$
$W_{k,0}$	$x^4 + bx^2y^{2k+1} + axy^{3k+2} + y^{4k+2}$	$k \geqslant 1,\ b_0^2 \neq 4$	$12k+3$	$3k-1$
$W_{k,i}$	$x^4 + ax^3y^{k+1} + x^2y^{2k+1} + by^{4k+2+i}$	$i > 0,\ b_0 \neq 0$	$12k+3+i$	$3k-1$
$W^{\#}_{k,2q-1}$	$(x^2 + y^{2k+1})^2 + bxy^{3k+1+q} + ay^{4k+2+q}$	$q > 0,\ b_0 \neq 0$	$12k+2+2q$	$3k-1$
$W^{\#}_{k,2q}$	$(x^2 + y^{2k+1})^2 + bx^2y^{2k+1+q} + axy^{3k+2+q}$	$q > 0,\ b_0 \neq 0$	$12k+3+2q$	$3k-1$
W_{12k+5}	$x^4 + xy^{3k+2} + ax^2y^{2k+2} + by^{4k+3}$	$k \geqslant 1$	$12k+5$	$3k-1$
W_{12k+6}	$x^4 + y^{4k+3} + axy^{3k+3} + bx^2y^{2k+2}$	$k \geqslant 1$	$12k+6$	$3k-1$

In these formulae $b = b_0 + \ldots + b_{2k-1}y^{2k-1}$, $c = c_0 + \ldots + c_{2k-2}y^{2k-2}$; as elsewhere, $a = a_0 + \ldots + a_{k-2}y^{k-2}$ for $k > 1$ and $a = 0$ for $k = 1$.

III. SINGULARITIES OF CORANK 3 WITH A GIVEN 3-JET

Besides the unimodal singularities of the series T (see I.1) there are 3 infinite series of classes of such singularities, Q, S and U.

III.1. SERIES Q. Singularities with 3-jet $x^3 + yz^2$ form a single infinite series of classes

$$Q = \leftarrow Q_1 \leftarrow Q_2 \leftarrow \ldots, \text{ where } Q_1 = \leftarrow Q_{10} \leftarrow Q_{11} \leftarrow Q_{12} \leftarrow,$$

$$\underset{k>1}{Q_k} = \leftarrow Q_{k,0} \leftarrow \begin{matrix} Q_{k,1} \\ \uparrow \\ Q_{k,2} \leftarrow Q_{k,3} \leftarrow \ldots \end{matrix} \leftarrow Q_{6k+4} \leftarrow Q_{6k+5} \leftarrow Q_{6k+6} \leftarrow (Q_{k+1}).$$

Notation	Normal form	Restrictions	Multiplicity μ	Modality m
$Q_{k,0}$	$\varphi+\boldsymbol{b}x^2y^k+xy^{2k}$	$k>1,\ b_0^2\neq 4$	$6k+2$	k
$Q_{k,i}$	$\varphi+x^2y^k+\boldsymbol{b}y^{3k+i}$	$k>1,\ b_0\neq 0$	$6k+2+i$	k
Q_{6k+4}	$\varphi+y^{3k+1}+\boldsymbol{b}xy^{2k+1}$	$k\geqslant 1$	$6k+4$	k
Q_{6k+5}	$\varphi+xy^{2k+1}+\boldsymbol{b}y^{3k+2}$	$k\geqslant 1$	$6k+5$	k
Q_{6k+6}	$\varphi+y^{3k+2}+\boldsymbol{b}xy^{2k+2}$	$k\geqslant 1$	$6k+6$	k

In these formulae $\varphi = x^3 + yz^2$, $b = b_0 + \ldots + b_{k-1}y^{k-1}$.

III.2. SERIES S. Singularities with 3-jet $x^2z + yz^2$ form a single infinite series of classes

$$S = \leftarrow S_1 \leftarrow S_2 \leftarrow \ldots,$$

where

$$\begin{array}{ccccccccc} & & & & S_{k,1} \leftarrow \ldots & & & & \\ & & & \swarrow & \nwarrow & & & & \\ \boldsymbol{S}_{\mathbf{k}} = \leftarrow S_{12k-1} \leftarrow S_{12k} & \leftarrow & S_{k,0} & & & S_{12k+4} \leftarrow S_{12k+5} \leftarrow \boldsymbol{S}^*_{\mathbf{k+1}} \leftarrow (\boldsymbol{S}_{\mathbf{k+1}}). \\ & & & \nwarrow & \nearrow & & & & \\ & & & & S^{\#}_{k,1} \leftarrow \ldots & & & & \end{array}$$

Notation	Normal form	Restrictions	Multiplicity μ	Modality m
S_{12k-1}	$\varphi+y^{4k}+\boldsymbol{a}xy^{3k}+\boldsymbol{c}zy^{2k+1}$	—	$12k-1$	$3k-2$
S_{12k}	$\varphi+xy^{3k}+\boldsymbol{c}y^{4k+1}+\boldsymbol{a}zy^{2k+1}$	—	$12k$	$3k-2$
$S_{k,0}$	$\varphi+y^{4k+1}+\boldsymbol{a}xy^{3k+1}+\boldsymbol{b}zy^{2k+1}$	$b_0^2\neq 4$	$12k+2$	$3k-1$
$S_{k,i}$	$\varphi+x^2y^{2k}+\boldsymbol{a}x^3y^k+\boldsymbol{b}y^{4k+1+i}$	$i>0,\ b_0\neq 0$	$12k+2+i$	$3k-1$
$S^{\#}_{k,2q-1}$	$\varphi+zy^{2k+1}+\boldsymbol{b}xy^{3k+q}+\boldsymbol{a}y^{4k+q+1}$	$q>0,\ b_0\neq 0$	$12k+2q+1$	$3k-1$
$S^{\#}_{k,2q}$	$\varphi+zy^{2k+1}+\boldsymbol{b}x^2y^{2k+q}+\boldsymbol{a}xy^{3k+q+1}$	$q>0,\ b_0\neq 0$	$12k+2q+2$	$3k-1$
S_{12k+4}	$\varphi+xy^{3k+1}+\boldsymbol{a}zy^{2k+2}+\boldsymbol{b}y^{4k+2}$	—	$12k+4$	$3k-1$
S_{12k+5}	$\varphi+y^{4k+2}+\boldsymbol{a}xy^{3k+2}+\boldsymbol{b}zy^{2k+2}$	—	$12k+5$	$3k-1$

In these formulae $\varphi = x^2z + yz^2$, $a = a_0 + \ldots + a_{k-2}y^{k-2}$ for $k > 1$, $a = 0$ for $k = 1$; $b = b_0 + \ldots + b_{2k-1}y^{2k-1}$, $c = c_0 + \ldots + c_{2k-2}y^{2k-2}$.

The singularities $S^*_{\mathrm{k}}(k > 1)$ are divided into classes as follows:

$$\begin{array}{ccccc} & & \boldsymbol{SQ}_{\mathbf{k}} & & \\ & \swarrow & & \nwarrow & \\ \boldsymbol{S}^*_{\mathbf{k}} = \leftarrow S^*_{k,0} \leftarrow \boldsymbol{SP}_{\mathbf{k}} & & & & (\boldsymbol{S}_{\mathbf{k}}), \\ & \nwarrow & & \swarrow & \\ & & \boldsymbol{SR}_{\mathbf{k}} & & \end{array}$$

$$\mu(S^*_{k,0}) = 12k-4, \quad m(S^*_{k,0}, \ldots, \boldsymbol{SR}_{\mathbf{k}}) = 3k-2, \quad \operatorname{codim} \boldsymbol{S}^*_{\mathbf{k}} = 9k-3.$$

III. THE SERIES U. The singularities with 3-jet $x^3 + xz^2$ form a single infinite series of classes

$$U = \leftarrow U_1 \leftarrow U_2 \leftarrow \ldots,$$

where

$$\begin{array}{l} U_k = \leftarrow U_{12k} \leftarrow U_{k,0} \leftarrow U_{k,1} \leftarrow U_{12k+4} \leftarrow U^*_{k+1} \leftarrow (U_{k+1}). \\ \qquad\qquad\qquad\qquad\qquad \uparrow \\ \qquad\qquad\qquad\qquad\quad U_{k,2} \leftarrow \ldots \end{array}$$

Notation	Normal form	Restrictions	Multiplicity μ	Modality m
U_{12k}	$\varphi+y^{3k+1}+axy^{2k+1}+bzy^{2k+1}+$ $+dx^2y^{k+1}$	—	$12k$	$4k-3$
$U_{k,2q}$	$\varphi+xy^{2k+1}+ax^2y^{k+1}+by^{3k+2+q}+$ $+czy^{2k+1+q}$	$q\geqslant 0,\ c_0\neq 0$	$12k+2+2q$	$4k-2$
$U_{k,2q-1}$	$\varphi+xy^{2k+1}+ax^2y^{k+1}+$ $+bzy^{2k+1+q}+cz^2y^{k+q}$	$q>0,\ c_0\neq 0$	$12k+1+2q$	$4k-2$
U_{12k+4}	$\varphi+y^{3k+2}+axy^{2k+2}+bzy^{2k+2}+$ $+cx^2y^{k+1}$	—	$12k+4$	$4k-2$

In these formulae $\varphi = x^3 + xz^2$; $c_0^2 + 1 \neq 0$ for $q = 0$; everywhere,

$$a = a_0 + \ldots + a_{k-2}y^{k-2} \quad \text{for} \quad k>1, \quad a=0 \quad \text{for} \quad k=1;$$

$$b = b_0 + \ldots + b_{k-2}y^{k-2} \quad \text{for} \quad k>1, \quad b=0 \quad \text{for} \quad k=1;$$

$$c = c_0 + \ldots + c_{2k-1}y^{2k-1}, \quad d = d_0 + \ldots + d_{2k-2}y^{2k-2}.$$

The singularities U_k^* $(k > 1)$ are divided into classes:

$$\begin{array}{l} U^*_k = \leftarrow U^*_{k,0} \leftarrow UP_k \leftarrow UR_k \leftarrow UT_k \\ \qquad\qquad\qquad\qquad \uparrow \qquad\quad \uparrow \qquad\quad \uparrow \\ \qquad\qquad\qquad\quad UQ_k \leftarrow US_k \leftarrow (U_k), \end{array}$$

$$\mu(U^*_{k,0}) = 12k-4, \quad m(U^*_{k,0}, \ldots, UT_k) = 4k-3,$$

$$\operatorname{codim}(U^*_k) = 8k-2.$$

IV. THE SERIES V.

The singularities of corank 3 with 3-jet x^2y are divided into classes

$$\begin{array}{ccc} & V_{1,1} \leftarrow V_{1,2} \leftarrow \ldots & \\ & \swarrow \quad \nwarrow & \\ V = V_{1,0} & & V^*, \\ & \nwarrow \quad \swarrow & \\ & V^\#_{1,1} \leftarrow V^\#_{1,2} \leftarrow \ldots & \end{array},$$

where

Notation	Normal form	Restrictions	Multiplicity μ	Modality m
$V_{1,0}$	$x^2y+z^4+az^3y+bz^2y^2+zy^3$	$\Delta(a, b_0)\neq 0$	15	3
$V_{1,p}$	$x^2y+z^4+bz^3y+z^2y^2+ay^{4+p}$	$b^2\neq 4,\ a_0\neq 0$	$15+p$	3
$V^{\#}_{1,2q-1}$	$x^2y+z^3y+ay^2z^2+y^4+bxz^{2+q}$	$4a^3+27\neq 0$, $b_0\neq 0$	$14+2q$	3
$V^{\#}_{1,2q}$	$x^2y+z^3y+ay^2z^2+y^4+bz^{4+q}$	$4a^3+27\neq 0$, $b_0\neq 0$	$15+2q$	3

Here $p > 0$, $q > 0$, $a = a_0 + a_1y$, $b = b_0 + b_1z$. The singularities of the class V^* satisfy the conditions $\mu(V^*) \geqslant 17$, $m(V^*) \geqslant 3$, $c(V^*) = 13$.

V. OTHER SINGULARITIES

All the singularities whose normal forms are not given above belong to the following seven classes:

Notation	Corank	Adjacencies	Determination	$c \geqslant$	$\mu \geqslant$	$m \geqslant$	Theorem
N	2	$N \to W_{13}$	$j_4=0$	12	16	3	47—49
S^*	3	$S^*_k \to S_{12k-7}$	see Theorem 77	15	20	4	77—81
U^*	3	$U^*_k \to U_{12k-8}$	see Theorem 90	14	20	5	90—96
V^*	3	$V^* \to V_{1,1} \cup V^{\#}_{1,1}$	see Theorem 98	13	17	3	97—102
V'	3	$V' \to V$	$j_3=x^3$	13	18	4	103
V''	3	$V'' \to V'$	$j_3=0$	16	27	10	104
O	>3	$O \to T_{4,4,4}$	corank $\geqslant 4$	10	16	5	105

Here $k \geqslant 2$. The normal form for singularities of the class O, excepting a set of codimension $c = 11$, is indicated in [7].

§ 14. Determination of singularities

To shorten the statements of the theorems in § 14 we use the following *notation*:

f is the germ of a holomorphic function at an *isolated* critical point O of multiplicity μ, or its Taylor series at O, or a formal series in the variables x, y or x, y, z having finite μ.

$f \sim g$ means that the germs f and g at O are equivalent (that is, there exists a germ of a diffeomorphism or a formal series h such that $f = g\circ h$).

$\Rightarrow$ means "implies".

$\mapsto$ means "see" (references of the form $\mapsto i$ are not parts of the statements of the theorems; they indicate the number of the theorem where the singularities of that class are classified).

j_kf means the k-jet of f at O (or the Taylor polynomial of order k at O).

$A, \ldots, Z$ are the stable equivalence classes of germs of functions that are defined in §13.

$m(f)$ is the modality of the germ f at O.

$c(f)$ is the codimension of the stratum μ = const of the germ of f in the space of germs of functions with critical point O and critical value 0.

$c(\mathbf{K})$ is the codimension of the class $\mathbf{K}$ in that space.

$j_{\{x^{m_i}\}}f$ is the quasijet of f at O, defined by the monomials x^{m_i} (or the corresponding Taylor polynomial).

$j_{\{x^{m_i}\}}f \approx g$ denotes quasihomogeneous equivalence of jets or Taylor polynomials.

j^*, φ – the meaning of these symbols for Theorems **58–65**, **66–81**, **82–89**, **98–102** is explained before the first theorem of each group.

Δ is the discriminant. In Theorems **36**, **37**, **47**, **48**, **98**, and **99**, $\Delta = 4(a^3 + b^3) + 27 - a^2b^2 - 18ab$.

Clarifications. A system of n monomials $\{x^{m_i}\}$ in $x_1, \ldots, x_n$ with independent exponents $m_i \in \mathbf{Z}^n \subset \mathbf{R}^n$ defines a hyperplane $\Gamma \subset \mathbf{R}^n$, $\Gamma = \{m\colon (\alpha, m) = 1\}$. If all the components α_i of the vector α are positive, then α is called the *quasihomogeneity type*, and the number (α, m) is the *order* of the monomial x^m. The polynomial $f = \Sigma f_m x^m$ is quasihomogeneous of degree d and type α if $(\alpha, m) = d \, \forall \, m{:}f_m \neq 0$.

The quasihomogeneity type defines a grading and a decreasing ring filtration $\mathscr{E}_0 \supset \ldots$, where

$$\mathscr{E}_d = \{f\colon (\alpha, m) \geqslant d \, \forall m\colon f_m \neq 0\}.$$

The factor-space $\mathscr{E}_0/\bigcup \mathscr{E}_d$, $d > 1$, is called a *space of quasijets* defined by the monomials $\{x^{m_i}\}$ (or defined by the quasihomogeneity type α). For a fixed coordinate system quasijets can be identified with polynomials whose monomials are all of order at most 1 (that is, whose exponents lie on Γ or on the same side of Γ as O).

Quasihomogeneous diffeomorphisms are diffeomorphisms of $\mathbf{C}^n$ that preserve the grading of the ring $\mathbf{C}[[x_1, \ldots, x_n]]$. The Lie group of quasihomogeneous diffeomorphisms acts on the spaces of quasijets and on the spaces of quasihomogeneous polynomials. *Quasihomogeneous equivalence* refers to membership in a single orbit of this action.

DETERMINATION

1. $\mu(f) < \infty \Rightarrow$ one of four cases:

$$\begin{aligned} \text{corank } f \leqslant 1 &\models\!\Rightarrow \mathbf{2};\\ = 2 &\models\!\Rightarrow \mathbf{3};\\ = 3 &\models\!\Rightarrow \mathbf{50};\\ > 3 &\models\!\Rightarrow \mathbf{105}.\end{aligned}$$

2. corank $f \leqslant 1 \Rightarrow f \in A_k$ $(k \geqslant 1)$.

In Theorems 3–49, $f \in \mathbf{C}[[x, y]]$.

3. $j_2 f = 0 \Rightarrow$ one of four cases:

$$\begin{array}{ll} j_3 f \approx x^2y+y^3 & \models\Rightarrow \mathbf{4}; \\ \approx x^2y & \models\Rightarrow \mathbf{5}; \\ \approx x^3 & \models\Rightarrow \mathbf{6_1}; \\ =0 & \models\Rightarrow \mathbf{13}. \end{array}$$

4. $j_3 f = x^2y + y^3 \Rightarrow f \in D_4$.

5. $j_3 f = x^2y \Rightarrow f \in D_k$ $(k > 4)$.

In Theorems 6–9 the number $k \geqslant 1$.

6_k. $j_{x^3, y^{3k}} f(x, y) = x^3 \Rightarrow$ one of four cases:

$$\begin{array}{ll} j_{x^3,\, y^{3k+1}} f \approx x^3+y^{3k+1} & \models\Rightarrow \mathbf{7_k}; \\ j_{x^3,\, xy^{2k+1}} f \approx x^3+xy^{2k+1} & \models\Rightarrow \mathbf{8_k}; \\ j_{x^3,\, y^{3k+2}} f \approx x^3+y^{3k+2} & \models\Rightarrow \mathbf{9_k}; \\ j_{x^3,\, y^{3k+2}} f = x^3 & \models\Rightarrow \mathbf{10_{k+1}}. \end{array}$$

7_k. $j_{x^3, y^{3k+1}} f = x^3 + y^{3k+1} \Rightarrow f \in E_{6k}$.

8_k. $j_{x^3, y^{2k+1}} f = x^3 + xy^{2k+1} \Rightarrow f \in E_{6k+1}$.

9_k. $j_{x^3, y^{3k+2}} f = x^3 + y^{3k+2} \Rightarrow f \in E_{6k+2}$.

In Theorems 10–12, $k > 1$.

10_k. $j_{x^3, y^{3k-1}} f = x^3 \Rightarrow$ one of three cases:

$$\begin{array}{ll} j_{x^3,\, y^{3k}} f \approx x^3+ax^2y^k+y^{3k},\ 4a^3+27\neq 0 & \models\Rightarrow \mathbf{11_k}; \\ \approx x^3+x^2y^k & \models\Rightarrow \mathbf{12_k}; \\ \approx x^3 & \models\Rightarrow \mathbf{6_k}. \end{array}$$

11_k. $j_{x^3, y^{3k}} f = x^3 + ax^2y^k + y^{3k}$, $4a^3 + 27 \neq 0 \Rightarrow f \in J_{k,0}$.

12_k. $j_{x^3, y^{3k}} f = x^3 + x^2y^k \Rightarrow f \in J_{k,p}$ $(p > 0)$.

The series X_1

13. $j_3 f(x, y) = 0 \Rightarrow$ one of six cases:

$$\begin{array}{ll} j_4 \approx x^4+ax^2y^2+y^4,\ a^2\neq 4 & \models\Rightarrow \mathbf{14}; \\ \approx x^4+x^2y^2 & \models\Rightarrow \mathbf{15}; \\ \approx x^2y^2 & \models\Rightarrow \mathbf{16}; \\ \approx x^3y & \models\Rightarrow \mathbf{17}; \\ \approx x^4 & \models\Rightarrow \mathbf{25}; \\ =0 & \models\Rightarrow \mathbf{47}. \end{array}$$

14. $j_4 f = x^4 + ax^2y^2 + y^4$, $a^2 \neq 4 \Rightarrow f \in X_9 = X_{1,0} = T_{2,4,4}$.

15. $j_4 f = x^4 + x^2y^2 \Rightarrow f \in X_{1,p} = T_{2,4,4+p}$ $(p > 0)$.

16. $j_4 f = x^2y^2 \Rightarrow f \in Y^1_{p,q} = T_{2,4+p,4+q}$ $(p \geqslant q > 0)$.

17. $j_4 f = x^3y \Rightarrow j_{x^3y, y^4} f = x^3y \Rightarrow 18_1$.

In Theorems 18–21, $p \geqslant 1$.

18_p. $j_{x^3y, y^{3p+1}} f = x^3y \Rightarrow$ one of four cases:

$$\begin{aligned}
&j_{x^3y,\, y^{3p+2}}f \approx x^3y + y^{3p+2} &&\models 19_p;\\
&j_{x^3y,\, xy^{2p+2}}f \approx x^3y + xy^{2p+2} &&\models 20_p;\\
&j_{x^3y,\, y^{3p+3}}f \approx x^3y + y^{3p+3} &&\models 21_p;\\
&j_{x^3y,\, y^{3p+3}}f = x^3y &&\models 22_{p+1}.
\end{aligned}$$

19_p. $j_{x^3y, y^{3p+2}} f = x^3y + y^{3p+2} \Rightarrow f \in Z_{6p+5}$.

20_p. $j_{x^3y, xy^{2p+2}} f = x^3y + xy^{2p+2} \Rightarrow f \in Z_{6p+6}$.

21_p. $j_{x^3y, y^{3p+3}} f = x^3y + y^{3p+3} \Rightarrow f \in Z_{6p+7}$.

In Theorems 22–24, $p > 1$.

22_p. $j_{x^3y, y^{3p}} f = x^3y \Rightarrow$ one of three cases:

$$\begin{aligned}
j_{x^3y,\, y^{3p+1}}f &\approx y\,(x^3 + bx^2y^p + y^{3p}), \quad 4b^3 + 27 \neq 0 &&\models 23_p;\\
&\approx y\,(x^3 + x^2y^p) &&\models 24_p;\\
&\approx x^3y &&\models 18_p.
\end{aligned}$$

23_p. $j_{x^3y, y^{3p+1}} f = y(x^3 + bx^2y^p + y^{3p}),\ 4b^3 + 27 \neq 0 \Rightarrow f \in Z_{p-1,0}$.

24_p. $j_{x^3y, y^{3p+1}} f = y(x^3 + x^2y^p) \Rightarrow f \in Z_{p-1,r} (r > 0)$.

The series W.

25. $j_4 f(x, y) = x^4 \Rightarrow j_{x^4, y^4} f = x^4 \Rightarrow 26_1$.

In Theorems 26–35, $k \geqslant 1$.

26_k. $j_{x^4, y^{4k}} f = x^4 \Rightarrow$ one of three cases:

$$\begin{aligned}
&j_{x^4,\, y^{4k+1}}f \approx x^4 + y^{4k+1} &&\models 27_k;\\
&j_{x^4,\, xy^{3k+1}}f \approx x^4 + xy^{3k+1} &&\models 28_k;\\
&j_{x^4,\, xy^{3k+1}}f = x^4 &&\models 29_k.
\end{aligned}$$

27_k. $j_{x^4, y^{4k+1}} f = x^4 + y^{4k+1} \Rightarrow f \in W_{12k}$.

28_k. $j_{x^4, xy^{3k+1}} f = x^4 + xy^{3k+1} \Rightarrow f \in W_{12k+1}$.

29_k. $j_{x^4, xy^{3k+1}} f = x^4 \Rightarrow$ one of four cases:

$$\begin{aligned}
j_{x^4,\, y^{4k+2}}f &\approx x^4 + bx^2y^{2k+1} + y^{4k+2}, \quad b^2 \neq 4 &&\models 30_k;\\
&\approx x^4 + x^2y^{2k+1} &&\models 31_k;\\
&\approx (x^2 + y^{2k+1})^2 &&\models 32_k;\\
&= x^4 &&\models 33_k.
\end{aligned}$$

30_k. $j_{x^4, y^{4k+2}} f = x^4 + bx^2y^{2k+1} + y^{4k+2},\ b^2 \neq 4 \Rightarrow f \in W_{k,0}$.

31_k. $j_{x^4, y^{4k+2}} f = x^4 + x^2y^{2k+1} \Rightarrow f \in W_{k,i}\ (i > 0)$.

32_k. $j_{x^4, y^{4k+2}} f = (x^2 + y^{2k+1})^2 \Rightarrow f \in W_{k,i}\ (i > 0)$.

33_k. $j_{x^4, y^{4k+2}} f = x^4 \Rightarrow$ one of three cases:

$$\begin{aligned}
&j_{x^4,\, xy^{3k+2}}f \approx x^4 + xy^{3k+2} &&\models 34_k;\\
&j_{x^4,\, y^{4k+3}}f \approx x^4 + y^{4k+3} &&\models 35_k;\\
&j_{x^4,\, y^{4k+3}}f = x^4 &&\models 36_{k+1}.
\end{aligned}$$

34_k. $j_{x^4, xy^{3k+2}} f = x^4 + xy^{3k+2} \Rightarrow f \in W_{12k+5}$.

35_k. $j_{x^4, y^{4k+3}} f = x^4 + y^{4k+3} \Rightarrow f \in W_{12k+6}$.

In Theorems **36**–**46**, $k > 1$.

The series X_k

36$_k$. $j_{x^4, y^{4k-1}} f = x^4 \Rightarrow$ one of five cases:

$$j_{x^4, y^{4k}} f \approx x^4 + bx^3y^k + ax^2y^{2k} + xy^{3k}, \ \Delta \neq 0, \ ab \neq 9 \models\Rightarrow \mathbf{37_k};$$
$$\approx x^2(x^2 + axy^k + y^{2k}), \ a^2 \neq 4 \models\Rightarrow \mathbf{38_k};$$
$$\approx x^2(x + y^k)^2 \models\Rightarrow \mathbf{39_k};$$
$$\approx x^3(x + y^k) \models\Rightarrow \mathbf{40_k};$$
$$\approx x^4 \models\Rightarrow \mathbf{26_k}.$$

37$_k$. $j_{x^4, y^{4k}} f = x^4 + bx^3y^k + ax^2y^{2k} + xy^{3k}, \ \Delta \neq 0 \Rightarrow f \in X_{k,0}$.

38$_k$. $j_{x^4, y^{4k}} f = x^2(x^2 + axy^k + y^{2k}), a^2 \neq 4 \Rightarrow f \in X_{k,p}$ $(p > 0)$.

39$_k$. $j_{x^4, y^{4k}} f = x^2(x^2 + y^k)^2 \Rightarrow f \in Y^k_{r,s}$ $(1 \leqslant s \leqslant r)$

40$_k$. $j_{x^4, y^{4k}} f = (x + y^k)x^3 \Rightarrow f \sim f_1 f_2$, where

$$j_{x, y^k} f_1 \approx x + y^k, \quad j_{x^3, y^{3k}} f_2 = x^3 \models\Rightarrow \mathbf{41_k}.$$

In Theorems **41**–**44**, $i \geqslant 0, p > 0$.

41$_k$. $j_{x^3, y^{3k}} f_2 = x^3 \Rightarrow$ one of five cases:

$$f_2 \in E_{6(k+i)} \models\Rightarrow \mathbf{42_{k,\,i}};$$
$$f_2 \in E_{6(k+i)+1} \models\Rightarrow \mathbf{43_{k,\,i}};$$
$$f_2 \in E_{6(k+i)+2} \models\Rightarrow \mathbf{44_{k,\,i}};$$
$$f_2 \in J_{k+i+1,\,0} \models\Rightarrow \mathbf{45_{k,\,i+1}};$$
$$f_2 \in J_{k+i+1,\,p} \models\Rightarrow \mathbf{46_{k,\,i+1,\,p}}.$$

In Theorems **42**–**46**, $f(x, y) = f_1 f_2$, where $j_{x,y^k} f_1 \approx x + y^k$ and $j_{x^3, y^{3k}} f_2 = x^3$.

42$_{k,i}$. $f_2 \in E_{6(k+i)} \Rightarrow f \in Z^k_{12k+6i-1}$.

43$_{k,i}$. $f_2 \in E_{6(k+i)+1} \Rightarrow f \in Z^k_{12k+6i}$.

44$_{k,i}$. $f_2 \in E_{6(k+i)+2} \Rightarrow f \in Z^k_{12k+6i+1}$.

In Theorems **45**–**46**, $i \geqslant 1$, $p > 0$.

45$_{k,i}$. $f_2 \in J_{k+i,0} \Rightarrow f \in Z^k_{i,0}$.

46$_{k,i,p}$. $f_2 \in J_{k+i,p} \Rightarrow f \in Z^k_{i,p}$.

47. $j_4 f = 0 \Rightarrow$ one of two cases:

$$j_5 f \approx x^4y + ax^3y^2 + bx^2y^3 + xy^4, \ \Delta \neq 0, \ ab \neq 9 \models\Rightarrow \mathbf{48};$$
$$j_5 f \text{ is degenerate } \models\Rightarrow \mathbf{49}.$$

48. $j_5 f = x^4y + ax^3y^2 + bx^2y^3 + xy^4, \Delta \neq 0 \Rightarrow f \in N_{16}$, i.e. $f \sim x^4y + ax^3y^2 + bx^2y^3 + xy^4 + cx^3y^3, \Delta \neq 0, ab \neq 9; \mu(f) = 16, m(f) = 3, c(f) = 12$.

49. $j_5 f$ is degenerate $\Rightarrow \mu(f) > 16, m(f) > 2, c(f) > 12$.

SINGULARITIES OF CORANK 3.

In Theorems **50**–**104**, $f \in \mathbf{C}[[x, y, z]]$.

50. $j_2 f(x, y, z) = 0 \Rightarrow$ one of ten cases:

$$\begin{aligned} j_3f &\approx x^3+y^3+z^3+axyz, \quad a^3+27\neq 0 \models\Rightarrow 51;\\ &\approx x^3+y^3+xyz \models\Rightarrow 52 \text{ (series } \mathbf{P});\\ &\approx x^3+xyz \models\Rightarrow 54 \text{ (series } \mathbf{R});\\ &\approx xyz \models\Rightarrow 56 \text{ (series } \mathbf{T});\\ &\approx x^3+yz^2 \models\Rightarrow 58 \text{ (series } \mathbf{Q});\\ &\approx x^2z+yz^2 \models\Rightarrow 66 \text{ (series } \mathbf{S});\\ &\approx x^3+xz^2 \models\Rightarrow 82 \text{ (series } U);\\ &\approx x^2y \models\Rightarrow 97 \text{ (class } V);\\ &\approx x^3 \models\Rightarrow 103;\\ &=0 \models\Rightarrow 104. \end{aligned}$$

Series T

51. $j_3f=x^3+y^3+z^3+axyz,\ a^3+27\neq 0 \Rightarrow f\in P_8=T_{3,3,3}$.

52. $j_3f=x^3+y^3+xyz \Rightarrow f\sim x^3+y^3+xyz+\alpha(z),\ j_3(\alpha)=0 \mapsto 53$.

53. $f=x^3+y^3+xyz+\alpha(z),\ j_3(\alpha)=0 \Rightarrow f\in P_{p+5}=T_{3,3,3+p}\ (p>3)$.

54. $j_3f=x^3+xyz \Rightarrow f\sim x^3+xyz+\alpha(y)+\beta(z),\ j_3(\alpha,\beta)=0 \mapsto 55$.

55. $f=x^3+xyz+\alpha(y)+\beta(z),\ j_3(\alpha,\beta)=0 \Rightarrow f\in R_{p,q}=T_{3,p,q}\ (q\geqslant p>3)$.

56. $j_3f=xyz \Rightarrow f\sim xyz+\alpha(x)+\beta(y)+\gamma(z),\ j_3(\alpha,\beta,\gamma)=0 \mapsto 57$.

57. $f=xyz+\alpha(x)+\beta(y)+\gamma(z),\ j_3(\alpha,\beta,\gamma)=0 \Rightarrow f\in T_{p,q,r}\ (r\geqslant q\geqslant p>3)$.

Series Q. In Theorems 58–65, $\varphi = x^3+yz^2,\ j^*_\lambda=j_{yz^2,x^3,\lambda}$ (λ is a monomial).

58. $j_3 f=\varphi \Rightarrow f=\varphi+\alpha(y)+x\beta(y),\ j_3(\alpha,x\beta)=0 \mapsto 59_1$.

In Theorems 59–62, $k\geqslant 1$.

59_k. $f=\varphi+\alpha(y)+x\beta(y),\ j^*_{y^{3k}}f=\varphi \Rightarrow$ one of four cases:

$$\begin{aligned} j^*_{y^{3k+1}}f &\approx \varphi+y^{3k+1} \models\Rightarrow 60_k;\\ j^*_{xy^{2k+1}}f &\approx \varphi+xy^{2k+1} \models\Rightarrow 61_k;\\ j^*_{y^{3k+2}}f &\approx \varphi+y^{3k+2} \models\Rightarrow 62_k;\\ j^*_{y^{3k+2}}f &= \varphi \models\Rightarrow 63_{k+1}. \end{aligned}$$

60_k. $j^*_{y^{3k+1}}f=\varphi+y^{3k+1} \Rightarrow f\in Q_{6k+4}$.

61_k. $j^*_{xy^{2k+1}}f=\varphi+xy^{2k+1} \Rightarrow f\in Q_{6k+5}$.

62_k. $j^*_{y^{3k+2}}f=\varphi+y^{3k+2} \Rightarrow f\in Q_{6k+6}$.

In Theorems 63–65, $k>1$.

63_k. $f=\varphi+\alpha(y)+\beta(y),\ j^*_{y^{3k-1}}f=\varphi \Rightarrow$ one of three cases:

$$\begin{aligned} j^*_{y^{3k}}f &\approx \varphi+ax^2y^k+xy^{2k},\quad a^2\neq 4 \models\Rightarrow 64_k;\\ &\approx \varphi+x^2y^k \models\Rightarrow 65_k;\\ &=\varphi \models\Rightarrow 59_k. \end{aligned}$$

64_k. $j^*_{y^{3k}}f=\varphi+ax^2y^k+xy^{2k},\ a^2\neq 4 \Rightarrow f\in Q_{k,0}$.

65_k. $j^*_{y^{3k}}f=\varphi+x^2y^k \Rightarrow f\in Q_{k,i}\ (i>0)$.

Series S. In Theorems 68–81, $\varphi=x^2z+yz^2,\ j^*_\lambda=j_{x^2z,yz^2,\lambda}$ (λ is a monomial).

66. $j_3 f = \varphi \Rightarrow f \sim \varphi + \alpha(y) + x\beta(y) + z\gamma(y), j_3(\alpha, x\beta, z\gamma) = 0 \Rightarrow$ **67₁**.

In Theorems **67**–**76**, $k \geqslant 1$.

$\mathbf{67_k}$. $f = \varphi + \alpha(y) + x\beta(y) + z\gamma(y), j^*_{y^{4k-1}} f = \varphi \Rightarrow$ one of 3 cases:

$$j^*_{y^{4k}} f \approx \varphi + y^{4k} \quad |\Rightarrow \mathbf{68_k};$$
$$j^*_{xy^{3k}} f \approx \varphi + xy^{3k} \quad |\Rightarrow \mathbf{69_k};$$
$$j^*_{xy^{3k}} f = \varphi \quad |\Rightarrow \mathbf{70_k}.$$

$\mathbf{68_k}$. $j^*_{y^{4k}} f = \varphi + y^{4k} \Rightarrow f \in S_{12k-1}$.

$\mathbf{69_k}$. $j^*_{xy^{3k}} f = \varphi + xy^{3k} \Rightarrow f \in S_{12k}$.

$\mathbf{70_k}$. $f = \varphi + \alpha(y) + x\beta(y) + z\gamma(y), j^*_{xy^{3k}} f = \varphi \Rightarrow$ one of four cases:

$$j^*_{y^{4k+1}} f \approx \varphi + y^{4k+1} + bzy^{2k+1}, \quad b^2 \neq 4 \; |\Rightarrow \mathbf{71_k};$$
$$\approx \varphi + x^2y^{2k} \quad |\Rightarrow \mathbf{72_k};$$
$$\approx \varphi + zy^{2k+1} \quad |\Rightarrow \mathbf{73_k};$$
$$= \varphi \quad |\Rightarrow \mathbf{74_k}.$$

$\mathbf{71_k}$. $j^*_{y^{4k+1}} f = \varphi + y^{4k+1} + bzy^{2k+1}, b^2 \neq 4 \Rightarrow f \in S_{k,0}$.

$\mathbf{72_k}$. $j^*_{y^{4k+1}} f = \varphi + x^2y^{2k} \Rightarrow f \in S_{k,i}$ $(i > 0)$.

$\mathbf{73_k}$. $j^*_{y^{4k+1}} f = \varphi + zy^{2k+1} \Rightarrow f \in S^{\#}_{k,i}$ $(i > 0)$.

$\mathbf{74_k}$. $f = \varphi + \alpha(y) + x\beta(y) + z\gamma(y), j^*_{y^{4k+1}} f = \varphi \Rightarrow$ one of three cases:

$$j^*_{xy^{3k+1}} f = \varphi + xy^{3k+1} \quad |\Rightarrow \mathbf{75_k};$$
$$j^*_{y^{4k+2}} f = \varphi + y^{4k+2} \quad |\Rightarrow \mathbf{76_k};$$
$$j^*_{y^{4k+2}} f = \varphi \quad |\Rightarrow \mathbf{77_{k+1}}.$$

$\mathbf{75_k}$. $j^*_{xy^{3k+1}} f = \varphi + xy^{3k+1} \Rightarrow f \in S_{12k+4}$.

$\mathbf{76_k}$. $j^*_{y^{4k+2}} f = \varphi + y^{4k+2} \Rightarrow f \in S_{12k+5}$.

In Theorems 77–**81**, $k > 1$.

$\mathbf{77_k}$. $f = \varphi + \alpha(y) + x\beta(y) + z\gamma(y), j^*_{y^{4k+2}} f = \varphi \Rightarrow$ one of five cases:

$$j^*_{y^{4k-1}} f \approx \varphi + ax^2y^{k-1} + bxy^kz + xy^{3k-1}, \quad \Delta \neq 0 \; |\Rightarrow \mathbf{78_k};$$
$$\approx \varphi + xy^kz + ax^3y^{k-1}, \quad a^2 \neq a \; |\Rightarrow \mathbf{79_k};$$
$$\approx \varphi + x^3y^{k-1} \quad |\Rightarrow \mathbf{80_k};$$
$$\approx \varphi + xy^kz \quad |\Rightarrow \mathbf{81_k};$$
$$= \varphi \quad |\Rightarrow \mathbf{67_k}.$$

$\mathbf{78_k}$. $j^*_{y^{4k-1}} f = \varphi + ax^2y^{k-1} + bxy^kz + xy^{3k-1}, \Delta \neq 0 \Rightarrow f \in S^*_{k,0}$; $\mu(f) = 12k-4$, $m(f) = 3k-2$, $c(S^*_{k,0}) = 9k-3$.

$\mathbf{79_k}$. $j^*_{y^{4k-1}} f = \varphi + xy^kz + ax^3y^{k-1}, a^2 \neq a \Rightarrow f \in \boldsymbol{SP}_{\mathbf{k}}$; $\mu(f) \geqslant 12k-3$, $m(f) = 3k-2$, $c(\boldsymbol{SP}_{\mathbf{k}}) = 9k-2$.

$\mathbf{80_k}$. $j^*_{y^{4k-1}} f = \varphi + x^3y^{k-1} \Rightarrow f \in \boldsymbol{SQ}_{\mathbf{k}}$; $\mu(f) \geqslant 12k-2$, $m(f) = 3k-2$, $c(\boldsymbol{SQ}_{\mathbf{k}}) = 9k-1$.

$\mathbf{81_k}$. $j^*_{y^{4k-1}} f = \varphi + xy^kz \Rightarrow f \in \boldsymbol{SR}_{\mathbf{k}}$; $\mu(f) \geqslant 12k-2$, $m(f) = 3k-2$, $c(\boldsymbol{SR}_{\mathbf{k}}) = 9k-1$.

The *series U.*

In Theorems 82–89, $\varphi = x^3 + xz^2$, $j_\lambda^* = j_{x^3z^3,\lambda}$ (λ is a monomial).

82. $j_3 f = \varphi \Rightarrow f \sim \varphi + \alpha(y) + x\beta(y) + z\gamma(y) + x^2\delta(y)$, $j_3(\alpha, x\beta, z\gamma, x^2\delta) = 0$ $\Rightarrow$ **83_1**.

In Theorems 83–89, $k \geqslant 1$.

83_k. $f = \varphi + \alpha(y) + x\beta(y) + z\gamma(y) + x^2\delta(y)$, $j^*_{y^{3k}} f = \varphi \Rightarrow$ one of 2 cases:

$$j^*_{y^{3k+1}}f \approx \varphi + y^{3k+1} \models\Rightarrow 84_k,$$
$$= \varphi \models\Rightarrow 85_k.$$

84_k. $j^*_{y^{3k+1}} f = \varphi + y^{3k+1} \Rightarrow f \in U_{12k}$.

85_k. $f = \varphi + \alpha(y) + x\beta(y) + z\gamma(y) + x^2\delta(y)$, $j^*_{y^{3k+1}}f = \varphi \Rightarrow$ one of 3 cases:

$$j^*_{xy^{2k+1}}f \approx \varphi + xy^{2k+1} + czy^{2k+1}, \quad c(c^2+1) \neq 0 \models\Rightarrow 86_k;$$
$$\approx \varphi + xy^{2k+1} \models\Rightarrow 87_k;$$
$$= \varphi \models\Rightarrow 88_k.$$

86_k. $j^*_{xy^{2k+1}}f = \varphi + xy^{2k+1} + czy^{2k+1}$, $c(c^2+1) \neq 0 \Rightarrow f \in U_{k,0}$.

87_k. $j^*_{xy^{2k+1}}f = \varphi + xy^{2k+1} \Rightarrow f \in U_{k,p}$ $(p > 0)$.

88_k. $f = \varphi + \alpha(y) + x\beta(y) + z\gamma(y) + x^2\delta(y)$, $j^*_{xy^{2k+1}}f = \varphi \Rightarrow$ one of 2 cases:

$$j^*_{y^{3k+2}}f \approx \varphi + y^{3k+2} \models\Rightarrow 89_k;$$
$$= \varphi \models\Rightarrow 90_{k+1}.$$

89_k. $j^*_{y^{3k+2}}f = \varphi + y^{3k+2} \Rightarrow f \in U_{12k+4}$.

In Theorems 90–96, $k \geqslant 2$; $\varphi = x^2z + xz^2$, $j^*_\lambda = j_{x^3,z^3,\lambda}$ (λ is a monomial).

90_k. $f = x^3 + xz^2 + \alpha(y) + x\beta(y) + z\gamma(y) + x^2\delta(y)$, $j^*_{y^{3k-1}}f = x^3 + xz^2 \Rightarrow$ one of seven cases:

$$j^*_{y^{3k}}f \approx \varphi + ax^2y^k + bxy^kz + y^kz^2 + cxy^{2k}, \quad \Delta \neq 0 \models\Rightarrow 91_k;$$
$$\approx \varphi + ax^2y^k + bxy^kz + y^kz^2, 4a \neq b^2, \quad a(a+1-b) \neq 0 \models\Rightarrow 92_k;$$
$$\approx x^3 + ax^2z + xz^2 + y^kz^2, \quad a^2 \neq 4 \models\Rightarrow 93_k;$$
$$\approx \varphi + x^2y^k + axy^kz, \quad a^2 \neq a \models\Rightarrow 94_k;$$
$$\approx \varphi + x^2y^k \models\Rightarrow 95_k;$$
$$\approx \varphi + xy^kz \models\Rightarrow 96_k;$$
$$= \varphi \models\Rightarrow 83_k.$$

91_k. $j^*_{y^{3k}}f = \varphi + ax^2y^k + bxy^kz + y^kz^2 + cxy^{2k}$, $\Delta \neq 0 \Rightarrow f \in U^*_{k,0}$; $\mu(f) = 12k - 4$, $m(f) = 4k - 3$, $c(U^*_{k,0}) = 8k - 2$.

92_k. $j^*_{y^{3k}}f = \varphi + ax^2y^k + bxy^kz + y^kz^2$, $4a \neq b^2$, $a(a+1-b) \neq 0 \Rightarrow f \in \boldsymbol{UP}_k$; $\mu(f) \geqslant 12k - 3$, $m(f) = 4k - 3$, $c(\boldsymbol{UP}_k) = 8k - 1$.

93_k. $j^*_{y^{3k}}f = x^3 + ax^2z + xz^2 + y^kz^2$, $a^2 \neq 4 \Rightarrow f \in \boldsymbol{UQ}_k$; $\mu(f) \geqslant 12k - 2$, $m(f) = 4k - 3$, $c(\boldsymbol{UQ}_k) = 8k$.

94_k. $j^*_{y^{3k}}f = \varphi + x^2y^k + axy^kz$, $a^2 \neq a \Rightarrow f \in \boldsymbol{UR}_k$; $\mu(f) \geqslant 12k - 2$, $m(f) = 4k - 3$, $c(\boldsymbol{UR}_k) = 8k$.

95_k. $j^*_{y^{3k}}f = \varphi + x^2y^k \Rightarrow f \in \boldsymbol{US}_k$, $\mu(f) \geqslant 12k - 2$, $m(f) = 4k - 3$, $c(\boldsymbol{US}_k) = 8k + 1$.

96$_k$. $j^*_{y^{3k}} f = \varphi + xy^k z \Rightarrow f \in UT_k$, $\mu(f) \geqslant 12k - 1$, $m(f) = 4k - 3$, $c(UT_k) = 8k + 1$.

The *Class V*.

97. $j_3 f(x, y, z) = x^2 y \Rightarrow f \sim x^2 y + \alpha(y, z) + x\beta(z) \Rightarrow$ **98.**

In Theorems **98** and **102**, φ is one of the following 10 polynomials

$$z^4 + z^3 y,\ z^3 y + z^2 y^2,\ z^2 y^2 + z y^3,\ z^4 + z^2 y^2,\ z^4,\ z^3 y,\ z^2 y^2,\ z y^3,\ y^4,\ 0.$$

98. $f = x^2 y + \alpha(y, x) + x\beta(z)$, $j_3 f = x^2 y \Rightarrow$ one of four cases:

$$j_{x^2y,\, y^4,\, z^4} f \approx x^2 y + z^4 + az^3 y + bz^2 y^2 + zy^3,\ \Delta \neq 0,\ ab \neq 9 \mid\Rightarrow \mathbf{99};$$
$$\approx x^2 y + z^4 + bz^3 y + z^2 y^2,\ b^2 \neq 4 \mid\Rightarrow \mathbf{100};$$
$$\approx x^2 y + z^3 y + az^2 y^2 + y^4,\ 4a^3 + 27 \neq 0 \mid\Rightarrow \mathbf{101};$$
$$\approx x^2 y + \varphi \mid\Rightarrow \mathbf{102}.$$

99. $j_{x^2y,y^4,z^4} f = x^2 y + z^4 + az^3 y + bz^2 y^2 + zy^3$, $\Delta \neq 0 \Rightarrow f \in V_{1,0}$.

100. $j_{x^2y,y^4,z^4} f = x^2 y + z^4 + bz^3 y + z^2 y^2$, $b^2 \neq 4 \Rightarrow f \in V_{1,p}$ $p > 0$.

101. $j_{x^2y,y^4,z^4} f = x^2 y + z^3 y + az^2 y^2 + y^4$, $4a^3 + 27 \neq 0 \Rightarrow f \in V^{\#}_{1,p}$ $p > 0$.

102. $j_{x^2y,y^4,z^4} f = x^2 y + \varphi \Rightarrow \mu(f) \geqslant 17$, $m(f) \geqslant 3$, $c(f) \geqslant 13$.

103. $j_3 f(x, y, z) = x^3 \Rightarrow \mu(f) \geqslant 18$, $m(f) \geqslant 4$, $c(f) \geqslant 13$.

104. $j_3 f(x, y, z) = 0 \Rightarrow \mu(f) \geqslant 27$, $m(f) \geqslant 10$, $c(f) \geqslant 16$.

105. corank $f > 3 \Rightarrow \mu(f) \geqslant 16$, $m(f) \geqslant 5$, $c(f) \geqslant 10$.

REMARKS. Proofs of Theorems **1–105** can be found in the following places:

§15–**6, 18, 26, 33, 59, 67, 74, 83, 88.**

§16–**3, 10, 13, 22, 29, 36, 47, 50, 52, 54, 56, 58, 63, 66, 70, 77, 82, 85, 90, 97, 98.**

§17–**2, 4, 5, 7, 8, 9, 11, 14, 19, 20, 21, 23, 27, 28, 30, 34, 35, 37, 48, 51, 60, 61, 62, 64, 68, 69, 71, 75 ,76, 78, 84, 86, 89, 91, 99.**

§18–**53, 55, 57.**

Theorems **1, 17** and **25** are obvious. Theorems **12$_1$, 15, 16** are proved in [8].

Theorems **49** and **102–105** are proved by the methods of [8], but we do not give the proofs.

The proofs of Theorems **12$_k$** ($k > 1$), **24, 31, 32, 38–46, 65, 72, 73, 79, 80, 81, 87, 92–96, 100, 101** are based on new methods compared with those of [8] (using a spectral sequence) and are not carried out in the present article (see [91], [92]). In the classification of unimodal singularities these theorems are not used. A proof of the theorem on the classification of unimodal singularities is obtained by a juxtaposition of the following theorems:

1–5, 6$_{1,2}$–9$_{1,2}$, 10$_2$, 11$_2$, 13–17, 18$_1$–21$_1$, 22$_2$, 23$_2$, 25, 26$_1$–30$_1$, 36$_1$, 37$_1$, 47, 48, 50–58, 59$_1$–62$_1$, 63$_2$, 64$_2$, 66, 67$_1$–71$_1$, 82, 83$_1$–86$_1$. 97, 98, 105.

Here Theorems **10, 22, 29, 36, 47, 63, 79, 85, 98** are not used in their entirety (only the first case of each of them is needed).

The classification of bimodal singularities uses, in addition, Theorems 6_3–11_3, 12_2, 18_2–21_2, 22_3, 23_3, 24_2, 31_1–35_1, 36_2, 37_2, 59_2–62_2, 63_3, 64_3, 65_2, 72_1–76_1, 77_2, 78_2, 87_1, 89_1, 90_2, 91_2, 99.

Here Theorems 10_3, 22_3, 36_2, 63_3, 77_2, 90_2 are not used in their entirety (only the first case of each of them is needed).

§ 15. Proofs of the theorems on the rotation of a ruler (Theorems 6, 18, 26, 33, 59, 67, 74, 83, 88)

The method of the rotation of a ruler, by which these theorems are proved, is due to Newton [59]. Let us prove, for example, Theorem 6_k According to the hypothesis of this theorem, the support of the function lies strictly above a line joining the points (3,0) and $(0,3k)$, excluding (3,0). We place a ruler on that line, and, holding one end at (3,0), we move the other end to the side, until we meet the exponent of a monomial occurring with a non-zero coefficient (Fig. 7).

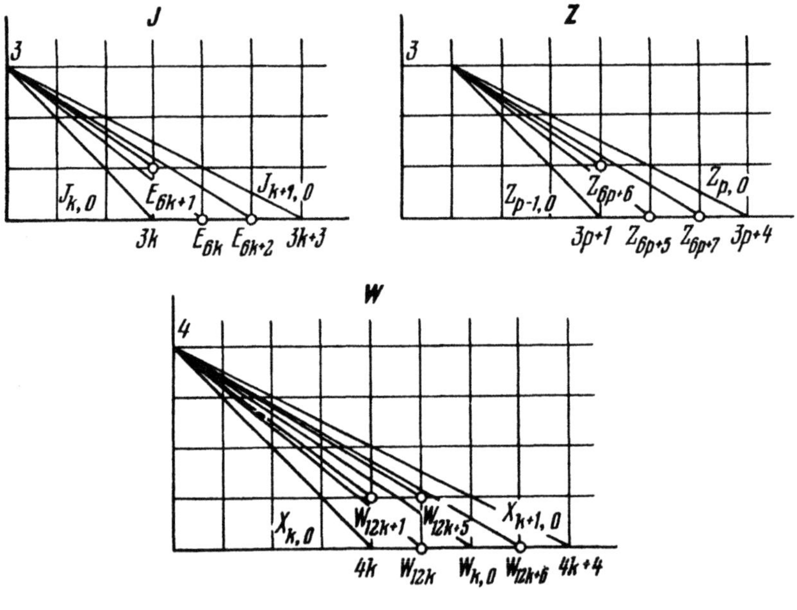

Fig. 7

Inside the triangle between the lines joining (3,0) to $(0,3k)$ and to $(0,3(k+1))$ there are in all three lattice points (in the positive quadrant), namely, $(0,3k+1)$, $(1,2k+1)$ and $(0,3k+2)$.

If the coefficient of y^{3k+1} is different from 0, then we have the first case of Theorem 6_k. If the coefficient of y^{3k+1} is 0, but that of xy^{2k+1} is different from zero, then we have the second case. If both monomials y^{3k+1} and xy^{2k+1} do not occur in f, but y^{3k+2} does, then we have the third case. If the coefficients of all three monomials y^{3k+1}, xy^{2k+1}, y^{3k+2}

are 0, then we have the fourth case.

In exactly the same manner the remaining theorems also reduce to the enumeration of the lattice points of one kind or another in the corresponding domains. The edges of the domains in the cases of corank 2 are as follows:

Theorem	Series	Near edge	Far edge	Monomials
6_k	J	(3, 0) (0, $3k$)	(3, 0) (0, $3k+3$)	y^{3k+1}, xy^{2k+1}, y^{3k+2}
18_p	Z	(3, 1) (0, $3p+1$)	(3, 1) (0, $3p+4$)	y^{3p+2}, xy^{2p+2}, y^{3p+3}
26_k	W	(4, 0) (0, $4k$)	(4, 0) (0, $4k+2$)	y^{4k+1}, xy^{3k+1}
33_k	W	(4, 0) (0, $4k+2$)	(4, 0) (0, $4k+4$)	xy^{3k+2}, y^{4k+3}

(see Fig. 7; following Newton,. we put the exponent of x on the vertical, and that of y on the horizontal axis).

In the cases of corank 3 (Theorems **59**, **67**, **74**, **83** and **88**) we assume that f only contains monomials of definite types. In the following table we indicate the near and far edges for each type of monomial, and also monomials of those types with exponents in the prescribed limits.

Theorem	Series	Type	Edges	Monomials
59	Q	y^q xy^q	$3k<q<3k+3$ $2k<q<2k+2$	y^{3k+1}, y^{3k+2} xy^{2k+1}
67	S	y^q xy^q zy^q	$4k-1<q<4k+1$ $3k-1<q<3k+\frac{1}{2}$ $2k<q<2k+1$	y^{4k} xy^{3k} —
74	S	xy^q y^q zy^q	$3k+\frac{1}{2}<q<3k+2$ $4k+1<q<4k+3$ $2k+1<q<2k+2$	xy^{3k+1} y^{4k+2} —
83	U	y^q xy^q, zy^q x^2y^q	$3k<q<3k+\frac{3}{2}$ $2k<q<2k+1$ $k<q<k+\frac{1}{2}$	y^{3k+1} — —
88	U	y^q xy^q, zy^q x^2y^q	$3k+\frac{3}{2}<q<3k+3$ $2k+1<q<2k+2$ $k+\frac{1}{2}<q<k+1$	y^{3k+2} — —

Let us prove, for example, Theorem 67_k. According to the hypothesis of the theorem, on the plane passing through the exponents of the monomials x^2z, yz^2 and y^{4k-1} and closer to it than to the origin there are no points of the support of f, except two, namely the exponents of the monomials x^2z and yz^2. We join these two points by a line and we rotate the plane around this line, moving the point t of intersection with the axis of y-exponents to the side until we meet the exponent of the support of f. According to the hypothesis of the theorem, f contains only monomials of one of the three types y^q, xy^q and zy^q (excluding x^2z, yz^2). When t moves from $4k - 1$ to $4k + 1$, the plane (which we denote by Π_t) successively passes through the exponents of precisely two monomials of our types, y^{4k} and xy^{3k} (Fig. 8 shows the plane of x- and y-exponents; the line around which we rotate the plane Π_t, intersects it at $(4, -1)$).

If the coefficient of y^{4k} in f is not zero, then we have the first case of Theorem **67**; if it is zero, but the coefficient of xy^{3k} is not 0, then we have the second case; and if the coefficients are both zero, then we have the third case. This proves Theorem **67**; the other theorems (**59**, **74**, **83**, **88**) are proved in the same way.

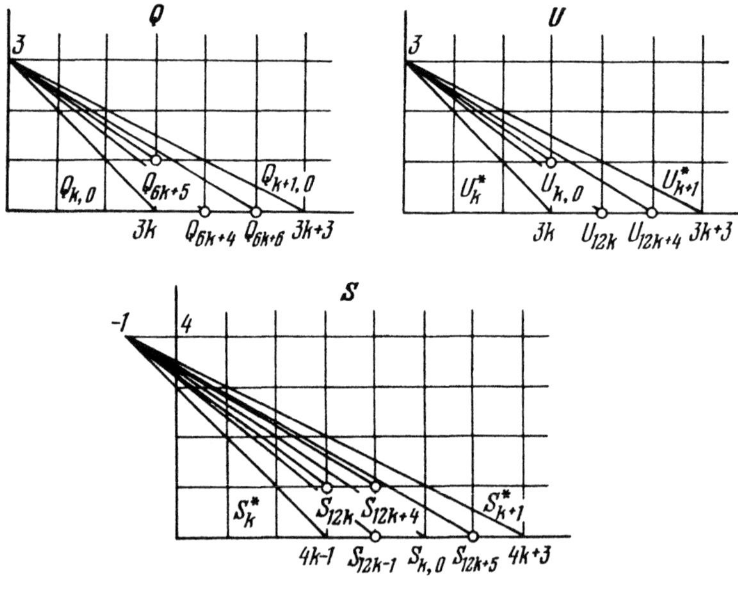

Fig. 8

REMARK. Actually, we have proved somewhat more than is stated in Theorems **6**, . . . , **88**. Namely, in each of these theorems there is a *last case*; in this case we prove that the *support of f is located no closer than the far edge* indicated in the table. (Explanation. In the cases of corank 2 in the table we have indicated two points on the far edge. In the cases of corank 3 the far edge is the plane Π_t, where $t = q_{\max}$ is the upper bound for the exponent q of the monomial y^q indicated in the table.)

EXAMPLE. In proving Theorem **67** we have proved that in the third case the support of f has no points lying on the same side as the origin of the plane passing through the exponents of the monomials x^2z, yz^2, y^{4k+1}

In accordance with this remark, the proofs of Theorems **10**, **22**, **29**, **36**, **63**, **70**, **77**, **85**, **90** lead to the classification of quasihomogeneous singularities, for which all the exponents of the monomials lie on the corresponding far edge.

Analogous (but somewhat simpler) arguments with moving a ruler lead to the classification of quasihomogeneous singularities of the proof of Theorems 3, **49**, **50**, **98**.

§16. Normal forms of quasihomogeneous singularities (proofs of Theorems 3, 10, 13, 22, 29, 36, 47, 50, 52, 54, 56, 58, 63, 66, 70, 77, 82, 85, 90, 97, 98)

The remark at the end of §15 reduces the proofs of these theorems to the classification of orbits of the action of the group of quasihomogeneous diffeomorphisms on the space of quasihomogeneous functions with fixed quasihomogeneity type and with fixed coefficients for certain *distinguished* monomials. The run of the calculations is as follows. First we compute the exponents of homogeneity and with respect to them the monomial vector fields that generate over **C** the Lie algebra of the group of all quasihomogeneous diffeomorphisms. Fixing the coefficients of the distinguished monomials gives an affine plane in the space of quasihomogeneous functions. We find the isotropy algebra of this plane. The calculations show (unexpectedly) that the actions of the Lie algebras within the limits of each of our series of singularities are affinely equivalent to each other. Thus, our classification theorems reduce to a finite number of problems. These problems can be identified with simple geometric classification problems, as indicated by the table on p. 48.

In the cases of Theorems 3, 13, **47** and **50** the classification problems coincide by their very definition. The linear classifications of 3-forms in $\mathbf{C}^2$ and $\mathbf{C}^3$ and of 4-forms in $\mathbf{C}^2$ are well known (see, for example, [45], [64]), therefore, the proofs of Theorems 3, 13 and **50** are not given here. I only remark that Theorem **50** describes the stratification by singularities shown in Fig.9 on p.48 of the 10-dimensional space of cubic forms.

The stratifications of the spaces of cubic curves in Theorems **63**, **70**, **77**, **85** and **90** use different fragments of this diagram.

The identification of our classification problems with the geometric problems is based on two general theorems on quasihomogeneous functions (Theorems A and B below). To state these theorems we introduce some definitions and notation.

Theorem No.	Series	Geometric problem
3	D	Linear classification of 3-forms in $\mathbf{C}^2$
10, 22	J, Z	Affine classification of triples of points in $\mathbf{C}^1$
13	X	Linear classification of 4-forms in $\mathbf{C}^2$
29	W	Linear classification of pairs of points in $\mathbf{C}^1$
36	X	Affine classification of quadruples of points in $\mathbf{C}^1$
47	N	Linear classification of 5-forms in $\mathbf{C}^2$
50	P	Linear classification of 3-forms in $\mathbf{C}^3$
63	Q	Affine classification of cubic curves in $\mathbf{C}^2$ with a finite cusp
70	S	Affine classification of reducible cubic curves in $\mathbf{C}^2$ having a finite double point
77	S^*	Affine classification of cubic curves in $\mathbf{C}^2$ having a simple tangent at infinity
85	U	Affine classification of centrally-symmetric cubic curves in $\mathbf{C}^2$ with precisely three points at infinity
90	U^*	Affine classification of cubic curves in $\mathbf{C}^2$ with precisely three points at infinity
98	V	Affine classification of polynomials of degree ≤ 4 in one variable in $\mathbf{C}^1$

Let $\alpha = (\alpha_1, \ldots, \alpha_n)$ be a set of positive rational numbers. We consider the arithmetic space $\mathbf{C}^n$ with fixed coordinates $x_1, \ldots, x_n$.

DEFINITION. A *quasihomogeneous function* of degree d and type $\boldsymbol{\alpha}$ is a polynomial $\Sigma a_m x^m \in \mathbf{C}[x_1, \ldots, x_n]$ for which $(\boldsymbol{m}, \boldsymbol{\alpha}) = d$ for all $\boldsymbol{m}$.

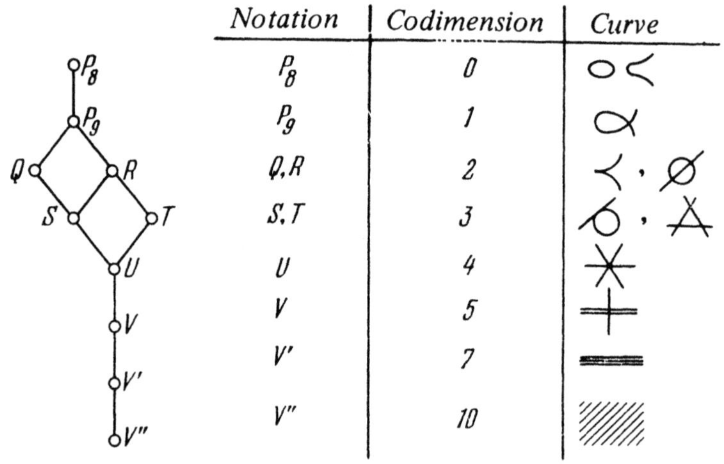

Fig. 9

DEFINITION. The *group of quasihomogeneous diffeomorphisms* of type $\boldsymbol{\alpha}$ is the group of germs of diffeomorphisms of $\mathbf{C}^n$ at O taking any quasihomogeneous function of type $\boldsymbol{\alpha}$ and degree d into a quasihomogeneous function of the same degree. The Lie algebra of this group is called the *quasihomogeneous algebra* and is denoted by $\mathfrak{a}(\boldsymbol{\alpha})$. For example, $\mathfrak{a}(1, 1) = \mathfrak{gl}(2, \mathbf{C})$.

DEFINITION. The *space of exponents* of functions of $x_1, \ldots, x_n$ is the arithmetic space $\mathbf{C}^n$ whose points $\boldsymbol{m}$ with non-negative integer coordinates are the exponents of the monomials $x^{\boldsymbol{m}}$.

DEFINITION. The *support of quasihomogeneous functions of degree d and type* $\boldsymbol{\alpha}$ is the set of all non-negative integer points $\boldsymbol{m}$ on the plane $(m, \alpha) = d$. The support is said to be *complete* if it is not contained in an affine subspace of $\mathbf{C}^n$ of dimension less than $n - 1$.

Quasihomogeneous functions can be regarded as functions given on the support ($\Sigma\, a_m x^m$ has the value a_m at m). All such functions form a linear space $\mathbf{C}^\nu$, where ν is the number of points of the support. The group of quasihomogeneous diffeomorphisms and the quasihomogeneous algebra $\mathfrak{a}(\alpha)$ act on this space $\mathbf{C}^\nu$. From the definitions it follows immediately that *the Lie algebra* $\mathfrak{a}(\alpha)$ *is generated as a* **C**-*linear space by all the monomial vector fields* $x^{\boldsymbol{p}}\, \partial_i$ *for which* $(p, \alpha) = \alpha_i$ (here and later on, $\partial_i = \partial/\partial x_i$). For example, the n monomials $x_i \partial_i$ belong to $\mathfrak{a}(\boldsymbol{\alpha})$ for any $\boldsymbol{\alpha}$.

DEFINITION. The *roots* of the quasihomogeneous algebra $\mathfrak{a}(\alpha)$ are all the non-zero vectors $\boldsymbol{m}$ of the space of exponents that lie in the plane $(m, \alpha) = 0$ and have the form $\boldsymbol{m} = \boldsymbol{p} - \mathbf{1}_i$ (where $\mathbf{1}_i$ is the vector whose i-th component is equal to 1 and the remainder equal to 0, and the vector p has non-negative integer components).

In other words, $\boldsymbol{m}$ is a root if $x^{\boldsymbol{p}}\, \partial_i$ is a monomial vector field from $\mathfrak{a}(\alpha)$ other than $x_i \partial_i$.

We observe that i can be regained from a root $\boldsymbol{m}$, since $\boldsymbol{m}$ has precisely one negative coordinate $m_i = -1$ (not all the components of $\boldsymbol{m}$ can be non-negative because $(m, \alpha) = 0$).

THEOREM A. *We assume that the support is complete. Then the action of the Lie algebra* $\mathfrak{a}(\alpha)$ *on the space of functions on the support is uniquely determined with respect to the affine equivalence class of the pair (support, root system).*

THEOREM B. *The quasihomogeneous Lie algebra* $\mathfrak{a}(\alpha)$ *is determined, up to finitely many variants, by its root system (as a subspace of the linear space spanned by the roots) and by its dimension.*

In other words, if we do not distinguish algebras that are obtained from each other by direct addition of a trivial (commutative) algebra, then *there are only finitely many non-isomorphic Lie algebras* $\mathfrak{a}(\boldsymbol{\alpha})$ *with linearly equivalent root systems.*

In our examples this finite number is equal to 1, and I do not know whether it can be greater than 1.

REMARK. We must emphasize that the affine equivalences of supports and linear equivalences of root systems in Theorems A and B are not at all necessary to preserve either the coordinate simplex $m_i \geqslant 0$ on the plane $(m, \alpha) = d$ or the lattice of non-negative integral $\boldsymbol{m}$ in $\mathbf{C}^n$. We also observe that under the hypotheses of Theorems A and B the *groups* of quasihomogeneous diffeomorphisms and their orbits in spaces of quasihomogeneous functions do not necessarily coincide; however, the connected components of the orbits coincide.

PROOF OF THEOREM B. Let $M \subset \mathbf{C}^r \subset \mathbf{C}^n$ be a root system of $\mathfrak{a}(\alpha)$, generating the plane $\mathbf{C}^r$ in $\mathbf{C}^n$ $(0 \leqslant r \leqslant n-1)$. With each $m \in M$ we associate a basis vector e_m in $\mathbf{C}^\nu$ (where ν is the number of roots). We consider the r-dimensional linear space $H = \mathrm{Hom}(\mathbf{C}^r, \mathbf{C})$ and the direct sum $\mathfrak{b} = H \oplus \mathbf{C}^\nu$.

LEMMA. *The space* $\mathfrak{b} = H \oplus \mathbf{C}^\nu$ *can be given the following Lie algebra structure*:

(1) $[h_1, h_2] = 0$ $(\forall h_1, h_2 \in H)$;

(2) $[h, e_m] = (h, m)e_m$ $(\forall h \in H,\ m \in M)$;

(3) $[e_{m_1}, e_{m_2}] = N_{m_1, m_2} e_{m_1+m_2}$, *where* $m_1 + m_2 \neq 0$,

$$N_{m_1,m_2}\begin{cases} = 0 \textit{ if } m_1 + m_2 \textit{ is not a root};\\ = -\max\{\lambda\colon m_1 + \lambda m_2 \text{ is a root}\} \textit{ if this maximum is} > 1;\\ = +\max\{\lambda\colon m_2 + \lambda m_1 \text{ is a root}\} \textit{ if this maximum is} > 1;\\ = \pm 1 \textit{ if both maxima} = 1 \textit{ (the case when both maxima are} > 1 \textit{ is}\end{cases}$$

impossible).

(4) $[e_m, e_{-m}] = h_m$, *where the function* $h_m \in H$ *changes sign under the reflection of* $\mathbf{C}^r$ *that preserves* M *and takes* m *into* $-m$, *normalized by the condition* $h_m(m) = 2$ (*such a reflection exists and is unique for any pair of opposite roots*).

The quasihomogeneous Lie algebra $\mathfrak{a}(\alpha)$ *is isomorphic to the direct sum of the Lie algebra* $\mathfrak{b}$ (*for some choice of the sign* $\pm$ *in* (3)) *and a trivial (commutative) algebra*: $\mathfrak{a}(\alpha) \cong \mathfrak{b} \oplus \mathbf{C}^{n-r}$.

PROOF OF THE LEMMA. We consider a monomial basis for $\mathfrak{a}(\alpha)$ over $\mathbf{C}$ and write the basis monomials as follows:

$$h_i = x_i\partial_i, \qquad e_m = x^m x_i \partial_i$$

(I recall that i is uniquely determined by the root m). We claim that these generators satisfy the commutation relations (1)–(4).

We regard the differentiations h_i and e_m as linear operators in the space of all functions on the lattice $\mathbf{Z}^n$ in the space of exponents $\mathbf{C}^n$. Then h_i is the operator of multiplication by the i-th coordinate. We denote the operator of multiplication by a function like that function. Thus,

$$(ha)(k) = h(k)\,a(k), \text{ where } k \in \mathbf{Z}^n,\ a\colon \mathbf{Z}^n \to \mathbf{C}.$$

The relation (1) is now proved, because multiplications by functions commute.

We denote by σ_m the action of translation by $m \in \mathbf{Z}^n$ on the function:

$$(\sigma_m a)(k) = a(k - m), \text{ where } k \in \mathbf{Z}^n,\ a\colon \mathbf{Z}^n \to \mathbf{C}.$$

Then $e_m = \sigma_m h_i$. Computing the commutator of multiplication by a linear function and a translation, we obtain

$$[h, \sigma_m] = h(m)\sigma_m.$$

From this we immediately get (2). Next, computing the commutator of e_{m_1} and e_{m_2} we obtain

$$e_{m_1}e_{m_2} = \sigma_{m_1}h_{i_1}\sigma_{m_2}h_{i_2} = \sigma_{m_1}\sigma_{m_2}(h_{i_1}h_{i_2} + h_{i_1}(m_2)h_{i_2}),$$
$$[e_{m_1}, e_{m_2}] = \sigma_{m_1+m_2}(h_{i_1}(m_2)h_{i_2} - h_{i_2}(m_1)h_{i_1}).$$

If $m_1 + m_2$ is not a root and not 0, then the operator on the right-hand side can belong to $\mathfrak{a}(\alpha)$ only if it is 0. In this case $[e_{m_1}, e_{m_2}] = 0$.

If $m_1 + m_2$ is a root, then this vector has precisely one negative component, which is equal to -1. The vectors m_1 and m_2 also have one negative component, equal to -1. Therefore $m_1 + m_2$ has either the same negative component as m_1 or as m_2. For example, let us assume that $m_1 + m_2$ has the same negative component as m_1.

Then the i_1-th and i_2-th components of the vectors m_1, m_2, and $m_1 + m_2$ have the form

	m_1	m_2	m_1+m_2	$m_1+\lambda m_2$
h_{i_1}	-1	0	-1	-1
h_{i_2}	$p>0$	-1	$p-1\geqslant 0$	$p-\lambda$

Consequently, $m_1 + \lambda m_2$ is a root for $\lambda \leqslant p = h_{i_2}(m_1)$, that is, $h_{i_2}(m_1) = \max\{\lambda : m_1 + \lambda m_2 \text{ is a root}\}$.

Thus, in case $h_{i_1}(m_1 + m_2) = -1$, we have

$$[e_{m_1}, e_{m_2}] = -h_{i_2}(m_1)\sigma_{m_1+m_2} = N_{m_1, m_2}e_{m_1+m_2},$$

where $N_{m_1,m_2} = h_{i_2}(m_1)$; and so we have proved (3).

Suppose that m and $-m$ are roots. Both these vectors have precisely one negative coordinate, equal to -1, so that $e_m = x_i\partial_j$, $e_{-m} = x_j\partial_i$. Consequently, the weights α_i and α_j are equal. The interchange of the coordinates i and j in $\mathbf{C}^n$ is a reflection taking the system of all roots into itself and changing the places of m and $-m$. Further, $[e_m, e_{-m}] = h_i - h_j$. If we regard $h_m = h_i - h_j$ as a function in $\mathbf{C}^n$, then it changes sign under the reflection switching i and j, and is equal to $+2$ on m. Thus, the operators h and e_m satisfy the relations (1)–(4).

Now we consider the subspace $\mathbf{C}^r$ spanned by the roots in the coordinate space $\mathbf{C}^n$ with the usual Hermitian metric, $\langle k, l\rangle = \Sigma k_i\bar{l}_i$. The space $\mathbf{C}^r$ and the metric are invariant relative to all permutations of the coordinates with equal weights ($\alpha_i = \alpha_j$). Therefore, the orthogonal complement $\mathbf{C}^{n-r}$ to $\mathbf{C}^r$ in $\mathbf{C}^n$ is also invariant. We represent the linear space of $\mathfrak{a}(\alpha)$ as $\mathfrak{a}(\alpha) = H^r \oplus K^\nu \oplus H^{n-r}$, where H^r consists of those linear functions h on $\mathbf{C}^n$ that vanish on $\mathbf{C}^{n-r}$; H^{n-r} of the linear functions h on $\mathbf{C}^n$ that vanish on $\mathbf{C}^r$; K^ν of linear combinations of the vectors e_m.

It follows from these commutation relations that H^{n-r} lies in the centre of $\mathfrak{a}(\alpha)$ and that $H^r \oplus K^\nu$ is an ideal isomorphic to $\mathfrak{b}$. Consequently,

$\mathfrak{a}(\alpha) \cong \mathfrak{b} \oplus H^{n-r}$, and the lemma is proved.

We can now complete the proof of Theorem B. The relations (1)–(4) express the commutators in $\mathfrak{b}$ and $\mathfrak{a}$ by the geometry of the roots, without reference to coordinates (excluding the choice of sign ± in one of the relations (3)). Thus, the set of roots, as vectors in the **C**-linear space $\mathbf{C}^r$, determines $\mathfrak{b}$ up to a finite number of possibilities, and this proves Theorem B.

REMARK. I do not know whether different choices of signs in (3) yield non-isomorphic algebras.

PROOF OF THEOREM A. We extend functions on the support onto the lattice $\mathbf{Z}^{n-1}$ of all integer points in the support plane, by setting them equal to 0 outside the support. The operators h_i, e_m, σ_m, h_m, defined in the proof of Theorem B on the functions on $\mathbf{Z}^n$, also act on the space of functions on the lattice $\mathbf{Z}^n$ in the support plane. These operators in the space of functions on $\mathbf{Z}^{n-1}$ will be denoted by the same letters. Thus, h_i and h_m are operators of multiplication by affine functions in the support plane, σ_m is the operator of translation by the root m and $e_m = \sigma_m h_i$.

Suppose that m is a root. We call a point k of the support a *base point* for m if $k + m$ does not belong to the support. The set of all base points of m is called the *base of the root* m (in the given support).

The base of the root $m = p - 1_i$ is formed by all the points of the support whose i-th coordinate is 0. Consequently, the entire base lies in an affine hyperplane $\mathbf{C}^{n-2}$ of the support plane.

The base of each root of the complete support belongs to precisely one affine hyperplane $\mathbf{C}^{n-2} \subset \mathbf{C}^{n-1}$. For, each point of the support is obtained from a point of the base by subtracting a non-negative integral multiple of m, so that if the base were contained in $\mathbf{C}^{n-3}$, the support would be contained in $\mathbf{C}^{n-2}$ and would not be complete.

Thus, there exists precisely one affine function in the support plane that is equal to 0 on the base of m and whose increment along m is -1. This function is h_i (restricted to the support plane). Consequently, the restriction of h_i to the support plane can be reconstructed uniquely from the support and the root m.

The action of e_m on the space of functions on the support can now be described in terms only of the geometry of the support and the roots: $e_m = \sigma_m h_i$, that is, $(e_m a)(k) = h_i(k - m)a(k - m)$ for any function a. We note that the operator e_m leaves invariant the space of functions that vanish outside the support, since h_i vanishes at base points.

The algebra $\mathfrak{a}(\alpha)$ acts on the space of functions on a lattice in the support plane, so that we have a representation $\varphi\colon \mathfrak{a}(\alpha) \to \mathscr{E}$, where $\mathscr{E}$ is the algebra of endomorphisms of this (infinite-dimensional) space of functions. We consider the image of the representation φ. *This image is determined by the geometry of the support and the roots.* More precisely, let $\mathfrak{a}_1 = \mathfrak{a}(\alpha_1)$ and $\mathfrak{a}_2 = \mathfrak{a}(\alpha_2)$ be two quasihomogeneous algebras, and

$S_1 \subset \mathbf{C}_1^{n-1}$, $S_2 \subset \mathbf{C}_2^{n-1}$ be complete supports. Let ψ: $\mathbf{C}_1^{n-1} \to \mathbf{C}_2^{n-1}$ be an affine mapping that takes S_1 into S_2 bijectively and the root system for $\mathfrak{a}_1$ into that for $\mathfrak{a}_2$. Then *the isomorphism ψ^*, induced by ψ, of the space of functions on $\mathbf{C}_2^{n-1}$ into the space of functions on $\mathbf{C}_1^{n-1}$ takes the Lie algebra $\varphi(\mathfrak{a}_2)$ isomorphically into the Lie algebra $\varphi(\mathfrak{a}_1)$.*

For $\mathfrak{a}(\alpha)$ is generated over $\mathbf{C}$ by the monomials $x_i\partial_i$ and $x^m x_i \partial_{i(m)}$. The images of the fields $\Sigma\, c_i x_i \partial_i$ in $\mathscr{E}$ are the operators of multiplication by all possible affine functions in the support plane. The images of the fields $x^m x_i \partial_{i(m)}$ are the operators e_m, defined by the geometry of the support and the roots. Consequently, $\psi^* e_{\psi(m)} = e_m \psi^*$ and, hence, ψ induces an isomorphism Ψ: $\varphi(\mathfrak{a}_2) \to \varphi(\mathfrak{a}_1)$.

The kernel of the Lie algebra homomorphism φ: $\mathfrak{a}(\alpha) \to \mathscr{E}$ is 0. For suppose that $(h + \Sigma\, c_m e_m)\, a = 0$ for all functions a. We choose a point k at which all the h_{i_m} are different from 0, and we apply $h + \Sigma\, c_m e_m$ to the function δ_k, which is equal to 1 only at k. We obtain $h(k)\delta_k + \Sigma c_m h_{i_m}(k)\, \delta_{k+m} = 0$, hence $c_m = 0$ and $h(k) = 0$. Consequently, $h \equiv 0$. Thus, the Lie algebra $\varphi(\mathfrak{a}(\alpha))$ is isomorphic to $\mathfrak{a}(\alpha)$.

The isomorphisms $\mathfrak{a}_2 \to \varphi(\mathfrak{a}_2) \xrightarrow{\Psi} \varphi(\mathfrak{a}_1) \to \mathfrak{a}_1$ show that both the Lie algebras $\mathfrak{a}_1$ and $\mathfrak{a}_2$, and their actions on the spaces of functions on S_1 and S_2 are isomorphic, and Theorem A is proved.

COROLLARY 1. *Suppose that the set of weights α and the degree d are such that there exists a quasihomogeneous function with an isolated critical point O with zero 2-jet. Then the root system and the support uniquely determine the Lie algebra $\mathfrak{a}(\alpha)$ and its action φ.*

PROOF. Under the assumptions the support is complete. For the isolation implies that for each i there is a monomial of the form $x_i^{a_i} x_{j(i)}$ ($j \in \{1, \dots, n\}$, and by hypothesis $a_i > 1$). We claim that the exponents of these n monomials that belong to the support are linearly independent. The system of equations $z_{j(i)} + a_i z_i = 0$ relative to z has only the zero solution for $a_i > 1$ (it is easy to see this by considering the cycles of an endomorphism of a finite set, $i \mapsto j$). Consequently, the determinant of its matrix is not 0, and hence, our support is complete. The assertion of the corollary now follows from Theorem A.

COROLLARY 2. *Let γ: $\mathbf{C}_1^{n-1} \to \mathbf{C}_2^{n-1}$ be an affine isomorphism of a complete support plane S_1 into a complete support plane S_2, taking S_1 into part of S_2, the roots of $\mathfrak{a}_1$ into part of the roots of $\mathfrak{a}_2$ and the bases of the roots of $\mathfrak{a}_1$ into part of the bases of the corresponding roots of $\mathfrak{a}_2$. Then γ induces an isomorphism of the action φ of $\mathfrak{a}_1$ onto functions on S_1 and an isomorphism of the action of a subalgebra of $\mathfrak{a}_2$ onto the space of functions on S_2 that vanish outside $\gamma(S_1)$.*

PROOF. The subalgebra indicated in the corollary is generated over $\mathbf{C}$ by the operators h of multiplication by all affine functions and by the operators

e_m, where m is the image of a root of the first algebra. The actions e_m are defined by the roots and the bases, and thus commute with the action of γ.

COROLLARY 3. *Suppose that under the hypotheses of Corollary 2 several points in the support S_1 are distinguished and the values of the functions are fixed at them. All the functions on S_1 with fixed values at these points form an affine plane P in the space of functions on S_1. Let $\mathfrak{a}_P$ be the isotropy algebra of this plane P. Then the isomorphism γ of Corollary 2 induces an isomorphism of the action of $\mathfrak{a}_P$ on P with the action of some subalgebra of the Lie algebra $\mathfrak{a}_2$, preserving the plane $\gamma^{-1*}P$ in the space of functions on S_2.*

PROOF. The mapping of the first action into the second commutes with γ^{-1*}, as required.

The *reduction of the classification Theorems* **10**, ..., **98** *to standard classification problems* is carried out by means of Corollary 3. In the table given below, for each theorem we have indicated: the weight α_i and the

Theorem	Type	Weights $\alpha; d$	Monomials of the support	e_m	Monomials of the homogeneous support	Images of e_m
10	J_k	$k, 1; 3k$	$x_*^3, x^2y^k, xy^{2k}, y^{3k}$	$y^k\partial_x$	X_*^3, X^2Y, XY^2, Y^3	$Y\partial_X$
22	Z_{l-1}	$l, 1; 3l+1$	$x^3y_*, x^2y^{l+1}, xy^{2l+1}, y^{3l+1}$	$y^l\partial_x$	X_3^*, X^2Y, XY^2, Y^3	$Y\partial_X$
29	W_k	$2k+1,2; 8k+4$	$x_*^4, x^2y^{2k+1}, y^{4k+2}$	—	X_*^2, XY, Y^2	—
36	X_k	$k, 1; 4k$	$x_*^4, x^3y^k, x^2y^{2k}, xy^{3k}, y^{4k}$	$y^k\partial_x$	$X_*^4, X^3Y, X^2Y^2, XY^3, Y^4$	$Y\partial_X$
63	Q_k	$2k, 2, 3k-1; 6k$	$x_*^3, yz_*^2, x^2y^k, xy^{2k}, y^{3k}$	$y^k\partial_x$	$X_*^3, Y^2Z_*, X^2Y, XY^2, Y^3$	$Y\partial_X$
70	S_k	$2k+1, 2, 4k; 8k+2$	$x^2z_*, yz_*^2, x^2y^{2k}, y^{2k+1}z, y^{4k+1}$	$y^{2k}\partial_z$	$XYZ_*, YZ_*^2, XY^2, Y^2Z, Y^3$	$Y\partial_Z$
77	S_k^*	$k,1, 2k-1; 4k-1$	$x^2z_*, yz_*^2, x^3y^{k-1}, xy^kz, x^2y^{2k-1}, zy^{2k}, xy^{3k-1}, y^{4k-1}$	$y^k\partial_x$ $y^{2k-1}\partial_z$ $xy^{k-1}\partial_z$	$X^2Z_*, YZ_*^2, X^3, XYZ, X^2Y, Y^2Z, XY^2\ Y^3$	$Y\partial_X$ $Y\partial_Z$ $X\partial_Z$
85	U_k	$2k+1,2, 2k+1; 6k+3$	$x_*^3, xz_*^2, x^2z_\circ, z_\circ^3, xy^{2k+1}, zy^{2k+1}$	$x\partial_z$ $z\partial_x$	$X_*^3, XZ_*^2, X^2Z_\circ, Z_\circ^3, XY^2, Y^2Z$	$X\partial_Z$ $X\partial_X$
90	U_k^*	$k, 1, k; 3k$	$x^2z_*, xz_*^2, x_\circ^3, z_\circ^3, x^2y^k, xy^kz, z^2y^k, xy^{2k}, zy^{2k}, y^{3k}$	$x\partial_z$ $z\partial_x$ $y^k\partial_x$ $y^k\partial_z$	$X^2Z_*, XZ_*^2, X_\circ^3, Z_\circ^3, X^2Y, XYZ, Z^2Y, XY^2, ZY^2, Y^3$	$X\partial_Z$ $Z\partial_X$ $Y\partial_X$ $Y\partial_Z$
98	V	$3, 2, 2; 8$	$x^2y_*, z^4, y^4, y^3z, y^2z^2, yz^3$	$y\partial_z$	$XY_*^2, Z^4, Y^4, Y^3Z, Y^2Z^2, YZ^3$	$Y\partial_Z$

degree of quasihomogeneity d, the monomials of the support (those distinguished by the sign $*$ have the coefficient 1 and those by the sign $\circ$ have the coefficient 0), the root vector fields e_m for each quasihomogeneity type, and then their images in the support of homogeneous forms (all of weight 1) and in the Lie algebra of the general linear group (Fig. 10).

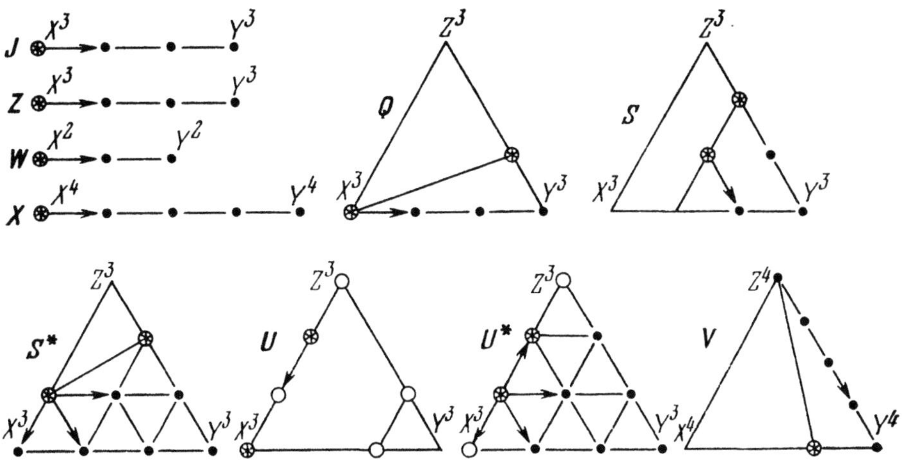

Fig. 10

According to Corollary 3, the orbits of the identity component of the group of quasihomogeneous diffeomorphisms that fix the plane P (consisting of functions on the support with the values 0 and 1 at the points distinguished by the signs $\circ$ and $*$) are transformed under a mapping onto the support of the homogeneous functions into orbits of the corresponding group of *linear* transformations. (This circumstance is a nice property of the roots in the relevant cases in Theorems **10–98**; in general, the images of the roots for $\mathfrak{a}(\alpha)$ in the homogeneous support need not also be roots of the linear group.)

The Lie algebra of this group of linear transformations is easy to describe: it is generated by a torus part, acting on a function on the support as multiplication by affine functions that are zero at the points of the support distinguished by the sign $*$ (at which a non-zero value of the function is fixed) and by the images of those root vectors e_m that do not take other points of the support into points distinguished by the sign $\circ$ (the last condition excludes only $X\partial_Z$ and $Z\partial_X$ in the cases U and U^*).

PROOF OF THEOREMS **10** AND **22**. By Corollary 3 and the table, the problem reduces to the classification of the polynomials $p = X^3 + AX^2Y + BXY^2 + CY^3$ relative to the substitutions $X = X' + \lambda Y$, $Y = \gamma Y'$. By the substitution $X = X' + \lambda Y$ the polynomial p reduces to a form in which $B = 0$. If $C \neq 0$, the substitution $Y = \gamma Y'$ takes p to the form $X^3 + aX^2Y + Y^3$. The roots of the polynomial $X^3 + aX^2 + 1$ are distinct if and only if $4a^3 + 27 \neq 0$. If the roots have multiplicities 2 and 1, then the substitution $X = X' + \lambda Y$ takes p to the form $X^3 + AX^2Y$, $A \neq 0$. Using the substitution $Y = \gamma Y'$ we carry p to the form $X^3 + X^2Y$. If the multiplicity is 3, then $p = X'^3$. Theorems **10** and **22** are now proved.

PROOF OF THEOREM **29**. The problem reduces to the classification of the polynomials $X^2 + AXY + BY^2$ relative to the substitution $Y = \gamma Y'$. The four cases of Theorem **29** correspond to distinct non-zero roots, to one zero root, to a multiple non-zero root and to a multiple zero root of the polynomial $X^2 + AX + B$.

PROOF OF THEOREM **36**. The problem reduces to the classification of the polynomials $p = X^4 + AX^2Y + BX^2Y^2 + CXY^2 + DY^4$ with respect to the substitutions $X = X' + \lambda Y$ and $Y = \gamma Y'$. The first substitution changes the coefficients to

$$A' = 4\lambda + A, \quad B' = 6\lambda^2 + 3\lambda A + B, \quad C' = 4\lambda^3 + 3\lambda^2 A + 2\lambda B + C,$$
$$D' = \lambda^4 + A\lambda^3 + B\lambda^2 + C\lambda + D.$$

For λ we choose one of the roots of the polynomial D'. If these roots are simple, then $C' \neq 0$, so that the substitution $Y = \gamma Y'$ takes p to the form $X^4 + bX^3Y + aX^2Y^2 + XY^3$. The discriminant of $X^4 + bX^3 + aX^2 + X$ is $4(a^3 + b^3) - a^2b^2 - 18ab + 27$.

We observe that $ab - 9 = \dfrac{A'B' - 9C}{C'} = \dfrac{-12\lambda^3 + \ldots}{4\lambda^3 + \ldots}$ is a rational function with numerator of degree three in λ.

Consequently, if the roots of D' are simple, then at least one of the four roots λ of D' does not make the number $ab - 9$ vanish. Thus, in the case of simple roots, p reduces to the form $X^4 + bX^3Y + aX^2Y^2 + XY^3$, where $4(a^3 + b^3) + 27 \neq a^2b^2 + 18ab$, $ab \neq 9$.

If there is one root of multiplicity two, then the substitution $X = X' + \lambda Y$ allows us to reach $D' = C' = 0$, $B' \neq 0$, after which $Y = \gamma Y'$ gives $X^2(X^2 + aXY + Y^2)$, $a^2 \neq 4$.

If there are two double roots, one triple or one quadruple root we obtain $X^2(X + Y)^2$, $X^3(X + Y)$ and X^4, respectively. Theorem **36** is now proved.

REMARK. The restriction $ab \neq 9$ is necessary for the reduction to normal form of non-quasihomogeneous functions with principal quasi-homogeneous part of the form indicated here (that is, for the proof of Theorem **37**). The possibility of reduction to a normal form with $ab \neq 9$ in Theorems **47** and **98** is proved in exactly the same way (and follows from the calculations given here and from Corollary 3).

PROOF OF THEOREM **63**. The problem reduces to the classification of the polynomials $p = X^3 + Y^2Z + AX^2Y + BXY^2 + CY^3$ relative to the substitution $X = X' + \lambda Y$, $Y = \gamma Y'$, $Z = \gamma^2 Z'$. We consider the roots λ of the polynomial $\lambda^3 + A\lambda^2 + B\lambda + C$. If they are simple, the substitution $X = X' + \lambda Y$ yields $C = 0$, $B \neq 0$, after which a second substitution transforms p to $X^3 + Y^2Z + aX^2Y + XY^2$, $a^2 \neq 4$. Then we have the first case of Theorem **63**. If λ is a double root, we obtain $X^3 + Y^2Z + X^2Y$. Under the conditions of the theorem ($A = 0$) a triple root can only occur if $B = C = 0$, and Theorem **63** is proved.

REMARK. It is not hard to check that we have classified plane cubic curves with a finite cusp ($X = 0$, $Y = 0$, $Z = 1$) relative to the affine group. The three cases we have met are distinguished by the nature of the points of intersection with the line at infinity ($Z = 0$), which can be 3, 2, or 1 in number.

PROOF OF THEOREM **70**. The problem reduces to the classification of the polynomials $p = XYZ + YZ^2 + AXY^2 + BY^2Z + CY^3$ relative to the substitution $Z = Z' + \lambda Y$; $X = \gamma X'$, $Y = \gamma^{-2}Y'$, $Z = \gamma Z'$. Instead of (A, B, C) it is convenient to consider in the space of polynomials the coordinates $(A, E = B - 2A, D = 4C - B^2)$. The first substitution takes the point (A, E, D) to $(A + \lambda, E, D)$. Consequently, it suffices to consider the action of the second substitution on the plane $A = 0$, which carries the point (E, D) to $(\gamma^{-3}E, \gamma^{-6}D)$. Consequently, the orbits are 0, half the parabola $D = \varkappa E^2$ without 0, and the line $E = 0$ without 0. Also let $4\Delta = D + E^2$ (that is, $\Delta = A^2 + C - AB$). Four cases are possible:

	Condition	The orbit passes through the point with the coordinates	Consequently, the polynomial reduces to the form
1°	$D\neq 0,\ \Delta\neq 0$	$A=0,\ B=b,\ C=1$	$XYZ+YZ^2+bY^2Z+Y^3,\ b^2\neq 4$
2°	$D=0,\ \Delta\neq 0$	$A=1,\ B=C=0$	$XYZ+YZ^2+XY^2$
3°	$D\neq 0,\ \Delta=0$	$A=C=0,\ B=1.$	$XYZ+YZ^2+Y^2Z$
4°	$D=0,\ \Delta=0$	$A=B=C=0$	$XYZ+YZ^2$

In the last case from the very start the polynomial has the form $XYZ + YZ^2$, since under the hypotheses of Theorem **70** the original polynomial has no monomial x^2y^{2k}, and hence $A = 0$. Theorem **70** is now proved.

REMARK. It is not hard to check that we have classified reducible plane cubic curves (with Y splitting off as factor), having at least two finite (with $X \neq 0$) double points ($Y = 0$, $XZ + Z^2 = 0$) relative to the affine group. The condition $D = 0$ corresponds to tangency with infinity, and $\Delta = 0$ is the complete degeneration of the cubic into 3 lines (Fig.11 on p.58).

PROOF OF THEOREM **77**. The problem reduces to the classification of cubic forms $X^2Z + YZ^2 + 0XZ^2 + 0Z^3 + \dots$ relative to the substitutions $X = X' + \alpha Y$, $Z = Z' + \beta Y + \gamma X$; $X = \lambda X'$, $Y = \lambda^4 Y'$, $Z = \lambda^{-2}Z'$.

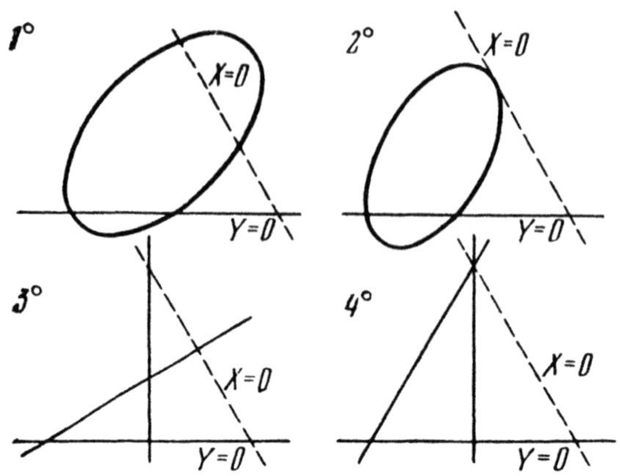

Fig. 11

A form of this kind gives a cubic curve with a simple tangent $Y = 0$ at the non-singular point $X = Y = 0$. Depending on the kind of singularities of this curve five cases are possible.

1°. If the curve is non-singular, then we pass a tangent to the curve different from $Y = 0$ through the point $X = Y = 0$, and set $X = X' + \alpha Y$, where $X' = 0$ is the equation of the new tangent. Then, joining the second point of intersection of the line $Y = 0$ with the curve to the point of contact of the line $Y = 0$ with the line $Z' = 0$ (where $Z = Z' + \beta Y + \gamma X$). Now the form is $X^2Z + YZ^2 + aX^2Y + bXYZ + cXY^2$. The point $X = Z = 0$ is non-singular, so that $c \neq 0$. By dilations we can achieve that $c = 1$. This is the first case of Theorem 77.

2°. Suppose that the curve has a singular point. This cannot lie on the line "at infinity" $Y = 0$. Moving $X' = Z' = 0$ to the singular point by the translations $X = X' + \alpha Y$, $Z = Z' + \beta Y$, we take the form into $X^2Z + YZ^2 + aX^2Y + bXYZ$.

Now by the substitution $Z = Z' + \gamma X$ we annihilate the term with X^2Y; and then X^3 arises and the form becomes $X^2Z + YZ^2 + bXYZ + aX^3$.

If the singular point is non-degenerate, then $b \neq 0$ and by dilations we may take $b = 1$. This is the second case of Theorem 77 (the case *SB*). The condition $a^2 \neq a$ guarantees the uniqueness of the singular point.

3°. If the singular point is degenerate, then $b = 0$. If it is a cusp, then $a \neq 0$ and by dilations we may take $a = 1$. This is the third case of Theorem 77 (case *SQ*).

4°. If there are two singular points, then both are non-degenerate and the curve splits. By the substitution $Z = Z' + \beta Y + \gamma X$ we reduce the equation of the linear factor to the form Z. By the substitution $X = X' + \alpha Y$ we make the point $X' = Z' = 0$ singular. The form is taken to $X^2Z + YZ^2 + aXYZ$. Since there are two singular points, we have

$a \neq 0$ and we may make $a = 1$. This is the fourth case of Theorem 77 (the case *SR*).

5°. The curve cannot completely split (since the axis $Y = 0$ is a *simple* tangent). Therefore, what remains to be analyzed is the case when the curve consists of an irreducible conic and a tangent to it. But by the hypothesis of Theorem 77, the equation of the curve was $X^2Z + YZ^2 + AY^3 + BXY^2 + CZY^2 = 0$. The line $Z = 0$ has third order contact at $Y = Z = 0$ with such a curve, and hence, belongs to the curve. Consequently, $A = B = 0$. The condition of tangency with the conic gives $C = 0$. Thus, in the last case $A = B = C = 0$, and Theorem 77 is proved.

REMARK. It is not hard to check that we have given an affine classification of all plane cubic curves having the line "at infinity" $Y = 0$ as a simple tangent.

PROOF OF THEOREM 85. The problem reduces to the classification of the forms $\varphi + Y^2\psi$ (where φ is a non-degenerate cubic form in X and Z, and ψ is an arbitrary linear form in X and Z) relative to linear substitutions of X and Z and dilations of Y.

Let $\psi \neq 0$. Then, depending on whether φ is divisible by ψ or not, by a linear transformation taking φ to the form $X^3 + XZ^2$, we can transform ψ either to the form $AX + CZ$, where $C^3 + A^2C \neq 0$, $A \neq 0$ or to the form AX. The substitution $Y = \lambda Y'$ makes $A = 1$ and we obtain either the first ($C^3 + C \neq 0$) or the second ($C = 0$) case of Theorem 85. Finally, if $\psi = 0$, then we have the third case.

REMARK. It is not hard to check that we have given an affine classification of all centrally-symmetric plane cubic curves having precisely three points at infinity ($Y = 0$). Here the three cases of Theorem 85 correspond to a non-degenerate curve, a curve split into a line and a conic, and a curve completely split (into three lines).

PROOF OF THEOREM 90. The problem reduces to the classification of the forms $\varphi_3 + Y\varphi_2 + Y^2\varphi_1 + Y^3\varphi_0$, where φ_i is a form of degree i with respect to (X, Z), and φ_3 is not degenerate. We admit linear substitutions of (X, Z), dilations of Y and the substitutions $X = X' + \alpha Y$, $Z = Z' + \beta Y$. In other words, the question is one of the classification of the equations of all cubic curves having exactly three points on the line at infinity $Y = 0$. Depending on the singularities of the curve, seven cases are possible.

1°. We assume that the curve is non-singular. We choose one of the points of intersection of the curve with the line at infinity and we pass a tangent to the curve through this point, so that the point of contact is finite (which, as is easily seen, is always possible). Then the contact is of second order (simple). We choose an origin of the affine coordinate system $(X/Y, Z/Y)$ at the point of contact. We direct the axis $X = 0$ along the tangent and the axis $Z = 0$ into one of the two remaining points at infinity.

In writing our form in this coordinate system, the terms

Y^3, Y^2Z, Z^3, X^3 are missing and YZ^2, X^2Z, XZ^2 are present with non-zero coefficients. By dilations we make the non-zero coefficients equal to 1, after which the form becomes $X^2Z + XZ^2 + aX^2Y + bXYZ + YZ^2 + cXY^2$. This form is non-degenerate if its discriminant, the polynomial $\Delta(a, b, c)$, is different from 0. This is the first case of Theorem **90**.

2°. We assume that the curve is the folium of Descartes. We choose an origin of the affine coordinate system $(X/Y, Z/Y)$ at the singular point and direct the axes $(X = 0, Z = 0)$ to the two points of the curve at infinity. Then in writing down the form, all the terms divisible by Y^2 are missing, as are the terms X^3 and Z^3, and X^2Z, XZ^2, X^2Y, Z^2Y are present with non-zero coefficients.

By dilatations we make the first three non-zero coefficients equal to 1, after which the form becomes $X^2Z + XZ^2 + aX^2Y + bXYZ + YZ^2$. This form yields the folium of Descartes if $b^2 \neq 4a$, $a(a + 1 - b) \neq 0$. This is the second case of Theorem **90** (case *UP*).

3°. If the curve has a cusp, then we choose the origin of the system $(X/Y, Z/Y)$ there and direct the axis $Z = 0$ along the tangent to the cusp, and the axis $X = 0$ to one of the points at infinity of the curve. After dilatations we reduce the form to $X^3 + aX^2Z + XZ^2 + Z^2Y$. This form yields a curve with a cusp $X = Z = 0$. It has three distinct points at infinity $(Y = 0)$ for $a^2 \neq 4$. This is the third case of Theorem **90** (case *UQ*).

4°. If the curve splits into a line and a non-singular conic, then we choose the origin of the coordinate system $(X/Y, Z/Y)$ at one of their points of intersection, and we direct the axis $X = 0$ along a line and the axis $Z = 0$ to one of the points of the conic at infinity. After dilatations the form becomes $X^2Z + XZ^2 + X^2Y + aXYZ$.

The conic does not split if $a \neq 1$. The conic is not tangent to the line $X = 0$ for $a \neq 0$. This is the fourth case of Theorem **90** (case *UR*).

5°. If the conic is tangent to the line under the conditions of 4°, then $a = 0$. This is the fifth case of Theorem **90** (case *US*).

6°. If the curve splits into three lines, then we make two of them the coordinate axes $X = 0$ and $Z = 0$. Then the form becomes $XZ(aX + bY + cZ)$, $ac \neq 0$. If the lines do not intersect at one point, then $b \neq 0$ and we reduce the form to $X^2Z + XZ^2 + XYZ$ by dilatations. This is the sixth case of Theorem **90** (case *UT*).

7°. If the curve given by the form $X^3 + XZ^2 + \alpha Y^3 + \beta XY^2 + \gamma ZY^2 + \delta YX^2$ splits into three lines intersecting at the same point, then necessarily $\alpha = \beta = \gamma = \delta = 0$. For let us consider the point $X = Y = 0$. The tangent to the curve at this point $(X = 0)$ must be contained in the curve. Consequently, $\gamma = \alpha = 0$. The curve $X^2 + Z^2 + \beta Y^2 + \delta XZ$ must have a singular point on the axis $X = 0$. Consequently, $\beta = \delta = 0$. This is the last case of Theorem **90**, which is therefore proved, since we have exhausted all the possible cases (by Theorem **50**).

PROOF OF THEOREM **98**. The problem reduces to the classification of the polynomials $aZ^4 + bZ^3Y + cZ^2Y^2 + dZY^3 + eY^4$ relative to the transformations $Z = \alpha Z' + \beta Y'$, $Y = \gamma Y'$. This classification in the case $a \neq 0$ is carried out in the proof of Theorem **36**, and in the case $a = 0$, $b \neq 0$ in the proof of Theorem **10**. The case $a = b = 0$ is trivial, and Theorem **98** is proved.

PROOFS OF THEOREMS **52**, **54**, **56**, **58**, **66**, **82**, **97**. In [8], 7.3 we proved the following lemma.

LEMMA 1. *Let f_0 be a quasihomogeneous function of degree d (not necessarily non-degenerate) and suppose that the monomials e_k form a basis of the local ring of the gradient mapping f_0. Then every series $f = f_0 + f_1$, where the series f_1 has a higher order of quasihomogeneity than d, reduces by a formal change of variables to the form $f_0 + \Sigma\, c_k e_k$, where the c_k are constants, and the e_k are monomials of the basis whose degree is greater than d.*

We verify the next result directly.

LEMMA 2. *The ideals generated by the partial derivatives of the cubic forms of Theorems* **52**–**97** *contain the following elements*:

Theorem No.	f_0	Elements of the ideal
52	x^3+y^3+xyz	$x^3,\ y^3,\ x^2z,\ y^2z,\ xz^2,\ yz^2$
54	x^3+xyz	$xy,\ xz,\ x^3,\ y^2z,\ yz^2$
56	xyz	$xy,\ xz,\ yz$
58	x^3+yz^2	$x^2,\ z^2,\ yz$
66	x^2z+yz^2	$xz,\ z^2,\ x^2+2yz$
82	x^3+xz^2	$xz,\ 3x^2+z^2,\ x^3,\ z^3$
97	x^2y	$xy,\ x^2$

The proofs of all these theorems follow from Lemmas 1 and 2.

§ 17. Proofs of theorems on the normal form of semiquasihomogeneous functions (Theorems 2, 4, 5, 7, 8, 9, 11, 14, 19, 20, 21, 23, 27, 28, 30, 34, 35, 37, 48, 51, 60, 61, 62, 64, 68, 69, 71, 75, 76, 78, 84, 86, 89, 91, 99)

In these theorems we are given a non-degenerate quasihomogeneous function f_0 of degree N and we find a normal form for the function $f_0 + f_1$, where f_1 has order greater than N. The normal form is like $f_0 + \Sigma\, c_s e_s$, where the e_s are monomials that generate the space of all functions of order greater than N over $\mathbf{C}$ together with the gradient ideal $I = \{(\partial f_0/\partial x_i)\}$ (see Lemma 1 in § 16 and Theorem 7.2 in [8]).

Therefore, for the proof it suffices to verify that for each (weighted) degree $d > N$ the space of quasihomogeneous polynomials of degree d is generated over $\mathbf{C}$ by the polynomials $x^m\, \partial f_0/\partial x_i$ (where $x^m\, \partial_i$ are the "leading monomials" of degree $d - N$) and monomials e_s of degree d included in the normal form of § 13. To verify the independence of the monomials e_s mod I it suffices to compare the number of these with the number p_d of basis monomials of degree d, which is given by the Poincaré polynomial [8]

$$p(t) = \Sigma p_d t^d = \Pi (t^{N-A_i} - 1)/\Pi (t^{A_i} - 1),$$

where the A_i are weights (so that the degree of x^m is equal to $\Sigma\, m_i A_i$, that of f_0 is N).

As an example we consider the proof of Theorem 37. In this case $f_0 = x^4 + bx^3y^k + ax^2y^{2k} + xy^{3k}$, so that $A_1 = k$, $A_2 = 1$; $N = 4k$. The degree of $x^p y^q$ is $d = kp + q$. The degree of the Poincaré polynomial is $6k - 2$. Consequently, the condition for f_0 to be non-degenerate is the same as that for all the monomials of degree $d = 6k - 1$ to belong to I.

In the following table, for the values of d in which we are interested (that is, for $4k < d \leqslant 6k - 1$) we have indicated the number $\nu(d)$ of monomials of degree d, the leading monomials $x^m \partial_i$, and the basis monomials e_s.

Here in the notation of the basis and leading monomials we do not indicate the degree of y, for example, in place of $xy^u \partial_y$ we write $x\partial_y$ (the quantity u can be recovered from d).

d	$4k \leqslant d \leqslant 5k-2$	$5k-1$	$5k \leqslant d \leqslant 6k-2$	$6k-1$
$\nu(d)$	5	5	6	6
$\{x^m \partial_i\}$	$x\partial_x, \partial_x, \partial_y$	$x\partial_x, \partial_x, x\partial_y, \partial_y$	$x^2\partial_x, x\partial_x, \partial_x, x\partial_y, \partial_y$	$x^2\partial_x, x\partial_x, \partial_x, x^2\partial_y, x\partial_y, \partial_y$
$\{e_s\}$	x^2, x^3	x^3	x^3	—

To check that $\{e_s\}$ and $\{x^m \partial_i f_0\}$ generate $\mathbf{C}^{\nu(d)}$, it suffices to verify that the "quasidiscriminant" is different from 0 (this is a determinant formed from the components of the expansions of these $\nu(d)$ functions in the basis of monomials of degree d). The four cases of the preceding table correspond to 4 determinants $\Delta_1 - \Delta_4$, which are, respectively,

$$\begin{vmatrix} 4 & & & & \\ 3b & 4 & kb & & 1 \\ 2a & 3b & 2ka & 1 & \\ 1 & 2a & 3k & & \\ & 1 & & & \end{vmatrix}, \quad \begin{vmatrix} 4 & & kb & & \\ 3b & 4 & 2ka & kb & 1 \\ 2a & 3b & 3k & 2ka & \\ 1 & 2a & & 3k & \\ & 1 & & & \end{vmatrix}, \quad \begin{vmatrix} 4 & & & & & \\ 3b & 4 & & kb & & \\ 2a & 3b & 4 & 2ka & kb & 1 \\ 1 & 2a & 3b & 3k & 2ka & \\ & 1 & 2a & & 3k & \\ & & 1 & & & \end{vmatrix}, \quad \begin{vmatrix} 4 & & & & & \\ 3b & 4 & & kb & & \\ 2a & 3b & 4 & 2ka & kb & \\ 1 & 2a & 3b & 3k & 2ka & kb \\ & 1 & 2a & & 3k & 2ka \\ & & 1 & & & 3k \end{vmatrix}$$

Thus, $\Delta_1 = 12k$, $\Delta_2 = 4k^2(ab - 9)$, $\Delta_3 = 4\Delta_2$, $\Delta_4 = -64k^3\Delta_5$, where Δ_5 is the discriminant of the polynomial $x^3 + bx^2 + ax + 1$. The non-degeneracy condition $\Delta_5 \neq 0$ has the form $4(a^3 + b^3) + 27 \neq a^2b^2 + 18ab$.

If $\Delta_5 \neq 0$, $ab \neq 9$, then all four determinants are different from 0, and hence, f reduces to the normal form

$$f_0+\sum c_s e_s = x^4 + bx^3y^k + ax^2y^{2k} + xy^{3k},$$

where $a = a_0 + \ldots + a_{k-2}y^{k-2}$, $b = b_0 + \ldots + b_{2k-2}y^{2k-2}$.

But by Theorem 36 every non-degenerate quasihomogeneous function $x^4 + bx^3y^k + ax^2y^{2k} + xy^{3k}$ is equivalent to one for which $ab \neq 9$. Theorem 37 is now proved.

The difficulty in the proof of this and the remaining theorems of the present section is that we have to guess the monomials e_s that are suitable simultaneously for all the values of the parameters of the quasihomogeneous principal part. The existence of such monomials is not a priori obvious.

It is also appropriate to remark that the classification of semiquasihomogeneous functions, like that of quasihomogeneous polynomials, reduce to calculations that actually do not depend on k. However, in contrast to the quasihomogeneous case, in the semiquasihomogeneous case (and a fortiori in the general case of an arbitrary Newton diagram), this independence from k has not yet been explained satisfactorily.

In many cases we may avoid the determinants and guess the basis monomials with the help of the technique of "crosswords" from [8]. For example, let us prove Theorem 71.

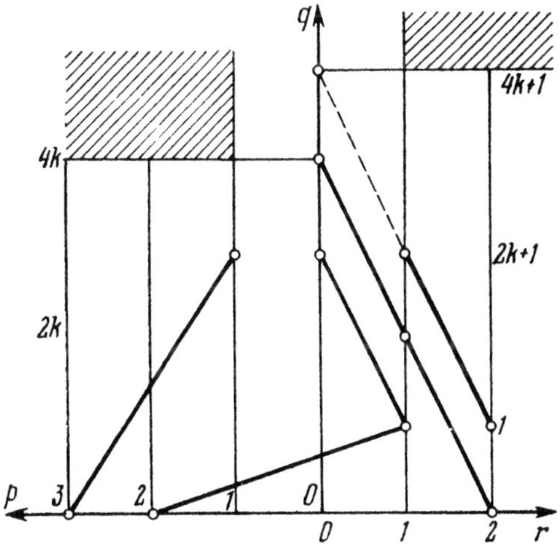

Fig. 12

In this theorem, $f_0 = x^2z + yz^2 + y^{4k+1} + bzy^{2k+1}$, $b^2 \neq 4$.

The gradient ideal I contains xz, so that the exponents of all the basis monomials $e_s = x^py^qz^r$ lie on the planes (p, q) and (q, r), where it is also convenient to draw the crossword. The derivatives

$$\partial f_0/\partial y = z^2 + \alpha zy^{2k} + \beta y^{4k}, \quad \partial f_0/\partial z = x^2 + 2yz + by^{2k+1}$$

(where $\alpha = (2k+1)b$, $\beta = 4k+1$) correspond to chains of three points. Translating the chains, we can see that all the monomials on the planes

(p, q) and (q, r) are congruent mod I to combinations of monomials of the form xy^q, y^q, zy^q. Looking at the crossword we successively find (Fig. 12) on p. 63 $xy^4 \in I$, $x^3 + bxy^{2k+1} \in I$, $2yz^2 + bzy^{2k+1} \in I$, $yz^2 + \alpha zy^{2k+1} + \beta y^{4k+1} \in I$, $bzy^{2k+1} + 2y^{4k+1} \in I$, $2zy^{4k+1} + by^{6k+1} \in I$. For $b^2 \neq 4$, from the last two relations we conclude $y^{6k+1} \in I$, $zy^{4k+1} \in I$. Consequently, the monomials $xy^q (3k+1 \leqslant q \leqslant 4k-1)$ and $zy^q (2k+1 < q \leqslant 4k)$ together with I generate the whole space of functions whose order is greater than the degree of f_0. Hence, for $b^2 \neq 4$ the quasihomogeneous function f_0 is non-degenerate.

To prove the independence of these new monomials mod I, we observe that $12k+2$ of the monomials $xy^q (0 \leqslant q \leqslant 4k-1)$, $y^q (0 \leqslant q \leqslant 4k)$, $zy^q (0 \leqslant q \leqslant 4k)$ together with I generate the whole ring $\mathbf{C}[[x, y, z]]$ over $\mathbf{C}$. On the other hand, the weights A_i and the degree N are equal to $(2k+1, 2, 4k)$ and $8k+2$. Hence, the dimension of the local ring $\mathbf{C}[[x, y, z]]/I$ over $\mathbf{C}$ is equal to

$$\mu = \Pi(N - A_i)/\Pi A_i = 12k + 2.$$

Consequently, the listed $12k+2$ monomials are independent, as required.

The remaining theorems on semiquasihomogeneous singularities are verified similarly; the determinants only must be considered for the proof of Theorems **37**, **48**, **51**, **64**, **71**, **78**, **86**, **91**, **99**. Here the crossword gives not only the normal forms, but also all the generators of the local ring (and, consequently, a formula for the versal deformation).

§ 18. Proofs of theorems on the normal forms of singularities of the series T (Theorems 53, 55, 57)

We use the terminology of [8].

PROPOSITION. *Let* $f_0 = ax_1x_2x_3 + x_1^{p_1} + x_2^{p_2} + x_3^{p_3}$, *where* $3 \leqslant p_1 \leqslant p_2 \leqslant p_3 > 3, a \neq 0$. *Then*:

1) *the function* f_0 *has a critical point* O *of finite multiplicity* $\mu = p_1 + p_2 + p_3 - 1$;

2) *the system of monomials* $1, x_1x_2x_3, x_i^{s_i}$ $(0 < s_i < p_i,\ i = 1, 2, 3)$ *defines a regular basis of the local ring*;

3) *condition* A *holds for the Newton filtration defined by the four monomials of* f_0.

PROOF. We apply the technique of crosswords described in [8], §9.7.

LEMMA 1. *All the cycles of the admissible segments for* f_0 *are trivial.*

PROOF. This follows from the linear independence of the three basic segments, which is guaranteed by the conditions on p_i.

LEMMA 2. *The maximal admissible chains are the following*:

1) *each of the points* $x_i^{s_i}$ $(0 \leqslant s_i \leqslant p_i - 2)$ *has a trivial admissible chain from one of these points*;

2) *there are four finite maximal chains*

$$x_1x_2 - x_3^{p_3-1}, \quad x_2x_3 - x_1^{p_1-1}, \quad x_3x_1 - x_2^{p_2-1}, \quad x_1x_2x_3 - \{x_i^{p_i}\};$$

3) *infinite admissible chains issue from all the remaining points.*

Lemma 2 is geometrically obvious, but we give a formal proof.

PROOF OF LEMMA 2. We consider an arbitrary monomial $\Pi x_i^{q_i}$. It can be written in the form $x^\xi y^\eta z^\zeta$, where $\xi \leqslant \eta \leqslant \zeta$; in this notation (depending on q_i), $f_0 = axyz + x^p + y^q + z^r$. Then there are eight mutually exclusive possibilities:

0. $0 = \xi < \eta = \zeta = 1.$

I. $\xi < \eta \leqslant \zeta \neq 1.$

II. $0 < \xi = \eta < \zeta.$

III. $0 < \xi = \eta = \zeta > 1.$

IV. $0 < \xi = \eta = \zeta = 1.$

V. $0 = \xi = \eta, \quad \zeta > r.$

VI. $0 = \xi = \eta, \quad \zeta = r.$

VII. $0 = \xi = \eta, \quad \zeta < r.$

Corresponding to this we distinguish eight types of monomials. If for a monomial $x^\xi y^\eta z^\zeta$ the condition $\zeta \leqslant \eta \leqslant \xi$ does not hold, then we define its type as that of a monomial with regular order of the exponents obtained by a permutation of the names of the variables.

The vector-exponent of a monomial of type i is called a point of type i. From each point of the types I, II, III, IV there issues an admissible x-, y- or z-segment, taking into one point the same of a higher Newton filtration. In the following table we indicate the ends of these segments and their type:

Type of start	Start	Type of segment	End	Type of end	Filtration
I	ξ, η, ζ	x	$\xi+p-1, \eta-1, \zeta-1$	I, II or III	$>$
II	ξ, ξ, ζ	y	$\xi-1, \xi+q-1, \zeta-1$	I	$=$
III	ξ, ξ, ξ	z	$\xi-1, \xi-1, \xi+r-1$	II	$=$
V	$0, 0, \zeta$	z	$1, 1, \zeta-r+1$	II	$=$

Consequently, from each point of each of the types I, II, III, V there issues an admissible chain of at most three segments, ending at a point of a strictly larger filtration of one of the same types. Thus, the maximal chains of points of these types are infinite.

The admissible chains of points $x_i^{s_i}$ $(0 \leqslant s_i \leqslant p_i - 2)$ of type VII are obviously trivial. There remain ten points, corresponding to monomials of the types 0, IV, VI, VII:

$$x_i^{p_i}, \quad x_i^{p_i-1}, \quad x_1x_2, \quad x_2x_3, \quad x_3x_1, \quad x_1x_2x_3.$$

The six admissible segments passing through them have no other vertices. Therefore the maximal chains of them are finite. Lemma 2 is now proved.

The fact that of the critical point O of f_0 is isolated can be verified directly. In each finite maximal chain of monomials we choose a higher filtration. We obtain a regular basis of $\mu = p_1 + p_2 + p_3 - 1$ monomials. Assertions 1 and 2 of our proposition are proved (see [8], §9.7).

All strictly upper monomials for f_0 have types I, II, III, V. Consequently, all of them belong to the ideal $\{\partial f_0/\partial x_i\}$. We check that condition A holds (see [8], §9.2).

We denote by φ the Newton filtration (φ assigns to each function the cardinal number d of the smallest term $\mathscr{E}_d$, of the filtration to which the function belongs; $\varphi(f_0) = \varphi(x_i^{p_i}) = \varphi(x_1x_2x_3) = 1$). We denote by ψ the filtration induced by in the space of vector fields:

$$\psi(\gamma) = \max\{\delta : \gamma\mathscr{E}_d \subset \mathscr{E}_{d+\delta} \forall d\}.$$

LEMMA 3. *Each upper monomial α admits the decomposition*

$$\alpha = \gamma f_0 + \alpha', \quad \varphi(\alpha') > \varphi(\alpha), \quad \psi(\gamma) = \varphi(\alpha) - 1.$$

PROOF OF LEMMA 3. In the proof of Lemma 2 we constructed an admissible chain (of at most three segments), joining α to a monomial of the higher filtration α' ($\alpha = \alpha_1; \alpha_2; \ldots; \alpha'$). By the definition of an admissible chain

$$\alpha_i = a_i\alpha_{i+1} + b_iv_i\partial f_0/\partial w_i, \tag{1}$$

where a_i and b_i are constants, v_i is a monomial, and w_i is one of the coordinates x_1, x_2, x_3. By construction, $\varphi(\alpha_{i+1}) \geqslant \varphi(\alpha_i)$, and at least one of the inequalities is strict.

We claim that $\psi(v_i\partial/\partial w_i) = \varphi(\alpha_i) - 1$. For we can extend a piecewise-linear function φ of the exponents of the monomials into a domain where one of the exponents is equal to -1, linearly through a plane on which this exponent is 0. Then

$$\psi(v_i\partial/\partial w_i) = \varphi(v_i/w_i).$$

The vector-exponents of the monomials $x_1x_2x_3$ and v_i/w_i lie together in one of three convex cones, in which φ depends linearly on the exponents. Therefore,

$$\varphi(v_ix_1x_2x_3/w_i) = \varphi(v_i/w_i) + \varphi(x_1x_2x_3) = \psi(v_i\partial/\partial w_i) + 1.$$

On the other hand, we compute v_i and w_i for each of the groups constructed in the proof of Lemma 2. If α_i is of type I, II or III, we obtain $\varphi(v_ix_1x_2x_3/w_i) = \varphi(\alpha_i)$, and if of type V, $\varphi(v_ix_1x_2x_3/w_i) = \varphi(\alpha_{i+1}) = \varphi(\alpha_i)$. Thus, in all cases $\psi(v_i\partial/\partial w_i) = \varphi(\alpha_i) - 1$.

Since $\varphi(\alpha_i) \geqslant \varphi(\alpha_1)$, we obtain $\psi(v_i\partial/\partial w_i) \geqslant \varphi(\alpha_1) - 1$. Now we express according to (1) α_1 by α_2, α_2 by α_3 etc. to α', with $\varphi(\alpha') > \varphi(\alpha)$. This gives us the required decomposition and proves the proposition.

Now we can deduce Theorems 53, 55, 57. We claim that if *f has a critical point O of finite multiplicity and*

$$f = xyz + \alpha(x) + \beta(y) + \gamma(z), \quad j_2(\alpha, \beta) = 0, \quad j_3(\gamma) = 0,$$

then in a neighbourhood of O,

$$f \sim f_0 = axyz + x^p + y^q + z^r, \quad a \neq 0, \quad 3 \leqslant p \leqslant q \leqslant r > 3.$$

For suppose that the first non-zero terms of the Taylor series for α, β, γ have the exponents p, q, r. Changing the notation, if necessary, we obtain $3 \leqslant p \leqslant q \leqslant r > 3$. By a dilatation of the axes we reduce f to the form $f_0 + f_1$, where $\varphi(f_1) > 1$. By the proposition proved above, all the monomials of f_1 lie in the ideal $\{\partial f_0/\partial x_i\}$ and f_0 satisfies condition A. By Theorem **9.5** of [8], $f \sim f_0$, as required.

References[1]

[1] N. A'Campo, Sur la monodromie des singularités isolées d'hypersurfaces complexes, Invent. Math. **20** (1973), 147–169. MR **49** # 3201.

[2] N. A'Campo, Le nombre de Lefschetz d'une monodromie, Indag. Math. **35** (1973), 113–118. MR **47** # 8903.

[3] V. I. Arnol'd, Singularities of smooth mappings, Uspekhi Mat. Nauk **23**:1 (1968), 3–44. MR **37** # 2243.
= Russian Math. Surveys **23**:1 (1968), 1–43.

[4] V. I. Arnol'd, On some topological invariants of algebraic functions, Trudy Moskov. Mat. Obshch. **21** (1970), 27–46. MR **43** # 225.
= Trans. Moscow Math. Soc. **21** (1970), 30–52 (1972).

[5] V. I. Arnol'd, Integrals of rapidly oscillating functions and singularities of the projections of Lagrangian manifolds, Funktsional. Anal. i Prilozhen. **6**:3 (1972), 61–62. MR **50** # 8594.
= Functional Anal. Appl. **6** (1973), 222–224.

[6] V. I. Arnol'd, Normal forms of functions close to degenerate critical points, the Weyl groups A_k, D_k, E_k, and Lagrangian singularities, Funktsional. Anal. i Prilozhen. **6**:4 (1972), 3–25. MR **50** # 8595.
= Functional Anal. Appl. **6** (1973), 254–272.

[7] V. I. Arnol'd, Remarks on the stationary phase method and Coxeter numbers, Uspekhi Mat. Nauk **28**:5 (1973), 17–44. MR **53** # 1635.
= Russian Math. Surveys **28**:5 (1973), 19–48.

[8] V. I. Arnol'd, Normal forms of functions in a neighbourhood of a degenerate critical point, Uspekhi Mat. Nauk **29**:2 (1974), 11–49.
= Russian Math. Surveys **29**:2 (1974), 10–50.

[9] V. I. Arnol'd, Critical points of functions and the classification of caustics, Uspekhi Mat. Nauk **29**:3 (1974), 243–244 (*summary of a lecture, with errata to [6] and [8]).

[10] V. I. Arnol'd, *Matematicheskoi metody klassicheskoi mekhaniki,* Izdat. Nauka, Moscow 1974.
*Translation: *Mathematical methods of classical mechanics,* Springer-Verlag, Berlin-Heidelberg-New York and Izdat. Mir, Moscow, to appear.

[1] The references prefixed by an asterisk were added by the translator, while references [91] through [132] were added by the author for this translation.

[11] V. I. Arnol'd, Contact manifolds, Legendrian mappings, and singularities of wave fronts, Uspekhi Mat. Nauk **29**:4 (1974), 153–154 (*summary of a lecture)
[12] V. I. Arnol'd, Lectures on bifurcations in versal families, Uspekhi Mat. Nauk **27**:5 (1972), 119–184.
= Russian Math. Surveys, **27**:5 (1972), 54–123.
[13] V. I. Arnol'd, I. M. Gel'fand, Yu. I. Manin, B. G. Moishezon, S. P. Novikov and I. R. Shafarevich, Galina Nikolaevna Tyurina (obituary), Uspekhi Mat. Nauk **26**:1 (1971), 207–211. MR **44** # 6430.
= Russian Math. Surveys **26**:1 (1971), 193–197.
[14] M. Artin, On the solutions of analytic equations, Invent. Math. **5** (1968), 277–291. MR **38** # 344.
[15] M. F. Atiyah, Resolution of singularities and division of distributions, Comm. Pure Appl. Math. **23** (1970), 145–150. MR **41** # 815.
[16] I. N. Bernstein and S. I. Gel'fand, Meromorphy of the functions P^λ, Funktsional. Anal. i Prilozhen. **3**:1 (1969), 84–85. MR **40** # 723.
= Functional Anal. Appl. **3** (1969), 68–69.
[17] I. N. Bernstein, Analytic continuation of generalized functions with respect to a parameter, Funktsional. Anal. i Prilozhen. **6**:4 (1972), 26–40. MR **47** # 9269.
= Functional Anal. Appl. **6** (1973), 273–285.
[18] E. Brieskorn, Rationale Singularitäten komplexer Flächen, Invent. Math. **4** (1968), 336–358. MR **36** # 5136.
[19] E. Brieskorn, Beispiele zur Differentialtopologie von Singularitäten, Invent. Math. **2** (1966), 1–14. MR **34** # 6788.
[20] E. Brieskorn, Singular elements of semisimple algebraic groups, Actes Congres Internat. Mathématiciens, Nice 1970, Tome 2, 279–284, Gauthier-Villars, Paris 1971.
[21] E. Brieskorn, Die Fundamentalgruppe des Raumes der regulären Orbite einer endlichen komplexen Spiegelungsgruppe, Invent. Math. **12** (1971), 57–61. MR **45** # 2692.
= Uspekhi Mat. Nauk **30**:6 (1975), 147–151.
[22] E. Brieskorn, Sur les groupes des tresses (d'après V. I. Arnol'd), Séminaire Bourbaki, 1971–72, exp. 401, Lecture Notes in Math. **317**, Springer-Verlag, Berlin-Heidelberg-New York 1973.
[23] E. Brieskorn and K. Saito, Artin-Gruppen unde Coxeter-Gruppen, Invent. Math. **17** (1972), 245–271. MR **48** # 2263.
= Matematika **18**:6 (1974), 56–79.
[24] E. Brieskorn, Die Monodromie der isolierten Singularitäten von Hyperflächen, Manuscripta Math. **2** (1970), 103–106. MR **42** # 2509.
= Matematika **15**:4 (1971), 130–160.
[25] J. Cerf, La stratification naturelle des espaces de fonctions différentiables réelles et le théorème de la pseudo-isotopie, Inst. Hautes Études Sci. Publ. Math. **39** (1970), 5–173. MR **45** # 1176.
[26] P. Deligne, Les immeubles des groupes de tresses généralisés, Invent. Math. **17** (1972), 273–302.
[27] M. Demazure, Sousgroupes algébriques de rang maximum du groupe de Cremona, Ann. Sci. Ecole Normale Sup. (4) **3** (1970), 507–588. MR **44** # 1672.
[28] I. V. Dolgachev, Factor-conical singularities of complex hypersurfaces, Funktsional. Anal. i Prilozhen. **8**:2 (1974), 75–76. MR **49** # 10700.
= Functional Anal. Appl. **8** (1974), 160–161.

[29] I. V. Dolgachev, Automorphic forms and quasihomogeneous singularities, Funktsional. Anal. i Prilozhen. **9**:2 (1975), 67–68.
= Functional Anal. Appl. **9** (1975), 149–151.
[30] J. J. Duistermaat, Oscillatory integrals, Lagrange immersions, and unfolding of singularities, Comm. Pure Appl. Math. **27** (1974), 207–281.
[31] A. Fresnel, Mémoire sur la diffraction de la lumière, Mém. Acad. Roy. Sci. Paris **5** (1821–22), 339–475 (*published in 1826, submitted in 1818 – the usual citation is incorrect).
[32] D. B. Fuks, Quillenization and bordism, Funktsional. Anal. i Prilozhen. **8**:1 (1974), 36–42. MR **49** # 8043.
= Functional Anal. Appl. **8** (1974), 31–36.
[33] D. B. Fuks, Cohomology of braid groups mod 2, Funktsional. Anal. i Prilozhen. **4**:2 (1970), 62–73; MR **43** # 226.
= Functional Anal. Appl. **4** (1970), 143–151.
[34] A. M; Gabrielov, Intersection matrices for certain singularities, Funktsional. Anal. i Prilozhen. 7:3 (1973), 18–32. MR **48** # 2418.
= Functional Anal. Appl. 7 (1974), 182–193.
[35] A. M. Gabrielov, Dynkin diagrams of unimodal singularities, Funktsional. Anal. i Prilozhen. **8**:3 (1974), 1–6. MR **51** # 3516.
= Functional Anal. Appl. **8** (1975), 192–196.
[36] A. M. Gabrielov, Bifurcations, Dynkin diagrams, and the modality of isolated singularities, Funktsional. Anal. i Prilozhen. **8**:2 (1974), 7–12.
= Functional Anal. Appl. **8** (1974), 94–98.
[37] A. M. Gabrielov and A. G. Kushnirenko, Description of the deformations with constant Milnor number for homogeneous functions, Funktsional. Anal. i Prilozhen. **9**:4 (1975), 67–68.
[38] J. Guckenheimer, Catastrophes and partial differential equations, Ann. Inst. Fourier (Grenoble) **23**:2 (1973), 31–59. MR **51** # 1879.
[39] S. M. Gusein-Zade, Intersection matrices for some singularities of functions of two variables, Funktsional. Anal. i Prilozhen. **8**:1 (1974), 11–15. MR **49** # 3202.
= Functional Anal. Appl. **8** (1974), 10–13.
[40] S. M. Gusein-Zade, Dynkin diagrams for singularities of functions of two variables, Funktsional. Anal. i Prilozhen. **8**:4 (1974), 23–30.
= Functional Anal. Appl. **8** (1975), 295–300.
[41] A. E. Hatcher, Parametrized h-cobordism theory, Ann. Inst. Fourier (Grenoble) **23**:2 (1973), 61–74. MR **50** # 1267.
[*41a] A. E. Hatcher and J. Wagoner, *Pseudo-isotopies of compact manifolds*, Astérisque No. 6, Offilib, Paris 1973. MR **50** # 5821.
[42] L. Hörmander, Fourier integral operators. I, Acta Math. **127** (1971), 79–183. MR **52** # 9399.
= Matematika **16**:1 (1972), 17–61; **16**:2 (1972), 67–136.
[43] F. Hirzebruch, The topology of normal singularities of an algebraic surface (d'après Mumford), Séminaire Bourbaki, 15ème année, Exp. No. 250, 1962/1963, Benjamin, New York.
[44] F. Hirzebruch and K. H. Mayer, *O(n)-Mannigfaltigkeiten, exotische Sphären und Singularitäten*, Lecture Notes in Math. **57**, Springer-Verlag, Berlin-Heidelberg-New York, 1968. MR **37** # 4825.

[45] F. Klein, *Vorlesungen über die Entwicklung der Mathematik im 19. Jahrhundert*, Springer-Verlag, Berlin 1926; reprinted, Chelsea, New York 1967.

[46] N. Kuiper, Algebraic equations for non-smoothable 8-manifolds, Inst. Hautes Etudes Scient. Publ. Math. **33** (1967), 139–155. MR **39** # 3494.

[47] F. Latour, Stabilité des champs d'applications differentiables, généralisation d'un théorème de J. Mather, C. R. Acad. Sci. Paris **268** (1969), A1331–1334. MR **39** # 7617.

[48] F. Lazzeri, A theorem on the monodromy of isolated singularities (Singularités à Cargèse, 1972), Astérisque **7–8** (1973), 269–275, Soc. Math. France, Paris 1973.

[49] Lê Dũng Tráng, Les théorèmes de Zariski de type Lefschetz, preprint, Centre Maths. École Polytechnique, Paris 1971.

[*49a] (published version of [49]) Lê Dũng Tráng and Helmut A. Hamm, Un théorème de Zariski du type de Lefschetz, Ann. Sci. Ecole Norm. Sup. (4) **6** (1973), 317–355.

[50] Lê Dũng Tráng and C. P. Ramanujam, The invariance of the Milnor number implies the invariance of the topological type, preprint, Centre Maths. École Polytechnique, Paris 1973, Amer. J. Math. **98** (1976), 67–68.

[51] E. Locijenga, The complement of the bifurcation variety of a simple singularity, Invent. Math. **23** (1974), 105–116.

[52] B. Malgrange, Intégrales asymptotiques et monodromie, Sém. Leray au Collège de France, 1972/1973; *published version: Ann. Sci. Ecole Norm. Sup. (4) **7** (1974), 405–430. MR **51** # 8459.

[53] B. Malgrange, Sur les polynômes de I. N. Bernstein, Sém. Goulaouic-Schwartz 1973/74, École Polytechnique, Paris, Exp. No. 20.
= Uspekhi Mat. Nauk **29**:4 (1974), 81–88.
= Russian Math. Surveys **29**:4 (1974), 81–88.

[*53a] B. Malgrange, Le polynôme de Bernstein d'une singularité isolée, in: Fourier integral operators and partial differential equations, Lecture Notes in Math. **459**, 98–119, Springer-Verlag, Berlin-Heidelberg-New York 1975.

[54] J. N. Mather, Stability of C^∞-mappings I–VI: Ann. of Math.(2) **87** (1968), 89–104. MR **38** # 726; ibid. **89** (1969), 254–291. MR **41** # 4582; Inst. Hautes Etudes Sci. Publ. Math. **35** (1969), 127–156 (279–308). MR **43** # 1215a; ibid. **37** (1970), 223–248. MR **43** # 1215b; Advances in Math. **4** (1970), 301–335. MR **43** # 1215c; Proc. Liverpool Conf. on Singularities, Vol. I, Lecture Notes in Math. **192**, 207–253, Springer-Verlag, Berlin-Heidelberg-New York 1971. MR **45** # 2747.
= (I) (translation of a preliminary version). In the collection: *Osobennosti differentsiruemykh otobrazhenii* (Singularities of differentiable mappings), 198–215, Izdat. Mir, Moscow 1968. MR **39** # 3511.
= (II) (translation of a preliminary version) In the collection *Osobennosti differentsiruemykh otobrazhenii* (Singularities of differentiable mappings), 217–267, Izdat. Mir, Moscow 1968. MR **40** # 4966.
= (III) Matematika **14**:1 (1970), 146–175.
= (IV) Uspekhi Mat. Nauk **28**:6 (1973), 165–190.
= (V) Uspekhi Mat. Nauk **29**:1 (1974), 99–128.
= (VI) Uspekhi Mat. Nauk **29**:1 (1974), 129–158.

[55] V. P. Maslov, *Teoriya vozmushchenii i asimptoticheskie metody* (Perturbation theory and asymptotic methods), Izdat. Moskov. Gos. Univ., Moscow 1965. Translation: *Théorie des perturbations et méthodes asymptotiques*, Dunod, Paris, 1972.

[56] V. P. Maslov, On the focusing of energy in a crystal lattice, Uspekhi Mat. Nauk **27**:6 (1972), 224 (*summary of a lecture).

[57] J. Milnor, *Singular points of complex hypersurfaces,* Annals of Math. Studies **61**, Princeton Univ. Press, Princeton, N.J., 1968. MR **39** # 969.
Translation: *Osobye tochki kompleksnykh giperpoverkhnostei*, Izdat. Mir, Moscow 1971. MR **50** # 8560.

[58] J. Milnor and P. Orlik, Isolated singularities defined by weighted homogeneous polynomials, Topology **9** (1970), 385–393. MR **45** # 2757.

[59] I. Newton, The method of fluxions (1671), Mathematical papers, Vol. 3, Cambridge University Press, Cambridge 1969, 49 ff.

[60] P. Orlik and P. Wagreich, Isolated singularities of algebraic surfaces with C^*-action, Ann. of Math. (2) **93** (1971), 205–228. MR **44** # 1662.

[61] V. P. Palamodov, The multiplicity of a holomorphic mapping, Funktsional. Anal. i Prilozhen. **1**:3 (1967), 54–65. MR **38** # 4720.
= Functional Anal. Appl. **1** (1967), 218–226.

[62] F. Pham, Formules de Picard-Lefschetz généralisées et ramifications des intégrales, Bull. Soc. Math. France **93** (1968), 333–367. MR **33** # 4064.

[63] F. Pham, Remarque sur l'equisingularité universelle, Preprint, Faculté des Sciences, Nice (1970), 1–24.

[64] H. Poincaré, Sur les formes cubiques ternaires et quaternaires. I, J. École Polytechnique **50** (1881), 190–253 = Oeuvres V, 28–72.
= *Izbrannye trudy*, Tom II, Izdat. Nauka, Moscow 1973, 829–836 (*These pages correspond to §4 of the original paper.)

[65] K. Saito, Einfach-elliptische Singularitäten, Invent. Math., **23** (1974), 289–325. MR **50** # 7147.

[66] K. Saito, Quasihomogene isolierte Singularitäten von Hyperflächen, Invent. Math. **14** (1971), 123–142. MR **45** # 3767.

[67] A. M. Samoilenko, On the equivalence of a smooth function with a Taylor polynomial in a neighbourhood of a critical point of finite type. Funktsional. Anal. i Prilozhen. 2:4 (1968), 63–69. MR **42** # 5132.
= Functional Anal. Appl. **2** (1969), 318–323.

[68] G. Segal, Configuration-spaces and iterated loop-spaces, Invent. Math. **21** (1973), 213–222. MR **48** # 9710.

[69] D. Siersma, The singularities of C^∞-functions of right codimension smaller or equal than eight, Indag. Math. **35**:1 (1973), 31–37. MR **51** # 11574.

[*69a] D. Siersma, *Classification and deformation of singularities,* Academisch Proefschrift, Univ. of Amsterdam, 1974. MR **50** # 3267.

[70] *Singularités à Cargèse,* Astérisque 7/8, Soc. Math. France, Paris 1973. MR **48** # 11548 (table of contents review).

[71] B. Teissier, Cycles évanescents, sections planes et conditions de Whitney, Astérisque 7/8, 285–362, Soc. Math. France, Paris 1973. MR **51** # 10682.

[72] R. Thom and H. Levine, Stability of differentiable mappings. I, Bonner Math. Schriften **6**, Univ. Bonn 1959.
= Proc. Liverpool Singularities Symposium. I, 1–89, Lecture Notes in Math. **192**, Springer-Verlag, Berlin-Heidelberg-New York 1971.
= in the collection: *Osobennosti differentsiruemykh otobrazhenii* (Singularities of differentiable mappings), 9–101, Izdat. Mir, Moscow 1968. MR **40** # 896.

[73] R. Thom and M. Sebastiani, Un résultat sur la monodromie, Invent. Math. **13** (1971), 90–96. MR **45** # 2201.

[74] R. Thom, *Stabilité structurelle et morphogénèse*, Benjamin, Reading, Mass., 1972. Translation: *Structural stability and morphogenesis*, Benjamin, Reading, Mass., 1974.

[75] R. Thom, The bifurcation subset of a space of maps, in: *Manifolds*, Amsterdam, 1970, Lecture Notes in Math. **197**, 202–208, Springer-Verlag, Berlin-Heidelberg-New York 1971. MR **43** # 6941.
= Uspekhi Mat. Nauk **27**:5 (1972), 51–58.

[76] G. N. Tyurina, The topological properties of isolated singularities of complex spaces of codimension one, Izv. Akad. Nauk SSSR Ser. Mat. **32** (1968), 605–620. MR **37** # 3053.
= Math. USSR-Izv. **2** (1968), 557–571.

[77] G. N. Tyurina, Locally semi-universal flat deformations of isolated singularities of complex spaces, Izv. Akad. Nauk SSSR Ser. Mat. **33** (1969), 1026–1058. MR **40** # 5903.
= Math. USSR-Izv. **3** (1969), 967–999.

[78] G. N. Tyurina, Resolution of singularities of flat deformations of rational double points, Funktsional. Anal. i Prilozhen. **4**:1 (1970), 77–83. MR **42** # 2031.
= Functional Anal. Appl. **4** (1970), 68–73.

[79] J. C. Tougeron, Idéaux de fonctions différentiables, Ann. Inst. Fourier (Grenoble) **18**:1 (1968), 177–240. MR **39** # 2171.

[80] A. N. Varchenko, Theorems of topological equisingularity of families of algebraic varieties and a family of polynomial mappings, Izv. Akad. Nauk SSSR Ser. Mat. **36** (1972), 957–1019. MR **49** # 2725.
= Math. USSR-Izv. **6** (1972), 949–1008.

[81] I. M. Vinogradov, *Metod trigonometricheskikh summ v teorii chisel* (The method of trigonometric sums in number theory), Izdat. "Nauka", Moscow 1971.

[82] I. A. Volodin, Generalized Whitehead groups and pseudo-isotopies, Uspekhi Mat. Nauk **27**:5 (1972), 229–230. MR **52** # 15501.

[83] P. Wagreich, Singularities of complex surfaces with solvable fundamental groups, Topology **11** (1972), 51–72. MR **44** # 2754.

[84] J. B. Wagoner, Algebraic invariants for pseudoisotopies, Proc. Liverpool Singularities Symposium. II, 164–190. Lecture Notes in Math. **209**, Springer-Verlag, Berlin-Heidelberg–New York 1971. MR **50** # 1266.

[85] A. Weinstein, Singularities of families of functions, in: *Differentialgeometrie im Grossen, Oberwolfach* 1971, Ber. Math. Forschungsinst. Oberwolfach, Heft 4, BI, Mannheim 1971.

[86] H. Whitney, On singularities of mappings of Euclidean spaces. I, Ann. of Math. (2) **62** (1955), 374–410. MR **17**–518.

[87] V. M. Zakalyukin, A theorem of versality, Funktsional. Anal. i. Prilozhen. **7**:2 (1973), 28–32. MR **47** # 9670.
= Functional Anal. Appl. 7 (1973), 110–112.

[88] O. Zariski, On the Poincaré group of a projective hypersurface, Ann. of Math. (2) **38** (1937), 131–141.

[89] V. S. Kulikov, Degenerations of elliptic curves and resolutions of unimodal and bimodal singularities, Funktsional. Anal. i Prilozhen. **9**:1 (1975), 72–73.
= Functional Anal. Appl. **9** (1975), 69–70.

[90] A. G. Kushnirenko, Newton polygon and Milnor numbers, Funktsional. Anal. i Prilozhen. **9**:1 (1975), 74–75.
= Functional Anal. Appl. **9** (1975), 71–72.

[91] V. I. Arnol'd, A spectral sequence to reduce functions to a normal form, Funktsional. Anal. i Prilozhen. **9**:3 (1975), 81–82. MR **52** #4332.

[92] V. I. Arnol'd, A spectral sequence to reduce functions to a normal form, I. G. Petrovskii memorial volume, Acad. Sci. USSR, 1976, to appear.

[93] V. I. Arnol'd, On the theory of envelopes, Uspekhi Mat. Nauk **31**:5 (1976), to appear.

[94] V. I. Arnol'd, Evolution of wave fronts and equivariant Morse lemma, Comm. Pure Appl. Math. 1976, to appear in a volume dedicated to Siegel.

[95] V. I. Arnol'd, Some unsolved problems in the theory of singularities, Trudy Sobolev Seminar **1**, to appear (1976).

[96] A. G. Kushnirenko, The Newton polyhedron and the number of solutions of a system of *k* equations in *k* unknowns, Uspekhi Mat. Nauk **30**:2 (1975), 266–267 (*summary of a lecture)

[97] A. G. Kushnirenko, The Newton polyhedron and Bezout's theorem, Funktsional. Anal. i Prilozhen. **10**:3 (1976), to appear

[98] A. G. Kushnirenko (Kouchnirenko), Polyhèdres de Newton et nombres de Milnor, Invent. Math., **32** (1976), 1–31.

[99] A. G. Kushnirenko (Kouchnirenko), Recepte pour diviner le nombre de solutions d'un systeme d'équations polynomiales à polyhèdre de Newton donnée, C. R. Acad. Sci., to appear (1976).

[100] D. N. Bernstein, The number of solutions of a system of equations, Funktsional. Anal. i Prilozhen. **9**:3 (1975), 1–4.

[101] D. N. Bernstein, The number of lattice points in polyhedra with integer vertices, Funktsional. Anal. i Prilozhen. **10**:3 (1976), to appear.

[102] V. M. Zakalyukin, On Lagrangian and Legendre singularities, Funktsional. Anal. i Prilozhen. **10**:1 (1976), 37–45.

[103] V. M. Zakalyukin, Rearrangements in wave fronts depending on a parameter, Funktsional. Anal. i Prilozhen. **10**:2 (1976), 69–70.

[104] A. N. Varchenko, ζ-functions of monodromy and the Newton diagram, Invent. Math., to appear (1976).

[105] A. N. Varchenko, Oscillatory integrals and the Newton polyhedron, Funktsional. Anal. i Prilozhen. **10**:3 (1976), to appear.

[106] A. N. Varchenko, Oscillatory integrals and the Newton polyhedron, Notes prepared for Bonner Arbeitstagung, June 1975.

[107] S. M. Gusein-Zade, On the characteristic polynomial of the monodromy for series of singularities, Funktsional. Anal. i Prilozhen. **10** (1976), to appear.

[107a] S. M. Gusein-Zade, Monodromy of isolated singularities, Uspekhi Mat. Nauk **31** (1976), to appear.

[108] O. V. Lyashko, Decompositions of simple singularities of functions, Funktsional. Anal. i Prilozhen. **10**:2 (1976), 49–56.

[109] S. I. Kaliman, The holomorphic universal covering of the space of polynomials without multiple roots, Funktsional. Anal. i Prilozhen. **9**:1 (1975), 71.

[110] V. S. Kulikov, Degenerations of elliptic curves and pseudotaut singularities, Uspekhi Mat. Nauk **30**:2 (1975), 215–216.

[111] V. S. Kulikov, Degenerations of elliptic curves and surface singularities, Ph.D. thesis, Moscow State University, Moscow 1975.

[112] J. Milnor, On the 3-dimensional Brieskorn manifolds $M(p, q, r)$, In: *Knots, groups, and 3-manifolds* (Papers dedicated to the memory of R.H. Fox), 175–225, Annals of Math. Study **84**, Princeton University Press, Princeton, N.J., 1975.

[113] J. Steenbrink, Limits of Hodge structures, Notes prepared for the Bonner Arbeitstagung, June 1975.

[*113a] J. H. M. Steenbrink, Limits of Hodge structures, Invent. Math. **31** (1976), 229–257.

[114] J. H. M. Steenbrink, Intersection form for quasi-homogeneous singularities, Univ. of Amsterdam, Mathematisch Instituut, Report No. 75–09.

[115] M. Golubitsky, Contact equivalence for Lagrangian submanifolds, in *Dynamical Systems – Warwick* 1974, Lecture Notes in Math. **468**, 71–72, Springer-Verlag, Berlin-Heidelberg-New York 1975.

[*115a] M. Golubitsky and V. W. Guillemin, Contact equivalence for Lagrangian manifolds, Advances in Math. **15** (1975), 375–387. MR **51** # 1868 (detailed version of above).

[116] K. Jänich, Caustics and catastrophes, in: *Dynamical Systems – Warwick* 1974, Lecture Notes in Math. **468**, 100–101, Springer-Verlag, Berlin-Heidelberg-New York 1975.

[117] E. C. Zeeman, Catastrophe theory in biology, in: *Dynamical Systems – Warwick* 1974, Lecture Notes in Math. **468**, 101–104, Springer-Verlag, Berlin-Heidelberg-New York 1975.

[118] John Guckenheimer, Solving a single conservation law, in: *Dynamical Systems – Warwick* 1974, Lecture Notes in Math. **468**, 108–134, Springer-Verlag, Berlin-Heidelberg-New York 1975.

[119] John Guckenheimer, Bifurcation and catastrophe, in: *Dynamical Systems* – Salvador 1971, 95–109, Academic Press, New York-London 1973. MR **49** # 9878.

[120] Gordon Wassermann, *Stability of unfoldings,* Lecture Notes in Math. **393**, Springer-Verlag, Berlin-Heidelberg-New York 1974.

[121] Gordon Wassermann, Stability of unfoldings in space and time, Acta Math. **135** (1975), 57–128.

[122] Gordon Wassermann, Stability of caustics, Math. Ann. **216** (1975), 43–50. MR **51** # 9116.

[123] John Guckenheimer, Caustics, in: Global analysis and its applications. II, 281–289, Internat. Atomic Energy Agency, Vienna 1974.

[124] John Guckenheimer, Caustics and non-degenerate Hamiltonians, Topology **13** (1974), 127–133. MR **50** # 11316.

[125] John Guckenheimer, Isochrons and phaseless sets, J. Math. Biol. **1** (1974/75), 259–273.

[126] John Guckenheimer, Shocks and rarefactions in two space dimensions, Arch. Rational Mech. Anal. **59** (1975), 281–291. MR **52** # 8668.

[127] E. C. Zeeman, Levels of structure in catastrophe theory illustrated by applications in the social and biological sciences, Proc. Internat. Congress Mathematicians, Vancouver 1974. II, 533–546, Canadian Math. Congress 1975.

[128] K. Jänich, Caustics and catastrophes, Math. Ann. **209** (1974), 161–180. MR **50** # 3273.

[129] K. Lamotke, Die Homologie isolierter Singularitäten, Math. Z. **143** (1975), 27–44. MR **52** # 11104.

[130] N. A'Campo, Le groupe de monodromie du déploiement des singularités isolées des courbes planes. I, Math. Ann. **213** (1975), 1–32. MR **51** # 13282.

[131] N. A'Campo, Le groups de monodromie du déploiement des singularités isolées des courbes planes. II, Proc. Internat. Congress Mathematicians, Vancouver 1974, Vol. I, 395–404, Canadian Math. Congress 1975.

[132] A. Durfee, Fibered knots and algebraic singularities, Topology **13** (1974), 47–59. MR **49** # 1523.

[*133] P. Gabriel, Unzerlegbare Darstellungen. I, Manuscripta Math. **6** (1972), 71–103. MR **48** # 11212.

[*134] P. Gabriel, Problèmes actuels de théorie de représentations, Enseigement Math. **20** (1974), 323–332. MR **51** # 3229.

[*135] P. Gabriel, Représentations indécomposables, Sém. Bourbaki, 1973/74, Exp. 444, Lecture Notes in Math. **431**, Springer-Verlag, Berlin-Heidelberg-New York 1975.

[*136] L. A. Nazarova and A. V. Roiter, Representations of partially ordered sets, Zap. Nauchn. Sem. Leningrad. Otdel. Mat. Inst. Steklov (LOMI) **28** (1972), 5–31. MR **49** # 4877.
= J. Soviet Math. **3** (1975), 585–606.

[*137] L. A. Nazarova, Representations of quivers of infinite type, Izv. Akad. Nauk SSSR Ser. Mat. **37** (1973), 752–791. MR **49** # 2785.
= Math. USSR-Izv. **7** (1973), 749–792.

[*138] I. M. Gel'fand and V. A. Ponomarev, Problems of linear algebra and classification of quadruples of subspaces in a finite-dimensional vector space. In: Hilbert space operators and operator algebras, Colloq. Math. Soc. János Bolyai **5**, 163–237, North-Holland, Amsterdam 1972. MR **50** # 9896.

[*139] I. N. Bernstein, I. M. Gel'fand and V. A. Ponomarev, Coxeter functors and a theorem of Gabriel, Uspekhi Mat. Nauk **28**:2 (1973), 19–33.
= Russian Math. Surveys **28**:2 (1973), 17–32.

[*140] I. M. Gel'fand and V. A. Ponomarev, Free modular lattices and their representations, Uspekhi Mat. Nauk **29**:6 (1974), 3–58.
= Russian Math. Surveys **29**:6 (1975), 1–56.

[*141] V. Dlab and C. M. Ringel, On algebras of finite representation type, J. Algebra **33** (1975), 306–394. MR **50** # 9974.

[*142] V. Dlab and C. M. Ringel, Representations of graphs and algebras, Mem. Amer. Math. Soc. No. **173** (1976).

[*143] W. V. D. Hodge, The isolated singularities of an algebraic surface, Proc. London Math. Soc. (2) **36** (1929), 133–143.

[*144] M. V. Berry, Waves and Thom's theorem, Advances in Physics **25**:1 (1976), 1–26.

[*145] N. A'Campo, La fonction zêta d'une monodromie, Comment. Math. Helvet. **50** (1975), 233–248. MR **51** # 8106.

[*146] D. Siersma, Periodicities in Arnold's lists of singularities, Univ. of Amsterdam, Mathematisch Instituut, Report No. 76–01.

[*147] A. H. Durfee, Fourteen characterizations of rational double points and simple critical points, Inst. for Adv. Study, Princeton, N.J., 1976, preprint.

Received by the Editors, 26 December 1974

Translated by J. S. Joel

Inventiones math. 35, 87 – 109 (1976)

Inventiones mathematicae

Pudet dicere, ad quot figurarum loca has computationes, otiosus eo tempore, produxi. Nam tunc sane nimis delectabar inventis hisce.

I. Newton, Letter to Oldenburg, October 24, 1676

Local Normal Forms of Functions

V. I. Arnold (Moscow)

To Jean-Pierre Serre

Introduction

This article contains lists of normal forms of functions in the neighbourhoods of critical points (the classification of all singularities with the number of modules $m=0$, 1 and 2 or with multiplicity $\mu \leqq 16$ included) and the proofs of most of these classification theorems (the classification of unimodular singularities included).

For instance, the number ν of stable μ-equivalence classes (see the definitions below), is given for $\mu \leqq 16$ by the following table:

μ	**1**	**2**	3	**4**	5	6	7	**8**	9	10	11	12	13	14	15	**16**
ν	**1**	**1**	1	**2**	2	3	3	**4**	4	7	11	15	14	17	22	**32**

The motivations for and the applications of the classification of critical points of functions, and their relation to Coxeter groups, braids, automorphic functions, platonics, caustics, wavefronts and stationary phase method are discussed in [1].

The technical results are formulated in §§ 1 and 2. We begin with some general remarks, arising from these empirical facts.

1. Normal Forms

For many classes of singularities there exist simple normal forms.

The following definitions give an exact meaning to these words.

The group of germs (or jets) of diffeomorphisms of $\mathbb{C}^n$, leaving O invariant, acts on the space of germs (jets) of functions at O. A *singularity class* is a subset of the space of germs of functions, invariant under this action. Each orbit is such a class. Two germs (jets) belonging to the same orbit are *equivalent*.

Another example of a class is the *μ-equivalence class*. The multiplicity (or the Milnor number) μ of a critical point $O \in \mathbb{C}^n$ of a function f is the Poincaré index of the vector field grad f at the singular point O. Germs of functions f_0, f_1 at a critical point O are *μ-equivalent*, if there exist a holomorphical familie of functions f_t with critical point O of constant multiplicity μ (independent of t, $0 \leqq t \leqq 1$).

Originally publ. in: Invent. Math. 35(1), 87-109, © Springer 1976
Reprint by Springer-Verlag Berlin Heidelberg 2016 with kind permission.

To define the *normal forms*, let us consider the space of polinomials, $M = \mathbb{C}[x_1, \ldots, x_n]$, as a subset in the space of germs of functions $f(x_1, \ldots, x_n)$ at O.

The *normal form for a given class K of functions* is a mapping $\Phi\colon B \to M$ from a finite dimensional linear "parameter space" B to the space of polynomials, satisfying three conditions:

1. $\Phi(B)$ intersects all the orbits, belonging to K;
2. the counterimage of every orbit in B is a finite set;
3. the counterimage of the whole complement to K lies in some hypersurface in B.

The normal form is *polynomial* (resp. *affine*) if Φ is a polynomial (resp. affine) mapping. An affine normal form is *simple*, if Φ is of the form:

$$\Phi(b_1, \ldots, b_r) = \varphi + b_1 \mathbf{x}^{\mathbf{m}_1} + \cdots + b_r \mathbf{x}^{\mathbf{m}_r},$$

where φ is a fixed polynomial, b_i are complex numbers and $\mathbf{x}^{\mathbf{m}_i}$ are monomials. In most applications the polynomial φ itself is "simple", that is – a sum of *few* monomials.

The existence of a unique normal form (even a polynomial one) for the whole μ-equivalence class is by no mean evident a'priori. It is a rather surprising empirical fact, that such normal forms exist for all the singularities in the lists of § 1 (and hence for all the singularities with modules number $m = 0, 1$ and 2).

Most of our normal forms are simple; perhaps, all singularities of § 1 have simple normal forms. It is not known, how vast is the class of singularities, whose μ-equivalence classes admit simple (or polynomial) normal forms (here it is natural to consider stable equivalence classes, see 5).

2. *The Series of Singularities*

All singularities in our lists are divided into some *series*, such as A_k or $\mathbf{Z}_k$. We shall use light characters $A, \ldots, Z$ to denote μ-equivalence classes and bold characters $\mathbf{A}, \ldots, \mathbf{Z}$ to denote singularity classes which contain more than one μ-equivalence classes.

While the series undoubtely exist, they have yet no formal definition. Let us consider, for instance, series **A** and **D**, formed by the orbits of germs

$$A_k = x^{k+1} + y^2 \,(k \geqq 1), \qquad D_k = x^2 y + y^{k-1} \,(k \geqq 4).$$

The A_k and D_k classes adjacences are

$$\begin{array}{ccccccccc} A_1 & \leftarrow & A_2 & \leftarrow & A_3 & \leftarrow & A_4 & \leftarrow & \cdots \\ & & & & \uparrow & & \uparrow & & \\ & & & & D_4 & \leftarrow & D_5 & \leftarrow & \cdots . \end{array}$$

[A class L of singularities is *adjacent* to class K, $K \leftarrow L$, if every germ (sufficient jet) of $f \in L$ can be deformed into a germ (sufficient jet) in K by an arbitrary small deformation.]

It is clear, that A_k and D_k are two different series. What is, however, the exact meaning of this?

To define series **A** is to define such a rule of reversing adjacency arrows, that leads from A_k to A_{k+1} and not to D_{k+1}. In the case of **A** singularities the rule is easy to find (**A** singularities have corank 1). In more complicated cases (see §§ 1, 2) one can also find the rules of reversing arrows (different in different cases). So, series with one or more subscripts and parameters occur (*e.g.*, $T_{k,l,m}=axyz+x^k+y^l+z^m$).

In all these cases one can find the definition of the series *after* one have found the series, while no a'priori general definition is known. It is clear, however, that the series are connected with singularities of infinite multiplicity (*e.g.* $\mathbf{D}\sim x^2y$, $\mathbf{T}\sim xyz$). So the hierarchy of series reflects the hierarchy of nonisolated singularities.

3. *Periodicity*

The decomposition of many singularity classes into μ-equivalence classes is in some sence periodic. The whole decomposition (stratification) can be regarded as a chain of similar fragments ("animals"). Each animal is built of points (μ-equivalence classes), two of which are distinguished (the "head" and the "tail"). An animal can also contain some classes whose adjacency arrows join the tail to the head and some "legs" (infinite series of classes). The tail of each animal in the chain is adjacent to the head of the following one. For example, the stratification of the corank 2 singularities with the 3-jet equal to x^3 is a chain of animals each consisting of 5 points and 1 infinite leg:

$$\begin{array}{l} \circ \leftarrow \bullet \leftarrow \bullet \leftarrow \bullet \leftarrow \circ \\ \quad\;\;\uparrow \\ \quad\;\;\bullet \leftarrow \cdots \end{array} \qquad (J_k \text{ in the list of } \S\,1).$$

The phenomenon of periodicity is only partialy explained and for quasihomogenous singularities only (the explanation is based upon some root technique for the quasihomodenous Lie algebra, related to that of Enriques and Demazure [2], see § 3). However this periodicity arises in all the calculations, connected with our classification theorems (so that it is sufficient to consider only first animal from every animal chain). This empirical periodicity have at present no satisfactory explanation. Like the existence of series, it suggests some algebraic structure of the set of all μ-equivalence classes.

4. *Small Modules Number Classes*

The main characteristic of a singularity class is from the point of view of applications its codimension c in the space of germs of functions with critical point O and critical value 0. Indeed, a generic function has only nondegenerate singularities (singularities with codimension $c=0$). Degenerate singularities appears inevitably only when *families* of functions depending upon parameters are considered. A class **K** of codimension c is unavoidable by small deformation if the number of parameters l is $\geqq c$.

So in all the applicatious to l-parametrical problems we need classification of all singularities *up to codimension* l (that is, we need classify such a set of classes, that the complement to their union be of codimension larger than l).

This problem is not to be confused with the classification of singularities with *orbit codimension* $\leqq l$ (that is, with $\mu \leqq l+1$). The only value of the last problem for applications is its relation with the former one.

From the topological point of view the main characteristic of a singularity is the *multiplicity* μ of the critical point (equal to the number of nondegenerate critical points, to which a given point bifurcate after a small variation; not to be confused with the algebraico-geometrical multiplicity of the hypersurface $f^{-1}(0)$ at O).

An unexpected result of our calculation is the fact, that *algebraically the most natural classification theorems are not those of singularities of small codimension c or of small multiplicity* μ, *but the classification of singularities with small modules number, m.*

The *modules number m* of a germ of function $f: (\mathbb{C}^n, O) \to (\mathbb{C}, O)$ at critical point O is the minimal number M, such that some neighbourhood of a sufficient k-jet of the function f at O is covered by a finite number of no more than M-parametrical families of the orbits of the group of diffeomorphisms germs $(\mathbb{C}^n, O) \to (\mathbb{C}^n, O)$ acting on the space of functions with critical point O and critical value 0.

The modules number m is equal to the dimension of the $\mu=$const stratum in the base space of the versal deformation minus 1 (see Gabrielov [3]). So the codimension c of this stratum in the space of germs of functious with critical value 0 at the critical point O, the multiplicity μ and the modules number m are related by the formula

$$\mu = c + m + 1.$$

At present, the complete classifications are known for:

(i) all singularities with $c \leqq 10$;
(ii) all singularities with $\mu \leqq 16$;
(iii) all singularities with $m \leqq 2$.

The singularities with the modules number $m=0$, 1 and 2 are called *simple*, *unimodular* and *bimodular* respectively. Lists of them are given in § 1. A study of these lists leads to the following 4 conclusions:

1. The simple singularities are classified by the Coxeter groups A_k, D_k, E_6, E_7, E_8 (that is, by the platonic in the usual 3-space, see [4]).

2. The unimodular singularities form one infinite series $T_{k,l,m}$ and 14 "exceptional" families, generated by 14 quasihomogenous singularities ([5]).

All the quasihomogenous unimodular singularities can be constructed from 14 remarkable Lobatchevski triangles and 3 remarkable Euclidean triangles like simple singularities are constructed from platonics (Dolgatchev [6]).

3. The bimodular singularities form 8 infinite series and 14 "exceptional" families, generated by quasihomogenous singularities.

All the quasihomogenous bimodular singularities can be constructed from 6 Lobatchevski quadrangles and 14 Lobatchevski triangles. In the last case one has to consider 2-, 3- and 5-fold coverings of the surface, constructed in the canonical way (Dolgatchev).

4. All the singularities with the modules number $m=1$ and 2 are classified by the degenerations of elliptic curves, studied by Kodaira (Koulikov [7]). The Koulikov construction begins with a blowing up of 1, 2 or 3 points of the surface on the minimal resolution of the degenerate fiber; then one has to blow down this (resolved) fiber.

Unfortunately, all these coincidence of different classifications have only a' posteriori proofs and thus depend on our calculations.

5. *Stabilization*

Two germs $f: (\mathbb{C}^n, O) \to (\mathbb{C}, 0)$ and $g: (\mathbb{C}^m, O) \to (\mathbb{C}, 0)$ are *stably equivalent*, if they become equivalent after a direct addition to both of nondegenerate quadratic forms (*e.g.*, $f(x)=x^3$ is stably equivalent to $g(x, y)=x^3+y^2$, but not to $h(x, y)=x^3$). Two stably equivalent germs on equidimensional spaces are equivalent. In § 1 we shall not show explicitely the arguments of functions when all the arguments are presented in the formulae.

The *corank* of a germ of a function at a critical point is the nullity of its second differential. Every germ of corank r is stably equivalent to a germ of a function in r variables. Classifying functions up to the stable equivalence one normally chooses a normal form with a minimal number of variables (equal to the corank).

Our calculations show, however, that sometimes *one can obtain more natural normal forms with more variables.*

The simplest case is the inclusion of the singularities $x^p+ax^2y^2+y^q$ in the series T; series $\mathbf{W}^{\#} \sim (x^2+y^3)^2$ provides more interesting examples.

To formulate the general stabilisation problem we need some definitions. Let $f \in \mathbb{C}[[x_1, \ldots, x_n]]$. The *support* of f is the set $\operatorname{supp} f \subset \mathbb{Z}^n \subset \mathbb{R}^n$ of all indexes $\mathbf{m}$ of the monomials $\mathbf{x}^{\mathbf{m}}$ with nonzero coefficients in f. The *Newton polyhedron* $\Gamma = \Gamma(f)$ is the boundary of the convex hull of the union of all positive ortants $\mathbf{m} + \mathbb{R}^n_+$, $\mathbf{m} \in \operatorname{supp} f$.

A function f is *Γ-nondegenerate* if the multiplicity μ of its critical point O has the minimal possible value ν, for functions with a given Newton polyhedron Γ. An explicite formula for this *Newton number* $\nu(\Gamma)$ is given by A.G. Kouchnirenko (see [1]).

Most of our normal forms are Γ-nondegenerate functions. It is not known, how vast is the class of functions stably equivalent to Γ-nondegenerate ones. For Γ-nonedegenerate functions classification and reducing to normal forms can be done in a very explicit way: these problems are rather stereometrical (see [8]).

Our lists of normal forms are given in § 1. They contain all the singularities with the modules number $m=0$, 1 and 2, all the singularities with $\mu \leqq 16$, all the singularities of corank 2 with nonzero 4-jet, all the singularities of corank 3 with a 3-jet, which determine an irreducible cubic, and some other singularities.

§ 2 is an outline of the proofs of these results. It contains the statements of 105 theorems. These theorems together with the lists of § 1 form a kind of singularity determinator like plant determinators: an algorithm is described for finding every singularity's place in the lists of § 1.

Most of the theorems of § 2 are proved in § 3. All these proofs depend heavily on the geometrical technique, described in [8]. From this geometrical point of view, the formulae of §§ 1 and 2 are to be considered as descriptions of supports and Newton polyhedrons, and to prove the theorems one has rather to manipulate with these polyhedrons than to calculate.

§ 1. Lists of Normal Forms

Letters $A, \ldots, \mathbf{Z}$ stand here for stable equivalence classes of function germs (or families of function germs).

I. Singularities with Modules Number m = 0, 1 and 2

I.0. Simple Singularities ($m=0$). There are 2 infinite series **A**, **D**, and 3 "exceptional" singularities E_6, E_7, E_8:

A_k	$x^{k+1},\ k \geqq 1$
D_k	$x^2 y + y^{k-1},\ k \geqq 4$
E_6	$x^3 + y^4$
E_7	$x^3 + x y^3$
E_8	$x^3 + y^5$

The adjacency diagram

$$\begin{array}{ccccccccc} \mathbf{A} \leftarrow \mathbf{D} \leftarrow & E_6 & \leftarrow & E_7 & \leftarrow & E_8 \\ & \uparrow & & \uparrow & & \uparrow \\ & (\mathbf{P}) & & (\mathbf{X}) & & (\mathbf{J}) \end{array} \qquad \begin{array}{l} \mathbf{A} = A_1 \leftarrow A_2 \leftarrow A_3 \leftarrow A_4 \leftarrow \cdots \\ \qquad\qquad\qquad\quad \uparrow \\ \qquad\qquad\qquad\quad (D) \\ \mathbf{D} = D_4 \leftarrow D_5 \leftarrow D_6 \leftarrow \cdots. \\ \qquad\qquad\quad \uparrow \\ \qquad\qquad\quad (\mathbf{E}) \end{array}$$

The definitions of **P**, **X**, **J** are to be found below (see II).

I.1. Unimodular Singularities ($m=1$). There are 3 families of parabolical singularities, one series of hyperbolical singularities (with 3 subscripts), and 14 families of exceptional singularities.

The parabolical singularities

P_8	$x^3 + y^3 + z^3 + axyz$	$a^3 + 27 \neq 0$
X_9	$x^4 + y^4 + ax^2 y^2$	$a^2 \neq 4$
J_{10}	$x^3 + y^6 + ax^2 y^2$	$4a^3 + 27 \neq 0$

The hyperbolical singularities

$$T_{p,q,r}\colon x^p + y^q + z^r + axyz, \qquad a \neq 0,\ \frac{1}{p} + \frac{1}{q} + \frac{1}{r} < 1.$$

The 14 exceptional families

E_{12}	$x^3+y^7+axy^5$	E_{13}	$x^3+xy^5+ay^8$
E_{14}	$x^3+y^8+axy^6$	Z_{11}	$x^3y+y^5+axy^4$
Z_{12}	$x^3y+xy^4+ax^2y^3$	Z_{13}	$x^3y+y^6+axy^5$
W_{12}	$x^4+y^5+ax^2y^3$	W_{13}	$x^4+xy^4+ay^6$
Q_{10}	$x^3+y^4+yz^2+axy^3$	Q_{11}	$x^3+y^2z+xz^3+az^5$
Q_{12}	$x^3+y^5+yz^2+axy^4$	S_{11}	$x^4+y^2z+xz^2+ax^3z$
S_{12}	$x^2y+y^2z+xz^3+az^5$	U_{12}	$x^3+y^3+z^4+axyz^2$

Some of the adjacencies between the unimodular singularities

$$\begin{array}{ccccccccccccc}
 & & & & J_{10}=T_{2,3,6} & \rightarrow & (E_8) & & & & & & \\
 & & & & \uparrow & & & & & & & & \\
\cdots\rightarrow & T_{2,3,8} & \rightarrow & & T_{2,3,7} & \leftarrow & E_{12} & \leftarrow & E_{13} & \leftarrow & E_{14} & \leftarrow & (\mathbf{J}_3)
\end{array}$$

$$\begin{array}{cccccccccccc}
 & & & X_9=T_{2,4,4} & \rightarrow & (E_7) & & & & & & \\
 & & & \uparrow & & & & & & & & \\
\cdots\rightarrow & T_{2,4,6} & \rightarrow & T_{2,4,5} & \leftarrow & Z_{11} & \leftarrow & Z_{12} & \leftarrow & Z_{13} & \leftarrow & (\mathbf{Z}_1^1) \\
\cdots\rightarrow & \uparrow & & \uparrow & & \uparrow & & \uparrow & & & & \\
\cdots\rightarrow & T_{2,5,6} & \rightarrow & T_{2,5,5} & \leftarrow & W_{12} & \leftarrow & W_{13} & \leftarrow & (\mathbf{W}_{1,0},\mathbf{N}) & &
\end{array}$$

$$\begin{array}{cccccccccccc}
 & & & P_8=T_{3,3,3} & \rightarrow & (E_6) & & & & & & \\
 & & & \uparrow & & & & & & & & \\
\cdots\rightarrow & T_{3,3,5} & \rightarrow & T_{3,3,4} & \leftarrow & Q_{10} & \leftarrow & Q_{11} & \leftarrow & Q_{12} & \leftarrow & (\mathbf{Q}_2) \\
\cdots\rightarrow & \uparrow & & \uparrow & & \uparrow & & \uparrow & & & & \\
\cdots\rightarrow & T_{3,4,5} & \rightarrow & T_{3,4,4} & \leftarrow & S_{11} & \leftarrow & S_{12} & \leftarrow & (\mathbf{S}_{1,0}) & & \\
\cdots\rightarrow & \uparrow & & \uparrow & & \uparrow & & & & & & \\
\cdots\rightarrow & T_{4,4,5} & \rightarrow & T_{4,4,4} & \leftarrow & U_{12} & \leftarrow & (\mathbf{U}_{1,0},\mathbf{V}) & & & & \\
\cdots\rightarrow & & & \uparrow & & & & & & & & \\
 & & & (\mathbf{O}) & & & & & & & &
\end{array}$$

The singularity classes enclosed in brackets are not unimodular.

I.2. Bimodular Singularities ($m=2$). There are 8 infinite series and 14 exceptional families. In all the formulae of this section $\mathbf{a}=a_0+a_1y$.

The 4 infinite series of bimodular singularities of corank 2

Notation	Normal form	Restrictions	Multiplicity μ
$J_{3,0}$	$x^3+bx^2y^3+y^9+cxy^7$	$4b^3+27\neq 0$	16
$J_{3,p}$	$x^3+x^2y^3+\mathbf{a}y^{9+p}$	$p>0, a_0\neq 0$	$16+p$
$Z_{1,0}$	$y(x^3+dx^2y^2+cxy^5+y^6)$	$4d^3+27\neq 0$	15
$Z_{1,p}$	$y(x^3+x^2y^2+\mathbf{a}y^{6+p})$	$p>0, a_0\neq 0$	$15+p$
$W_{1,0}$	$x^4+\mathbf{a}x^2y^3+y^6$	$a_0^2\neq 4$	15
$W_{1,p}$	$x^4+x^2y^3+\mathbf{a}y^{6+p}$	$p>0, a_0\neq 0$	$15+p$
$W^{\#}_{1,2q-1}$	$(x^2+y^3)^2+\mathbf{a}xy^{4+q}$	$q>0, a_0\neq 0$	$15+2q-1$
$W^{\#}_{1,2q}$	$(x^2+y^3)^2+\mathbf{a}x^2y^{3+q}$	$q>0, a_0\neq 0$	$15+2q$

The 4 infinite series of bimodular singularities of corank 3

Notation	Normal form	Restrictions	Multiplicity μ
$Q_{2,0}$	$x^3+yz^2+\mathbf{a}x^2y^2+xy^4$	$a_0^2 \neq 4$	14
$Q_{2,p}$	$x^3+yz^2+x^2y^2+\mathbf{a}z^{6+p}$	$p>0, a_0 \neq 0$	$14+p$
$S_{1,0}$	$x^2z+yz^2+y^5+\mathbf{a}zy^3$	$a_0^2 \neq 4$	14
$S_{1,p}$	$x^2z+yz^2+x^2y^2+\mathbf{a}y^{5+p}$	$p>0, a_0 \neq 0$	$14+p$
$S^{\#}_{1,2q-1}$	$x^2z+yz^2+zy^3+\mathbf{a}xy^{2+q}$	$q>0, a_0 \neq 0$	$14+2q-1$
$S^{\#}_{1,2q}$	$x^2z+yz^2+zy^3+\mathbf{a}x^2y^{2+q}$	$q>0, a_0 \neq 0$	$14+2q$
$U_{1,0}$	$x^3+xz^2+xy^3+\mathbf{a}y^3z$	$a_0(a_0^2+1) \neq 0$	14
$U_{1,2q-1}$	$x^3+xz^2+xy^3+\mathbf{a}y^{1+q}z^2$	$q>0, a_0 \neq 0$	$14+2q-1$
$U_{1,2q}$	$x^3+xz^2+xy^3+\mathbf{a}y^{3+q}z$	$q>0, a_0 \neq 0$	$14+2q$

The 14 exceptional families

E_{18}	$x^3+y^{10}+\mathbf{a}xy^7$	E_{19}	$x^3+xy^7+\mathbf{a}y^8$
E_{20}	$x^3+y^8+\mathbf{a}xy^8$	Z_{17}	$x^3y+y^8+\mathbf{a}xy^6$
Z_{18}	$x^3y+xy^6+\mathbf{a}y^9$	Z_{19}	$x^3y+y^9+\mathbf{a}xy^7$
W_{17}	$x^4+xy^5+\mathbf{a}y^7$	W_{18}	$x^4+y^7+\mathbf{a}x^2y^4$
Q_{16}	$x^3+yz^2+y^7+\mathbf{a}xy^5$	Q_{17}	$x^3+yz^2+xy^5+\mathbf{a}y^8$
Q_{18}	$x^3+yz^2+y^8+\mathbf{a}xy^6$	S_{16}	$x^2z+yz^2+xy^4+\mathbf{a}y^6$
S_{17}	$x^2z+yz^2+y^6+\mathbf{a}zy^4$	U_{16}	$x^3+xz^2+y^5+\mathbf{a}x^2y^2$

All the functions of all these families are bimodular (when the restrictions hold).

Some of the adjacencies between the bimodular singularities

$$J_{3,0} \leftarrow J_{3,1} \leftarrow E_{18} \leftarrow E_{19} \leftarrow E_{20}, \qquad Z_{1,0} \leftarrow Z_{1,1} \leftarrow Z_{17} \leftarrow Z_{18} \leftarrow Z_{19},$$

(with $\uparrow \cdots$ below $J_{3,1}$ and below $Z_{1,1}$)

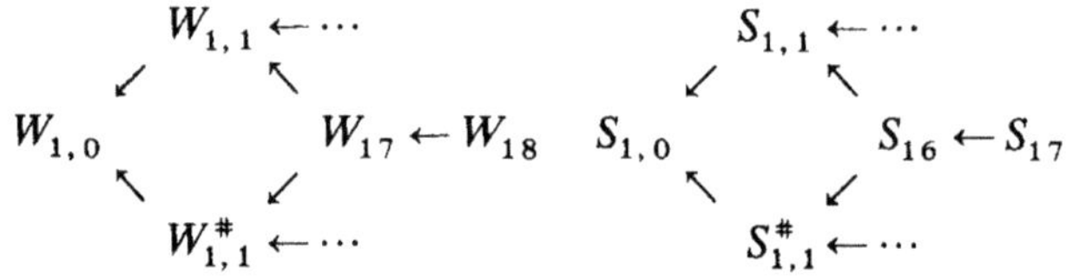

$$Q_{2,0} \leftarrow Q_{2,1} \leftarrow Q_{16} \leftarrow Q_{17} \leftarrow Q_{18}, \qquad U_{1,0} \leftarrow U_{1,1} \leftarrow U_{16}.$$

(with $\uparrow \cdots$ below $Q_{2,1}$ and below $U_{1,1}$)

The pyramids of exceptional singularities with modules numbers $m=1$ and 2

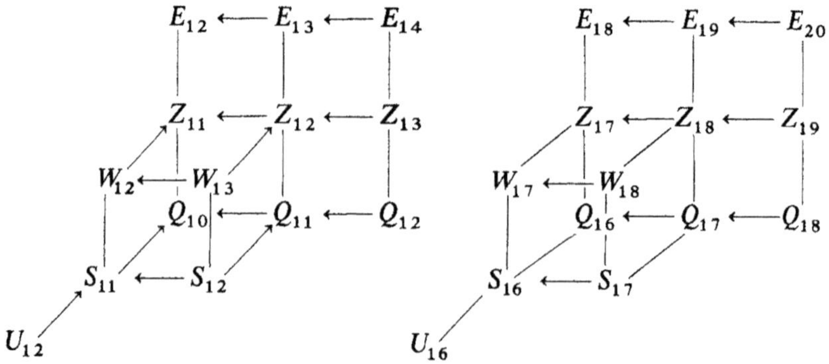

Each vertical joins the singularities of the same Kodaira class in the Koulikov's construction [7].

II. The Corank 2 Singularities with Nonzero 4-Jets

Through this section $\mathbf{a}=a_0+\dots+a_{k-2}y^{k-2}$ for $k>1$ and $\mathbf{a}=0$ for $k=1$.

II_1. The Corank 2 Singularities with Nonzero 3-Jets. These are the simple singularities **A**, **D**, E_6, E_7, E_8 and the singularities of the following infinite series of classes.

$$\mathbf{J}=\mathbf{J}_2\leftarrow\mathbf{J}_3\leftarrow\cdots,\quad\text{where } \mathbf{J}_k=\begin{array}{l}\leftarrow J_{k,0}\leftarrow J_{k,1}\leftarrow E_{6k}\leftarrow E_{6k+1}\leftarrow E_{6k+2}\leftarrow(\mathbf{J}_{k+1})\\ \qquad\qquad\quad\uparrow\\ \qquad\qquad\quad J_{k,2}\leftarrow J_{k,3}\leftarrow\cdots.\end{array}$$

Notation	Normal form	Restrictions	Multiplicity μ	Modules number m
$J_{k,0}$	$x^3+bx^2y^k+y^{3k}+\mathbf{c}xy^{2k+1}$	$k>1,\ 4b^3+27\neq0$	$6k-2$	$k-1$
$J_{k,i}$	$x^3+x^2y^k+\mathbf{a}y^{3k+i}$	$k>1,\ i>0,\ a_0\neq0$	$6k+i-2$	$k-1$
E_{6k}	$x^3+y^{3k+1}+\mathbf{a}xy^{2k+1}$	$k\geqq1$	$6k$	$k-1$
E_{6k+1}	$x^3+xy^{2k+1}+\mathbf{a}y^{3k+2}$	$k\geqq1$	$6k+1$	$k-1$
E_{6k+2}	$x^3+y^{3k+2}+\mathbf{a}xy^{2k+2}$	$k\geqq1$	$6k+2$	$k-1$

Here $\mathbf{c}=c_0+\dots+c_{k-3}y^{k-3}$ for $k>2$; $\mathbf{c}=0$ for $k=2$.

II_2. The Corank 2 Singularities with Zero 3-Jets and Nonzero 4-Jets. These singularities form one infinite series of classes

$$\mathbf{X}=\mathbf{X}_1\leftarrow\mathbf{X}_2\leftarrow\cdots$$

where

$$\begin{array}{ccccc} & & & \mathbf{Y}_k & \\ & & \swarrow & & \nwarrow \\ \mathbf{X}_k=X_{k,0}\leftarrow & \mathbf{X}_k^* & & & \mathbf{W}_k\leftarrow; \\ & & \nwarrow & & \swarrow \\ & & & \mathbf{Z}_k & \end{array}$$

here

$$\begin{array}{ll} \qquad\ \ \downarrow\downarrow & \qquad\quad\ \downarrow \\ \mathbf{X}_k^*=\leftarrow X_{k,1}\leftarrow X_{k,2}\leftarrow\cdots, & \mathbf{Y}_k=\leftarrow Y_{1,1}^k\longleftarrow Y_{2,1}^k\Leftarrow\cdots, \\ \mathbf{Z}_k=\leftarrow\mathbf{Z}_0^k\longleftarrow\mathbf{Z}_1^k\longleftarrow\cdots, & \mathbf{Z}_0^k=\leftarrow Z_{12k-1}^k\leftarrow Z_{12k}^k\leftarrow Z_{12k+1}^k, \\ \qquad\quad\uparrow & \end{array}$$

$$\begin{array}{l}\mathbf{Z}_i^k=\leftarrow Z_{i,0}^k\leftarrow Z_{i,1}^k\leftarrow Z_{12k+6i-1}^k\leftarrow Z_{12k+6i}^k\leftarrow Z_{12k+6i+1}^k\leftarrow(\mathbf{Z}_{i+1}^k),\qquad i>0\\ \qquad\qquad\qquad\ \ \uparrow\\ \qquad\qquad\qquad Z_{i,2}^k\leftarrow Z_{i,3}^k\leftarrow\cdots\end{array}$$

$$\begin{array}{lcccl} & & W_{k,1}\leftarrow W_{k,2}\leftarrow\cdots & & \\ & \swarrow & & \nwarrow & \\ \mathbf{W}_k=\leftarrow W_{12k}\leftarrow W_{12k+1}\leftarrow W_{k,0} & & & & W_{12k+5}\leftarrow W_{12k+6}\leftarrow(\mathbf{X}_{k+1}) \\ & \nwarrow & & \swarrow & \\ & & W_{k,1}^{\#}\leftarrow W_{k,2}^{\#}\leftarrow\cdots. & & \end{array}$$

Classes **X** *and* **Y**

Notation	Normal Form	Restrictions	Multiplicity μ	Modules number m
$X_{k,0}$	$x^4+\mathbf{b}x^3y^k+\mathbf{a}x^2y^{2k}+xy^{3k}$	$\Delta\neq 0, a_0b_0\neq 9$ $k>1$	$12k-3$	$3k-2$
$X_{k,p}$	$x^4+\mathbf{a}x^3y^k+x^2y^{2k}+\mathbf{b}y^{4k+p}$	$a_0^2\neq 4, b_0\neq 0$ $p>0, k>1$	$12k-3+p$	$3k-2$
$Y^k_{r,s}$	$[(x+\mathbf{a}y^k)^2+\mathbf{b}y^{2k+s}](x^2+y^{2k+r})$	$1\leqq s\leqq 7, k>1$ $a_0\neq 0\neq b_0$	$12k-3+r+s$	$3k-2$

The case $k=1$:

$X_{1,0}$	$x^4+a_0x^2y^2+y^4$	$a_0^2\neq 4$	9	1
$X_{1,p}$	$x^4+x^2y^2+a_0y^{4+p}$	$a_0\neq 0$	$9+p$	1
$Y^1_{r,s}$	$x^{4+r}+a_0x^2y^2+y^{4+s}$	$a_0\neq 0$	$9+r+s$	1

Of course, $X_{1,0}=X_9$, $X_{1,p}=T_{2,4,4+p}$, $Y^1_{r,s}=T_{2,4+r,4+s}$. Here

$$\Delta=4(a_0^3+b_0^3)-a_0^2b_0^2-18a_0b_0+27,$$

$$\mathbf{b}=b_0+\cdots+b_{2k-2}y^{2k-2}.$$

Class **Z**

For singularities $Z^k_{i,0}$ and Z^k_μ $(k>1)$ normal forms are $f=(x+\mathbf{a}y^k)f_2$, where $a_0\neq 0$ and f_2 is given in the following table:

Notation	f_2	Restrictions	Multiplicity, μ	Modules number, m
$Z^k_{i,0}$	$x^3+dx^2y^{k+1}+\mathbf{c}xy^{2k+2i+1}+y^{3k+3i}$	$4d^3+27\neq 0$ $k>1, i\geqq 0$	$12k-3+6i$	$3k+i-2$
$Z^k_{12k+6i-1}$	$x^3+\mathbf{b}xy^{2k+2i+1}+y^{3k+3i+1}$	$k>1, i\geqq 0$	$12k+6i-1$	$3k+i-2$
Z^k_{12k+6i}	$x^3+xy^{2k+2i+1}+\mathbf{b}y^{3k+3i+2}$	$k>1, i\geqq 0$	$12k+6i$	$3k+i-2$
$Z^k_{12k+6i+1}$	$x^3+\mathbf{b}xy^{2k+2i+2}+y^{3k+3i+2}$	$k>1, i\geqq 0$	$12k+6i+1$	$3k+i-2$

Singularity $Z^k_{i,p}$ $(k>1,\ i>0,\ p>0)$ admits the normal form

$$(x^2+\mathbf{a}xy^k+\mathbf{b}y^{2k+i})\cdot(x^2+y^{2k+2i+p}),\quad a_0\neq 0,\ b_0\neq 0,$$

its multiplicity is $\mu=12k+6i+p-3$, the modules number $m=3k+i-2$.

For $k=1$ one modifies the above formulae in the following way:

1. One omits the index $k=1$ in the notations.

2. Singularities $Z_{i,0}, Z_{6i+11}, Z_{6i+12}, Z_{6i+13}$ $(i>0)$ have normal forms $f=yf_2$, where f_2 is given by the table above.

3. $Z_{i,p}$: $y(x^3+x^2y^{i+1}+\mathbf{b}y^{3i+p+3})$, $b_0\neq 0$, $i>0$, $p>0$.

Through this section

$$\mathbf{b}=b_0+\cdots+b_{2k+i-2}y^{2k+i-2},$$

$$\mathbf{c}=c_0+\cdots+c_{2k+i-3}y^{2k+i-3}.$$

Class **W**

Notation	Normal form	Restriction	Multiplicity μ	Modules number m
W_{12k}	$x^4+y^{4k+1}+\mathbf{a}xy^{3k+1}+\mathbf{c}x^2y^{2k+1}$	$k\geqq 1$	$12k$	$3k-2$
W_{12k+1}	$x^4+xy^{3k+1}+\mathbf{a}x^2y^{2k+1}+\mathbf{c}y^{4k+2}$	$k\geqq 1$	$12k+1$	$3k-2$
$W_{k,0}$	$x^4+\mathbf{b}x^2y^{2k+1}+\mathbf{a}xy^{3k+2}+y^{4k+2}$	$k\geqq 1, b_0^2\neq 4$	$12k+3$	$3k-1$
$W_{k,i}$	$x^4+\mathbf{a}x^3y^{k+1}+x^2y^{2k+1}+\mathbf{b}y^{4k+2+i}$	$i>0, b_0\neq 0$	$12k+3+i$	$3k-1$
$W^{\#}_{k,2q-1}$	$(x^2+y^{2k+1})^2+\mathbf{b}xy^{3k+1+q}+\mathbf{a}y^{4k+2+q}$	$q>0, b_0\neq 0$	$12k+2+2q$	$3k-1$
$W^{\#}_{k,2q}$	$(x^2+y^{2k+1})^2+\mathbf{b}x^2y^{2k+1+q}+\mathbf{a}xy^{3k+2+q}$	$q>0, b_0\neq 0$	$12k+3+2q$	$3k-1$
W_{12k+5}	$x^4+xy^{3k+2}+\mathbf{a}x^2y^{2k+2}+\mathbf{b}y^{4k+3}$	$k\geqq 1$	$12k+5$	$3k-1$
W_{12k+6}	$x^4+y^{4k+3}+\mathbf{a}xy^{3k+3}+\mathbf{b}x^2y^{2k+2}$	$k\geqq 1$	$12k+6$	$3k-1$

Here $\mathbf{b}=b_0+\cdots+b_{2k-1}y^{2k-1}$, $\mathbf{c}=c_0+\cdots+c_{2k-2}y^{2k-2}$; as always, $\mathbf{a}=a_0+\cdots+a_{k-2}y^{k-2}$, for $k>1$ and $\mathbf{a}=0$ for $k=1$.

III. The Corank 3 Singularities with Reduced 3-Jets

Besides the unimodular singularities T (see $\mathbf{I}_1$), there are 3 infinite series of classes **Q**, **S**, **U** of such singularities.

III$_1$. Series Q. The singularities with the 3-jet x^3+yz^2 form one infinite series of classes:

$$\mathbf{Q}=\leftarrow\mathbf{Q}_1\leftarrow\mathbf{Q}_2\leftarrow\cdots,\qquad \text{where } \mathbf{Q}_1=\leftarrow Q_{10}\leftarrow Q_{11}\leftarrow Q_{12},$$

$$\begin{array}{lllllll} \mathbf{Q}_k=\leftarrow Q_{k,0}\leftarrow & Q_{k,1} & \leftarrow Q_{6k+4}\leftarrow Q_{6k+5}\leftarrow Q_{6k+6}\leftarrow(\mathbf{Q}_{k+1}) \\ & \uparrow & \\ k>1 & Q_{k,2}\leftarrow Q_{k,3}\leftarrow\cdots. & \end{array}$$

Notation	Normal form	Restrictions	Multiplicity μ	Modules number m
$Q_{k,0}$	$\varphi+\mathbf{b}x^2y^k+xy^{2k}$	$k>1, b_0^2\neq 4$	$6k+2$	k
$Q_{k,i}$	$\varphi+x^2y^k+\mathbf{b}y^{3k+i}$	$k>1, b_0\neq 0$	$6k+2+i$	k
Q_{6k+4}	$\varphi+y^{3k+1}+\mathbf{b}xy^{2k+1}$	$k\geqq 1$	$6k+4$	k
Q_{6k+5}	$\varphi+xy^{2k+1}+\mathbf{b}y^{3k+2}$	$k\geqq 1$	$6k+5$	k
Q_{6k+6}	$\varphi+y^{3k+2}+\mathbf{b}xy^{2k+2}$	$k\geqq 1$	$6k+6$	k

Here $\varphi=x^3+yz^2$, $\mathbf{b}=b_0+\cdots+b_{k-1}y^{k-1}$.

III$_2$. Series S. The singularities with the 3-jet x^2z+yz^2 form one infinite series of classes:

$$\mathbf{S}=\leftarrow\mathbf{S}_1\leftarrow\mathbf{S}_2\leftarrow\cdots,$$

where

$$\begin{array}{ccccc} & & S_{k,1}\leftarrow\cdots & & \\ & \swarrow & & \nwarrow & \\ \mathbf{S}_k=\leftarrow S_{12k-1}\leftarrow S_{12k}\leftarrow S_{k,0} & & & & S_{12k+4}\leftarrow S_{12k+5}\leftarrow \mathbf{S}^{*}_{k+1}\leftarrow(\mathbf{S}_{k+1}) \\ & \nwarrow & & \swarrow & \\ & & S^{\#}_{k,1}\leftarrow\cdots. & & \end{array}$$

Notation	Normal form	Restrictions	Multiplicity μ	Modules number m
S_{12k-1}	$\varphi+y^{4k}+\mathbf{a}xy^{3k}+\mathbf{c}zy^{2k+1}$	–	$12k-1$	$3k-2$
S_{12k}	$\varphi+xy^{3k}+\mathbf{c}y^{4k+1}+\mathbf{a}zy^{2k+1}$	–	$12k$	$3k-2$
$S_{k,0}$	$\varphi+y^{4k+1}+\mathbf{a}xy^{3k+1}+\mathbf{b}zy^{2k+1}$	$b_0^2\neq 4$	$12k+2$	$3k-1$
$S_{k,i}$	$\varphi+x^2y^{2k}+\mathbf{a}x^3y^k+\mathbf{b}y^{4k+1+i}$	$i>0,\ b_0\neq 0$	$12k+2+i$	$3k-1$
$S^{\#}_{k,2q-1}$	$\varphi+zy^{2k+1}+\mathbf{b}xy^{3k+q-1}+\mathbf{a}y^{4k+q}$	$q>0,\ b_0\neq 0$	$12k+2q+1$	$3k-1$
$S^{\#}_{k,2q}$	$\varphi+zy^{2k+1}+\mathbf{b}x^2y^{2k+q}+\mathbf{a}xy^{3k+q+1}$	$q>0,\ b_0\neq 0$	$12k+2q+2$	$3k-1$
S_{12k+4}	$\varphi+xy^{3k+1}+\mathbf{a}zy^{2k+2}+\mathbf{b}y^{4k+2}$	–	$12k+4$	$3k-1$
S_{12k+5}	$\varphi+y^{4k+2}+\mathbf{a}xy^{3k+2}+\mathbf{b}zy^{2k+2}$	–	$12k+5$	$3k-1$

Here $\varphi=x^2z+yz^2$;

$\mathbf{a}=a_0+\cdots+a_{k-2}y^{k-2}$ for $k>1$, $\mathbf{a}=0$ for $k=1$;

$\mathbf{b}=b_0+\cdots+b_{2k-1}y^{2k-1}$;

$\mathbf{c}=c_0+\cdots+c_{2k-2}y^{2k-2}$.

The $\mathbf{S}_{\mathbf{k}}^*$ $(k>1)$ singularities are:

$$\mathbf{S}_{\mathbf{k}}^* = \leftarrow S^*_{k,0} \leftarrow \mathbf{SP}_{\mathbf{k}} \quad \begin{matrix} \swarrow \mathbf{SQ}_{\mathbf{k}} \nwarrow \\ \nwarrow \mathbf{SR}_{\mathbf{k}} \swarrow \end{matrix} \quad (\mathbf{S}_{\mathbf{k}})$$

$\mu(S^*_{k,0})=12k-4$, $m(S^*_{k,0},\ldots,\mathbf{SR}_{\mathbf{k}})=3k-2$, $\operatorname{codim}\mathbf{S}_{\mathbf{k}}^*=9k-3$.

III$_3$. Series U. The singularities with the 3-jet x^3+xz^2 form one infinite series of classes

$$\mathbf{U}=\leftarrow\mathbf{U}_1\leftarrow\mathbf{U}_2\leftarrow\cdots,$$

where

$$\mathbf{U}_{\mathbf{k}}=\leftarrow U_{12k}\leftarrow U_{k,0}\leftarrow U_{k,1}\leftarrow U_{12k+4}\leftarrow\mathbf{U}^*_{\mathbf{k+1}}\leftarrow(\mathbf{U}_{\mathbf{k+1}})$$
$$\uparrow$$
$$U_{k,2}\leftarrow\cdots.$$

Notation	Normal form	Restrictions	Multiplicity μ	Modules number m
U_{12k}	$\varphi+y^{3k+1}+\mathbf{a}xy^{2k+1}+\mathbf{b}zy^{2k+1}+\mathbf{d}x^2y^{k+1}$	–	$12k$	$4k-3$
$U_{k,2q}$	$\varphi+xy^{2k+1}+\mathbf{a}x^2y^{k+1}+\mathbf{b}y^{3k+2+q}+\mathbf{c}zy^{2k+1+q}$	$q\geqq 0;\ c_0\neq 0$	$12k+2+2q$	$4k-2$
$U_{k,2q-1}$	$\varphi+xy^{2k+1}+\mathbf{a}x^2y^{k+1}+\mathbf{b}zy^{2k+1+q}+\mathbf{c}z^2y^{k+q}$	$q>0;\ c_0\neq 0$	$12k+1+2q$	$4k-2$
U_{12k+4}	$\varphi+y^{3k+2}+\mathbf{a}xy^{2k+2}+\mathbf{b}zy^{2k+2}+\mathbf{c}x^2y^{k+1}$	–	$12k+4$	$4k-2$

Here $\varphi=x^3+xz^2$; $c_0^2+1\neq 0$ for $q=0$; and trough the table

$\mathbf{a}=a_0+\cdots+a_{k-2}y^{k-2}$ for $k>1$, $\mathbf{a}=0$ for $k=1$,

$\mathbf{b}=b_0+\cdots+b_{k-2}y^{k-2}$ for $k>1$, $\mathbf{b}=0$ for $k=1$,

$$\mathbf{c}=c_0+\cdots+c_{2k-1}y^{2k-1},$$
$$\mathbf{d}=d_0+\cdots+d_{2k-2}y^{2k-2}.$$

The $\mathbf{U}_\mathbf{k}^*$ singularities ($k>1$) are

$$\begin{array}{llll} \mathbf{U}_\mathbf{k}^* = \leftarrow U_{k,0}^* \leftarrow & \mathbf{UP}_\mathbf{k} \leftarrow & \mathbf{UR}_\mathbf{k} \leftarrow & \mathbf{UT}_\mathbf{k} \\ & \uparrow & \uparrow & \uparrow \\ & \mathbf{UQ}_\mathbf{k} \leftarrow & \mathbf{US}_\mathbf{k} \leftarrow & (\mathbf{U}_\mathbf{k}), \end{array}$$

$\mu(U_{k,0}^*)=12k-4$, $m(U_{k,0}^*,\ldots,\mathbf{UT}_\mathbf{k})=4k-3$, $\operatorname{codim}(\mathbf{U}_\mathbf{k}^*)=8k-2$.

IV. Series V. The corank 3 singularities with the 3-jet x^2y belong to the classes

$$\begin{array}{ccc} & V_{1,1} \leftarrow V_{1,2} \leftarrow \cdots & \\ \swarrow & \nwarrow & \\ \mathbf{V}=V_{1,0} & & \mathbf{V}^* \\ \nwarrow & \swarrow & \\ & V_{1,1}^{\#} \leftarrow V_{1,2}^{\#} \leftarrow \cdots. & \end{array}$$

Notation	Normal form	Restrictions	Multiplicity μ	Modules number m
$V_{1,0}$	$x^2y+z^4+az^3y+\mathbf{b}z^2y^2+zy^3$	$\Delta(a,b_0)\neq 0$	15	3
$V_{1,p}$	$x^2y+z^4+bz^3y+z^2y^2+\mathbf{a}y^{4+p}$	$b^2\neq 4$, $a_0\neq 0$	$15+p$	3
$V_{1,2q-1}^{\#}$	$x^2y+z^3y+ay^2z^2+y^4+\mathbf{b}xz^{2+q}$	$4a^3+27\neq 0$, $b_0\neq 0$	$15+2q-1$	3
$V_{1,2q}^{\#}$	$x^2y+z^3y+ay^2z^2+y^4+\mathbf{b}z^{4+q}$	$4a^3+27\neq 0$, $b_0\neq 0$	$15+2q$	3

Here $p>0$, $q>0$, $\mathbf{a}=a_0+a_1y$, $\mathbf{b}=b_0+b_1z$. For the $\mathbf{V}^*$ singularities $\mu(\mathbf{V}^*)\geqq 17$, $m(\mathbf{V}^*)\geqq 3$, $c(\mathbf{V}^*)=13$.

V. Other Singularities. All the singularities, which normal forms are not given in this article, belong to the following 7 classes:

Notation	Corank	Adjacency	Definition	$c\geqq$	$\mu\geqq$	$m\geqq$	Theorem
$\mathbf{N}$	2	$\mathbf{N}\to W_{13}$	$j_4=0$	12	16	3	**47–49**
$\mathbf{S}^*$	3	$\mathbf{S}_\mathbf{k}^*\to S_{12k-9}$	th. **77**	15	20	4	**77–81**
$\mathbf{U}^*$	3	$\mathbf{U}_\mathbf{k}^*\to U_{12k-8}$	th. **90**	14	20	5	**90–96**
$\mathbf{V}^*$	3	$\mathbf{V}^*\to V_{1,1}\cup V_{1,1}^{\#}$	th. **98**	13	17	3	**97–102**
$\mathbf{V}'$	3	$\mathbf{V}'\to\mathbf{V}$	$j_3=x^3$	13	18	4	**103**
$\mathbf{V}''$	3	$\mathbf{V}''\to\mathbf{V}'$	$j_3=0$	16	27	10	**104**
$\mathbf{O}$	>3	$\mathbf{O}\to T_{4,4,4}$	corank $\geqq 4$	10	16	5	**105**

Here $k>1$.

Remark. For every arrow $\mathbf{K}\leftarrow\mathbf{L}$ in §1 (say, $\mathbf{J_2}\leftarrow\mathbf{J_3}$), every singularity in class $\mathbf{K}$ has a neighbourhood which does not intersect $\mathbf{L}$ (*e.g* there is no adjacency $J_{2,i}\to J_{3,0}$ for any i).

§ 2. Singularity Determinator

Notations

f	a germ of a holomorphic function at an *isolated* critical point O, with finite multiplicity μ; its Taylor series at O; a formal series in x, y or in x, y, z, with multiplicity μ.
$f \sim g$	germs (or series) f and g at O are equivalent (there exists a germ of a diffeomorphism h (or a formal series h) such that $f = g \circ h$.
$\Rightarrow$	"implies"
$\mapsto$	"see" (references $\mapsto$ **i** are not parts of the theorem statements; they give the number of the theorem containing the classification of the given singularity classes).
$j_k f$	k-jet of f at O (or the Taylor k-polynomial).
$A, \ldots, Z$	stable equivalence classes defined in § 1.
$m(f)$	modules number of the germ of f at O.
$c(f)$	codimension of the μ=const stratum of a function germ f in the space of germs of functions having critical value 0 at the critical point O.
$c(\mathbf{K})$	codimension of the singularity class $\mathbf{K}$ in the same space.
$j\{\mathbf{x}^{\mathbf{m}_i}\} f$	quasijet (or quasi Taylor polynomial of f at O determined by the monomials $\mathbf{x}^{\mathbf{m}_i}$, see the explanation below).
$j\{\mathbf{x}^{\mathbf{m}_i}\} f \approx g$	quasihomogenous equivalence of two quasijets or of two quasi Taylor polynomials, see the explanation below.
j^*, φ	abbreviations used in theorems **58–65**, **66–81**, **82–89**, **98–102**; their meaning is to be found above the first theorem of each group.
Δ	discriminant. In theorems **36**, **37**, **47**, **48**, **98**, **99**, $\Delta = 4(a^3+b^3)+27-a^2b^2-18ab$.

Explanation. A system $\{\mathbf{x}^{\mathbf{m}_i}\}$ of n monomials in n variables $x_1, \ldots, x_n$ with linearly independent exponents $\mathbf{m}_i \in \mathbb{Z}^n \subset \mathbb{R}^n$ defines an hyperplane $\Gamma \subset \mathbb{R}^n$, $\Gamma = \{\mathbf{m}: (\boldsymbol{\alpha}, \mathbf{m}) = 1\}$. If all the components α_i of the vector $\boldsymbol{\alpha}$ are positive, $\boldsymbol{\alpha}$ is called the *quasihomogenity type*. The number $(\boldsymbol{\alpha}, \mathbf{m})$ is then called the *order* of the monomial $\mathbf{x}^{\mathbf{m}}$.

A polynomial $f = \sum f_{\mathbf{m}} \mathbf{x}^{\mathbf{m}}$ is *quasihomogenous, of degree d and of type* $\boldsymbol{\alpha}$ if $(\boldsymbol{\alpha}, \mathbf{m}) = d$ for all $\mathbf{m}$ with $f_{\mathbf{m}} \neq 0$.

The type $\boldsymbol{\alpha}$ defines a graded ring structure in $\mathbb{C}[[x_1, \ldots, x_n]]$ and a ring filtration $\mathscr{E}_0 \supset \ldots, \mathscr{E}_d = \{f: (\boldsymbol{\alpha}, \mathbf{m}) \geqq d \, \forall \mathbf{m}: f_{\mathbf{m}} \neq 0\}$.

The factor space $\mathscr{E}_0/\mathscr{E}_d$, $d > 1$ is by definition the *quasijet space defined by the monomials* $\{\mathbf{x}^{\mathbf{m}_i}\}$ (or by the type $\boldsymbol{\alpha}$). For a fixed coordinate system one can indentify quasijets with the polynomials having only monomials of order $\leqq 1$ (that is, their exponents lie on Γ or the same side of Γ as O).

Quasihomogenous diffeomorphisms are diffomorphisms of $\mathbb{C}^n$ which preserve the degrees in the graded ring $\mathbb{C}[[x_1, \ldots, x_n]]$. The Lie group of the quasihomogenous diffeomorphisms acts on the space of quasijets and on the space of quasihomogenous polynomials. Two quasijets (or two polynomials) are *quasihomogenously equivalent* if they belong to the same orbit of this action.

Determinator

1. $\mu(f)<\infty \Rightarrow$ one of the four possibilities holds:
corank $f \leqq 1 \mapsto$ **2**,
$=2 \mapsto$ **3**,
$=3 \mapsto$ **50**,
$>3 \mapsto$ **105**.

2. corank $f \leqq 1 \Rightarrow f \in A_k$ $(k \geqq 1)$.

Through theorems **3–49**, $f \in \mathbb{C}[[x, y]]$.

3. $j_2 f=0 \Rightarrow$ one of the four possibilities holds:
$j_3 f \approx x^2 y+y^3 \mapsto$ **4**,
$\approx x^2 y \mapsto$ **5**,
$\approx x^3 \mapsto$ **6$_1$**,
$=0 \mapsto$ **13**.

4. $j_3 f=x^2 y+y^3 \Rightarrow f \in D_4$.

5. $j_3 f=x^2 y \Rightarrow f \in D_k (k>4)$.

Through theorems **6–9**, $k \geqq 1$.

6$_k$. $j_{x^3, y^{3k}} f(x, y)=x^3 \Rightarrow$ one of the four possibilities holds:
$j_{x^3, y^{3k+1}} f \approx x^3+y^{3k+1} \mapsto$ **7$_k$**,
$j_{x^3, xy^{2k+1}} f \approx x^3+x y^{2k+1} \mapsto$ **8$_k$**,
$j_{x^3, y^{3k+2}} f \approx x^3+y^{3k+2} \mapsto$ **9$_k$**,
$j_{x^3, y^{3k+2}} f =x^3 \mapsto$ **10$_{k+1}$**.

7$_k$. $j_{x^3, y^{3k+1}} f=x^3+y^{3k+1} \Rightarrow f \in E_{6k}$.

8$_k$. $j_{x^3, xy^{2k+1}} f=x^3+x y^{2k+1} \Rightarrow f \in E_{6k+1}$.

9$_k$. $j_{x^3, y^{3k+2}} f=x^3+y^{3k+2} \Rightarrow f \in E_{6k+2}$.

Through theorems **10–12**, $k>1$.

10$_k$. $j_{x^3, y^{3k-1}} f=x^3 \Rightarrow$ one of the three possibilities holds:
$j_{x^3, y^{3k}} f \approx x^3+a x^2 y^k+y^{3k}, \quad 4a^3+27 \neq 0 \mapsto$ **11$_k$**;
$\approx x^3+x^2 y^k \mapsto$ **12$_k$**;
$\approx x^3 \mapsto$ **6$_k$**.

11$_k$. $j_{x^3, y^{3k}} f=x^3+a x^2 y^k+y^{3k}, \quad 4a^3+27 \neq 0 \Rightarrow f \in J_{k,0}$.

12$_k$. $j_{x^3, y^{3k}} f=x^3+x^2 y^k \Rightarrow f \in J_{k,p}$ $(p>0)$.

Series X

13. $j_3 f(x, y)=0 \Rightarrow$ one of the six possibilities holds:
$j_4 \approx x^4+a x^2 y^2+y^4, \quad a^2 \neq 4 \mapsto$ **14**;
$\approx x^4+x^2 y^2 \mapsto$ **15**;
$\approx x^2 y^2 \mapsto$ **16**;
$\approx x^3 y \mapsto$ **17**;
$\approx x^4 \mapsto$ **25**;
$=0 \mapsto$ **47**;

14. $j_4 f=x^4+a x^2 y^2+y^4, \quad a^2 \neq 4 \Rightarrow f \in X_9 = X_{1,0}=T_{2,4,4}$.

15. $j_4 f=x^4+x^2 y^2 \Rightarrow f \in X_{1,p}=T_{2,4,4+p}$ $(p>0)$.

16. $j_4 f = x^2 y^2 \Rightarrow f \in Y^1_{p,q} = T_{2,4+p,4+q}$ $(p>0, q>0)$.

17. $j_4 f = x^3 y \Rightarrow j_{x^3 y, y^4} f = x^3 y \mapsto$ **18**$_1$.

Through theorems **18–21**, $p \geqq 1$.

18$_p$. $j_{x^3 y, y^{3p+1}} f = x^3 y \Rightarrow$ one of the four possibilities holds:

$j_{x^3 y, y^{3p+2}} f \approx x^3 y + y^{3p+2} \mapsto$ **19$_p$**,

$j_{x^3 y, x y^{2p+2}} f \approx x^3 y + x y^{2p+2} \mapsto$ **20$_p$**,

$j_{x^3 y, y^{3p+3}} f \approx x^3 y + y^{3p+3} \mapsto$ **21$_p$**,

$j_{x^3 y, y^{3p+3}} f = x^3 y \mapsto$ **22$_{p+1}$**.

19$_p$. $j_{x^3 y, y^{3p+2}} f = x^3 y + y^{3p+2} \Rightarrow f \in Z_{6p+5}$.

20$_p$. $j_{x^3 y, x y^{2p+2}} f = x^3 y + x y^{2p+2} \Rightarrow f \in Z_{6p+6}$.

21$_p$. $j_{x^3 y, y^{3p+3}} f = x^3 y + y^{3p+3} \Rightarrow f \in Z_{6p+7}$.

Through theorems **22–24**, $p > 1$.

22$_p$. $j_{x^3 y, y^{3p}} f = x^3 y \Rightarrow$ one of the three possibilities holds:

$j_{x^3 y, y^{3p+1}} f = y(x^3 + b x^2 y^p + y^{3p})$, $\quad 4b^3 + 27 \neq 0 \mapsto$ **23$_p$**;

$y(x^3 + x^2 y^p) \mapsto$ **24$_p$**;

$x^3 y \mapsto$ **18$_p$**.

23$_p$. $j_{x^3 y, y^{3p+1}} f = y(x^3 + b x^2 y^p + y^{3p})$, $\quad 4b^3 + 27 \neq 0 \Rightarrow f \in Z_{p-1,0}$.

24$_p$. $j_{x^3 y, y^{3p+1}} f = y(x^3 + x^2 y^p) \Rightarrow f \in Z_{p-1,r}$ $(r>0)$.

Series W

25. $j_4 f(x,y) = x^4 \Rightarrow j_{x^4, y^4} f = x^4 \mapsto$ **26$_1$**.

Through theorems **26–34**, $k \geqq 1$.

26$_k$. $j_{x^4, y^{4k}} f = x^4 \Rightarrow$ one of the three possibilities holds:

$j_{x^4, y^{4k+1}} f \approx x^4 + y^{4k+1} \mapsto$ **27$_k$**,

$j_{x^4, x y^{3k+1}} f \approx x^4 + x y^{3k+1} \mapsto$ **28$_k$**,

$j_{x^4, x y^{3k+1}} f = x^4 \mapsto$ **29$_k$**.

27$_k$. $j_{x^4, y^{4k+1}} f = x^4 + y^{4k+1} \Rightarrow f \in W_{12k}$.

28$_k$. $j_{x^4, x y^{3k+1}} f = x^4 + x y^{3k+1} \Rightarrow f \in W_{12k+1}$.

29$_k$. $j_{x^4, x y^{3k+1}} f = x^4 \Rightarrow$ one of the four possibilities holds:

$j_{x^4, y^{4k+2}} f \approx x^4 + b x^2 y^{2k+1} + y^{4k+2}$, $\quad b^2 \neq 4 \mapsto$ **30$_k$**;

$\approx x^4 + x^2 y^{2k+1} \mapsto$ **31$_k$**;

$\approx (x^2 + y^{2k+1})^2 \mapsto$ **32$_k$**;

$= x^4 \mapsto$ **33$_k$**.

30$_k$. $j_{x^4, y^{4k+2}} f = x^4 + b x^2 y^{2k+1} + y^{4k+2}$, $\quad b^2 \neq 4 \Rightarrow f \in W_{k,0}$.

31$_k$. $j_{x^4, y^{4k+2}} f = x^4 + x^2 y^{2k+1} \Rightarrow f \in W_{k,i,(i>0)}$.

32$_k$. $j_{x^4, y^{4k+2}} f = (x^2 + y^{2k+1})^2 \Rightarrow f \in W^{\#}_{k,i,(i>0)}$.

33$_k$. $j_{x^4, y^{4k+2}} f = x^4 \Rightarrow$ one of the three possibilities holds:

$j_{x^4, x y^{3k+2}} f \approx x^4 + x y^{3k+2} \mapsto$ **34$_k$**;

$j_{x^4, y^{4k+3}} f \approx x^4 + y^{4k+3} \mapsto$ **35$_3$**;

$j_{x^4, y^{4k+3}} f = x^4 \mapsto$ **36$_{k+1}$**.

34$_k$. $j_{x^4, x y^{3k+2}} f = x^4 + x y^{3k+2} \Rightarrow f \in W_{12k+5}$.

35$_k$. $j_{x^4, y^{4k+3}} f = x^4 + y^{4k+3} \Rightarrow f \in W_{12k+6}$.

Through theorems **36–46,** $k > 1$.

Series X$_k$

36$_k$. $j_{x^4, y^{4k-1}} f = x^4 \Rightarrow$ one of the five possibilities holds:

$j_{x^4, y^{4k}} f \approx x^4 + b x^3 y^k + a x^2 y^{2k} + x y^{3k}, \quad \Delta \neq 0, \quad ab \neq 9 \mapsto$ **37$_k$**;

$\approx x^2 (x^2 + a x y^k + y^{2k}), \quad a^2 \neq 4 \mapsto$ **38$_k$**;

$\approx x^2 (x + y^k)^2 \mapsto$ **39$_k$**;

$\approx x^3 (x + y^k) \mapsto$ **40$_k$**;

$\approx x^4 \mapsto$ **26$_k$**.

37$_k$. $j_{x^4, y^{4k}} f = x^4 + b x^3 y^k + a x^2 y^{2k} + x y^{3k}, \quad \Delta \neq 0 \Rightarrow f \in X_{k,0}$.

38$_k$. $j_{x^4, y^{4k}} f = x^2 (x^2 + a x y^k + y^{2k}), \quad a^2 \neq 4 \Rightarrow f \in X_{k,p} (p > 0)$.

39$_k$. $j_{x^4, y^{4k}} f = x^2 (x^2 + y^k)^2 \Rightarrow f \in Y_{r,s} (1 \leqq s \leqq r)$.

40$_k$. $j_{x^4, y^{4k}} f = (x + y^k) x^3 \Rightarrow f = f_1 f_2, \quad$ where $j_{x, y^k} f_1 \approx x + y^k$;

$j_{x^3, y^{3k}} f_2 = x^3 \mapsto$ **41$_k$**.

Through theorems **41–44,** $i \geqq 0$, $p > 0$.

41$_k$. $j_{x^3, y^{3k}} f_2 = x^3 \Rightarrow$ one of the five possibilities holds:

$f_2 \in E_{6(k+i)} \mapsto$ **42$_{k,i}$**;

$f_2 \in E_{6(k+i)+1} \mapsto$ **43$_{k,i}$**;

$f_2 \in E_{6(k+i)+2} \mapsto$ **44$_{k,i}$**;

$f_2 \in J_{k+i+1,0} \mapsto$ **45$_{k,i+1}$**;

$f_2 \in J_{k+i+1,p} \mapsto$ **46$_{k,i+1,p}$**.

Through theorems **42–46**, $f(x, y) = f_1 f_2$, where

$j_{x,y^k} f_1 \approx x + y^k, \quad j_{x^3, y^{3k}} f_2 = x^3$.

42$_{k,i}$. $f_2 \in E_{6(k+i)} \Rightarrow f \in Z^k_{12k+6i-1}$.

43$_{k,i}$. $f_2 \in E_{6(k+i)+1} \Rightarrow f \in Z^k_{12k+6i}$.

44$_{k,i}$. $f_2 \in E_{6(k+i)+2} \Rightarrow f \in Z^k_{12k+6i+1}$.

Through theorems **45–46,** $i \geqq 1$, $p > 0$.

45$_{k,i}$. $f_2 \in J_{k+i,0} \Rightarrow f \in Z^k_{i,0}$.

46$_{k,i,p}$. $f_2 \in J_{k+i,p} \Rightarrow f \in Z^k_{i,p}$.

47. $j_4 f = 0 \Rightarrow$ one of the two possibilities holds:

$j_5 f \approx x^4 y + a x^3 y^2 + b x^2 y^3 + x y^4, \quad \Delta \neq 0, \quad ab \neq 9 \mapsto$ **48**;

$j_5 f$ is degenerate $\mapsto$ **49**.

48. $j_5 f = x^4 y + a x^3 y^2 + b x^2 y^3 + x y^4, \quad \Delta \neq 0 \Rightarrow f \in N_{16}$, i.e.

$f \sim x^4 y + a x^3 y^2 + b x^2 y^3 + x y^4 + c x^3 y^3, \quad ab \neq 9, \quad \Delta(a, b) \neq 0$;

$\mu(f) = 16, \quad m(f) = 3, \quad c(f) = 12$.

49. $j_5 f$ is degenerate $\Rightarrow \mu(f) > 16$, $m(f) > 2$, $c(f) > 12$.

Corank 3 Singularities

Through theorems **50–104,** $f \in \mathbb{C}[[x, y, z]]$.

50. $j_2 f(x, y, z) = 0 \Rightarrow$ one of the ten possibilities holds.

$j_3 f \approx x^3 + y^3 + z^3 + a x y z, \quad a^3 + 27 \neq 0 \mapsto$ **51**;

$\approx x^3+y^3+xyz \mapsto$ **52** (series **P**);
$\approx x^3+xyz \mapsto$ **54** (series $\mathbf{R}=T_{3,p,q}$);
$\approx xyz \mapsto$ **56** (series **T**),
$\approx x^3+yz^2 \mapsto$ **58** (series **Q**);
$\approx x^2z+yz^2 \mapsto$ **66** (series **S**);
$\approx x^3+xz^2 \mapsto$ **82** (series **U**);
$\approx x^2y \mapsto$ **97** (class **V**);
$\approx x^3 \mapsto$ **103**;
$=0 \mapsto$ **104**.

Series T

51. $j_3 f = x^3+y^3+z^3+axyz, \quad a^3+27 \neq 0 \Rightarrow f \in P_8 = T_{3,3,3}$.
52. $j_3 f = x^3+y^3+xyz \Rightarrow f \sim x^3+y^3+xyz+\alpha(z), \quad j_3\alpha=0, \quad \mapsto$ **53.**
53. $f = x^3+y^3+xyz+\alpha(z), \quad j_3\alpha=0 \Rightarrow f \in P_{p+5} = T_{3,3,p}\,(p>3)$.
54. $j_3 f = x^3+xyz \Rightarrow f = x^3+xyz+\alpha(y)+\beta(z), \quad j_3\alpha=j_3\beta=0, \quad \mapsto$ **55.**
55. $f = x^3+xyz+\alpha(y)+\beta(z), \quad j_3(\alpha,\beta)=0 \Rightarrow f \in R_{p,q} = T_{3,p,q} \quad (q \geqq p>3)$.
56. $j_3 f = xyz \Rightarrow f \sim xyz+\alpha(x)+\beta(y)+\gamma(z), \quad j_3(\alpha,\beta,\gamma)=0, \quad \mapsto$ **57.**
57. $f = xyz+\alpha(x)+\beta(y)+\gamma(z), \quad j_3(\alpha,\beta,\gamma)=0 \Rightarrow f \in T_{p,q,r} \quad (r \geqq q \geqq p>3)$.

Series Q

Through theorems **58–65** $\varphi = x^3+yz^2$, $j^*_\lambda = j_{yz^2,x^3,\lambda}$ (λ is a monomial).
58. $j_3 f = \varphi \Rightarrow f \sim \varphi+\alpha(y)+x\beta(y), \quad j_3(\alpha, x\beta)=0, \quad \mapsto$ **59$_1$**.
Through theorems **59–62**, $k \geqq 1$.
59$_k$. $f = \varphi+\alpha(y)+x\beta(y), \quad j^*_{y^{3k}} f = \varphi \Rightarrow$ one of the four possibilities holds:
$j^*_{y^{3k+1}} f \approx \varphi+y^{3k+1} \mapsto$ **60$_k$**;
$j^*_{xy^{2k+1}} f \approx \varphi+xy^{2k+1} \mapsto$ **61$_k$**;
$j^*_{y^{3k+2}} f \approx \varphi+y^{3k+2} \mapsto$ **62$_k$**;
$j^*_{y^{3k+2}} f = \varphi \mapsto$ **63$_{k+1}$**.
60$_k$. $j^*_{y^{3k+1}} f = \varphi+y^{3k+1} \Rightarrow f \in Q_{6k+4}$.
61$_k$. $j^*_{xy^{2k+1}} f = \varphi+xy^{2k+1} \Rightarrow f \in Q_{6k+5}$.
62$_k$. $j^*_{y^{3k+2}} f = \varphi+y^{3k+2} \Rightarrow f \in Q_{6k+6}$.
Through theorems **63–65**, $k>1$.
63$_k$. $f = \varphi+\alpha(y)+x\beta(y), \quad j^*_{y^{3k-1}} f = \varphi \Rightarrow$ one of the three possibilities holds:
$j^*_{y^{3k}} f \approx \varphi+ax^2y^k+xy^{2k}, \quad a^2 \neq 4 \mapsto$ **64$_k$**;
$\approx \varphi+x^2y^k \mapsto$ **65$_k$**;
$= \varphi \mapsto$ **59$_k$**.
64$_k$. $j^*_{y^{3k}} f = \varphi+ax^2y^k+xy^{2k}, \quad a^2 \neq 4 \Rightarrow f \in Q_{k,0}$.
65$_k$. $j^*_{y^{3k}} f = \varphi+x^2y^k \Rightarrow f \in Q_{k,i} \quad (i>0)$.

Series S

Through theorems **66–81** $\varphi = x^2z+yz^2$, $j^* = j_{x^2z,yz^2,\lambda}$ (λ is a monomial).
66. $j_3 f = \varphi \Rightarrow f = \varphi+\alpha(y)+x\beta(y)+z\gamma(y), \quad j_3(\alpha, x\beta, z\gamma)=0, \quad \mapsto$ **67$_1$**.
Through theorems **67–76**, $k \geqq 1$.

67$_k$. $f=\varphi+\alpha(y)+x\beta(y)+z\gamma(y)$,
$j^*_{y^{4k-1}}f=\varphi$ $\Rightarrow$ one of the three possibilities holds:
$j^*_{y^{4k}}f \approx \varphi+y^{4k} \mapsto$ **68**$_k$;
$j^*_{xy^{3k}}f \approx \varphi+xy^{3k} \mapsto$ **69**$_k$;
$j^*_{xy^{3k}}f = \varphi \mapsto$ **70**$_k$.

68$_k$. $j^*_{y^{4k}}f=\varphi+y^{4k} \Rightarrow f\in S_{12k-1}$.

69$_k$. $j^*_{xy^{3k}}f=\varphi+xy^{3k} \Rightarrow f\in S_{12k}$.

70$_k$. $f=\varphi+\alpha(y)+x\beta(y)+z\gamma(y)$, $j^*_{xy^{3k}}f=\varphi \Rightarrow$ one of the four possibilities holds:
$j^*_{y^{4k+1}}f \approx \varphi+y^{4k+1}+bzy^{2k+1}$, $b^2 \neq 4 \mapsto$ **71**$_k$;
$\approx \varphi+x^2y^{2k} \mapsto$ **72**$_k$;
$\approx \varphi+zy^{2k+1} \mapsto$ **73**$_k$;
$= \varphi \mapsto$ **74**$_k$.

71$_k$. $j^*_{y^{4k+1}}f=\varphi+y^{4k+1}+bzy^{2k+1}$, $b^2\neq 4 \Rightarrow f\in S_{k,0}$.

72$_k$. $j^*_{y^{4k+1}}f=\varphi+x^2y^{2k} \Rightarrow f\in S_{k,i}$ $(i>0)$.

73$_k$. $j^*_{y^{4k+1}}f=\varphi+zy^{2k+1} \Rightarrow f\in S^{\#}_{k,i}$ $(i>0)$.

74$_k$. $f=\varphi+\alpha(y)+x\beta(y)+z\gamma(y)$,
$j^*_{y^{4k+1}}f=\varphi \Rightarrow$ one of the three possibilities holds:
$j^*_{xy^{3k+1}}f \approx \varphi+xy^{3k+1} \mapsto$ **75**$_k$;
$j^*_{y^{4k+2}}f \approx \varphi+y^{4k+2} \mapsto$ **76**$_k$;
$j^*_{y^{4k+2}}f = \varphi \mapsto$ **77**$_{k+1}$.

75$_k$. $j^*_{xy^{3k+1}}f=\varphi+xy^{3k+1} \Rightarrow f\in S_{12k+4}$.

76$_k$. $j^*_{y^{4k+2}}f=\varphi+y^{4k+2} \Rightarrow f\in S_{12k+5}$.

Through theorems **77**–**81**, $k>1$.

77$_k$. $f=\varphi+\alpha(y)+x\beta(y)+z\gamma(y)$, $j^*_{y^{4k-2}}f=\varphi \Rightarrow$ one of the five possibilities holds:
$j^*_{y^{4k-1}}f \approx \varphi+ax^2y^{k-1}+bxy^kz+xy^{3k-1}$, $\Delta\neq 0 \mapsto$ **78**$_k$.
$\approx \varphi+xy^kz+ax^3y^{k-1}$, $a^2\neq a \mapsto$ **79**$_k$.
$\approx \varphi+x^3y^{k-1} \mapsto$ **80**$_k$.
$\approx \varphi+xy^kz \mapsto$ **81**$_k$.
$= \varphi \mapsto$ **67**$_k$.

78$_k$. $j^*_{y^{4k-1}}f=ax^2y^{k-1}+bxy^kz+xy^{3k-1}$, $\Delta\neq 0 \Rightarrow f\in S^*_{k,0}$, $\mu(f)=12k-4$, $m(f)=3k-2$, $c(S^*_{k,0})=9k-3$.

79$_k$. $j^*_{y^{4k-1}}f=\varphi+xy^kz+ax^3y^{k-1}$, $a^2\neq a \Rightarrow f\in$ **SP**$_k$, $\mu(f)\geqq 12k-3$, $m(f)=3k-2$, $c($**SP**$_k)=9k-2$.

80$_k$. $j^*_{y^{4k-1}}f=\varphi+x^3y^{k-1} \Rightarrow f\in$ **SQ**$_k$, $\mu(f)\geqq 12k-2$, $m(f)=3k-2$, $c($**SQ**$_k)=9k-1$.

81$_k$. $j^*_{y^{4k-1}}f=\varphi+xy^kz \Rightarrow f\in$ **SR**$_k$, $\mu(f)\geqq 12k-2$, $m(f)=3k-2$, $c($**SR**$_k)=9k-1$.

Series U

Through theorems **82**–**89**, $\varphi=x^3+xz^2$, $j^*=j_{x^3,z^3,\lambda}$ (λ is a monomial).

82. $j_3f=\varphi \Rightarrow f\sim\varphi+\alpha(y)+x\beta(y)+z\gamma(y)+x^2\delta(y)$, $j_3(\alpha, x\beta, z\gamma, x^2\delta)=0 \mapsto$ **83**$_1$.

Through theorems **83**–**89**, $k\geqq 1$.

83$_k$. $f=\varphi+\alpha(y)+x\beta(y)+z\gamma(y)+x^2\delta(y)$, $j^*_{y^{3k}}f=\varphi\Rightarrow$ one of the two possibilities holds:
$j^*_{y^{3k+1}}f\approx\varphi+y^{3k+1}\mapsto$ **84$_k$**,
$=\varphi\mapsto$ **85$_k$**.

84$_k$. $j^*_{y^{3k+1}}f=\varphi+y^{3k+1}\Rightarrow f\in U_{12k}$.

85$_k$. $f=\varphi+\alpha(y)+x\beta(y)+z\gamma(y)+x^2\delta(y)$, $j^*_{y^{3k+1}}f=\varphi\Rightarrow$ one of the three possibilities holds:
$j^*_{xy^{2k+1}}f\approx\varphi+xy^{2k+1}+czy^{2k+1}$, $c(c^2+1)\neq 0\mapsto$ **86$_k$**;
$\approx\varphi+xy^{2k+1}\mapsto$ **87$_k$**;
$=\varphi\mapsto$ **88$_k$**.

86$_k$. $j^*_{xy^{2k+1}}f=\varphi+xy^{2k+1}+czy^{2k+1}$, $c(c^2+1)\neq 0\Rightarrow f\in U_{k,0}$.

87$_k$. $j^*_{xy^{2k+1}}f=\varphi+xy^{2k+1}\Rightarrow f\in U_{k,p}$ $(p>0)$.

88$_k$. $f=\varphi+\alpha(y)+x\beta(y)+z\gamma(y)+x^2\delta(y)$, $j^*_{xy^{2k+1}}f=\varphi\Rightarrow$ one of the two possibilities holds:
$j^*_{y^{3k+2}}f\approx\varphi+y^{3k+2}\mapsto$ **89$_k$**;
$=\varphi\mapsto$ **90$_{k+1}$**.

89$_k$. $j^*_{y^{3k+2}}f=\varphi+y^{3k+2}\Rightarrow f\in U_{12k+4}$.
Through theorems **90–96**, $k\geqq 2$;
$\varphi=x^2z+xz^2$; $j^*=j_{x^3,z^3,\lambda}$ (λ is a monomial).

90$_k$. $f=x^3+xz^2+\alpha(y)+x\beta(y)+z\gamma(y)+x^2\delta(y)$, $j^*_{y^{3k-1}}f=x^3+xz^2\Rightarrow$ one of the seven possibilities holds:
$j^*_{y^{3k}}f\approx\varphi+ax^2y^k+bxy^kz+y^kz^2+cxy^{2k}$, $\Delta\neq 0\mapsto$ **91$_k$**; $\mapsto$ **91$_k$**;
$\approx\varphi+ax^2y^k+bxy^kz+y^kz^2$, $4a\neq b^2$, $a(a+1-b)\neq 0\mapsto$ **92$_k$**;
$\approx x^3+ax^2z+xz^2+z^2y^k$, $a^2\neq 4\mapsto$ **93$_k$**;
$\approx\varphi+x^2y^k+axy^kz$, $a^2\neq a\mapsto$ **94$_k$**;
$\approx\varphi+x^2y^k\mapsto$ **95$_k$**;
$\approx\varphi+xy^kz\mapsto$ **96$_k$**;
$=\varphi\mapsto$ **83$_k$**.

91$_k$. $j^*_{y^{3k}}f=\varphi+ax^2y^k+bxy^kz+y^kz^2+cxy^{2k}$, $\Delta\neq 0\Rightarrow f\in U^*_{k,0}$,
$\mu(f)=12k-4$, $m(f)=4k-3$, $c(U^*_{k,0})=8k-2$.

92$_k$. $j^*_{y^{3k}}f=\varphi+ax^2y^k+bxy^kz+y^kz^2$, $4a\neq b^2$, $a(a+1-b)\neq 0\Rightarrow f\in \mathbf{UP}_k$,
$\mu(f)\geqq 12k-3$, $m(f)=4k-3$, $c(\mathbf{UP}_k)=8k-1$.

93$_k$. $j^*_{y^{3k}}f=x^3+ax^2z+xz^2+z^2y^k$, $a^2\neq 4\Rightarrow f\in\mathbf{UQ}_k$, $\mu(f)\geqq 12k-2$,
$m(f)=4k-3$, $c(\mathbf{UQ}_k)=8k$.

94$_k$. $j^*_{y^{3k}}f=\varphi+x^2y^k+axy^kz$, $a^2\neq a\Rightarrow f\in\mathbf{UR}_k$, $\mu(f)\geqq 12k-2$,
$m(f)=4k-3$, $c(\mathbf{UR}_k)=8k$.

95$_k$. $j^*_{y^{3k}}f=\varphi+x^2y^k\Rightarrow f\in\mathbf{US}_k$, $\mu(f)\geqq 12k-1$, $m(f)=4k-3$,
$c(\mathbf{US}_k)=8k+1$.

96$_k$. $j^*_{y^{3k}}f=\varphi+xy^kz\Rightarrow f\in\mathbf{UT}_k$, $\mu(f)\geqq 12k-1$, $m(f)=4k-3$,
$c(\mathbf{UT}_k)=8k+1$.

Class V

97. $j_3 f(x,y,z)=x^2y \Rightarrow f \sim x^2y+\alpha(y,z)+x\beta(z) \mapsto$ **98.**

In theorems **98** and **102** φ is one of the 10 polynomials:
$z^4+z^3y,\ z^3y+z^2y^2,\ z^2y^2+zy^3,\ z^4+z^2y^2,\ z^4,\ z^3y,\ z^2y^2,\ zy^3,\ y^4,\ 0.$

98. $f=x^2y+\alpha(y,z)+x\beta(z),\quad j_3 f=x^2y \Rightarrow$ one of the four possibilities holds:

$$\begin{aligned} j_{x^2y,y^4,z^4} f &\approx x^2y+z^4+az^3y+bz^2y^2+zy^3, \quad \Delta \neq 0, \quad ab \neq 9 \mapsto \mathbf{99};\\ &\approx x^2y+z^4+bz^3y+z^2y^2, \quad b^2 \neq 4 \mapsto \mathbf{100};\\ &\approx x^2y+z^3y+az^2y^2+y^4, \quad 4a^3+27 \neq 0 \mapsto \mathbf{101};\\ &\approx x^2y+\varphi \mapsto \mathbf{102}.\end{aligned}$$

99. $j_{x^2y,y^4,z^4} f=x^2y+z^4+az^3y+bz^2y^2+zy^3,\quad \Delta \neq 0 \Rightarrow f \in V_{1,0}.$

100. $j_{x^2y,y^4,z^4} f=x^2y+z^4+bz^3y+z^2y^2,\quad b^2 \neq 4 \Rightarrow f \in V_{1,p},\quad p>0.$

101. $j_{x^2y,y^4,z^4} f=x^2y+z^3y+az^2y^2+y^4,\quad 4a^3+27 \neq 0 \Rightarrow f \in V^{\#}_{1,p},\quad p>0.$

102. $j_{x^2y,y^4,z^4} f=x^2y+\varphi \Rightarrow \mu(f) \geqq 17,\quad m(f) \geqq 3,\quad c(f) \geqq 13.$

103. $j_3 f(x,y,z)=x^3 \Rightarrow \mu(f) \geqq 18,\quad m(f) \geqq 4,\quad c(f) \geqq 13.$

104. $j_3 f(x,y,z)=0 \Rightarrow \mu(f) \geqq 27,\quad m(f) \geqq 10,\quad c(f) \geqq 16.$

105. Corank $f>3 \Rightarrow \mu(f) \geqq 16,\quad m(f) \geqq 5,\quad c(f) \geqq 10.$

§ 3. The Proofs

Theorems **1**, **17** and **25** are obvious. Theorems **12_1**, **15**, **16** are proved in [8]. The proofs of theorems **12_k** ($k>1$), **24**, **31**, **32**, **38–46**, **65**, **72**, **73**, **79**, **80**, **81**, **87**, **92–96**, **100**, **101** are based on new techniques as compared with [8] (some spectral sequence) and are not given here.

The proofs of the classification theorems for unimodular singularities do not depend on these theorems.

The classification of unimodular singularities is obtained from the following theorems:

1–5, **$6_{1,2}$–$9_{1,2}$**, **10_2**, **11_2**, **13–17**, **18_1–21_1**, **22_2**, **23_2**, **25**, **26_1–30_1**, **36_1**, **37_1**, **47**, **48**, **50–58**, **59_1–62_1**, **63_2**, **64_2**, **66**, **67_1–71_1**, **82**, **83_1–86_1**, **97**, **98**, **105**.

Here one needs only the simplest cases (the first one) of theorems **10_2**, **22_2**, **29_1**, **36**, **47**, **63_2**, **70_1**, **85_1**, **98**.

To classify all bimodular singularities one needs in addition theorems:

6_3–11_3, **12_2**, **18_2–21_2**, **22_3**, **23_3**, **24_2**, **31_1–35_1**, **36_2**, **37_2**, **59_2–62_2**, **63_3**, **64_3**, **65_2**, **72_1–76_1**, **77_2**, **78_2**, **87_1–89_1**, **90_2**, **91_2**, **99**.

(In fact, only the simplest cases of theorems **10_3**, **22_3**, **36_2**, **63_3**, **77_2**, **90_2** are needed.)

Theorems **6**, **18**, **26**, **33**, **59**, **67**, **74**, **83**, **88** are proved by the Newton method [9] of a moving ruler (line, plane). This method reduces the proof to the counting of the integer points in triangles resp. polyhedrones on the exponent plane (resp. in the space).

The proofs of theorems **3**, **10**, **13**, **22**, **29**, **36**, **47**, **50**, **52**, **54**, **56**, **58**, **63**, **66**, **70**, **77**, **82**, **85**, **90**, **97**, **98** can be reduced to the classifications of orbits of the actions of some quasihomogenous diffeomorphism groups on the spaces of quasi-homenous polynomials.

Some quasihomogenous Lie algebra "roots" technique reduces all these classification problems to geometrical problems as shown in the table below.

Theorem	Series	Geometrical problem
3	D	Linear classification of the 3-forms in $\mathbb{C}^2$.
10, 22	J, Z	Affine classifications of the triples of points in $\mathbb{C}^1$.
13	X	Linear classification of the 4-forms in $\mathbb{C}^2$.
29	W	Linear classification of the couples of points in $\mathbb{C}^1$.
36	X	Affine classification of the quadruples of points in $\mathbb{C}^1$.
47	N	Linear classification of the 5-forms in $\mathbb{C}^2$.
50	P	Linear classification of the 3-forms in $\mathbb{C}^3$.
63	Q	Affine classification of the cubics with a finite cusp point in $\mathbb{C}^2$.
70	S	Affine classification of the cubics with at least 2 finite nodes in $\mathbb{C}^2$.
77	S*	Affine classification of the cubics in $\mathbb{C}^2$, having the infinity line a simple tangent.
85	U	Affine classification of the cubics in $\mathbb{C}^2$, admitting a centre of symmetry and having exactly 3 points at infinity.
90	U*	Affine classification of the cubics in $\mathbb{C}^2$, having exactly 3 points at infinity.
98	V	Affine classification of the polynomials of degree $\leqq 4$ in 1 variable in $\mathbb{C}^1$.

Theorems on the normal forms of the semiquasihomogenous functions (theorems **2, 4, 5, 7, 8, 9, 11, 14, 19, 20, 21, 23, 27, 28, 30, 34, 35, 37, 48, 51, 60, 61, 62, 64, 68, 69, 71, 75, 78, 84, 86, 89, 91, 99**) and theorems **52, 54, 56, 58, 66, 82, 97** follow from section **7.3** in [8]. Theorems **49** and **102** are corollaries of **48** and **100, 101**.

To prove theorems **53, 55, 57** we use the following lemmata (the terminology is that of [8]).

Lemma 1. *Let* $f_0 = a x_1 x_2 x_3 + x_1^{p_1} + x_2^{p_2} + x_3^{p_3}$, *where* $3 \leqq p_1 \leqq p_2 \leqq p_3 > 3$, $a \neq 0$. *Then*

1) *at* O f_0 *has an isolated critical point with* $\mu = p_1 + p_2 + p_3 - 1$;

2) *monomials* $1, x_1 x_2 x_3, x_i^{s_i}$ $(0 < s_i < p_i,\ i = 1, 2, 3)$ *define a regular basis of the local ring*;

3) *condition A holds for the Newton filtration, defined by the 4 monomials of* f_0.

To prove this lemma we use the crossword technique (see [8], section **9.7**). We find the following 2 geometrical facts:

Lemma 2. *All the cycles formed by the admissible segments for* f_0, *are trivial.*

This follows from the linear independence of the 3 main segments; this independence holds when p_i satisfies the above stated restrictions.

Lemma 3. *The maximal admissible chains are:*

1) *every point* $x_i^{s_i}$ $(0 \leqq s_i \leqq p_i - 2)$ *is a trivial chain formed by this point*;

2) *there exist 4 finite maximal chains* $x_1 x_2 - x_3^{p_3-1}, x_2 x_3 - x_1^{p_1-1}, x_3 x_1 - x_2^{p_2-1}, x_1 x_2 x_3 - x_i^{p_i}$;

3) *all other points have infinite admissible chains.*

Lemma 1 is an easy corollary of lemmata 2 and 3.

The proofs of theorems **53, 55, 57**.

These theorems can be stated as follows.

Theorem. *Let f have a critical point O of finite multiplicity and*

$$f = xyz + \alpha(x) + \beta(y) + \gamma(z), \quad j_2(\alpha, \beta) = 0, \quad j_3(\gamma) = 0.$$

Then in some neighbourhood of O

$$f \sim f_0 = axyz + x^p + y^q + z^r, \quad a \neq 0, \; 3 \leqq p \leqq q \leqq r > 3.$$

Proof. Let the first nonzero terms of the Taylor series for α, β, γ have exponents p, q, r. By changing if nesseserly the notations, we obtain $3 \leqq p \leqq q \leqq r > 3$. By dilatations of coordinates we reduce f to the form $f_0 + f_1$, where all the exponents of the monomials of f_1 are above the Newton polyhedron of f_0.

According to lemma 1, all the monomials of f_1 belong to the ideal $(\partial f_0 / \partial x_i)$, and f_0 satisfies the condition A of [8].

By theorem **9.5** of [8] we have $f \sim f_0$; theorems **53**, **55**, **57** are thus proved.

References

1. Arnold, V.I.: Critical points of smooth functions. In: The Vancouver International Congress of Mathematicians, 1974
2. Demazure, M.: Sousgroupes algébriques de rang maximum du group de Cremona. Annales Sc. de l'E.N.S., 4-e serie, t.3, fasc **4**, 507–588 (1970)
3. Gabrielov, A.M.: Bifurcations, Dynkin diagrams and the modules numbers of isolated singularities. Funct. Anal. and appl. **8**, N2 7–13 (1974)
4. Arnold, V.I.: Normal forms of functions near degenerate critical points, Weyl groups A_k, D_k, E_k and Lagrange singularities. Funct. Anal. and appl. **6**, N4 254–272 (1972)
5. Arnold, V.I.: Classification of unimodular critical points of functions. Funct. Anal. and appl. **7**, N3 75–76 (1973)
6. Dolgatchev, I.V.: Factor conical singularities of complex surfaces. Funct. Anal. and appl. **8**, N2 75–76 (1974)
7. Koulikov, V.S.: On some singularities of surfaces. Funct. Anal. and appl. **9**, N1 72–73 (1975)
8. Arnold, V.I.: Normal forms of functions in the neighbourhoods of degenerate critical points. Russian Math. surveys **29**, N2 11–49 (1974)
9. Newton, I.: La méthode des fluxions., trad. Buffon. Paris: Debure l'Ainé 1740
10. Klein, F.: Vorlesungen über die Entwicklung der Mathematik im 19 Jahrhundert. Berlin: Springer 1926
11. Poincaré, H.: Sur les formes cubiques ternaires et quaternaires. J. Ecole Polytechn. **50**, 199–253 (1881)

V.I. Arnold
Department of Mathematics
University of Moscow
Moscow 117234/USSR

Received April 14, 1975

Proceedings of Symposia in Pure Mathematics
Volume 40 (1983), Part 1

SOME OPEN PROBLEMS IN THE THEORY OF SINGULARITIES[1]

V. I. ARNOLD

Many conjectures in the theory of singularities were first suggested by experimental material. Several of these conjectures have been proved (see the survey [**1**]). Therefore it seems reasonable to give a list of some problems which are still unsolved. I shall use without further explanation the terminology of [**1**], where one can find the more complete list of references as well.

1. Is it true that the stratum $\mu = \text{const}$ is smooth? For the cases which appear in the tables, in particular for $\mu \leqslant 16$, this is true. Recently A. Shoshitaishivili proved that the functions with isomorphic local rings form a smooth manifold in the base of the versal deformation. The smoothness of the stratum $\mu = \text{const}$ would follow from the positive solution of the following problem. Consider the algebraic action of a complex algebraic group in a finite-dimensional affine space (e.g. linear representation). Let us consider the set of points for which the isotropy groups have a fixed dimension. Is this set a smooth manifold?

2. Does the real modality of a real function coincide with the complex modality? This is clearly true for the simple singularities, and V. Muravlev checked that this is the case for unimodal singularities as well. The number of real moduli does not exceed the number of complex ones. E. Vinberg pointed out to me that for the representations of real algebraic groups such coincidence does not necessarily take place. For example for the standard action of the group of quaternions in $\mathbf{R}^4$ the number of moduli is 0, and after complexification it is different from zero.

1980 *Mathematics Subject Classification*. Primary 57R45.

[1]The Russian original appeared in "Theory of cubic formulas and applications of functional analysis to problems in mathematical physics", Proceedings of the Sobolev Seminar 1, Novosibirsk, 1976. This translation was prepared by I. Dolgachev, A. Libgober and Y. Yomdin. It is reprinted here with the author's kind permission.

0082-0717/82/0000-0357/$04.00

Originally publ. in: Trudy Sem. S.L. Soboleva, No. 1, 5-15, © Akad. Nauk SSSR, Sibirsk Otdel, Inst. Mat., Novosibirsk 1976
English transl. in: Proc. Symp. Pure Math. 40, Part 1, 57-69, © Amer. Math. Soc., Providence, R.I. 1983
Reprint by Springer-Verlag Berlin Heidelberg 2016 with kind permission.

3. Is every function stably equivalent to a Γ-non-degenerate function (in a neighborhood of a critical point of finite multiplicity)?

4. For which weights $\alpha_s = A_s/N$ are there nondegenerate quasihomogeneous functions of degree 1? For the functions with number of variables less than four the necessary and sufficient condition is the following: the rational function

$$\prod_s \frac{1 - z^{N-A_s}}{1 - z^{A_s}}$$

should be a polynomial. For functions in four variables this is already false, as was shown by B. Ivlev [**2**, p. 22].

Experimentation shows that all reasonable characteristics of a Γ-non-degenerate singularity can be simply expressed via the geometry of the Newton polyhedron. Here are several examples making this conjecture more precise.

5. Compute the modality of a Γ-non-degenerate function from its Newton polyhedron. In particular prove for semiquasihomogeneous functions that the modality is equal to the number of monomials of the basis of the local ring in the diagram and above it. For functions in two variables the modality is apparently equal to the number of integer points in the polygon bounded by the coordinate rays from the point (2, 2) and the Newton diagram.

Partial results in this direction were obtained by A. Kouchnirenko and A. Gabrielov [**3**, **4**].

6. Compute the signature of the intersection form in the middle-dimensional homology group of the local nonsingular level set of a Γ-non-degenerate function depending on $n \equiv 3$ (4) variables. The answer is known only for the sums of powers (Hirzebruch [**5**]) and several simple cases (Gabrielov [**6**]). The experimental data shows that for quasihomogeneous functions the following is true. The number of the positive (resp. negative) squares is equal to the number of λ_s's in the upper (resp. lower) half plane, where λ_s are square roots of the eigenvalues of the monodromy operator given by the condition

$$\lambda_s = e^{\pi i[(m_s,\alpha)+\Sigma\,\alpha_k]}.$$

(This conjecture was proven by Steenbrink in the fall of 1975.)

Here m_s is the exponent of the monomial x^{m_s} defining an eigenvector of the monodromy by the formula

$$\omega_s = x^{m_s}\frac{dx_1 \wedge \cdots \wedge dx_n}{df},$$

and $(\alpha_1 \cdots \alpha_n)$ are the weights. (Note that the number ν_r of monomials x^{m_s} for which $(m_s, \alpha) = r$ is given by the formula

$$\sum \nu_r t^{rN} = \prod \frac{1 - t^{N-A_k}}{1 - t^{A_k}}$$

where $\alpha_k = A_k/N$.)

The numbers λ_s are the eigenvalues of the symplectic operator

$$W = \begin{pmatrix} 0 & \mathrm{Var} \\ -{}^t\mathrm{Var}^{-1} & 0 \end{pmatrix}$$

which acts on $H^* + H_*$. The square of this operator gives the monodromy.

7. Find a normal form to which all Γ-non-degenerate functions with given Newton-diagram can be reduced. The answer is partially known in the case when the diagram has only one face. Experiment shows that in other cases one can get rather simple forms as well. However even for homogeneous functions it is unknown whether there is a simple or even a polynomial normal form for all Γ-non-degenerate homogeneous functions of a given degree.

8. Find the Jordan normal form for the monodromy operator for Γ-non-degenerate functions with given Newton diagrams. The characteristic polynomial was computed by Varchenko. Recent work of A. Todorov and Gordon on "geometric monodromy" gives hope that this problem is close to solution.

9. What is the maximal number of components the local real level set of the morsification of a real function with given (say Γ-non-degenerate) singularity can have? In the case of functions in two variables the number of all components (ovals and components going out of the neighborhood of a given point) is less than or equal to $g + k(g + 1, k)$. Here g is the genus of the corresponding Riemann surface (i.e. the number of handles), and k is the number of components of the boundary. If the original function is Γ-non-degenerate, then the genus is exactly the number of integer points strictly between Γ and the coordinate axes, (this follows easily from the formula of Koushnirenko [3]) and $k + 1$ is the number of integer points on Γ.

However, already in the case of two variables it is unknown, whether there is always a morsification with $k + g$ components, and moreover how many ovals there are among these components and how their position depends on the singularity. In the case of a greater number of variables it is natural to look not only at the number of components but also at the sums of the Betti numbers, and furthermore at the Euler characteristics of the level sets or of the sets having smaller values (cf. the papers by Petrovskiĭ, Oleĭnik, Gudkov, Rohlin, Harlamov, Zvonilov, Utkin, Krahnov, Krasnov, Mišačev, etc. on the 16th problem of Hilbert [7]). In the papers by Harlamov it is shown what role the dimensions $h^{p,q}$ of the spaces of forms of type (p, q) play in these problems. In particular, we go over to the question how to find out the role played by the genus $h^{p,0}$ for a local fiber of a holomorphic function and how to compute this number from the Newton diagram.

10. Čebyšev polynomials of several variables. For each isolated critical point of a function there is a "Čebyšev polynomial"—the morsification with the minimal number of critical values. (The usual Čebyšev polynomials are obtained from the function of one variable of the form z^n.) It is asked which of the noteworthy

properties of Čebyšev polynomials of one variable go over to the polynomials of several variables defined in such a way.

11. Uniform estimates of oscillating integrals. An oscillating integral has the form

$$I(h, \lambda) = \int_{\mathbf{R}^n} e^{iF(x,\lambda)/h}\varphi(x)\,dx, \qquad h \to 0,$$

where φ is a smooth function concentrated in a sufficiently small neighborhood of the origin, F a real deformation of the function $f(x) = F(x, 0)$ which depends smoothly on the parameter λ, and h is a small parameter; $F(0, 0) = 0$.

The greatest lower bound β of all numbers γ such that for any deformation F

$$(1) \qquad |I(h, \lambda)| \leqslant C(\varphi)\,|h|^{n/2-\gamma}$$

holds for all $|\lambda|$ which are sufficiently small, is called the uniform exponent of the singularity of the function f at 0.

The problem is to compute the exponent β (let us say, for a Γ-non-degenerate function f).

In the examples, where β could be found (Vinogradov **[8]**, Duistermaat **[9]**), this exponent coincides with the exponent β_0 of the first term of the asymptotic expansion

$$I(h, 0) \sim Ch^{n/2-\beta_0} ln^{\kappa_0} h + \cdots .$$

For a Γ-non-degenerate function the exponent β_0 can presumably be calculated from the point $(z, \ldots, z)$ where the Newton diagram and the diagonal of the first octant intersect, $\beta_0 = n/2 - 1/z$. This conjecture has been proved by V. N. Karpuskin for the semiquasihomogeneous case, and by A. N. Varčenko for the Γ-non-degenerate case provided that $z > 1$.

For each pair of integers n and l one can define the universal uniform exponent $\beta(n, l)$ as the greatest lower bound of all γ such that the oscillating integral admits an estimate (1) which is uniform with respect to the parameter λ for all families F of functions of n variables x and l parameters λ, except for a coresidual set in the function space.

The problem of calculating the rational numbers $\beta(n, l)$ turns out to be very difficult, since it seems to be almost equivalent to the problem of a complete classification of all singularities. For a fixed l and for $n \to \infty$ the numbers $\beta(n, l)$ stabilize, $\beta(n, l) = \beta(\infty, l) = \beta(l)$. The first 10 numbers $\beta(l)$ are as follows

l	0	1	2	3	4	5	6	7	8	9	10
β	0	1/6	1/4	1/3	3/9	5/12	1/2	1/20	13/24	9/16	2/3

(from $l = 6$ on, the unproved conjecture $\beta = \beta_0$ is used). Presumably $\beta(l)$ grows like $\sqrt{2l}/6$ for $l \to \infty$ (presumably the maximal Milnor number and the number of moduli for an l-parameter family in general position grow like $2^{\sqrt{2l}}$).

12. Uniform estimates for the variation of the inverse image. Let g be the germ of a diffeomorphism of an n-dimensional real (or complex) space onto another of the same kind and let $g(O_1) = O_2$ be a point of multiplicity μ of the image space. Depending on the type of the singularity of g in the point O_1, the μ inverse images of the point O_2 can approach each other in different ways.

In order to describe the difference between these ways, it is useful to study the asymptotic geometric characteristics of the inverse image of a small ball with center in the point O_2.

Let D be a set in n-dimensional Euclidean space which is sufficiently good. Then one can define its volume $v_n(D)$, the number $v_0(D)$ of components and the so-called variations in all dimensions $v_k(D)$ (cf. Vituškin **[10]**). By definition, $v_k(D)$ is the average (taken over all k-dimensional subspaces of the n-dimensional space) of the k-dimensional volumes of the orthogonal projections D onto the k-dimensional subspaces, taking care of multiplicity, i.e. of the number of components which are projected on one point. If D is convex, the numbers $v_k(D)$ coincide, up to coefficients which are independent of D, with the coefficients of the formula for the volume of the ε-neighborhood

$$\operatorname{mes}(D + \varepsilon) = v_n(D) + c_1 v_{n-1}(D)\varepsilon + \cdots + c_n v_n(D)\varepsilon^n.$$

If the boundary of D is smooth, these coefficients can be calculated as integrals from the symmetric functions of the main curvatures of the boundary of D. The one-dimensional variation $v_1(D)$ characterizes the length D, but not the diameter!

The functions, the asymptotics of which interest us, are the variations of the inverse image of a ball of radius δ with center O_2 under the mapping g:

$$w_k(\delta) = v_k(g^{-1}\{y: |y| < \delta\})$$

(more exactly, of the part of the inverse image in a ball of sufficiently small radius ρ with center in O_1: $\delta < \delta_0(\rho)$).

As in the case of oscillating integrals, one can define individual exponents

$$\eta_k^{(0)}(g) = \inf\{\xi_k: w_k(\delta) \leqslant C\delta^{\xi_k}\}. \tag{2}$$

For instance, the exponents for the mapping $y_k = x_k^{a_k}$ are equal to $\eta_n^{(0)} = \Sigma\,\alpha_j$, $\eta_{n-1}^{(0)} = (\Sigma\,\alpha_j) - \alpha_{\max}, \ldots, \eta_1^{(0)} = \alpha_{\min}$ (where $\alpha_j = 1/a_j$ are the weights).

Let $G(x, \lambda)$ be a deformation of the mapping g, which depends on l parameters λ (such that $G(x, 0) = g(x)$).

The uniform exponents $\eta_k(g)$ are defined by formula (2), where w_k is changed into $w_k(\delta, \lambda) = v_k\{x: |G(x, \lambda)| < \delta\}$ and where the constant C is independent of λ which varies in a neighborhood of O. It is asked to try to calculate all these exponents (let us say, for the gradient of a Γ-non-degenerate function), and also the corresponding universal exponents $\eta_k(n, l)$ (where $\eta = \dim\{x\}$, $l = \dim\{\lambda\}$).

The only result in this direction known to me is the following theorem of N. N. Nehoroŝev **[11]** which he proved as part of a problem in celestial mechanics: the estimate for the deformation G of a mapping g near 0

$$\max_{0 \leqslant r \leqslant \kappa} \min_{|x| = r} |G(x, \lambda)| \geqslant C\kappa^{\rho} \tag{3}$$

which is uniform with respect to λ is violated only on a set whose codimension in the space of functions g grows with ρ. An estimate of this form would follow from the uniform estimate of the first variation above. In this way, the calculation of the universal exponent $\eta_1(n, l)$ for gradient mappings would even for small values of n and l give estimates for the speed of the evolution of the action variables in Hamiltonian systems which are almost integrable. For a function of one variable the estimate (3), together with the exact value of the constant C, is obtained from the theory of Čebyšev polynomials which have the smallest deviation from 0.

13. Problem A, D, E. The Dynkin diagrams

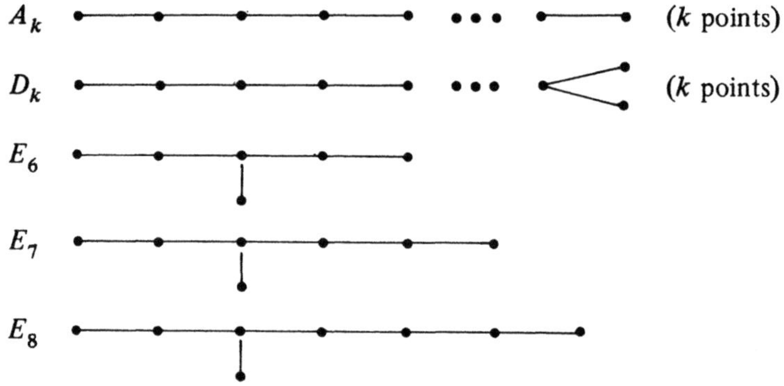

appear unexpectedly at the solution of so different classification problems, as the classifications (1) of critical points of functions, (2) of regular polyhedra in $\mathbf{R}^3$, (3) the category of linear spaces, (4) caustics, (5) wave fronts, (6) reflexion groups, and (7) simple Lie groups.

Some connections between these objects are known, but in the majority of cases (e.g. in the cases $1 \Leftrightarrow 2 \Leftrightarrow 3$) the coincidence of the answers to different problems has until now no explanation at all. The problem "A, D, E" consists in finding such a general classification theorem from which the solutions of all listed problems could be derived such that one could derive, let us say, the classification of simple stable singularities of caustics from the classificatin of regular polyhedra.

14. The problem $K(\pi, 1)$. Since 1972 it is known that the complements of the bifurcation varieties of functions and of the bifurcation varieties for their level sets are Eilenberg-Mac Lane space $K(\pi, 1)$ in the case of simple singularities A, D, E.

Does this result admit any generalization for the case of nonsimple singularities?

15. Singularities and reflexions. To each simple singularity two reflection groups are attached: the group generated by reflexions at the sides of a spherical triangle (with which the singularity can be built up as the spectrum of the ring of invariants), and the group generated by the Picard-Lefschetz reflexions (with which the monodromy group and the bifurcation diagram of the level sets can be described).

I. V. Dolgačev **[12, 13]** generalized the first construction and obtained all quasihomogeneous singularities with the help of Lobacevskiĭ polygons and rings of automorphic functions. It is asked whether the second construction does not admit a generalization for nonsimple singularities, too. In other words, one should complete the diagram:

Construction of simple singularities from regular polygons	→	Construction of quasihomogeneous 2-dimensional singularities from automorphic forms
↓		↓
Construction of bifurcation diagrams of simple singularities	→	?

to a commutative one.

For the modular group (and its generalization for the case of curves K of genus $g > 1$) the natural domain for the action is not the complex homology space $H_1(K, \mathbf{C})$ but the Siegel upper half plane (i.e. the Grassmann variety of Lagrange $\mathbf{C}$-subspaces in $H_1(K, \mathbf{C})$ with positive definite imaginary part of the action function).

Maybe, even for singularities the natural domain for the action of the monodromy group is not the complex homology space, but something like the upper half plane. Here we once more consider the Hodge filtration for singularities (compare problem 9). Recent work of Steenbrink is devoted to this question **[14]**.

16. Singularities and generalized Whitehead groups. Complete the following diagram to a commutative one:

Morse theory	→	Generalized Whitehead groups
↓		↓
Picard-Lefschetz theory	→	?

Here the upper horizontal arrow is the theory of Cerf-Wagoner-Volodin-Hatcher **[15–18]**, expressing homotopy properties of the space of Morse functions through the natural stratification of the space of smooth real functions. One can expect, that also by the natural stratification of the space of germs of holomorphic functions some kind of a complex theory of Cerf-Wagoner-Volodin-Hatcher can be built, which must relate to the real one as the Chern classes relate to Stiefel-Whitney classes.

17. Stable cohomology ring. To a critical point of a holomorphic function corresponds the cohomology ring $H^*(f)$ of the complement of the bifurcation diagram of the level sets in a base of the versal deformation.

Let f_2 be a germ of a function from the versal deformation of f_1. Then the transversal to the corresponding stratum in a base of the versal deformation gives an imbedding of complements and a homomorphism $H^*(f_1) \to H^*(f_2)$ of cohomology rings. For example, if $f_1 = x^n$, $f_2 = x^{n-1}$, then the H^* are the cohomology rings of braid groups of n and $n - 1$ strings, respectively, and the homomorphism induces a stabilization of the cohomology rings of braid groups as $n \to \infty$.

The question is whether such a stabilization occurs in a general case, and, if the answer is positive, then what is the stable cohomology ring? (It can be considered as the cohomology ring of the complement to the bifurcation diagram in a base of the versal deformation of the function $f = 0$.)

First of all one should clarify, whether the imbedding of the transversal section defines a canonical (not depending on the imbedding) ring homomorphism. It is so if (1) the stratum of f_2 in the base of f_1 is irreducible, (2) a holomorphic automorphism of a base of the versal deformation, which preserves the bifurcation diagram, also preserves the elements of the ring $H^*(f_2)$.

But, for example, the stratum $f_2 = A_3$ in a base of the versal deformation of $f_1 = D_4$ contains three irreducible components. However, for our purpose it is sufficient to prove, instead of the property (1), the stable irreducibility of the stratum f_2 (in a base of the versal deformation of a function, more degenerate than f_1).

Similar questions can be posed for complements to the bifurcation diagrams of functions. In this case also the following interesting objects arise:

(a) the subgroup of the braid group, which corresponds to the motion of the critical values, when the parameters go around the bifurcation diagram.

(b) the subring of the cohomology ring of the complement of the bifurcation diagram, induced from the cohomology of the braid group by the mapping, which associates to a point in a base of the versal deformation the unordered set of critical values.

These subgroups and subrings are unknown even for classes of singularities, which have been well studied. An interesting conjecture, describing these groups, was formulated by Looijenga [**19**].

Finally let us formulate some problems, concerning applications of the theory of singularities.

18. Converse of the Lagrange-Dirichlet theorem. Prove that the equilibrium position 0 of the Newton system $\ddot{x} = -\operatorname{grad} U$ is unstable if the critical point 0 of the polynomial $U(x_1, \ldots, x_n)$ is not a local minimum point. An incorrect proof was published by N. G. Chetaev. For the case $n = 2$ there is an unpublished proof by V. P. Palamodov.

I mention also the following problem of R. Thom: Prove that every phase curve of the system $\dot{x} = \operatorname{grad} U$ which approaches the critical point 0 has a definite tangent direction (for U a polynomial).

19. Algorithmic unsolvability of the stability problem. Is the problem of stability of the equilibrium position 0 of the system $\dot{x}_k = P_k(x)$ $(k = 1, \ldots, h)$, algorithmically solvable?

Here the P_k are polynomials with rational coefficients. Similar problem, the algorithmic unsolvability of which probably implies the unsolvability of the preceding one, are the following:

(1) The problem of existence of a limit cycle for the system $\dot{x} = P(x, y)$, $\dot{y} = Q(x, y)$, where P, Q are polynomials with rational coefficients.

(2) The problem of positivity of a real abelian integral $\oint R(x, y)\,dx$ over the oval of the curve $P(x, y) = 0$. (The coefficients of P and R are rational numbers.)

All of these problems at least do not admit an algebraic algorithm (see [**20**]).

20. Typical singularities of the solutions of variational problems.

It is known, that variational problems lead to discontinuities and singularities even in the case that all the data in the statements of the problems are smooth. The singularities arising here can be pathologically complicated because of degeneracies of an infinite multiplicity. Question: Is it possible to avoid pathologies by considering only generic cases?

As an example, consider the problem of the shortest path from a point A in Euclidean space to a point B, which avoids some domain C bounded by a smooth submanifold $\mathfrak{M}$.

We call a system of necessary conditions almost sufficient if for all generic manifolds $\mathfrak{M}$ and for each $\mathfrak{L} > 0$, there are only finitely many curves of length $< \mathfrak{L}$ joining A and B outside C, which satisfy the necessary conditions.

Can an almost sufficient system of necessary conditions be found explicitly?

As far as I know such a system has not been found even for the case that $\mathfrak{M}$ is a surface in three-dimensional space.

As another example consider the problem of the fastest path from a point A to a point B under the condition, that the velocity at each x belongs to a given smooth submanifold F_x of the tangent space of R^n at x. (F_x is called the indicatrix at x.)

If the indicatrix is not convex, one should use mixed strategies i.e. the indicatrix should be replaced by its convex hull. Hence we obtain the following problem: Describe the singularities of the convex hull of a generic k-dimensional submanifold F_x of $\mathbf{R}^n$.

Except of the trivial case $n = 2$, this problem has been solved only for $n = 3$, $k = 2$ (V. M. Zakaljukin) and $n = 3$, $k = 1$ (W. D. Sedych).

In order to find almost sufficient systems of necessary conditions one should also describe bifurcations of these singularities in n-parameter families (the parameter is the point x). The general question of the existence of explicit almost sufficiently necessary conditions can probably be solved without explicitly writing down the conditions themselves.

I mention here also the problem of describing the singularities of the set of points which are accessible from A (for generic F_x). In particular, let A be a point of a space with a generic relativistic metric. Consider the set of points which are accessible from A by time-like paths. What singularities can its boundary, as well as the set of points accessible by isotropic curves, have?

Related problems of describing the singularities of the function "the value of the extremum of the functional, depending on a finite point" for problems of general position are partly solved in the theory of Lagrangian and Legendrian singularities (see [**1**, **21**]).

21. Singularities in the theory of partial differential equations. In the theory of P.D.E. there are many cases of degeneration (when some determinants vanish). For example, there are cases of degeneration of ellipticity or hyperbolicity, the appearance of multiple characteristics, and so on. To analyze these degenerations one should find codimensions of corresponding submanifolds in jet spaces, and find out in what problems cases of general position have been completely investigated, as well as which cases of degeneracies which are unremovable by a small deformation of a one-parameter family of problems have been studied and so on.

As the simplest example consider the oblique derivative problem for the Laplace equation in the three-dimensional ball.

Degenerations appear since the vector field can be tangent to the surface of the space along some line, and can be tangent to this line at isolated points. In this problem (and in a corresponding n-dimensional problem) all the generic degenerations have been described (see [**22**]). Unfortunately, the influence of these degenerations on the behavior of the solutions near the variety of singularities on the surface of an n-dimensional ball have so far been investigated only for $n = 2$. In a general case only function spaces, in which the problem becomes Fredholm, are known [**23**]. The situation is similar for many problems of mathematical physics in which solutions are found by methods of functional analysis in Sobolev spaces or their generalizations.

The problem here is to describe the character of singularities on surfaces, curves and at points of the space, which really occur in solutions of equations for generic smooth initial or boundary conditions, and which are responsible for the fact that the solutions turn out to be not infinitely differentiable, but only to belong to the appropriate function space. Work in this direction was started by J. Gukenheimer [**24**].

Added in proof. In a letter to the editor dated October 5, 1981, the author provided the following comments.

Problem 5. The 2-variable case is solved; the formula conjectured there was proved by Kouchnirenko in 1976.

Problems 6, 8. These seem to be solved by Steenbrink and A. N. Kirillov, but only a conjectural (and wrong if no restrictions on Γ are supposed) formula by Steenbrink appears (in an Appendix to the Report 76-06, University of Amsterdam, 1977).

Problem 7. First negative answer seems to be that of Wahl, see also Vasiljev, Functional Anal. Appl. **12**: 3 (1978), 72.

Problem 9. See my paper in Functional Anal. Appl. **12**: 1 (1978), 1–14, *Index of vector field singular point, Petrovski-Oleĭnik inequalities and mixed Hodge structures*, and the discussion of problem 5 in the Bombay paper. These results are used by Nikoulin and Vinberg to prove the absence of reflection groups in Lobachevski spaces (dim ⩾ 30).

Problem 11. A counterexample to uniform estimations by Varchenko, Functional Anal. Appl. **10**: 3 (1976), 13–38; generalization to complex phase cases by Vasiljev, Functional Anal. Appl. **13**: 4 (1979); applications by Colinde Verdiere, Ann. Sci. École. Norm. Sup. Sec. 4 **10**: 4 (1977), 550–575; proof of uniform estimation for the two-variable case by A. N. Karpouchkin, Dokl. Akad. Nauk SSSR **254**: 1 (1980), 28–31. See also Varchenko, Funkcional Anal.i Prilozen **16**: 1 (1982), 1–12.

Problem 12. See E. E. Landis, Vestnik Mosk. Univ. Ser. 1 Math. Mech. (1980), no. 3, 24–28; *Tangential singularities*, Functional Anal. Appl. **15**: 2 (1981), 36–50.

Problems 13, 14. See Arnold, Uspehi Mat. Nauk SSSR **33**: 5 (1978), 91–107, **34**: 2 (1979), 3–38, Slodowy, L. N. M. **815** (1980), Ljashko, Uspehi Mat. Nauk SSSR **34**: 3 (1979), 205–206, Gorjunov Functional Anal. Appl. **15**: 2 (1981), 1–8; Knorrer 1981; Gorjunov 1982, Guivental, Functional Anal. Appl. **14**: 2 (1980), Varchenko and Guivental, Functional Anal. Appl. **16**: 2 (1982), 7–20. Guivental, Functional Anal. Appl. **82**, and others. Ljashko has found a singularity for H_3, but not yet for H_4.

Problem 15. Is there any relation of the 14 exceptional singularities to the 14 Lobachevski simplexes of Lanner (which are the only simplexes generating discrete reflection groups with compact fundamental domains)?

Problem 16. See Bombay, §2.

Problem 17. The Looijenga conjecture proved by P. Deligne (unpublished, but see Lifchitz, Functional Anal. Appl. **15**: 1 (1981), 38–42, for an account on the proof). See also: Arnold, Functional Anal. Appl. **14**: 4 (1980) and Vasiljev, Functional Anal. Appl. **15**: 3 (1981) for the characteristic class theories arising from bifurcation diagram stratifications.

Problem 18. See the Palamodov proof in Functional Anal. Appl. **11**: 4 (1977).

Problem 19. A survey book, "Additional chapters of ordinary differential equations theory", Moscow, 1978 (Springer translation, 1982, French edition by Mir, Moscow, 1980). See also a survey, Arnold and Oleĭnik, "Topology of real algebraic manifolds", Vestnik Mosk. Univ. Math. Mech. Ser. 1 (1979), no. 6, 7–17 (also for problem 9). Additional references are

Hovanski, A. G., Doklady **255** (1980), 804–807.

Hovanski, A. G., C. R. Acad. Sci. Paris (1981).

Hovanski and Guelfond, Functional Anal. Appl. **14**: 2 (1980), 52–53.

Kasarnovski, Doklady **257** (1981), 804–808.

Problem 20. A survey book, V. Arnold, "Catastrophe theory", Znanie, Moscow, 1981 (Ser. Math., Cibernetics, N9, Sept. 1981). See also Platonova, O. A., Uspehi Mat. Nauk SSSR **36**: 1 (1981), 221–222 and Platonova, O. A., "Singularities in the obstacle problem", Functional Anal. Appl. **15**: 2 (1981), 86–87; Arnold, "Asymptotic rays", Functional Anal. Appl. **15**: 4 (1981); Sedych, Functional Anal. Appl. **11**: 1 (1977), 81–82; Zakaljukin, Functional Anal. Appl. **11**: 3 (1977), 76–77; Sedych, Trudy Mosk. Math. Obšč. (1981); Brisgalova, Functional Anal.

Appl. **12**: 1 (1978), 66–67. Sedych, Proceedings of I. G. Petrovski Seminar (1981), pp. 239–256 and Uspehi Mat. Nauk SSSR **36**: 5 (1981), 191–192; in the Platonova list of smooth manifold germ projections from $\mathbf{R}P^3$ to $\mathbf{R}P^2$ in Uspehi Mat. Nauk SSSR **36**: 1 (1981), 221–222, Theorem 6 should contain 14 singularities: add $z = x^5 + xy$, $z = x^4 + x^2y + xy^3$, $z = x^3 - xy^4$ to the 11 there.

The author wishes to take this opportunity to comment on the problems which appeared in his paper "On some problems in singularity theory" in Geometry and Analysis (papers dedicated to the Memory of V. K. Patodi), Indian Academy of Sciences, Bombay, 1980, pp. 1–9.

Problem 1. See Gorjunov, Vestnik Mos. Univ. Ser. 1, Math. Mech. (1981), no. 4, 19–21.

Problem 2. The Γ_1 description is proved (see problem 17 above). The Looijenga paper [**11**] does not contain the assertion attributed to it in the Bombay text, and his mistake in [**11**] can be corrected.

Problem 3. See comments to problem 17 above—papers by Vasiljev and Arnold; see also Funkcional Anal. i Prilozen **16**: 2 (1982), 68–69.

Problem 4. The (*) conjecture is proved by A. G. Alexandrov, 1981. See also: Gorjunov, Uspehi Mat. Nauk SSSR **35**: 2 (1980), 205–206.

Problem 5. The Hovanski paper, Functional Anal. Appl. **13**: 1 (1979),

Problem 8. The conjecture is proved by V. I. Matov, 1981. The Levantovski papers: Vestnik Mosk. Univ. (1980), 20–22; Uspehi Mat. Nauk SSSR **35**: 2 (1980), 213–224; Funkcional Anal. i Prilozen **16**: 1 (1982), 44–48.

Addendum by the author (*March* 3, 1982).

(1) Recently (1981) Varchenko has proved the conjecture on the modality of quasihomogeneous functions given in problem 5. The result follows from his mixed Hodge structure theorem.

(2) Additional references for problem 20. Landis, E. E., *Tangential singularities*, Functional Anal. Appl. **15**: 2 (1981), 36–49; Arnold, V. I., *Surgery of singularities of potential flows in collision free media and caustic metamorphoses in 3-space*, Proceedings of I. G. Petrovski Seminar (1982), pp. 8–56.

References

1. V. I. Arnold, *Critical points of smooth functions and their normal forms*, Russian Math. Surveys **30** (1975) (5), 3–65.

2. ______, *Normal forms of functions near degenerate critical points*, Russian Math. Surveys **29** (1974) (2), 11–45.

3. A. G. Kushnirenko, *Newton polyhedron and Milnor numbers*, Functional Anal. Appl. **9** (1975) (1), 74–75.

4. A. M. Gabrielov and A. G. Kushnirenko, *Description of deformations with constant Milnor number for homogeneous functions*, Functional Anal. Appl. **9** (1975) (9), 67–68.

5. E. Brieskorn, *Beispiele zur Differential topologie von Singularitaten*, Invent. Math. **2** (1966) (1), 1–14.

6. A. M. Gabrielov, *Intersection matrices for certain singularities*, Functional Anal. Appl. **7** (1973) (3), 18—32.

7. D. M. Gudkov, *Topology of real projective algebraic varieties*, Russian Math. Surveys **29** (1974) (4), 3–79.

8. I. M. Vinogradov, *Trigonometric sums in number theory*, M. (1971).

9. J. J. Duistermaat, *Oscillatory integrals, Lagrange immersions and unfolding of singularities*, Comm. Pure Appl. Math. **27** (1974), 209–281.

10. A. G. Vitushkin, *On multidimensional variations.*

11. N. N. Nehoroshev, *Stable estimates for smooth mappings and for gradients of smooth functions*, Mat. Sb. **90** (1973) (3), 432–478.

12. I. V. Dolgačev, *Factorconical singularities of complex surfaces*, Functional Anal. Appl. **8** (1979) (2), 75–76.

13. ______, *Automorphic forms and quasihomogeneous singularities*, Functional Anal. Appl. **9** (1975) (2), 67–68.

14. J. J. M. Steenbrink, *Limits of Hodge structures*, Inst. Hautes Études Sci. Publ. Math., preprint, 1975.

15. J. Cerf, *La stratification naturelle des espaces de fonctions differentiables reels et le theoreme de la pseudo-isotopie*, Inst. Hautes Études Sci. Publ. Math. **39** (1970), 5–173.

16. J. B. Wagoner, *Algebraic invariants for pseudoisotopies*, Proc. Liverpool Singularities Symposium II, Lecture Notes in Math., vol. 209, Springer-Verlag, Berlin and New York, 1971, pp. 164–190.

17. A. E. Hatcher, *Parametrized h-cobordism theory*, Ann. Inst. Fourier (Grenoble) **23** (1973) (2), 61–74.

18. I. A. Volodin, *Generalized Whitehead groups and pseudoisotopies*, Russian Math. Surveys **27** (1972) (5), 229–230.

19. E. Looijenga, *The complement of the bifurcation variety of a simple singularity*, Invent. Math. **23** (1974) (2), 105–116.

20. V. I. Arnold, *Algebraic unsolvability of the stability problem*, Functional Anal. Appl. **4** (1970) (3), 1–9.

21. V. M. Zakaljukin, *On Lagrangian and Legendrian singularities*, Functional Anal. Appl. **10** (1976) (1).

22. S. M. Vishik, *Vectorfields in a neighborhood of the boundary*, Vestnik Moskov. Gos. Univ. Ser. Mat. (1972) (1), 21–28.

23. V. G. Mazia, *On degenerating problem with oblique derivative*, Mat. Sb. **87** (1972) (3), 417–454.

24. J. Gukenheimer, *Solving a single conservation law*, preprint, 1973.

DEPARTMENT OF MATHEMATICS, UNIVERSITY OF MOSCOW, MOSCOW, 117234 USSR

On the Theory of Envelopes

V.I. Arnold

(The Petrovskiĭ seminar on Differential equations and mathematical problems of physics; meeting on September 24, 1975.[1])

1°. Let $f : E \to M$ be a smooth mapping. Diffeomorphisms $r : E \to E$ and $l : M \to M$ are said to be *f-compatible*, if $fr = lf$. Here, l and r are called *f-raising* and *f-lowering* diffeomorphisms, respectively.

2°. Let f be the Whitney map, which takes an n-tuple $(x_1, \ldots, x_n)$ to the n-tuple $(y_1 = x_1^{n+1} + x_2 x_1^{n-1} + \ldots + x_n x_1, y_2 = x_2, \ldots, y_n = x_n)$. The set of critical values of the map f is called the *swallowtail.*

Theorem 1. *Let g be a germ at 0 of a smooth function of $(y_1, \ldots, y_n)$, $g(0) = 0$, $\partial g/\partial y_2(0) \neq 0$. Then g can be put into the form $\pm y_2$ by means of an f-raising (i.e. swallowtail-preserving) diffeomorphism.*

Theorem 2. *Let π be a germ at 0 of a smooth function of $(x_1, \ldots, x_n)$, $\pi(0) = 0$, $\partial \pi/\partial x_1(0) \neq 0$. Then π can be put into the normal form $x_1 + \varphi(f(x))$ by an f-lowering diffeomorphism.*

Let $F(x, x_{n+1}) = (f(x), y_{n+1})$, where $y_{n+1} = x_{n+1}$ and f is the Whitney map.

Theorem 3. *Let π be a germ at 0 of a smooth function of $(x_1, \ldots, x_{n+1})$, $\pi(0) = 0$, $\partial \pi/\partial x_{1,n+1}(0) \neq 0$. Then π can be put into the normal form $x_1 + x_{n+1}$ by an F-lowering diffeomorphism.*

3°. A diagram $B \leftarrow E \to M$, where $\pi : E \to B$ is a smooth bundle and $f : E \to M$ is a smooth mapping, is called a *family in M*. *Equivalence* of families is a commutative diagram whose two rows are the families and whose three columns are diffeomorphisms. The dimension of the base B is called the *number of parameters* of the family.

[1] Translation of Uspekhi Mat. Nauk, **31** (1976), no. 3, 248–249.

Originally publ. in: Usp. Mat. Nauk 31(3), 248-249, © Russ. Acad. Sciences 1976
Translated by Boris Khesin and Andrey Khesin.
Springer-Verlag Berlin Heidelberg 2016 with kind permission.

Denote by Ω_n the space of one-parameter families of hypersurfaces in $\mathbb{R}^n$ ($\dim B = 1$, $\dim E = \dim M = n$, and E is compact).

Theorem 4. *An open and everywhere dense subset of Ω_2 is constituted by the families that at every point of E are locally equivalent to one of the following*: 1) $f_1 = x_1$, $f_2 = x_2$; $\pi = x_1$; 2) $f_1 = x_1^2$, $f_2 = x_2$; *and either* $\pi = x_1 + x_2$, *or* $x_2 + x_1^3 + x_1x_2$, *or* $x_1 + x_2^2$; 3) $f_1 = x_1^3 + x_1x_2$, $f_2 = x_2$; $\pi = x_1 + \varphi(f(x))$.

Theorem 5. *An open and everywhere dense subset of Ω_3 is constituted by the families that at every point of E are locally equivalent to one of the following*: 1) $f_i = x_i$; $\pi = x_1$; 2) $f_1 = x_1^2$, $f_{2,3} = x_{2,3}$; *and either* $\pi = x_1 + x_3$, *or* $x_3 + x_1x_2$, *or* $x_3 + x_1x_3 + x_1^3 \pm x_1x_2^2$, *or* $x_1 \pm x_2^2 \pm x_3^2$; 3) $f_1 = x_1^3 + x_1x_2$, $f_{2,3} = x_{2,3}$; *and either* $\pi = x_1 + x_3$, *or* $(*)$ $x_1 \pm x_3^2 + \varphi(f_1(x), f_2(x))$, *or* $(*)$ $x_3 \pm x_1^2 + x_1\varphi(f_1(x), f_2(x))$; 4) $f_1 = x_1^4 + x_2x_1^2 + x_3x_1$, $f_{2,3} = x_{2,3}$; $\pi = x_1 + \varphi(f(x))$. (The results denoted by $(*)$ can possibly be improved.)

4°. Theorem 6 (Yu.A. Brodskiĭ). *The differential equation $\Phi(x, y, dy/dx) = 0$ for a typical Φ at a generic point of the discriminant manifold can be transformed into the normal form $(dy/dx)^2 = x$ by a local diffeomorphism of the (x, y)-plane.*

A qualitative study of this equation was recently carried out by R. Thom, although the normal form was not obtained.

5°. Proofs of Theorems 1 to 5 are based on the fact that symmetric vector fields on the plane $\sigma_1 = 0$ form a free module over the ring of functions of $(\sigma_2, \ldots, \sigma_n)$ with the generators $\mathrm{grad}(\sigma_i|_{\sigma_1=0})$.

Wave Front Evolution and Equivariant Morse Lemma

V. I. ARNOL'D
Moscow State University

1. Introduction

The simplest problem about bifurcation of wave front singularities is the following. Let us consider a plane semicubical parabola and a smooth curve, containing the cusp point of the parabola. We want to reduce this curve in a neighborhood of the cusp point to some simplest normal form by a parabola preserving local plane diffeomorphism.

Such problems appear naturally in a study of bifurcations of caustic and wave front singularities. In these cases normal forms are known for the varieties formed by the moving caustics or wave fronts in the space-time (in the example above this variety is the semicubical parabola and the space-time is the plane). The problem is to find normal forms for the time function (the smooth curve in the example above is the level set of the time function).

We give explicit solutions of these problems in the simplest (and most important) cases. For instance, every smooth plane curve, transversal to the tangent of the semicubical parabola at the cusp point, can locally be transformed into a straight line, orthogonal to this tangent, by a parabola preserving diffeomorphism. (See Figure 1.)

To formulate a more general result, we need the following

DEFINITION. The *swallowtail* $\Sigma \subset \mathbb{C}^k$ is the complex hypersurface

$$\Sigma = \{\lambda \in \mathbb{C}^k ; \Delta(\lambda) = 0\},$$

where Δ is the discriminant of the polynomial

$$p = z^{k+1} + \lambda_1 z^{k-1} + \cdots + \lambda_k,$$

the λ_j being coordinates in $\mathbb{C}^k$ (for $k = 2$ this is the plane semicubical parabola, for $k = 3$ the usual swallowtail surface in $\mathbb{C}^3$).

1.1. THEOREM. *Let f be a function, holomorphic at the origin $\lambda = 0$ of the space $\mathbb{C}^k \supset \Sigma$ such that $f(0) = 0$, $\partial f/\partial\lambda_1(0) \neq 0$. Then f can be reduced to the*

Originally publ. in: Comm. Pure Appl. Math. 29(6), 557-582, © Wiley Periodicals, Inc., A Wiley Company 1976
Reprint by Springer-Verlag Berlin Heidelberg 2016 with kind permission.

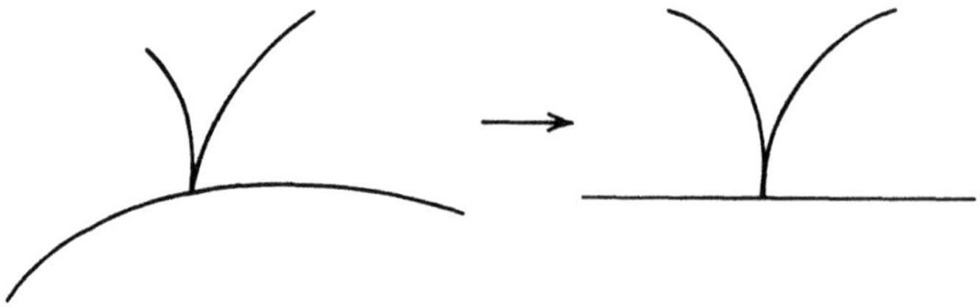

Figure 1

normal form $f=\lambda_1$ *by a swallowtail* Σ *preserving local holomorphic diffeomorphism of* $\mathbb{C}^k$ *at* 0.

The crucial point of the proof is to consider the coordinates λ_i as the symmetric group invariants. Theorem 1.1 is equivalent to a result on symmetric functions, which can be immediately extended to invariants of an arbitrary action of a compact group.

1.2. THEOREM. *Let G be a compact Lie group, acting linearly in* $\mathbb{C}^k$. *Then every G-invariant holomorphic function, having at* 0 *a nondegenerate critical point with critical value* 0, *can be reduced to its quadratic part by a G-invariant change of variables, biholomorphic at* 0.

Similar theorems hold in the real-analytic, formal and differentiable cases.

The proofs of Theorems 1.1 and 1.2 are given in Section 2. We give in Section 3 a different method for proving these theorems. These methods allow us to find normal forms for "simple" critical points of symmetric functions (Section 3), to deal with the real case (Section 4), and to study singularity bifurcations in one-parameter families of smooth mappings having corank 1 (Section 5). As other applications we list in Section 5 normal forms for the generic singularities of one-parameter curves and surface families at the envelopes, in Sections 6 and 7 we list normal forms for the singularity bifurcations of one-parameter families of Lagrange and Legendre mappings (that is, families of caustics and of wave fronts). We also discuss in Section 7 the Zeldovitch theory of "galaxy" generation in noninteracting particles media. All these theorems (Sections 4–7) are examples of applications of the general methods developed in Sections 2–3. In combination with the spectral sequence technique, described in [4], these methods imply many other results on the normal forms of singularity bifurcations.

Many publications have appeared in recent years both in the theory of singularity bifurcations and in that of equivariant singularities (see, for instance, [6], [7], [9], [10], [12], [13], [14], [15], [17], [19], [20], [23]).

The new idea of the present work is to use the relation between these two theories. This relation was unexpected (at least to the author), until Theorem 1.1 was proved by hard direct computations in [5].

While Theorem 1.1 is a simple corollary of the easy Theorem 1.2, it seems that it is impossible to find it in this way until the answer is known.

The proofs are elementary; that is, we use neither the esoteric parts of invariant theory, nor those of singularity theory.

The author thanks A. M. Gabrielov, A. M. Zakalyukin, Ya. B. Zeldovitch, V. Ya. Lin, D. B. Fuks, Ya. M. Eliasberg and H. Hamm for helpful discussions.

2. Normal Forms for G-Invariant Functions

2.1. Proof of Theorem 1.2: Let $H+F$ be a G-invariant function, H its nondegenerate quadratic part, F being of 3-rd order at 0. We want to construct a G-invariant (i.e., commuting with G) family g_t of local diffeomorphisms preserving 0 and satisfying

$$(1) \qquad (H+tF)(g_t x)=H(x), \qquad g_0(x)=x, \qquad t\in[0,1].$$

The time-dependent vector field ξ corresponding to g_t is, by definition,

$$\xi(g_t x, t)=\frac{\partial}{\partial t} g_t x .$$

Differentiating (1) with respect to t we obtain for ξ a linear (not a differential) equation

$$(2) \qquad \xi(H+tF)+F=0$$

(ξH is the derivative of H along ξ).

H being nondegenerate, this equation for ξ can be solved (in the class of possibly nonequivariant vector fields). To see this, let us consider local coordinates, such that $H=x_1^2+\cdots+x_k^2$. For the components η_i of the unknown vector field η equation (2) yields

$$-F=\sum \eta_i y_i, \qquad y_i=2x_i+t\frac{\partial F}{\partial x_i}.$$

The Jacobian $\det|\partial y_i/\partial x_j|_{x=0}$ is 1 for all t, since $dF(0)=d^2F(0)=0$. So (y_i, t) are local coordinates in some neighborhood of the t-axis in $\mathbb{C}^{k+1}=\{(x,t)\}$. Since $F(0)=0$, there exist, by the Hadamard lemma, η_i satisfying $-F=\sum \eta_i y_i$, $\eta_i(0)=0$. Thus we have constructed a (nonequivariant) solution η of equation (2).

To find an equivariant solution, let us integrate η, using the Haar measure on G:

$$\xi=\int_{g\in G} g\eta\, dg, \qquad (g\eta)(x)=g_*(g^{-1}x)\eta(g^{-1}x)$$

We obtain a G-invariant solution ξ of (2), vanishing at 0 identically in t. The corresponding diffeomorphisms g_t are defined in some neighborhood of 0 for all $t\in[0,1]$. It follows from (1) that g_1 is the desired diffeomorphism.

Remarks. (a) For the real analytic or differentiable cases the proof is the same. (b) It is not necessary to suppose that the action of G is linear: it becomes linear after a suitable change of variables (see [8]).

As an application of Theorem 1.2 let us consider the following situation. Let G be a finite irreducible group, generated by reflections and acting on a complex linear space $M\approx\mathbb{C}^k$. Let $\rho_1,\cdots,\rho_k$ be the basic homogeneous invariants. We define a mapping $\pi: M\to N\approx\mathbb{C}^k$ by the formulae

$$\lambda_1=\rho_1(x),\cdots,\lambda_k=\rho_k(x),$$

where $\lambda_1,\cdots,\lambda_k$ are the coordinates in N.

The *discriminant* $\Sigma\subset N$ is the set of critical values of π (that is, the mirror images).

Theorem 1.2 implies

2.2. COROLLARY. *Let f be holomorphic at $0\in N$ and let $f(0)=0$, $\partial f/\partial\lambda_1(0)\neq 0$, where the coordinate λ_1 corresponds to the invariant of the minimal degree.*

Then f can be reduced (at some neighborhood of 0) to a normal form $f=\lambda_1$ by a biholomorphical diffeomorphism preserving Σ.

Proof: Let $F(x)=f(\pi x)$. Then F satisfies the conditions of Theorem 1.2. (There exists, up to a constant multiple, only one invariant of degree 2, because the group is irreducible; all other invariants have degrees greater than 2.) Hence, the quadratic part H of F is nondegenerate. By Theorem 1.2,

$$F(gx)=H(x)$$

for some G-invariant diffeomorphism g. We have

$$H(x)=c\rho_1(x),$$

and hence

$$F(gc^{-1/2}x)=\rho_1(x).$$

Now the required diffeomorphism h is defined by

$$h(\pi x) = \pi(gc^{-1/2}x).$$

2.3. Proof of Theorem 1.1: Let us consider the space

$$M = \{Z \in \mathbb{C}^{k+1} : \sigma_1(Z)\} = 0,$$

where σ_i are the elementary symmetric functions in the coordinates $Z_1, \cdots, Z_{k+1}$. The coordinate permutation group G acts on M. We choose basic invariants

$$(\sigma_2 \mid \sigma_1 = 0), \quad (-\sigma_3 \mid \sigma_1 = 0), \quad \cdots, \quad (\sigma_{k+1} \mid \sigma_1 = 0).$$

Then the discriminant Σ is the swallowtail, and we obtain Theorem 1.1 from Corollary 2.2.

Now let us consider the same situation with parameters. Let $\tau \in \mathbb{C}^p$ be a parameter, and $N \times \mathbb{C}^p \subset_\tau \mathbb{C}^{k+p}$ the space containing the *discriminant* cylinder $\Sigma \times \mathbb{C}^p \subset N \times \mathbb{C}^p$. Theorem 1.2 implies

2.4. COROLLARY. *A generic function $\Phi : N \times \mathbb{C}^p \to \mathbb{C}$ can be reduced, at every point of the edge $O \times \mathbb{C}^p$, to one of the 2 normal forms*

$$\Phi = \tau_1 + C, \qquad \Phi = \lambda_1 + \tau_1^2 + \cdots + \tau_p^2 + C$$

by a diffeomorphism, preserving the discriminant cylinder (here $C =$ const., τ_i are coordinates in $\mathbb{C}^p$, λ_1 being the same as in Corollary 2.2).

Proof: Let us consider the restriction of Φ to the edge $O \times \mathbb{C}^p$. At a noncritical point of the restriction we can choose a coordinate $\tau_i = \Phi$. For generic Φ, all the critical points of $\Phi|_{O \times \mathbb{C}^p}$ are nondegenerate. At the neighborhood of such a point let us consider Φ as a family of functions on $\mathbb{C}^p$, depending on a parameter $\lambda \in N$. By the Morse lemma with parameters (the proof of which is similar to that of Theorem 1.2), we can choose coordinates $\tau_i = \tau_i(\tau, \lambda)$, reducing Φ to $\tau_1^2 + \cdots + \tau_p^2 + \alpha(\lambda)$.

Now for a generic Φ the function $\alpha - \alpha(0)$ satisfies the conditions of Corollary 2.2. By a change of variables $\Lambda = \Lambda(\lambda)$, preserving Σ, we reduce α to the form $\lambda_1 + C$.

3. Liftable and Lowerable Vector Fields and Diffeomorphisms

Let $h : M \to N$ be a smooth mapping.

DEFINITION. The diffeomorphisms $r : M \to M$ and $l : N \to N$ are said to be *h-agreeing*, if and only if $hr = lh$. In this case, l is *h-liftable* and r is *h-lowerable*.

The vector fields W on M and V on N are said to be *h-agreeing*, if and only if $hW = V$. In this case, the field V is *h-liftable*, and the field W is the *h-lifted* field of V; the field W is *h-lowerable*, and the field V is the *h-lowered* field of W.

The (Whitney) *fold mapping* $h : M \to N$ is defined by the formulae

$$y_1 = x_1^2, \quad y_2 = x_2, \quad \cdots, \quad y_n = x_n$$

for suitable local coordinates $(x_1, \cdots, x_n)$ on M and $(y_1, \cdots, y_n)$ on N. The critical values of h form *the apparent contour* of N. Let h be the fold mapping.

3.1. LEMMA. *A holomorphic vector field on N is locally h-liftable at a point of the apparent contour of a fold mapping* ($y_1 = 0$) *if and only if this field is tangent to the contour.*

Proof: Let $V = \sum V_i \, \partial/\partial y_i$, $W = \sum W_i \, \partial/\partial x_i$. The agreement conditions are

$$V_1(h(x)) = 2x_1 W_1(x), \qquad V_j(h(x)) = W_j(x), \qquad j > 1.$$

For an h-liftable field, $V_1 = 0$ for $y_1 = 0$. Hence $V_1(y) = y_1 U(y)$. Then the preceding formulae define a smooth lifted field W.

3.2. LEMMA. *Let $h : M \to N$ be a holomorphic mapping of spaces of equal dimensions, and let the set of all critical points of h which are not fold points be at least of codimension* 2. *Then a holomorphic vector field V on N is h-liftable if and only if it is tangent to the apparent contour.*

Proof: On some neighborhood of a noncritical point of h the lifted field is defined automatically, and on some neighborhood of a fold point—by Lemma 3.1. Hence by Hartog's theorem the resulting field can be extended to the set of all other critical points (which is of codimension greater than or equal to 2).

The lowered field is evidently tangent to the apparent contour.

3.3. EXAMPLE. The Vieta mapping $h : M \to N$ is defined by the formulae

$$\lambda_1 = (\sigma_2 \mid \sigma_1 = 0), \quad \lambda_2 = (-\sigma_3 \mid \sigma_1 = 0), \quad \cdots, \quad \lambda_k = (\pm\sigma_{k+1} \mid \sigma_1 = 0),$$

where $M = \{x \in \mathbb{C}^{k+1} : \sigma_1(x) = 0\}$ and λ_i are coordinates on $N \approx \mathbb{C}^k$.

The *Whitney mapping* $h : \mathbb{C}^k \to N$ is defined by the formulae

$$\lambda_1 = x_1, \cdots, \lambda_{k-1} = x_{k-1}, \quad \lambda_k = -(x_k^{k+1} + x_1 x_k^{k-1} + \cdots + x_{k-1} x_k).$$

Both mappings have the same appearance (the swallowtail in N). Non-fold critical points form sets of codimension 2 in both cases. Lemma 3.2 implies

3.4. Corollary. *The sets of all Vieta-liftable and Whitney-liftable holomorphic vector fields (or germs at 0) coincide with the set of all vector fields (respectively germs), tangent to the swallowtail; hence, every Whitney-liftable field (or germ) is Vieta-liftable.*

Proof: This follows from Theorem 1.1 and Lemma 3.2.

It is often useful to have explicit formulae for the liftable fields. We begin with an evident remark.

3.5. Lemma. *Let $\rho_i = (\sigma_i \mid \sigma_1 = 0)$, $i = 2, \cdots, k+1$, be the basic symmetric functions on M. Then the vector fields $W_i = \nabla \rho_{i+2}$ are Vieta-lowerable.*

Here ∇ is the gradient, defined by the invariant metric.

Definition. The vector fields W_i are the *basic symmetric vector fields.*

We choose the indices for the basic symmetric vector fields $W_0, \cdots, W_{k-1}$ in such a way that the differentiation along W_i raises the degrees by i.

The corresponding fields on N are the *basic liftable vector fields.* We denote them by V_i, $i = 0, \cdots, k-1$.

3.6. Theorem. *For every function F, holomorphic at $0 \in N$ and vanishing at 0, there exists a liftable vector field V in some neighborhood of $0 \in N$ such that $V\lambda_1 = F$.*

Proof: First let us calculate the derivatives of the basic function ρ_2 along the basic fields W_i.

3.7. Lemma. *For $i = 0, \cdots, k-1$,*

$$W_i \rho_2 = -(i+2)\rho_{i+2}.$$

Proof: The derivatives of basic functions along basic fields form a symmetric matrix:

$$W_i \rho_{j+2} = (\Delta \rho_{i+2}, \Delta \rho_{j+2}).$$

Now

$$\rho_2=(\sigma_2 \mid \sigma_1=0)=(-\tfrac{1}{2}(x_1^2+\cdots+x_{k+1}^2) \mid \sigma_1=0)\,.$$

Hence $-\nabla\rho_2$ is the Euler field, $-W_0=(\sum x_i\,\partial/\partial x_i \mid \sigma_1=0)$. By the Euler theorem on the homogeneous functions,

$$W_0\rho_i=-i\rho_i\,.$$

By symmetry, $W_0\rho_i=W_{i-2}\rho_2$.

To complete the proof of the theorem consider $F=\sum \lambda_i F_i$. Then the required field is $V=\sum f_i V_i$, $f_i=(-1)^i(i+2)^{-1}F_i$.

3.8. Remark. In the same way one can prove Theorem 1.2. To construct a G-invariant field ξ, satisfying $\xi H=\rho$ (for a given homogeneous invariant ρ and a nondegenerate quadratic form $H=\frac{1}{2}(Ax \mid x)$, $A:\mathbb{C}^k\to\mathbb{C}^{k*}$), we write

$$\xi=(\deg\rho)^{-1}A^{-1}(d\rho)\,.$$

Theorem 3.6 can also be deduced from Theorem 1.2 or can be proved like it. The advantage of the present calculations is that they allow us to consider degenerate cases as well.

For instance, we find the following beginning of the hierarchy of functions at the cusp point of the semicubical parabola $4\lambda_1^3+27\lambda_2^2=0$:

$$\begin{array}{c} \mathrm{I}\leftarrow\mathrm{II}\leftarrow\mathrm{III}_2\leftarrow\mathrm{III}_3\leftarrow\cdots \\ \uparrow \\ \mathrm{IV}\leftarrow\mathrm{V} \end{array}$$

Type	Normal form	m_f	m_h	codim
I	$\pm\lambda_1$	0	0	0
II	$\pm\lambda_2$	0	0	1
III_n	$\pm\lambda_1^2+a\lambda_2^n,\ a\neq 0$	1	0	n
IV	$\pm\lambda_1\lambda_2+a\lambda_1^3$	1	0	3
V	monomials $\lambda_1, \lambda_2, \lambda_1^2, \lambda_1\lambda_2$ are absent	$\geqq 2$	$\geqq 2$	4

Here m_f and m_h are the number of moduli for the function and for the hypersurface (defined by the action of the group of plane diffeomorphism germs preserving the cusp parabola, on the space of function germs vanishing at the cusp point, see [3]).

The above list can be considered as the beginning of the classification of S(3)-invariant functions. For instance, the critical points occuring in the generic three-parameter families of S(3) symmetric functions are only of type I, II, III_2, III_3 and IV.

The beginning of the classification of S(4)-invariant functions (or that of functions in $\mathbb{C}^3$ under the action of swallowtail preserving diffeomorphisms is:

$$\begin{array}{ccccc} I & \leftarrow & II_2 & \leftarrow II_3 \leftarrow \cdots \\ & & \uparrow & \\ & & III & \leftarrow IV \end{array}$$

Type	Normal form	m_f	m_h	codim
I	$\pm\lambda_1$	0	0	0
II_n	$\pm\lambda_2+a\lambda_1^n,\ a\neq 0$	1	0	$n-1$
III	$\pm\lambda_1^2+a\lambda_3,\ \Delta(a)\neq 0$	1	1	2

More complicated singularities (IV) occur only in families of functions depending on three or more parameters.

For S(5) we find $I\leftarrow II\leftarrow III$, $I=\pm\lambda_1$, $II=\pm\lambda_2+a\lambda_1^2$, $m_h(III)>1$. It seems that for $n>5$ the only $S(n)$ invariant functions with $m_h<2$ are $\pm\lambda_1$.

The formulae of this section also give normal forms for functions which are invariant under the adjoint representations of $SL(n)$ or $SU(n)$.

The proofs are based on the techniques of [3], [4], and on the use of the matrices of Remark 3.9.

3.9. Remark. The derivatives of the basic symmetric functions along the basic symmetric vector fields form a symmetric matrix. Its elements

$$L_{i,j}=(\nabla\rho_{i+1},\nabla\rho_{j+1})$$

are symmetric functions, and it is often useful to know their expressions in terms of the basic symmetric functions ρ_α. For instance, the rows of these matrices can be considered as being formed by the components of the basic Vieta- or Whitney-liftable vector fields V_i.

The first 5 matrices (for $k=1,\cdots,5$) are

$$(-2\rho_2),\quad \begin{pmatrix}-2\rho_2 & -3\rho_3\\ -3\rho_3 & \frac{2}{3}\rho_2^2\end{pmatrix},\quad \begin{pmatrix}-2\rho_2 & -3\rho_3 & -4\rho_4\\ -3\rho_3 & \rho_2^2-4\rho_4 & \frac{1}{2}\rho_2\rho_3\\ -4\rho_4 & \frac{1}{2}\rho_2\rho_3 & \frac{3}{4}\rho_3^2-2\rho_2\rho_4\end{pmatrix}$$

$$\begin{pmatrix} -2\rho_2 & -3\rho_3 & -4\rho_4 & -5\rho_5 \\ -3\rho_3 & \frac{6}{5}\rho_2^2-4\rho_4 & \frac{4}{5}\rho_2\rho_3-5\rho_5 & \frac{2}{5}\rho_2\rho_4 \\ -4\rho_4 & \frac{4}{5}\rho_2\rho_3-5\rho_5 & \frac{6}{5}\rho_3^2-2\rho_2\rho_4 & \frac{3}{5}\rho_3\rho_4-3\rho_2\rho_5 \\ -5\rho_5 & \frac{2}{5}\rho_2\rho_4 & \frac{3}{5}\rho_3\rho_4-3\rho_2\rho_5 & \frac{4}{5}\rho_4^2-2\rho_3\rho_5 \end{pmatrix},$$

$$\begin{pmatrix} -2\rho_2 & -3\rho_3 & -4\rho_4 & -5\rho_5 & -6\rho_6 \\ -3\rho_3 & \frac{4}{3}\rho_2^2-4\rho_4 & \rho_2\rho_3-5\rho_5 & \frac{2}{3}\rho_2\rho_4-6\rho_6 & \frac{1}{3}\rho_2\rho_5 \\ -4\rho_4 & \rho_2\rho_3-5\rho_5 & \frac{3}{2}\rho_3^2-2\rho_2\rho_4-6\rho_6 & \rho_3\rho_4-3\rho_2\rho_5 & \frac{1}{2}\rho_3\rho_5-4\rho_2\rho_6 \\ -5\rho_5 & \frac{2}{3}\rho_2\rho_4-6\rho_6 & \rho_3\rho_4-3\rho_2\rho_5 & \frac{4}{3}\rho_4^2-2\rho_3\rho_5-4\rho_2\rho_6 & \frac{2}{3}\rho_4\rho_5-3\rho_3\rho_6 \\ -6\rho_6 & \frac{1}{3}\rho_2\rho_5 & \frac{1}{2}\rho_3\rho_5-4\rho_2\rho_6 & \frac{2}{3}\rho_4\rho_5-3\rho_3\rho_6 & \frac{5}{6}\rho_5^2-2\rho_4\rho_6 \end{pmatrix}.$$

The general answer was communicated to me by D. B. Fuks: for $i \leqq j$ and for the case of $k+1=n$ variables,

$$L_{i,j} = \frac{i(n-j)}{n}\rho_i\rho_j(1-\delta_{1,i}) - \sum (\beta-\alpha)\rho_\alpha\rho_\beta$$

(summation is over α, β, such that $\alpha+\beta=i+j$, $0 \leqq \alpha < i$, $\beta < n$, $\alpha \neq 1$; $\delta_{1,i}=0$ for $i \neq 1$ and $\delta_{1,i}=1$ for $i=1$).

The proof is based on the identity

$$\sum_{s=1}^{n} \sigma_i(\hat{s})\sigma_j(\hat{s}) = (n-j)\sigma_i\sigma_j - \sum_{r=0}^{i-1} (i+j-2r)\sigma_r\sigma_{i+j-r},$$

where $\sigma_0 = \sigma_k = 0$ for $k > n$, $\sigma_i(\hat{s}) = \sigma_i(x_1, \cdots, x_{s-1}, x_{s+1}, \cdots, x_n)$.

D. B. Fuks has also found that for the Lie algebra A of symmetric vector fields

$$\operatorname{codim}_A [A, A] = 1, \qquad A = [A, A] \oplus \{W_0\}, \qquad H^q(A, \mathbb{R}) = 0 \quad \text{for} \quad q > 1.$$

3.10. The Whitney-Lowerable Fields. The Whitney mapping $h : \mathbb{C}^k \to \mathbb{C}^k$ is defined by the formulae

$$\lambda_1 = x_1, \cdots, \lambda_{k-1} = x_{k-1}, \qquad -\lambda_k = x_k^{k+1} + x_1 x_k^{k-1} + \cdots + x_{k-1}x_k,$$

where x_5 and λ_5 are the coordinates in the first and second $\mathbb{C}^k$. In this section we give explicit formulae for the Whitney-lowerable vector fields. These formulae are crucial for the envelope theory (see Section 5).

For a Whitney-lowerable field u and the lowered field hu we contract a row from the components of u and hu in coordinates x_s and λ_s. For $k = 1, 2, 3$, we find the following component matrices of the basic Whitney-lowerable fields u_i and lowered fields hu_i, $i = 0, \cdots, k-1$:

$$(x_1 \,|\, 2\lambda_1),$$

$$\left(\begin{array}{cc|cc} 2x_1 & x_2 & 2\lambda_1 & 3\lambda_2 \\ -3p & x_2^2 + \frac{2}{3}x_1 & 3\lambda_2 & -\frac{2}{3}\lambda_1^2 \end{array}\right), \qquad p = x_2^3 + x_1 x_2,$$

$$\left(\begin{array}{ccc|ccc} 2x_1 & 3x_2 & x_3 & 2\lambda_1 & 3\lambda_2 & 4\lambda_3 \\ 3x_2 & -x_1^2 - 4q & x_3^2 + \frac{1}{2}x_1 & 3\lambda_2 & -\lambda_1^2 + 4\lambda_3 & -\frac{1}{2}\lambda_1\lambda_2 \\ -4q & -\frac{1}{2}x_1 x_2 & x_3^3 + x_1 x_3 + \frac{3}{4}x_2 & 4\lambda_3 & -\frac{1}{2}\lambda_1\lambda_2 & \frac{3}{4}\lambda_3^2 + 2\lambda_1\lambda_3 \end{array}\right);$$

$$q = x_3^4 + x_1 x_3^2 + x_2 x_3.$$

For instance, the first row of the second matrix signifies the Whitney-lowerable field $u_0 = 2x_1\, \partial/\partial x_1 + x_2\, \partial/\partial x_2$.

We have normalized the basic Whitney-lowerable fields $u_0, \cdots, u_{k-1}$ by the condition

$$(hu_i)\lambda_1 = (i+2)\lambda_{i+1}, \qquad i = 0, \cdots, k-1.$$

Then the last component of u_i is

$$(u_i)_k = x_k^{i+1} + x_1 x_k^{i-1} + \cdots + x_{i-1} x_k + \frac{i+1}{k+1} x_i.$$

The values of the other components at a point x are equal to those of the corresponding components of hu_i at $h(x)$. The matrix of the components of hu_i can be calculated from that of Remark 3.9: one must substitute $(-1)^i \lambda_{i-1}$ for ρ_i in Remark 3.9 and then change signs at the places where the column and row number are both even or both odd.

All this follows from the relation $hu_i = (-1)^{i+1} V_i$ for the basic liftable fields V_i (see Lemma 3.5).

3.11. Remark. The basic symmetric holomorphic vector fields (germs) generate all symmetric holomorphic vector fields (germs) as a module over the algebra of holomorphic symmetric functions (germs). This module is free, and the basic fields form its free basis (see [16]). Hence every Vieta-liftable (or Whitney-lowerable) holomorphic vector field germ can be represented in the form of a linear combination of basic field germs with holomorphic coefficients in one and only one way.

4. The Real Case

4.1. Definition. The real discriminant $\Sigma \subset C\mathbb{R}^k$ is the hypersurface

$$\Sigma = \{\lambda \in \mathbb{R}^k : \Delta(\lambda) = 0\},$$

where Δ is the discriminant of the polynomial

$$p = z^{k+1} + \lambda_1 z^{k-1} + \cdots + \lambda_k.$$

The real Vieta mapping $h : M \to N$ is defined by

$$\lambda_1 = (\sigma_2 \mid \sigma_1 = 0), \cdots, \lambda_k = (\pm\sigma_{k+1} \mid \sigma_1 = 0),$$

where $M = \{x \in \mathbb{R}^{k+1} : \sigma_1(x) = 0\}$; λ_j are the coordinates on $N \approx \mathbb{R}^k$.
The real Whitney mapping $h : M \to N$ is defined by

$$\lambda_1 = x_1, \cdots, \lambda_{k-1} = x_{k-1}, \quad \lambda_k = -(x_k^{k+1} + x_1 x_k^{k-1} + \cdots + x_{k-1} x_k),$$

where x_s are the coordinates on $M \approx \mathbb{R}^k$, λ_s those on $N \approx \mathbb{R}^k$.

4.2. Remark (A. M. Gabrielov). Let Σ_V, Σ_W be the sets of critical values for the real Vieta and Whitney mappings. Then we have

$$\Sigma_V \subset \Sigma_W \subset \Sigma.$$

In contrast to the complex case, in the real case all these 3 sets are in general different.

Let us for instance consider the case $k = 3$. In this case, Σ is the set of all real points of the complex swallowtail (Figure 2). The swallowtail has a self intersecting line. Continuation of this line beyond the origin 0 belongs to Σ, but not to the set Σ_W of critical values of the Whitney mapping. All real analytic fields, tangent to Σ_W, are tangent to this continuation of the

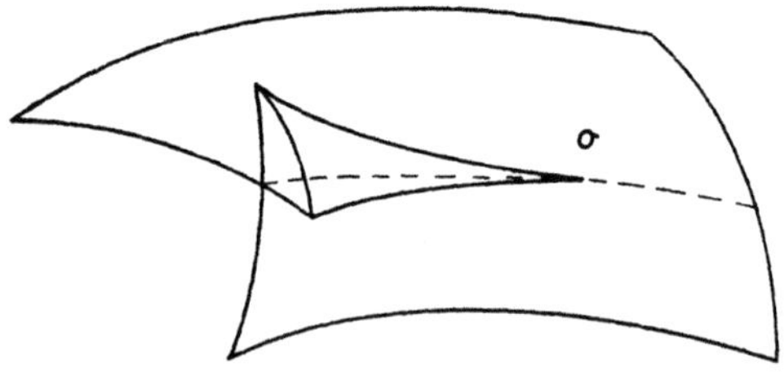

Figure 2

intersection line, since Σ is irreducible at 0. However a C^∞ Whitney-liftable field can have an arbitrary value at a point of this continuation.

The set Σ_V of critical values of the Vieta mapping consists only of those swallowtail points which form the boundary of the pyramidal part of the complement (this part corresponds to polynomials with 4 real roots).

Hence, in the C^∞-case there exist Vieta-liftable and Whitney-nonliftable fields. In addition, neither Vieta- (Whitney-) liftable C^∞-field germs nor C^∞-field germs tangent to Σ_V (Σ_W) form finite modules over the algebras of C^∞-function germs (while this is true for real analytic functions or formal power series).

In spite of these remarks, the following extension of Theorem 1.1 holds:

4.3. THEOREM. *Let f be a smooth function on some neighborhood of the origin $\lambda=0$ in $\mathbb{R}^k$, and let $f(0)=0$, $\partial f/\partial\lambda_1(0)\neq 0$. Then f can be reduced to a normal form $f=\pm\lambda_1$ by a diffeomorphism, which preserves the real discriminant variety Σ and is Whitney- and Vieta-liftable.*

Proof: Let $f=H+F$, where $H=c\lambda_1$, $c\neq 0$, $\partial F/\partial\lambda_1=0$. We want to construct a one-parameter family of diffeomorphisms g_t, preserving the origin $0\in\mathbb{R}^k$ and satisfying

$$(H+tF)(g_t(x))=H(x),\qquad g_0(x)=x,\qquad t\in[0,1]. \tag{1}$$

We define the time-dependent velocity field

$$\xi(g_t(x),t)=\frac{\partial}{\partial t}g_t(x). \tag{2}$$

By differentiation of (1), we obtain for ξ a linear (not a differential) equation

$$\xi(H+tF)+F=0. \tag{3}$$

To prove that the last equation is solvable, let us consider the basic liftable fields V_i (see Section 3). We define the functions $\Lambda_i=\Lambda_i(\lambda,t)$ by

$$\Lambda_i=V_i(H+tF),\qquad i=0,\cdots,k-1.$$

First we shall prove (a) *the mapping $(\lambda,t)\mapsto(\Lambda(\lambda,t),t)$ is a diffeomorphism of some neighborhood of the t-axis.*

To prove (a) we introduce weights: $\deg\lambda_j=j+1$. This defines a quasihomogeneous filtration. We find $\deg F>2=\deg H$; the differentiation along W_i raises the degrees by i. Now

$$\Lambda_i=P_i\lambda_i+t\omega_i(\lambda),\qquad \deg\omega_i>\deg\lambda_i,$$

(P_i are nonzero constants). Hence, on the t-axis, the Jacobian $\det(\partial\Lambda_i/\partial\lambda_j)$ is a nonzero constant; this implies (a).

By Hadamard's lemma, every smooth function Φ, vanishing on the t-axis, can be represented in the form

$$\Phi(\lambda, t) = \sum \Phi_i(\lambda, t)\Lambda_i(\lambda, t)$$

with smooth coefficients Φ_i. Hence, $F = \sum F_i \Lambda_i$. The field $\xi = -\sum F_i V_i$ satisfies (3), vanishes at $\lambda = 0$ for all t, is tangent to Σ, and is Vieta- and Whitney-liftable (since the same is true for the basic fields V_i).

We define g_t as the solutions of (2), and reduce f to $H = c\lambda_1$. Finally, we substitute $\theta^{i+1}\lambda_i$ for λ_i (this is a liftable diffeomorphism) to reduce $c\lambda_1 t \pm \lambda_1$.

REMARK. In the real analytic case we have constructed a real analytic diffeomorphism.

4.4. DEFINITION. The *cylindrical real discriminant* $\Sigma \times \mathbb{R}^p \subset N \times \mathbb{R}^p$ and the *cylindrical Vieta and Whitney mappings* $M^k \times \mathbb{R}^p \to N^k \times \mathbb{R}^p$ are the direct products of the real swallowtail $\Sigma \subset N$ and of the real Vieta and Whitney maps $M \to N$ by the parameter space $\mathbb{R}^p$ along with its identity mapping, respectively.

Theorem 4.3 implies

4.5. COROLLARY. *A generic smooth function $\Phi : N \times \mathbb{R}^p \to \mathbb{R}$ can be reduced, at every point of the edge $O \times \mathbb{R}^p$ of the real cylindrical discriminant, to one of the normal forms*

$$\Phi = \tau_1 + c \qquad \textit{or} \qquad \Phi = \pm\lambda_1 \pm \tau_1^2 \pm \cdots \pm \tau_p^2 + c$$

by a diffeomorphism which preserves the cylindric discriminant and is Vieta- and Whitney-liftable.

(Here c is a constant, τ_i are the coordinates in $\mathbb{R}^p$, and λ_1 is given in Definition 4.1.)

The proof is similar to that of Corollary 2.4.

4.6. EXAMPLE. In the three-space $\mathbb{R}^3$ with coordinates (x, y, z) let us consider a semicubical cylinder $x^2 = y^3$. At any point of the edge $(x = y = 0)$, a generic function $\Phi(x, y, z)$ can be reduced to

$$\Phi = z \qquad \text{or} \qquad \Phi = \pm y \pm z^2 + \text{const}.$$

If we consider z as the time coordinate, the last formula describes the evolution of the trace of the discriminant on the surface $\Phi = t$ as t varies.

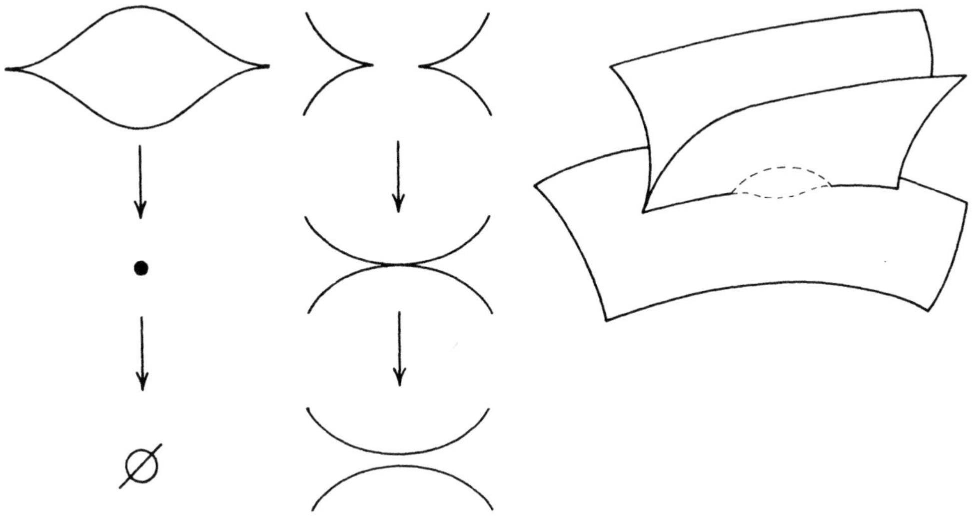

Figure 3

In accordance with the sign, the bifurcation consists of the generation or annihilation of two cusps (Figure 3). The trace curves can be turned towards each other (generation or annihilation of "lips" in Thom's notation) or away from each other ("bec á bec" in Thom's notation).

All functions Φ that can not be reduced to the above normal forms form a thin set in the space of all functions.

4.7. REMARK. The extension of Theorem 4.3 to all reflection groups is immediate. In this case, $\Sigma \subset \mathbb{R}^k$ is the real form of the discriminant, and λ_1 the real invariant of minimal (second) degree.

For applications the cases of A_k, D_k and E_k groups are particularly important. In these cases one can describe the discriminants Σ explicitly by the following construction.

A point $\lambda \in \mathbb{R}^k$ belongs to Σ if and only if 0 is a critical value of the *complex* mapping $\mathbb{C}^2 \to \mathbb{C}$, defined by the following function of 2 complex variables z_1, z_2:

$$A_k : z_1^{k+1} + z_2^2 + \lambda_1 z_1^{k-1} + \cdots + \lambda_k ,$$

$$D_k : z_1^2 z_2 \pm z_2^{k-1} + \lambda_1 z_2^{k-2} + \cdots + \lambda_{k-1} + \lambda_k z_1 ,$$

$$E_6 : z_1^3 + z_2^4 + \lambda_1 z_1 z_2^2 + \lambda_2 z_1 z_2 + \lambda_3 z_2^2 + \lambda_4 z_1 + \lambda_5 z_2 + \lambda_6 ,$$

$$E_7 : z_1^3 + z_1 z_2^3 + \lambda_1 z_2^4 + \lambda_2 z_2^3 + \lambda_3 z_1 z_2 + \lambda_4 z_2^2 + \lambda_5 z_1 + \lambda_6 z_2 + \lambda_7 ,$$

$$E_8 : z_1^3 + z_2^5 + \lambda_1 z_1 z_2^3 + \lambda_2 z_1 z_2^2 + \lambda_3 z_2^3 + \lambda_4 z_1 z_2 + \lambda_5 z_2^2 + \lambda_6 z_1 + \lambda_7 z_2 + \lambda_8 .$$

Only in the D_k case with even k are there two real forms of the discriminant.

In all these cases the normal forms for functions $f(\lambda)$ with $f(0)=0$, $\partial f/\partial\lambda_1(0)\neq 0$ will be $f=\pm\lambda_1$.

In the case with parameters (see Corollary 4.5) the normal forms will be

$$\Phi=\tau_1+C, \qquad \Phi=\pm\lambda_1\pm\tau_1^2\pm\cdots\pm\tau_p^2+C.$$

4.8. Remark (A. M. Gabrielov). Though the C^∞ fields which are tangent to Σ_W or Σ_V are not generated by the basic fields over the function algebra, they are Whitney- (or Vieta-) liftable.

4.9. Remark. It seems that Theorems 1.1, 1.2 and 4.3 admit extensions to the group actions on singular varieties, at least in the case when the quotients are $\mathbb{C}^k$ or $\mathbb{R}^k$.

Other extensions of Theorem 1.1 were found by V. M. Zakalyukin [22].

4.10. Remark. J. M. Eliashberg has used Corollary 4.5 in his very interesting paper [10] (Lemma 3.2.Б in [10]). However he has informed me that he cannot prove this lemma. Hence the results of this section provide some foundation for [10].

5. Applications to the Classification of Mapping Diagrams

Diagrams we shall deal with are $M\to N\to\mathbb{R}$ or $\mathbb{R}\leftarrow M\to N$.

An equivalence of two diagrams (of the same type) is a commutative diagram, e.g.

$$\begin{array}{ccccc} M_1 & \to & N_1 & \to & \mathbb{R}_1 \\ \downarrow & & \downarrow & & \downarrow \\ M_2 & \to & N_2 & \to & \mathbb{R}_2 \end{array} \qquad \begin{array}{ccccc} \mathbb{R}_1 & \leftarrow & M_1 & \to & N_1 \\ \downarrow & & \downarrow & & \downarrow \\ \mathbb{R}_2 & \leftarrow & M_2 & \to & N_2 \end{array}$$

vertical arrows being diffeomorphisms, $\mathbb{R}_1\approx\mathbb{R}_2\approx\mathbb{R}$.

We use similar definitions for germs.

If the mapping $\mathbb{R}_1\to\mathbb{R}_2$ in an equivalence is a translation ($t_2=t_1+\text{const.}$), we call this equivalence *strong equivalence.*

The information of Sections 2–4 implies many results on the classification of such diagrams up to equivalence (or strong equivalence) in the cases where $M\to N$ is the Whitney map.

5.1. Theorem (compare [9], [17]). *Let* $\dim M=\dim N=k<3$.

Then, for generic mappings $M\to N\to\mathbb{R}$, *the corresponding germ diagram is, for every point* $0\in M$, *strongly equivalent to one of the following:*

k	$h: M \to N, \quad y = h(x)$	$f: N \to \mathbb{R}, \quad f(y)$
1	$y = x$	$y, \quad \pm y^2$
	$y = x^2$	$\pm y$
2	$y_1 = x_1, \quad y_2 = x_2$	$y_1, \quad \pm y_1^2 \pm y_2^2$
	$y_1 = x_1, \quad y_2 = x_2^2$	$y_1, \quad \pm y_1^2 \pm y_2$
	$y_1 = x_1, \quad y_2 = x_2^3 + x_1 x_2$	$\pm y_1$
3	$y_1 = x_1, \quad y_2 = x_2, \quad y_3 = x_3$	$y_1, \quad \pm y_1^2 \pm y_2^2 \pm y_3^2$
	$y_1 = x_1, \quad y_2 = x_2^2, \quad y_3 = x_3$	$y_1, \quad \pm y_1^2 \pm y_3^2 \pm y_2$
	$y_1 = x_1, \quad y_2 = x_2^3 + x_1 x_2, \quad y_3 = x_3$	$y_3, \quad \pm y_3^2 \pm y_1$
	$y_1 = x_1, \quad y_2 = x_2, \quad y_3 = x_3^4 + x_1 x_3^2 + x_2 x_3$	$\pm y_1$

Proof: For $k \leqq 3$, a generic $h: M^k \to N^k$ has only standard Whitney singularities; they can be reduced to the normal forms of the "h" column in the table above (see, e.g. [11]). Then f can be reduced to the "if" column normal form by a Whitney-liftable diffeomorphism $N \to N$ (see Corollary 4.5).

Theorem 5.1 describes all bifurcations of singularities in generic one-parameter families of mappings $M^{k-1} \to N^{k-1}$ for $k \leqq 2$. Corollary 4.5 implies similar normal forms for mappings of corank less than 2, k being arbitrary.

5.2. THEOREM (compare [5]). *Let* $\dim M = \dim N = k \leqq 3$. *Then, for generic mappings* $\mathbb{R} \leftarrow M \to N$, *the corresponding germ diagram is, for every point* $0 \in M$, *strongly equivalent to one of the following:*

k	$h: M \to N, \quad y = h(x)$	$f: M \to \mathbb{R}, \quad f(x)$
1	$y = x$	$x \pm x^2$
	$y = x^2$	$x + \varphi(x^2)$
2	$y_1 = x_1, \quad y_2 = x_2$	$x_1, \quad \pm x_1^2 \pm x_2^2$
	$y_1 = x_1, \quad y_1 = x_2^2$	$x_1 + x_2, \pm x_1^2 + x_2, x_1 + x_2^3 \pm x_1 x_2$
	$y_1 = x_1, \quad y_2 = x_2^3 + x_1 x_2$	$\pm x_2 + \varphi(h(x))$
3	$y_1 = x_1, \quad y_2 = x_2, \quad y_3 = x_3$	$x_1, \pm x_1^2 \pm x_2^2 \pm x_3^2$
	$y_1 = x_1, \quad y_2 = x_2^2, \quad y_3 = x_3$	$x_1 + x_2, x_2 \pm x_1^2 \pm x_3^2, x_1 + x_2 x_3$ $x_1 \pm x_1 x_2 + x_2^3 \pm x_2 x_3^2$
	$y_1 = x_1, \quad y_2 = x_2^3 + x_1 x_2, \quad y_3 = x_3$	$x_2 + x_3, x_2 \pm x_3^2 + \varphi(h_{1,2}(x))$ $x_3 \pm x_2^2 + x_2 \varphi(h_{1,2}(x))$
	$y_1 = x_1, \quad y_2 = x_2, \quad y_3 = x_3^4 + x_1 x_3^2 + x_2 x_3$	$\pm x_3 + \varphi(h(x))$

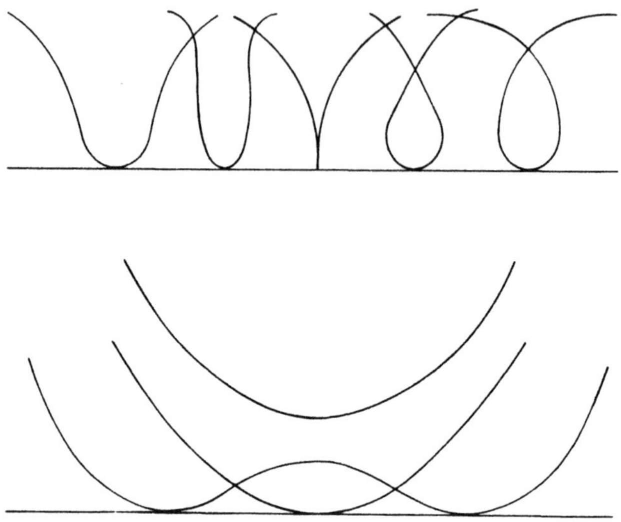

Figure 4

In the envelope theory for one-parameter families of hypersurfaces in N^k, $f: M^k \to \mathbb{R}$ is a fibration. In this case R represents the parameter space; to $t \in \mathbb{R}$ corresponds the hypersurface $h(f^{-1}(t))$.

Theorem 5.2 implies, for example, that in a generic family of plane curves only the following two bifurcations occur at a regular point of the envelope: (i) the "γ to U" transition through a cusp, and (ii) the "W to U" transition with annihilation of two contact points on the envelope. (See Figure 4.)

At a cusp of the envelope we find a "function-modulus" φ (probably, there exist as many topologically different types of smooth families as there exist different functions in two variables). The proof of Theorem 5.2 is based on the formulae in Section 3.11; it will appear elsewhere.

According to N. A. Nikishin, normal forms for generic f at fold points of h for $k \geqq 3$, are

$$x_1 + x_k, \qquad x_1 + x_1 x_k \pm x_2^2 \pm \cdots \pm x_k^2,$$

$$x_k \pm x_1^2 \pm \cdots \pm x_{k-1}^2, \qquad x_1 + x_2 x_k.$$

Here $y_1 = x_1, \cdots, y_k = x_k^2$.

6. Applications to the Evolution of Wave Fronts

Generic wave front singularities are described in [2], [3], [21]. There one can find a geometrical presentation in terms of contact manifolds, Legendre fibrations and Legendre singularities. During the time evolution of the front its singularities can bifurcate. Here we shall describe all bifurcations of singularities of generic one-parametric families of wave fronts provided that the dimension of the ambient space is less than 6.

We shall use the ancient coordinate notation. Let $F=F(x,\lambda)$ be a smooth function of $x\in\mathbb{R}^k$, $\lambda\in\mathbb{R}^l$.

6.1. Definition. *The front*, defined by the generating function F, is the following hypersurface $\mathbb{R}^{k+l+1}$:

$$\Sigma=\left\{(y,\lambda,z):\ \exists x:\ y_i=F_{x_i},\ z=F-\sum_i x_iF_{x_i}\right\}.$$

6.2. Example. The generating functions for the front singularities of type A_μ, D, E_μ are

$$A_\mu:\ F=\pm x_1^{\mu+1}+\lambda_1x_1^{\mu-1}+\cdots+\lambda_{\mu-2}x_1^2,$$

$$D_\mu:\ F=\pm x_1^2x_2\pm x_2^{\mu-1}+\lambda_1x_2^{\mu-2}+\cdots+\lambda_{\mu-3}x_2^2,$$

$$E_6:\ F=\pm x_1^3\pm x_2^4+\lambda_1x_1x_2^2+\lambda_2x_1x_2+\lambda_3x_2^2,$$

$$E_7:\ F=\pm x_1^3\pm x_1x_2^3+\lambda_1x_2^4+\lambda_2x_2^3+\lambda_3x_1x_2+\lambda_4x_2^2,$$

$$E_8:\ F=\pm x_1^3\pm x_2^5+\lambda_1x_1x_2^3+\lambda_2x_1x_2^2+\lambda_3x_2^3+\lambda_4x_1x_2+\lambda_5x_2^2.$$

Here $\mu\leqq k+l+1$, $k=1$, $\mu\geqq1$, for A_μ and $k=2$, $\mu\geqq4$ in the other cases.

It is known that all the singularities of generic wave fronts in spaces of dimension less than 7 are locally diffeomorphic to the A_μ, D_μ, E_μ singularities, $\mu<7$ (of course, different branches of a wave front can also intersect each other) (see [2], [3], [21]).

A one-parameter family of wave points is locally defined by a one-parameter family of generating functions $F=F(x,\lambda;t)$ (where $t\in\mathbb{R}$ is a parameter).

6.3. Definition. The big front Σ of a family, defined by $F=F(x,\lambda;t)$, $x\in\mathbb{R}^k$, $\lambda\in\mathbb{R}^1$, $t\in\mathbb{R}$, is the hypersurface in $\mathbb{R}^{k+l+2}$ defined by $\Sigma=\{y,\lambda,z,t:\exists x:\ y_i=F_{x_i},\ z=F-\sum_i x_iF_{x_i}\}$.

Hence the big front can be considered either as a disjoint union of temporary fronts or as a unique front in the space of greater dimension (in the last case, t is to be considered as λ_{l+1}).

We shall call the $\mathbb{R}^{k+l+2}$ space, containing the big front, the *space-time*, and we shall denote it by $\mathbb{R}^n$. Thus, the big front Σ is a hypersurface in the space-time $\mathbb{R}^n$, and the time $t:\ \mathbb{R}^n\to\mathbb{R}$ is a function on this space-time with no critical points.

Let us consider two one-parameter families of fronts, and let Σ_1, Σ_2 be their big fronts, in the space-times $\mathbb{R}_1^n$, and $\mathbb{R}_2^n$, with time functions $t_{1,2} : \mathbb{R}_{1,2}^n \to \mathbb{R}$, respectively.

DEFINITION. The bifurcations of wave front singularities at world points $O_1 \in \Sigma_1$ and $O_2 \subset \Sigma_2$ are *equivalent* if and only if there exists a local diffeomorphism $(\mathbb{R}_1^n, O_1) \to (\mathbb{R}_2^n, O_2)$ sending Σ_1 to Σ_2 and t_1 to $t_2 + \text{const}$.

To describe the bifurcations of wave front singularities up to equivalence we first consider some examples.

6.4. LIST OF EXAMPLES. The big front Σ will be a cylindrical discriminant set $\Sigma_0 \times \mathbb{R}^p \subset \mathbb{R}^{\mu+p}$, where Σ_0 is the discriminant set for A_μ, D_μ or E_μ, described in Remark 4.7.

The time function t will be one of the following functions:

$$t = \tau_1 \qquad \text{or} \qquad t = \pm\lambda_1 \pm \tau_1^2 \pm \cdots \pm \tau_p^2, \qquad p \geqq 0,$$

where τ_s are the coordinates in the parameter space $\mathbb{R}^p$.

6.5. THEOREM. *For generic one-parameter wave front families whose big fronts Σ have at 0 singularities of type A_μ, D_μ or E_μ, the pair (Σ, t) is locally equivalent at 0 to one of those in the list 6.4.*

6.6. THEOREM. *For generic one-parameter families of wave fronts in spaces of dimension q less than 6, all bifurcations of singularities are locally equivalent, at every point of the big front Σ, to one of those in the list 6.4, where $\mu + p \leqq q + 1 < 7$.*

Proofs: The big front is diffeomorphic to one of the real forms of the discriminant variety of the corresponding Coxeter group (see [2], [3], [21]). Hence Remark 4.7 implies Theorem 6.5.

Theorem 6.5 and the general results of Subsection 6.2 imply Theorem 6.6.

All the local bifurcations of singularities occuring in generic one-parameter families of wave fronts in $\mathbb{R}^3$ are presented in Figure 5 (compare also [3], [17]). Of course, only one wave front branch is presented.

7. Applications to the Evolution of Caustics and to Galaxy Generation

Generic caustic singularities are described in [1], [3], [21]. There one can find a geometric interpretation, in terms of symplectic manifolds, Lagrange fibrations and Lagrange singularity terms. During the time evolution of a

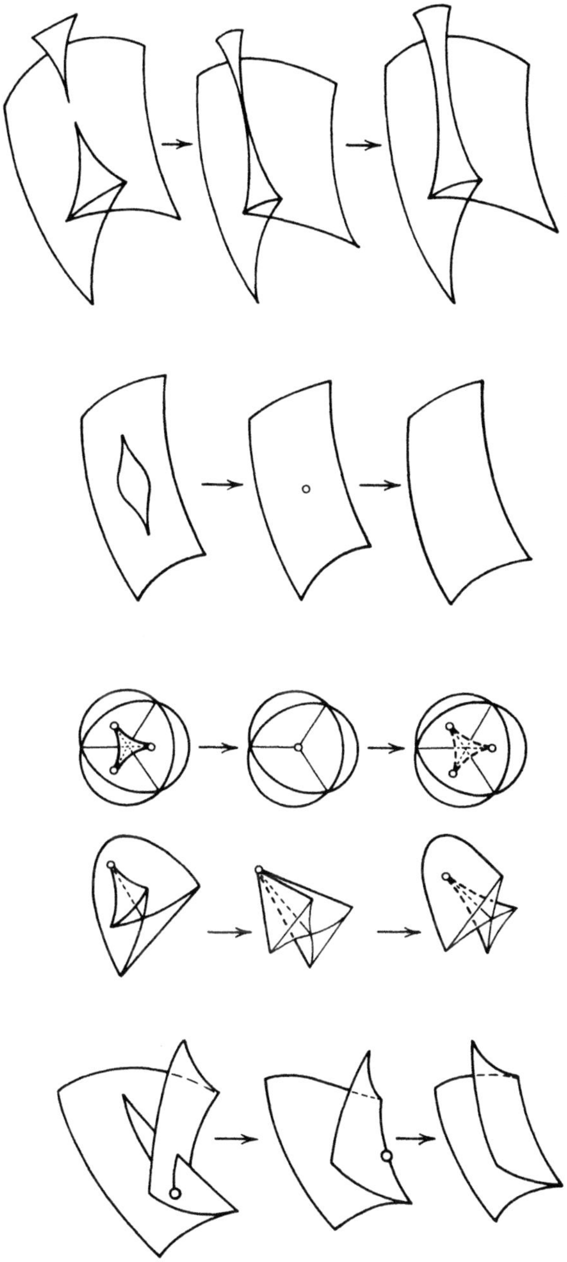

Figure 5

caustic its singularities can bifurcate. Here we shall describe all the singularity bifurcations for generic one-parameter families of caustics in spaces of dimension less than 4.

We shall again use the ancient coordinate notation. Let $F=F(x,\lambda)$ be a smooth function in $x\in\mathbb{R}^k$, $\lambda\in\mathbb{R}^l$.

7.1. DEFINITION. The *caustic*, defined by the generating function F, is the hypersurface $\Delta\subset R^{k+l}$ which consists of the critical values of the mapping $(x,\lambda)\mapsto(y=F_x,\lambda)$.

The generic caustic singularities in spaces of dimension less than 6 are locally diffeomorphic to those of the A_μ, D_μ, E_μ, $\mu<6$, caustic singularities. The corresponding functions F are given in Section 6 (see [1], [2], [3], [21]). (Different branches of a caustic can also intersect each other, but we shall deal with only one branch).

A one-parameter family of caustics is locally defined by a generating family of functions $F=F(x,\lambda;t)$ (where $t\in\mathbb{R}$ is a parameter).

7.2. DEFINITION. The *big caustic* Δ of a family, defined by $F=F(x,\lambda;t)$, $x\in\mathbb{R}^k$, $\lambda\in\mathbb{R}^l$, $t\in\mathbb{R}$, is the hypersurface in $\mathbb{R}^{k+l+1}$, which is the set of critical values for the mapping $(x,\lambda;t)\mapsto(y=F_x,\lambda;t)$.

Hence the big caustic can be considered either as a disjoint union of temporary caustics or as a unique caustic in the space-time: $\mathbb{R}^{k+l+1}=\mathbb{R}^n$. The time t is a function with no critical points on this $\mathbb{R}^n$.

7.3. DEFINITION. A *bifurcation* of caustic singularities is a pair of mapping germs $(\Delta,0)\to(\mathbb{R}^n,0)\to(\mathbb{R},0)$ (the first arrow being the inclusion of the big caustic, the second the time function). An *equivalence* of such bifurcations is a commutative diagram

$$\begin{array}{ccccc}\Delta_1 & \to & \mathbb{R}_1^n & \to & \mathbb{R}_1\\ \downarrow & & \downarrow & & \downarrow\\ \Delta_2 & \to & \mathbb{R}_2^n & \to & \mathbb{R}_2\end{array}$$

whose verticals are diffeomorphism germs. The equivalence is *strong* if the last one is a translation ($t_2=t_1+\text{const.}$).

7.4. LIST OF BIFURCATIONS OF CAUSTIC SINGULARITIES. The big caustic Δ will be the product $\Delta_0\times\mathbb{R}^p\subset\mathbb{R}^{\mu-1}\times\mathbb{R}^p$, where Δ_0 is the caustic of type A_μ or D_μ (the generating functions are listed in Subsection 6.2).

The time function t will be
Case A_μ:

$$t=\tau_1 \quad \text{or} \quad t=\pm\lambda_1\pm\tau_1^2\pm\cdots\pm\tau_p^2. \tag{1}$$

Case D_μ:

$$t=\tau_1 \quad \text{or} \quad t=\pm\lambda_1+y_1+a\lambda_2\pm\tau_1^2\pm\cdots\pm\tau_p^2, \tag{2}$$

$$t=\tau_1 \quad \text{or} \quad t=\pm\lambda_1+Py_1+Q\pm\tau_1^2\pm\cdots\pm\tau_p^2. \tag{3}$$

Here $P=1+a_1\lambda_2+\cdots+a_\nu\lambda_2^\nu$, $Q=b_1\lambda_2+\cdots+b_\nu\lambda_2^\nu$, $\mu=2\nu+2$, $2\nu+3$, the τ_s are the coordinates in $\mathbb{R}^p$, $p\geqq 0$; for $\mu=4$ one must replace $a\lambda_2$ by ay_2 in (2), (3) ($y_i=F_{x_i}$, see Example 6.2).

7.5. THEOREM. (V. M. Zakalyukin). *For a generic one-parameter family of caustics, at a big caustic singularity A_μ or D_μ, the bifurcation is locally equivalent to one given by* (1) *or* (2) *and strongly locally equivalent to one given by* (1) *or* (3).

7.6. COROLLARY. *All generic bifurcations of caustic singularities in one-parameter families in spaces of dimension $q<4$ are locally equivalent to those given by* (1) *or* (2) *and strongly equivalent to those given by* (1) *or* (3), *with $\mu+p\leqq q+2<6$.*

7.7. EXAMPLE. In the case of three space variables, generic caustic bifurcations are equivalent to the following:

A_3 : $F=x^4+\lambda_1x^2$, $\quad t=\tau_1$ or $\pm\lambda_1\pm\tau_1^2\pm\tau_2^2$,

A_4 : $F=x^5+\lambda_1x^3+\lambda_2x^2$, $\quad t=\tau_1$ or $\pm\lambda_1\pm\tau_1^2$,

D_4 : $F=x_1^2x_2\pm x_2^3\pm\lambda_1x_2^2$, $\quad t=\tau_1$ or $\pm\lambda_1+y_1+ay_2$,

A_5 : $F=x^6+\lambda_1x^4+\lambda_2x^3+\lambda_3x^2$, $\quad t=\pm\lambda_1$,

D_5 : $F=x_1^2x_2+x_2^4+\lambda_1x_2^3+\lambda_2x_2^2$, $\quad t=\pm\lambda_1+y_1+a\lambda_2$.

The proofs are based on the methods of Section 2 and on the spectral sequence techniques from [4]; they will appear elsewhere.

7.8. THE ZELDOVITCH THEORY (see [24], [25]). Let us consider a medium of noninteracting particles in $\mathbb{R}^3$. Let the initial velocity distribution be

$v = v(x)$ and the density distribution be $\rho = \rho(x)$; ρ and v are a smooth function and a vector field, respectively.

The inertial motion of particles defines a time dependent mapping $\mathbb{R}^3 \to \mathbb{R}^3$, $x \mapsto x + tv(x)$. This mapping is a diffeomorphism for small t (v being not too bad at infinity). But at some time t, singularities occur. The critical values of this mapping are points of condensation of particles (the density becomes infinite).

After that moment the physical assumption that interactions are absent is no longer realistic. Some physical shock conditions must be formulated. Here, however, we shall consider particles as noninteracting even after the generation of singularities.

In the Zeldovitch theory the initial vector field has a potential: $v = \nabla S$. Hence the critical values form the caustic of the Lagrange time depending mapping $\mathbb{R}^3 \to \mathbb{R}^3$,

$$x \mapsto x + \partial S/\partial x. \tag{4}$$

We shall study the bifurcations of these caustic singularities for generic S.

The formulae of Example 7.6 describe these bifurcations up to equivalence (that is, up to smooth change of variables in the space-time). For instance, the simplest formula

$$A_3 : F = x^4 + \lambda x^2, \qquad t = \lambda_1 + \tau_1^2 + \tau_2^2$$

describes in space-time with coordinates $(y, \lambda, \tau_1, \tau_2)$ a big caustic Δ, which is a semicubical cylinder ($27y^2 + 8\lambda^3 = 0$). The traces of this cylinder on the $t = \text{const.}$ manifolds are void for $t < 0$. For $t > 0$, a lens-like surface appears, similar to the surface of revolution of the lips about the short axis (see Figure 6).

7.9. Remark. Strictly speaking, to apply Theorem 7.5 to the galaxy generations, we must prove that generic functions S define generic families of caustics.

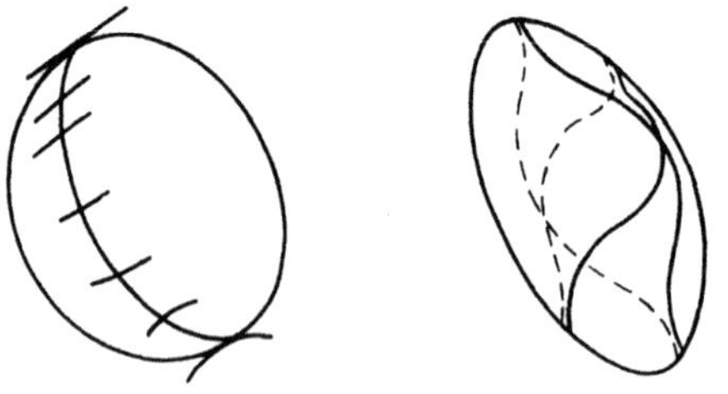

Figure 6

The most important result, namely, the fact that first a lens singularity of type A_3 appears was obtained by J. B. Zeldovitch from the following reasoning.

Let us consider the critical *points* of the mapping (4). A point x becomes critical at the moment t if and only if $-1/t$ is an eigenvalue for the Hessian $\partial^2 S/\partial x_i\, \partial x_j$ at x. The kernel of the derivative of our mapping at x is then generated by the eigenvectors with eigenvalue $-1/t$.

For a generic S the eigenvalues of the Hessian form a three-valued real function. Its branches intersect along curves (D_4 in our notations).

When t varies from 0 to $+\infty$ the number $-1/t$ varies from $-\infty$ to 0.

Hence the first singularity corresponds to the minimal (negative) eigenvalue of the Hessian.

This minimal eigenvalue corresponds to one of the three branches of the eigenvalue function. For generic S this branch is a smooth function at the minimum point, and the minimum is nondegenerate.

Hence after the first singularity generation the critical set looks like a small ellipsoid, centered at the minimum point. The kernels of the derivative of the mapping (4) at all its critical points will be almost parallel (since they are almost the same as at the minimum point).

It follows, that from the first moment of singularity generation on there will be a line on the ellipsoid of singular points, where the kernel is tangent to the ellipsoid. This line is close to an ellipse, and the singularities on this line are the Whitney cusp singularities. After the first generation of singularities there will be no other singularities.

From this description it is easy to see that the critical value set will be of lens form (Figure 6). Both longer axes (different generically) are of order $(t-t_0)^{1/3}$ while the shorter one is of order $(t-t_0)^{1/2}$.

Bibliography

[1] Arnol'd, V. I., *Normal forms for functions near degenerate critical points, the Weyl groups of A_k, D_k, E_k and Lagrangian singularities*, Functional Anal. i Prilozhen., 6:4, 1972, pp. 3-25; Functional Anal. Appl., 6, 1973, pp. 254-272.

[2] Arnol'd, V. I., *Mathematical Methods of Classical Mechanics*, Moscow, 1974.

[3] Arnol'd, V. I., *Critical points of smooth functions*, Vancouver Int. Congress of Math., 1974 or Russian Mathematical Surveys (Uspekhi), 1975, 30 No. 5, pp. 3-65.

[4] Arnol'd, V. I., *A spectral sequence to reduce functions to normal forms*, Funct. Anal. and Appl. 1975, 9, No. 3, pp. 81-82.

[5] Arnol'd, V. I., *On the theory of envelopes*, Russian Mathematical Surveys (Uspekhi), 1976, 31 No. 3.

[6] Baas, N., *Structural Stability of Composed Mappings*, Preprint.

[7] Bierstone, E., *Smooth functions, invariant under the action of a finite group*, Preprint, 1975.

[8] Bochner, S., *Compact groups of differentiable transformations*, Ann. Math. Ser. 2, 1945, 46:3, pp. 372-381.

[9] Chincaro, E., *Bifurcation of Whitney maps and critical Pareto sets*, Preprint, 1975.
[10] Eliashberg, J. M., *Surgery of smooth mapping singularities*, Isvestia A. N. U.S.S.R., 1972, No. 6.
[11] Morin, B., *Formes canoniques des singularités d'une application differentiable*, C.R. 260, 1965, pp. 5662–5665, pp. 6503–6506.
[12] Mather, J., *Differentiable invariants*, Preprint, 1975.
[13] Poenaru, V., *Stabilité structurelle équivariante, I*, Orsay, 1975.
[14] Ronga, F., *Stabilité locale des applications équivariantes*, Preprint 1975.
[15] Schwarz, G., *Smooth functions, invariant under the action of a compact Lie group*, to appear.
[16] Solomon, L., *Invariants of finite reflection groups*, Nagoya Math. J. 22, 1963, pp. 57–64.
[17] Sotomayor, J., *Bifurcation of Whitney Maps* $\mathbb{R}^n \to \mathbb{R}^2$ *and Critical Pareto Sets*, Springer Lecture Notes 468, 1975, pp. 75–77.
[18] Thom, R., *Stabilité Structurelle et Morphogenese*, Benjamin, 1972.
[19] Wassermann, G., *(r, s)-stability of unfolding*, Preprint, Regensburg, 1974.
[20] Wassermann, G., *Stability of Unfoldings*, Springer Lecture Notes 393, 1974.
[21] Zakalyukin, V. M., *On Lagrangian and Legendrian singularities*, Funct. Anal. and Applications, 10:1, 1976, pp. 37–45.
[22] Zakalyukin, V. M., *On one-parameter wave front families*, Funct. Anal. and Applications, 1976, 10:2.
[23] Zeeman, E. C., *Catastrophe Theory in Biology*, Springer Lecture Notes 468, 1975, pp. 101–104.
[24] Zeldovitch, J. B., and Mychkis, A. D., *Elements of Mathematical Physics*, Moscow, 1973.
[25] Zeldovitch, J. B., *The hydrodynamics of the universe*, Russian Mathematical Surveys (Uspekhi) 30:6, 1975, p. 204.

Received May, 1976.

A Correction to: Wave Front Evolution and Equivariant Morse Lemma

V. I. ARNOL'D

Communications on Pure and Applied Mathematics, Vol. 29, 1976, pp. 557–582

The following misprints occurred in the paper:

Page 559, esoteric should be theological,
Page 560, line 16: mirror images should be the image of the mirrors,
Page 561, line 8: σ_{k+1} should be $\pm\sigma_{k+1}$,
Page 561, line 12: $N\times\mathbb{C}^p \subset \tau\mathbb{C}^{k+p}$ should be $N\times\mathbb{C}^p \approx \mathbb{C}^{k+p}$,
Page 562, line 10: apparent contour *of N* should be $\cdots$ *in N*,
Page 563, line 3: the same appearance should be the same apparent contour,
Page 563, line −1: Δ should be ∇,
Page 566, line −3: x_5, λ_5 should be x_s, λ_s,
Page 567, line 1: contract should be construct,
Page 568, line 2: $\sum \subset C\mathbb{R}^k$ should be $\sum \subset \mathbb{R}^k$,
Page 568, line −3: intersecting should be intersections,
Page 570, line 10: $c\lambda_1 t \pm \lambda_1$ should be $c\lambda_1 t$ to $\pm\lambda_1$,
Page 573, line 13: "if" column should be "*f*" column
Page 574, line 2: R should be $\mathbb{R}$,
Page 575, line 7: hypersurface $\mathbb{R}^{k+l+1}$ should be hypersurface in $\mathbb{R}^{k+l+1}$,
Page 575, line 13: points should be fronts,
Page 576, line −5: [17] should be [18].

Originally publ. in: Comm. Pure Appl. Math. 30(6), 823, © Wiley Periodicals, Inc., A Wiley Company 1977
Reprint by Springer-Verlag Berlin Heidelberg 2016 with kind permission.

APPENDIX 2

A conjecture on the signature of the quadratic form of a quasihomogeneous singularity[1]

V. I. Arnol'd

A.2.1. DEFINITION. A function f: $\mathbf{C}^n \to \mathbf{C}$ is called *quasihomogeneous with exponents* $\alpha_1, \ldots, \alpha_n$ if

$$f(t^{\alpha_1}x_1, \ldots, t^{\alpha_n}x_n) = t \cdot f(x_1, \ldots, x_n) \quad \text{for any} \quad t \in \mathbf{C} - 0.$$

We assume that the α_i are rational with $0 < \alpha_i \leqslant \frac{1}{2}$.

A.2.2. CONJECTURE. *Let f be a quasihomogeneous function with an isolated critical point. Then the signature of the quadratic form of f is equal to the difference between the set of numbers λ_s (defined below) in the upper and lower half-planes.*

A.2.3. DEFINITION OF THE NUMBERS λ_s. As a basis for the cohomology we can choose the classes of differential $(n-1)$-forms

$$\omega_s = e_s \frac{dx_1 \wedge \ldots \wedge dx_n}{df},$$

where $e_1, \ldots, e_\mu$ are monomials defining a basis of the local ring ${}_nO/(\partial f/\partial x_1, \ldots, \partial f/\partial x_n)$ of f. These forms are eigenvectors of the classical monodromy operator with the eigenvalues

$$\Lambda_s = e^{2\pi i[(m_s, \alpha) + \sum \alpha_k]}, \quad \text{where } e_s = x^{m_s} \ (s = 1, \ldots, \mu).$$

Then $\lambda_s = \sqrt{\Lambda_s} = e^{\pi i[(m_s, \alpha) + \Sigma \alpha_k]}$.

REMARK. The number P_Q of values of s for which $(m_s, \alpha) + \Sigma\, \alpha_k$ has a given value Q/N (where N is the common denominator of the fractions $\alpha_k = A_k/N$) is given by the following formula:

$$\sum P_Q z^Q = \prod_{k=1}^{n} \frac{z^N - z^{A_k}}{z^{A_k} - 1}.$$

[1] This conjecture has now been proved by J. Steenbrinke.

Originally publ. in: Usp. Mat. Nauk 32(2), 60-61, © Russ. Acad. Sciences 1977
English transl.: Russ. Math. Surv. 32(2), 63-64, © London Mathematical Society 1977
Reprint by Springer Verlag Berlin Heidelberg 2016 with kind permission.

For sums of powers, the signature of the quadratic form has been calculated in [25]. The answer agrees with the conjecture A.2.2.

APPENDIX 3

Monodromy groups and bordering of singularities

A. M. Gabrielov

Let $f_\lambda(x)$ be a family of analytic functions defined in a neighbourhood of the origin in $\mathbf{C}^n$ and continuously depending on a parameter $\lambda \in [0, 1]$. We assume that $f_\lambda(x)$ for $\lambda \neq 0$, has an isolated critical point at the origin whose multiplicity μ is independent of λ, and that $f_0(x)$ has an isolated critical point at the origin of multiplicity $\mu_0 \geqslant \mu$. As is known (see [22]), such a family defines an embedding $i_*: H_{n-1}(V_\varepsilon(f_1)) \to H_{n-1}(V_\varepsilon(f_0))$ of the homology groups of the non-singular level sets of f_1 and f_0 compatible with the intersection indices in these groups. In addition, if $\Delta_1, \ldots, \Delta_\mu$ is a distinguished basis in $H_{n-1}(V_\varepsilon(f_1))$, then its image $\Delta_1' = i_*(\Delta_1), \ldots, \Delta_\mu' = i_*(\Delta_\mu)$ can be extended to a distinguished basis $\Delta_1', \ldots, \Delta_\mu', \Delta_{\mu+1}', \ldots, \Delta_{\mu_0}'$ (see [11], [21]). In particular, if $\mu = \mu_0$, then $i_*(\Delta_1), \ldots, i_*(\Delta_\mu)$ is a distinguished basis in $H_{n-1}(V_\varepsilon(f_0))$ with the same intersection matrix as $\Delta_1, \ldots, \Delta_\mu$. Thus, the monodromy group of the singularity is generated by the Picard–Lefschetz operators corresponding to the elements of a distinguished basis. From this the next result follows immediately:

A.3.1. PROPOSITION. *If $\mu = \mu_0$, then the monodromy groups of f_0 and f_1 are isomorphic.*

Now let $\mu < \mu_0$ and let G be the subgroup of the monodromy group of f_0 generated by the Picard–Lefschetz operators h_j' $(j = 1, \ldots, \mu)$ corresponding to the vanishing cycles $\Delta_1', \ldots, \Delta_\mu'$. The subgroup $i_*H_{n-1}(V_\varepsilon(f_1))$ of $H_{n-1}(V_\varepsilon(f_0))$ is G-invariant, and the action of the generators of G on this subgroup is the same as that of the Picard–Lefschetz operators h_j corresponding to the cycles $\Delta_1, \ldots, \Delta_\mu$. More precisely, for $a \in H_{n-1}(V_\varepsilon(f_1))$ we have $h_j'(i_*a) = i_*h_j(a)$. Thus, the correspondence $h_j' \mapsto h_j$ $(j = 1, \ldots, \mu)$ defines an epimorphism $\pi: G \to H$ of G onto the monodromy group H of f_1.

A.3.2. PROPOSITION (see [11]). *If the intersection matrix of f_1 is non-singular, then π is an isomorphism.*

PROOF. Let $\pi g = 1$, that is, let g be an operator in G that is the identity on the subgroup $i_*H_{n-1}(V_\varepsilon(f_1))$. Since the elements of the monodromy group preserve the intersection index, for the cycles $e \in H_{n-1}(V_\varepsilon(f_1))$ and $e' \in H_{n-1}(V_\varepsilon(f_0))$ we have $(ge' \circ i_*e) = (e' \circ i_*e)$, that is, $((ge' - e') \circ i_*e) = 0$. On the other hand, it follows from the Picard–Lefschetz formula that $(ge' - e') \in i_*H_{n-1}(V_\varepsilon(f_1))$, that is, $(ge' - e') = i_*e''$.

On Contemporary Developments of I.G. Petrovskiĭ's Works on Topology of Real Algebraic Varieties

V.I. Arnold

(The Petrovskiĭ seminar on Differential equations and mathematical problems of physics; meeting on January 18, 1977.[1])

The last several years witnessed a major progress in topology of real algebraic varieties, a theory where I.G. Petrovskiĭ had made cornerstone contributions. This progress is related to application of general ideas and methods of complex analytic and differential topological origin. In particular, one established the relation between I.G. Petrovskiĭ–O.A. Oleĭnik's estimates for the Euler characteristic of real manifolds and the middle Hodge number of their complexifications.

Let A be a real nonsingular hypersurface of even dimension n in the projective space of dimension $n+1$. The Petrovskiĭ–Oleĭnik inequality generalized by V.M. Kharlamov has the form $|\chi(A)-1| \leqslant h^{n/2,n/2}(\mathbb{C}A)-1$, where χ is the Euler characteristic, while $h^{p,q}$ is the Hodge number (i.e. the dimension of the bihomogeneous component of degree p in dz and degree q in $d\bar{z}$ in the complex cohomology group of the complexification $\mathbb{C}A$ of the surface A).

On the right-hand side of the Petrovskiĭ–Oleĭnik inequality, instead of the Hodge number, one encountered a certain polynomial in the degree m of the surface A. This "Petrovskiĭ–Oleĭnik polynomial" coincides with $h^{n/2,n/2}(\mathbb{C}A)-1$, as V.I. Zvonilov verified (by using the Hirzebruch formulas).

The definition of this polynomial proposed by I.G. Petrovskiĭ and O.A. Oleĭnik can be given as follows. In the real $(n+1)$-dimensional vector space let us consider a cube with integer lattice vertices and side length m. We draw hyperplanes through the cube vertices and orthogonally to the main diagonal. Such hyperplanes cut the cube into $n+1$ domains, which we will call *layers*.

The number of integer points strictly inside each layer is a polynomial in m. Consider the middle layer (the number of layers is odd, since n is even). The Petrovskiĭ–Oleĭnik polynomial is the number of integer points strictly inside this middle layer.

[1] Translation of Uspekhi Mat. Nauk, **32** (1977), no. 3, 215–216.

Originally publ. in: Usp. Mat. Nauk 32(3), 215-216, © Russ. Acad. Sciences 1977
Translated by Boris Khesin and Andrey Khesin.
Springer-Verlag Berlin Heidelberg 2016 with kind permission.

Thus, the results of I.G. Petrovskiĭ, O.A. Oleĭnik, V.M. Kharlamov, and V.I. Zvonilov imply the following theorem.

Theorem. *The middle Hodge number $h^{n/2,n/2}$ of an even dimensional complex nonsingular projective hypersurface of degree m is equal to one plus the number of lattice points strictly inside the middle layer in an $(n+1)$-dimensional cube with side length m.*

This theorem on topology of complex manifolds is closely related to recent works on Newton's polyhedra and toric manifolds (by A.G. Kushnirenko, D.N. Bernstein, A.G. Khovanskiĭ, D. Mumford, F. Ehlers, and others). The results of those works applied to a projective hypersurface express the outermost Hodge number $h^{n,0}$ as the number of lattice points strictly inside the outermost layer of the above-mentioned cube cutting.

Thus, there arises a natural conjecture that the remaining Hodge numbers are also equal to the numbers of lattice points in the corresponding layers. This conjecture was recently proven by V.I. Danilov (as a particular case of a general theory described in his survey being currently prepared for the Uspekhi Mat. Nauk [Russian Math. Surveys]).

The talk also described the current state of the art for the question on topology of a nonsingular surface of degree four in a three-dimensional projective space. This question constitutes a part of Hilbert's 16th problem. The efforts of D.A. Gudkov, G.A. Utkin, and V.M. Kharlamov allowed one to list all such surfaces up to a surface homeomorphism (so far the question on possible locations of the surface in space is still open). In addition to generalizations of the fundamental works of I.G. Petrovskiĭ, the works of Kharlamov completing these studies use the modern analytical and algebraic geometric techniques, namely, the works of G.N. Tyurina on $K3$ surfaces.

TOPOLOGY OF REAL ALGEBRAIC MANIFOLDS

V. I. Arnol'd and O. A. Oleinik

Vestnik Moskovskogo Universiteta. Matematika,
Vol. 34, No. 6, pp. 7-17, 1979

1. The issue of the topology of manifolds specified by algebraic equations is one of the fundamental and classical problems of mathematics; at the same time, it is still a long way from being solved and it is one of the problems whose solution determines to considerable extent the level of progress in mathematics. Some problems of this type are contained in Hilbert's 16th problem.

Plane second-order curves were investigated as early as the ancient Greeks, while third- and fourth-order curves were studied by Descartes and Newton. The topology of curves of higher order turned out to be much more difficult; the topology of nonsingular sixth-order curves was completed as recently as 1969, while all the possible positions of ovals of an eighth-order curve are unknown even today (see surveys [1,2]).

2. Harnack proved in 1876 [3] that the maximum number of connected components of a real algebraic curve of degree n on a projective plane does not exceed g + 1, where $g=\frac{(n-1)(n-2)}{2}$ is the genus of the curve. Curves with g + 1 ovals exist (examples were given as early as Harnack); I. G. Petrovskii proposed that they be called M-curves.

Hurwitz [4] and Klein [2] offered a new proof of Harnack's inequality. They employed a Riemannian surface formed by complex points of an irreducible curve. The complex conjugate defines the involution of a Riemannian surface of this type. The real points of the curve form the set of stationary points of this involution. If one of the ovals is removed, the remaining ones are homologically independent on the Riemannian surface (the film that effects homology together with its image under involution forms a connected component of the Riemannian surface, despite the assumed irreducibility). This implies that the number of ovals does not exceed g + 1. Thus, Harnack's inequality was not only proved but also extended to arbitrary (not necessarily plane) algebraic curves.

3. In a paper presented at the International Mathematical Congress in 1900, Hilbert included problems of the topology of real algebraic manifolds in his

Originally publ. in: Vestnik Mosk. Univ. Ser. I, Mat. Mekh. 6, 7-17, © Moscow Univ. Press, Moscow 1979
English transl.: Moscow Univ. Math. Bull. 34(6), 5-17, © Allerton Press Inc., New York 1979
Reprint by Springer-Verlag Berlin Heidelberg 2016 with kind permission.

16th problem. In particular, he specially singled out problems of the arrangement of ovals of sixth-order curves in RP^2 and of fourth-order surfaces in RP^3. The accompanying table gives the number N of "logically possible" arrangements of g + 1 ovals of M-curves of small degree n in RP^2.

n	$g+1$	N	ν
2	1	1	1
4	4	9	1
6	11	4766	3
8	22	268 282 855	$10 \leqslant ? \leqslant 144$

The last column of the table indicates how many of the "logically possible" arrangements are realized. Regarding sixth-order curves, Hilbert wrote that he was convinced that it was not possible for all 11 ovals to lie outside one another, but the proof did not appear. Hilbert [6], and subsequently, Briusotti and Wieman and others, proposed a variety of methods for constructing M-curves; in a series of papers, Roon attempted to prove Hilbert's hypothesis regarding sixth-order curves (see survey [1]).

4. In 1906, a paper by Ragsdale appeared [7], which contained extensive experimental material and a number of hypotheses. Each oval of a real nonsingular plane curve of even degree divides the projective plane into two parts, one of which is diffeomorphic to a circle and is called the interior of the oval (the other is diffeomorphic to a Moebius sheet). An oval is called positive (or even) if it lies inside an even number of other ones, and negative (or odd) if it lies inside an odd number of ovals.

Ragsdale's main hypothesis was the following. Assume that an M-curve of degree n = 2k consists of p even and m odd ovals. Then

$$p \leqslant \frac{3}{2} k(k-1)+1, \quad m \leqslant \frac{3}{2} k(k-1).$$

These bounds imply an inequality which was given by Ragsdale as a hypothesis:

$$|2(p-m)-1| \leqslant 3k^2-3k+1. \qquad (1)$$

This inequality was subsequently proved by Petrovskii, and we will call it the Petrovskii inequality. Ragsdale gave examples that demonstrated that the inequality cannot be improved. She also pointed out the relationship between the problem of the arrangement of the ovals and bounds for the Kronecker characteristics and indexes of singular points of vector fields.

5. In 1932, Comessatti proved [8] an inequality for the Euler characteristic χ of a nonsingular real algebraic surface; in modern notation, the inequality is

$$|\chi-2-t| \leqslant h^{1,1}-d,$$

where t is the trace of the involution of the complex conjugate on the space of homology classes generated by algebraic cycles; d is the dimension of this space; and $h^{1,1}$ is the Hodge complexification number. By applying this inequality to a suitable surface, it is possible to prove the Petrovskii inequality (as was recently pointed out by V. M. Kharlamov).

6. In 1933 and 1938, papers [9] and [10] of Petrovskii appeared, of which the second contained a proof of inequality (1) and its generalization to curves of odd degree. Petrovskii's proof was based on the Euler-Jacobi formula:

$$\Sigma g(x)/\det(\partial f/\partial x) = 0$$

(the sum is taken over all n^2 roots of the system of equations $f_1(x_1, x_2)=0, f_2(x_1, x_2)=0$, where f_1 and f_2 are polynomials of degree n; g is a polynomial whose degree is less than that of the denominator). Petrovskii proved, in particular, Hilbert's hypothesis regarding the impossibility of all 11 ovals of a sixth-degree curve being outside one another. This paper by Petrovskii initiated a lengthy series of studies in real algebraic geometry, and the entire subsequent development of the theory is related either directly or indirectly to the ideas, methods, and results of this pioneering study.

7. In 1949, Petrovskii and Oleinik proved [11] an inequality analogous to (1) for smooth real algebraic hypersurfaces in a space of any number m of measurements. For hypersurfaces of even dimension in RP^m, e.g., for algebraic surfaces in three-dimensional real projective space, they obtained an upper bound for their Euler characteristic as a function of the degree n of the hypersurface:

$$|\chi - 1| \leqslant \Pi_{m+1}(n),$$

where $\Pi_d(n)$ is the number of entire points strictly inside cube $(0, n)^d$, which lie on the hyperplane perpendicular to the principal diagonal of the cube and passing through the center of the cube.

For noneven-dimensional hypersurfaces of even degree, e.g., curves in RP^2, Petrovskii and Oleinik bounded the difference between the Euler characteristics of the parts into which this hypersurface divides a projective space. This bound has the form

$$|\chi_+ - \chi_-| \leqslant \Pi_{m+1}(n).$$

For m = 2, it becomes the Petrovskii inequality.

8. In 1951, Oleinik considered [12] spatial real algebraic curves cut out on a surface of degree n_1 by an algebraic surface of degree n_2. The bounds she obtained (see §15 below) are not improvable, at least for curves on surfaces of

7

degree 2. Oleĭnik also established separate bounds for the sums of even-dimensional and noneven-dimensional Betti numbers of algebraic hypersurfaces [13].

9. In 1964-1965, papers by Milnor [14] and Thom [15] appeared, containing bounds for the Betti numbers of real algebraic manifolds. These bounds are mostly weaker than those of Petrovskii and Oleinik. At the end of [15], Thom gave an important inequality—the sum of all Betti numbers of a smooth real algebraic manifold does not exceed the sum of all Betti numbers of the complexification of this manifold (all Betti numbers being modulo 2). This inequality is a direct corollary of Smith's theory applied to the involution of the complex conjugate. Thom ascribes it to Borel.

The Smith inequality precisely yields the Harnack inequality for algebraic curves. Manifolds for which the Smith inequality becomes an equality are called M-manifolds (at the suggestion of V. A. Rokhlin, who employed Petrovskii's term "M-curves" as a starting point).

10. By 1969, D. A. Gudkov had completed his investigation of the topology of plane nonsingular real curves of degree 6 (see [16]). For the 11 ovals of an M-curve of degree 6, only 3 arrangements turned out to be possible: exactly one oval contains other ones inside it, the number of the latter being only 1, 5, or 9. Gudkov's proofs are extremely complicated; they are an extension of the methods of Roon and of the Italian school of algebraic geometry.

By using Gudkov's results, G. A. Utkin obtained (see [16]) considerable information regarding the topology of surfaces of degree 4 in RP^3.

In considering curves of various degrees which he was able to construct, Gudkov arrived at the following "periodicity hypothesis": the difference between the number of even and odd ovals of an M-curve of even degree is congruent modulo 8 to the square of half the degree.

11. In 1971, V. I. Arnol'd (whom Gudkov had informed of his hypothesis) proved a weakened version of it (with 8 replaced by 4) [17]. Arnol'd's proof linked the study of real algebraic curves to the topology of four-dimensional manifolds and the arithmetic of quadratic forms. The use of these relationships led to progress in real algebraic geometry, facilitated by the employment of powerful techniques of modern differential topology and algebraic geometry.

The four-dimensional manifold considered in [17] was a double covering of a complex projective plane, branched along the curve being investigated.

12. Many real entities have complex analogs, sometimes quite unexpected ones. For example, the complex analog of group Z_2 is group Z; the complex analog

of a permutation group is an Artinian braid group; the complex analog of spaces $K(\pi,0)$ are spaces $K(\pi,1)$; and the complex analog of the Morse theory is the Picard-Lefschetz theory. (It would be very interesting to find a complex analog of algebraic K-theory using the Cerf-Hutcher-Wagoner-Volodin theory as a "real" starting point.) What is done upon complexification with a manifold with an edge? To answer this question, we specify a manifold with an edge by the inequality $f(x)\geqslant 0$. For complexification, we algebrize this inequality, i.e., we write it in the form $f(x)=z^2$. Thus, we arrive at a double covering which is branched along the complexification of the edge of the initial real manifold.

The parts into which an algebraic curve of even degree divides RP^2 give rise to two double coverings—manifolds of real dimension 4. The intersection index in two-dimensional integer homologies of this manifold specifies a unimodular bilinear form, while the complex conjugate defines a form-preserving involution.

Thus, the arithmetic of quadratic forms appear in the topology of real plane algebraic curves.

In addition to proofs of Gudkov's hypothesis modulo 4, Arnol'd's paper [17] contained a number of new constraints on the arrangement of ovals of plane algebraic curves.

13. Gudkov's hypothesis was fully proved by Rokhlin [18] in 1972. Rokhlin extended Gudkov's congruence to regular complete intersections [19], then to nonsingular algebraic manifolds [20]. For M-manifold A of even dimension, the Rokhlin congruence has the form

$$\chi(A)\equiv\sigma(CA) \bmod 16$$

(χ is the Euler characteristic, σ is the signature, and CA is the complexification of A). In the noneven-dimensional case, the Rokhlin congruence involves a signature of a double covering of a surface with branching along A. Rokhlin offered several proofs for Gudkov's congruence; the later proofs are simpler and more general, but the earlier ones contain some latent additional information.

14. The congruences of Gudkov and Rokhlin were soon generalized by Kharlamov [21,22], and also by Gudkov and Krakhnov [23] for M - 1- and M - 2-manifolds (for M - 1-manifolds the left side of the Smith inequality is 21 smaller than the right). Thus, for a plane curve of degree n = 2k consisting of $\frac{1}{2}(n-1)(n-2)$ ovals (a M - 1-curve), the difference between the number of even and odd ovals is congruent modulo 8 with $k^2\pm 1$.

15. Kharlamov obtained [24,25] new generalizations of the Petrovskii and Oleinik inequalities. In these generalizations, the Euler characteristics of

even-dimensional real algebraic manifolds are bounded in terms of the Hodge numbers of suitable complex manifolds. For the Euler characteristic of a real smooth manifold of dimension 2k, Kharlamov obtained the inequality

$$|\chi - 1| \leqslant h^{k,k} - 1, \tag{2}$$

where $h^{k,k}$ is the average Hodge number of the complexification of this manifold. Zvonilov proved in [26] that for a nonsingular hypersurface of degree n in CP^m the right side of inequality (2) coincides with the right side of the Petrovskii-Oleinik inequality. Thus, Kharlamov's inequality is a direct generalization of the Petrovskii-Oleinik inequality.

Kharlamov regards a noneven-dimensional manifold as the edge of a 2k-dimensional manifold with an edge, and proves the following inequality for the Euler characteristic of the latter:

$$|2\chi - 1| \leqslant h^{k,k} - 1,$$

where $h^{k,k}$ is the Hodge number of the double covering of the complexification of the manifold with edge which is branched along the complexification of the edge. For noneven-dimensional hypersurfaces in CP^m, he obtained precisely the Petrovskii-Oleinik inequality, while for curves on surfaces he obtained the Oleinik inequality.

16. Other generalizations of the Petrovskii-Oleinik inequality were obtained by Arnol'd in [27], on the basis of recent progress in the local theory of singularities of differentiable mappings. Arnol'd considered a finite-multiple critical point of a real analytic function of 2k variables, and bounded the index of the gradient of this function at the critical point in terms of the Hodge numbers of the mixed Hodge structure defined by Steenbrink [28] in homologies of the nonsingular local manifold of the level of the function. This bound has the form

$$|\mathrm{ind}| \leqslant h_1^{k,k} \tag{3}$$

(the subscript 1 indicating the Hodge number on the space adjoint to the eigenvalue, equal to unity, of the monodromy operator).

To bound the index of the gradient of a function of an odd number of variables, it suffices to increase the number of variables by 1, adding the square of the new variable to the function.

When the function is a homogeneous polynomial, inequality (3) becomes the Petrovskii-Oleinik inequality.

17. Progress in singularity theory has made it possible to bound from above not only the index of a singular point of a gradient vector field, but also the index of an isolated singular point of an arbitrary vector field. It was proved by Arnol'd in [27] that the modulus of the index of a singular point of a vector field in RP^m whose components are homogeneous polynomials of degree n - 1 is bounded from above by the Petrovskii-Oleinik number:

$$|\mathrm{ind}| \leqslant \Pi_{m+1}(n).$$

The proof is based on the Levine-Eisenbud-Khimshiashvili formula [29], which expresses the index of a singular point of a vector field in the form of the signature of a quadratic form on a local algebra corresponding to a finite-multiple singular point ("algebra of functions on merged singular points").

When the vector field is the gradient of a homogeneous polynomial of degree n, the inequality in question becomes the Petrovskii-Oleinik inequality. Therefore, Arnol'd called his inequality the generalized Petrovskii-Oleinik inequality.

18. The question of whether equality is attained in the Petrovskii-Oleinik inequality has been resolved only in the simplest cases: for plane curves (Ragsdale); for curves on second-order surfaces (Oleinik); for fourth-order surfaces. Equality is attained in all these cases. Khovanskii proved in [30] that equality is attained in the generalized Petrovskii-Oleinik inequality for vector fields; he constructed homogeneous polynomial vector fields of degree n - 1 in R^m with largest possible index (equal to $\Pi_m(n)$) . Moreover, he constructed fields with arbitrary indexes from $-\Pi_m(n)$ to $+\Pi_m(n)$ to $\Pi_m(n)$ congruent modulo 2 with $\Pi_m(n)$ (the indexes of all fields of fixed degree are identical modulo 2, since complex roots form pairs).

19. Khovanskii also obtained [30] bounds for the sum ind_+ of indexes of singular points of a polynomial vector field in a region bounded by an algebraic hypersurface. Assume that the components of the field are of degree $n_1-1, \dots, n_m-1$, while the hypersurface is of degree n_0. Consider the parallelepiped $\prod_{k=1}^{m}(0, n_k)$. We denote by $\Pi(n, n_0)$ the number of integral points z inside this parallelepiped for which $\Sigma n_k - n_0 \leqslant 2\Sigma z_k \leqslant \Sigma n_k + n_0$, and by $O(n, n_0)$ the number of integral points for which $\Sigma n_k - n_0 \leqslant 2\Sigma z_k \leqslant \Sigma n_k$. For some condition of nondegeneracy (which is usually satisfied)

$$|\mathrm{ind}_+| \leqslant O(n, n_0), \quad |\mathrm{ind}_+ - \mathrm{ind}_-| \leqslant \Pi(n, n_0), \quad |\mathrm{ind}_+ + \mathrm{ind}_-| \leqslant \Pi(n, 0).$$

These inequalities together with the obvious congruences modulo 2 comprise the complete system of equations of existence of the field with these data, in particular with sum of indexes ind_+ for singular points on one side of the hypersurface and ind_- on the other side of it. The proof employs an expression for

11

the indexes in the form of signatures of suitable forms, and the Eulier-Jacobi formula. It is a direct extension of the work of Petrovskii [10] and Petrovskii and Oleinik [11], and shows, incidentally, that Petrovskii's paper [10] already contained many elements of the proof of the signature formula for the index, which was obtained in [29].

20. Kharlamov extended generalized Petrovskii inequalities to algebraic manifolds with singularities. One of his generalizations involves a mixed Hodge structure of a nonsingular local manifold of level $\mathcal{A}$ of a homogeneous function constructed on the basis of a given hypersurface A of dimension 2k. In this case Kharlamov obtained the following generalization of the Petrovskii inequality:

$$|\chi(A)-1| \leqslant h_1^{k,k}(\mathcal{A}).$$

A double covering is adduced in the noneven-dimensional case. For plane curves of degree 2k with s double points and t return points, the Kharlamov inequality has the form

$$|\chi_+ - \chi_-| \leqslant 3k^2 - 3k + 1 - s - 2t,$$

where $\chi_\pm$ are the Euler characteristics of the parts into which the curve divides the projective plane.

In [31], Viro gave a number of other constraints which involve not only the type of the singular points but also their mutual disposition.

Extensive studies of disintegrating sixth-order curves were made by Polotovskii [32,33]. He discovered, in particular, the link between this issue and the "complex orientations" to be discussed below.

21. In addition to the issue of the topological classification proper of real algebraic curves, which was resolved by Harnack's theorem, and the issue of isotopic classification (Hilbert's question of mutual disposition of ovals), there is the no less important issue of the connectivity components of a manifold of real nonsingular plane curves of fixed degree.

The points of such a component constitute curves of the same isotopic type. Rokhlin pointed out in [34], however, that curves of the same isotopic type do not always belong to the same component; isotopy cannot always be realized in the class of smooth algebraic curves of fixed degree $n \geqslant 5$.

An invariant which makes it possible to distinguish connectivity components was essentially considered by Klein [35], who thoroughly investigated fourth-degree curves. The ovals of an M-curve divide a Riemannian complexification surface into two parts. Each part is an oriented real two-dimensional manifold, whose edge is made up of ovals. Consequently, each oval obtains an orientation (defined to within simultaneous variation of the orientation of all ovals). This orientation was called the complex orientation by Rokhlin. Two M-curves from one connectivity component of a manifold of curves of given degree are isotopic with allowance for complex orientations. Therefore, M-curves with nonequivalent complex orientations that are isotopic without allowance for orientations belong to different connectivity components.

A detailed survey of all findings concerning complex orientations of M-curves, and of complex characteristics of real curves in general, may be found in Rokhlin's paper [36], which also gives a number of known constraints on the isotopic type of a plane curve that do not appear in the present survey (strengthened Petrovskii and Arnol'd inequalities; inequalities of Mishachev and Zvonilov, Viro-Zvonilov, Rokhlin [34], and others).

22. The problem of classification of surfaces of fourth degree in RP^3 was definitively solved by Kharlamov as a result of a series of studies [37-39], which yielded both new constraints on the topological and isotopic types of the surfaces and new methods of constructing surfaces of specified types. Kharlamov was the first to employ the powerful resources of modern algebraic geometry in this area. A complex nonsingular surface of fourth degree in CP^3 is a manifold of type K - 3. Such manifolds have been studied in detail; the corresponding module spaces have been investigated; it has been determined how algebraic manifolds are arranged among all such manifolds. Kharlamov was able to use this information to study real surfaces. Thus, he obtained a completely new method of constructing examples of fourth-degree surfaces.

Kharlamov's results yield not only a topological but also an isotopic classification of nonsingular surfaces. His proofs made some use of arithmetic information obtained by Nikulin in [40].

23. Nikulin was able in [40] to completely arithmecize the problem of determining the connectivity components of manifolds of nonsingular real algebraic surfaces of type K - 3. He placed every even-dimensional real algebraic manifold in correspondence with a triple consisting of the following: 1) an integer-valued symmetrical bilinear unimodular form (form of intersections on space of integer-valued homologies of moderate dimension, reduced modulo the torsion); 2) an involution (induced by complex conjugacy); and 3) a "polarization," specified by an invariant or anti-invariant cycle (suitable degree of hyperplane section).

Nikulin constructed the complete system of invariants of such triples and, in a number of particularly interesting cases, he also found all the relationships among invariants. In particular, he established a one-to-one correspondence

13

between classes of triples with suitable values of the invariants and the connectivity components of a manifold of nonsingular real surfaces of type K - 3. He obtained the components, e.g., for a manifold of double coverings of a plane which are branched along plane curves of sixth degree (thus, for a manifold of most plane curves of sixth degree); for a manifold of nonsingular surfaces of fourth degree in RP^3; and for intersections of cubics and quadratics in RP^4 and intersections of three quadratics in RP^5.

24. Viro [41] developed entirely different methods of constructing examples of real algebraic manifolds with the given topology. In particular, he constructed M-surfaces and M-manifolds, i.e., he established the unimprovable nature of the inequality (a consequence of Smith's theory) for the sum of all Betti numbers of a real manifold modulo 2.

25. Studies of bounds of topological invariants of singularities in terms of Newton polygons led A. G. Kushnirenko to the hypothesis that what is essential for the topological complexity of an object is not the degree of the equation, but rather the number m of monomials that appear in the polynomial with nonzero coefficients. An elementary example of a theorem of this type is Descartes' theorem, which estimates the number of positive roots of a polynomial in terms of the number of sign changes in the sequence of its coefficients. If there are few monomials, then there are also few sign changes. Thus, we have the problem of "oligomials," i.e., the topology of objects specified by polynomials of arbitrarily large degree but with a restricted number of monomials.

It was recently shown by Sevast'yanov that the number of zeros of an m-omial on a plane algebraic curve of degree n is bounded by a constant C(m,n) which is independent of the degree of the m-omial. Viro improved his bound, obtaining

$$C(m, n) = \frac{n^2}{24}(m-1)\,m\,(m+1)\,(m+2) + n\left[\frac{(m+1)\,m\,(m-1)\,(m-2)}{24} + 3(m-1)\right].$$

Khovanskii and Gel'fond proved a number of corollaries of the oligomial hypothesis. Thus, they obtained a bound for the number of zeros of superpositions of oligomials. Their results can be carried over to the case of a field of real Liouville functions, obtained from a field of rational functions by attachment of a finite number of integrals and exponents of integrals. They proved that in the real case, the Liouville functions have a finite number of zeros. For example, cos x is not a Liouville function in the real sense, although it is in the complex sense. These results give hope that the oligomial hypothesis is also valid in the more general situation, e.g., that the number of ovals of curve $f = 0$ can be bounded in terms of the number of monomials of polynomial f.

26. Some unresolved issues.

1) Give an asymptotically exact bound for the number of connected components of a space of nonsingular real algebraic hypersurfaces of degree n.

2) Is equality attained in the Petrovskii-Oleinik inequality?

3) Is the Ragsdale hypothesis valid? This hypothesis can be reformulated as follows. Assume that $f(x, y, z)$ is a homogeneous polynomial of even degree; $F_{\pm}=f\pm t^2$; $RV_{\pm}$ is a local manifold of level $F_{\pm}=\pm\varepsilon$. Ragsdale's hypothesis involves bounding the numbers of components

$$b_0(RV_+) \leqslant h_1^{2,2}(F_+), \quad b_0(RV_-) \leqslant h_1^{2,2}(F_-)+1$$

in terms of a mixed Hodge structure (for suitable sign of f).

4) Give unimprovable bounds (in terms of the degree or the Hodge number) for the individual Betti numbers of real algebraic hypersurfaces, in particular for the number of components b_0. Perhaps it is easier to bound the numbers b_0, b_0-b_1, $b_0-b_1+b_2, \ldots$, as well as combinations of local Morse type numbers M_0, M_0-M_1, $M_0-M_1+M_2, \ldots$ (M_i is the number of critical points of index i that merge at zero for any Morseization of a homogeneous equation of a hypersurface).

5) What is the largest number of handles of a component of an algebraic surface of degree n in RP^3?

6) Bound the number of ovals of a curve whose equation is an oligomial, in terms of the number of its terms.

7) How many nonconvex ovals can a plane algebraic curve of degree n have?

8) Does the isotopic type of a pair (plane M-curve, its complex orientation) determine the connectivity component in space of nonsingular curves of fixed degree?

9) Investigate the fundamental complement group π_1 to the set of singular hypersurfaces in the complex projective space of all hypersurfaces of fixed degree in CP^m, and find the corresponding monodromy group (representation of π_1 by automorphisms of the homology group of the hypersurface).

10) Assume that P, Q, and H are polynomials of fixed degrees of x and y, and assume that $I(h)=\oint Pdx+Qdy$ over the oval H = h. What is the largest number of zeros of I(h) when I(h) is not identically zero?

11) Assume that P and Q in the system $\dot{x}$ = P, $\dot{y}$ = Q are polynomials of second degree, while H is the first integral (not necessarily polynomial). How many limit cycles can be generated from the components of the level lines of H for small perturbations of P and Q that leave them as second-degree polynomials?

12) Assume that P and Q in the system $\dot{x} = P$, $\dot{y} = Q$ are power series that begin with homogeneous polynomials P_n and Q_n of degree n. Is it true that for almost all pairs (P_n, Q_n) the number of limit cycles that are generated from zero under small perturbations of the system is bounded by a constant that depends only on n?

REFERENCES

1. D. A. Gudkov, "Topology of real projective algebraic manifolds," Uspekhi matem. nauk, vol. 29:4, pp. 3-79, 1974.
2. G. Wilson, "Hilbert's sixteenth problem," Topology, vol. 17:1, pp. 53-73, 1978.
3. A. Harnack, "Über Vielheiligkeit der ebenen algebraischen Kurven," Math. Ann., vol. 10, pp. 189-199, 1976.
4. A. Hurwitz, "Über Riemannsche Flächen mit gegebenen Verzweigungspunkten," Math. Ann., vol. 39, pp. 1-61, 1891.
5. F. Klein, Riemannsche Flächen (lit.), Vorles, I, II, Göttingen, 1892 (Neuedruck., 1906).
6. D. Hilbert, "Über die reellen Zuge algebraischer Kurven," Math. Ann., vol. 38, pp. 115-138, 1891.
7. V. Ragsdale, "On the arrangement of the real branches of plane algebraic curves," Amer. J. Math., vol. 28, pp. 377-404, 1906.
8. A. Comessatti, "Sulla connessione e sui numeri base delle superficie algebraiche reali," Rend. Semin. mat. Univ. Padova, vol. 3, pp. 141-162, 1932.
9. I. G. Petrovsky, "Sur la topologie des courbes reeles et algebriques," Compt. rend. Acad. Sci., Paris, vol. 197, pp. 1270-1273, 1933.
10. I. G. Petrovsky, "On the topology of real plane algebraic curves," Ann. Math., vol. 39, pp. 189-209, 1938.
11. I. G. Petrovskii and O. A. Oleinik, "Topology of real algebraic surfaces," Izv. AN SSSR. Ser. matem., vol. 13, pp. 389-402, 1949.
12. O. A. Oleinik, "Topology of real algebraic curves on an algebraic surface," Matem. sb., vol. 29(71), pp. 133-156, 1951.
13. O. A. Oleinik, "Bounds for Betti numbers of real algebraic hypersurfaces," Matem. sb., vol. 28(70), pp. 635-640, 1951.
14. J. Milnor, "On the Betti numbers of real varieties," Proc. Amer. Math. Soc., vol. 15, pp. 275-280, 1964.
15. R. Thom, "Sur l'homologie des variétés algébriques reelles," in collection: Differential and Combinatorial Topology: a Symposium in Honor of Marston Morse, Princeton University Press, pp. 255-265, 1965.
16. D. A. Gudkov and G. A. Utkin, "Topology of sixth-order curves and fourth order surfaces," Uch. zap. Gor'kovsk. un-ta, vol. 87, pp. 3-213, 1969.
17. V. I. Arnol'd, "Arrangement of ovals of real plane algebraic curves, involutions of four-dimensional smooth manifolds, and arithmetic of integer-valued quadratic forms," Funkts. analiz i ego prilozh., vol. 5:3, pp. 1-9, 1971.
18. V. A. Rokhlin, "Proof of Gudkov's hypothesis," Funkts. analiz i ego prilozh., vol. 6:2, pp. 62-64, 1972.
19. V. A. Rokhlin, "Congruences modulo 16 and Hilbert's sixteenth problem," Funkts. analiz i ego prilozh., vol. 6:4, pp. 58-64, 1972.
20. V. A. Rokhlin, "Congruences modulo 16 and Hilbert's sixteenth problem, II," Funkts. analiz i ego prilozh., vol. 7:2, pp. 91-92, 1973.
21. V. M. Kharlamov, "New congruences for Euler characteristic of real algebraic manifolds," Funkts. analiz i ego prilozh., vol. 7:2, pp. 74-78, 1973.
22. V. M. Kharlamov, "Additional congruences for Euler characteristics of four-dimensional real algebraic manifolds," Funkts. analiz i ego prilozh., vol. 9:2, pp. 51-60, 1975.

23. D. A. Gudkov and A. D. Krakhnov, "Periodicity of Euler characteristic of real algebraic M - 1-manifolds," Funkts. analiz i ego prilozh., vol. 7:2, pp. 13-19, 1973.
24. V. M. Kharlamov, "Generalized Petrovskii inequality," Funkts. analiz i ego prilozh., vol. 8:2, pp. 50-56, 1974.
25. V. M. Kharlamov, "Generalized Petrovskii inequality, II," Funkts. analiz i ego prilozh., vol. 9:3, pp. 93-94, 1975.
26. V. I. Zvonilov, "Kharlamov and Petrovskii-Oleinik inequalities," Funkts. analiz i ego prilozh., vol. 9:2, pp. 69-70, 1975.
27. V. I. Arnol'd, "Index of singular point of vector field, Petrovskii-Oleinik inequalities, and mixed Hodge structures," Funkts. analiz i ego prilozh., vol. 12:1, pp. 1-14, 1978.
28. J. H. M. Steenbrink, "Mixed Hodge structure on the vanishing cohomology," Report 76-06, University of Amsterdam, 1976.
29. D. Eisenbud and H. Levine, "The topological degree of finite C^∞ map germ," Ann. Math., no. 1, pp. 19-38, 1977.
30. A. G. Khovanskii, "Index of polynomial vector field," Funkts. analiz i ego prilozh., vol. 13:1, pp. 49-58, 1979.
31. O. Ya. Viro, "Generalized Petrovskii and Arnol'd inequalities and curves with singularities," Uspekhi matem. nauk, vol. 33:3, pp. 145-146, 1978.
32. G. M. Polotovskii, "Catalog of M-decaying curves of sixth order," Dokl. AN SSSR, vol. 236:3, pp. 548-551, 1977.
33. G. M. Polotovskii, "M - 1- and M - 2-decaying curves of sixth order," in collection: Methods of the Qualitative Theory of Differential Equations [in Russian], Gor'kii, pp. 130-148, 1978.
34. V. A. Rokhlin, "Complex orientations of real algebraic curves," Funkts. analiz i ego prilozh., vol. 8:4, pp. 71-75, 1974.
35. F. Klein, Gesammelte Mathematische Abhandlungen, Berlin, vol. 2, p. 155, 1922.
36. V. A. Rokhlin, "Complex topological characteristics of real algebraic curves," Uspekhi matem. nauk, vol. 33:5, pp. 77-89, 1978.
37. V. M. Kharlamov, "Maximum number of components of surface of degree 4 in RP^3," Funkts. analiz i ego prilozh., vol. 6:4, p. 101, 1972.
38. V. M. Kharlamov, "Topological types of nonsingular surfaces of degree 4 in RP^3," Funkts. analiz i ego prilozh., vol. 10:4, pp. 55-68, 1976.
39. V. M. Kharlamov, "Isotopic types of nonsingular surfaces of degree 4 in RP^3," Funkts. analiz i ego prilozh., vol. 12:1, pp. 86-87, 1978.
40. V. V. Nikulin, "Integer-valued symmetrical bilinear forms and some of their geometrical applications," Izv. AN SSSR. Ser. matem., vol. 3:1, pp. 111-177, 1979.
41. O. Ya. Viro, "Construction of M-manifolds," Uspekhi matem. nauk, vol. 34:4, p. 160, 1979.

BIFURCATIONS OF INVARIANT MANIFOLDS OF DIFFERENTIAL EQUATIONS AND NORMAL FORMS IN NEIGHBORHOODS OF ELLIPTIC CURVES

V. I. Arnol'd

§1. Description of the Results

An interesting connection exists between the invariant surfaces of a vector field in C^3 close to a singular point of Siegel type (or between the invariant curves of local hyperbolic automorphisms of C^2) on the one hand and the imbeddings of elliptic curves with zero self-intersection indices in complex-analytic surfaces on the other. This connection can be used in both directions.

A) An elliptic curve is constructed along an invariant manifold. This leads to the proof of the divergence of the Poincaré series linearizing a field or an automorphism in case the invariant manifolds accumulate to a singular point (which is observed under an "abnormal good commensurability" of the eigenvalues).

B) Let the elliptic curve be imbedded into a complex-analytic surface. It turns out that the problem of classifying such imbeddings is a problem with "small denominators" and its solution depends upon the theory of Diophantine approximations. We apply the standard methods of estimates of small denominators from the theory of normal forms of differential equations in the neighborhood of a singular point and the standard representations of this theory. The following results are obtained.

a) Linearization. A sufficiently small neighborhood of an elliptic curve imbedded in a surface with self-intersection index 0 is biholomorphically equivalent to a neighborhood of zero section of a normal bundle for almost all* normal bundles.

b) Materialization of Resonances. In the indicated space of modules there is picked out an everywhere-dense union of a denumerable set of hypersurfaces called resonance hypersurfaces. (A bundle is a resonance bundle if it becomes analytically trivial after passing to some finite-sheeted cyclic covering.)

Under a general continuous variation of a pair (elliptic curve, surface), at the instant of passing through resonance an elliptic curve is approximated by another elliptic curve covering it topologically.

c) Divergence of Series. Certain nonresonant points of the space of modules lie "abnormally close" to resonance hypersurfaces. As a consequence of this, in any neighborhood of the corresponding elliptic curve on the surface there are infinitely many elliptic curves corresponding to the different resonances and topologically covering the original curve with more and more sheets. In such a case the neighborhood of the elliptic curve is formally equivalent to a normal bundle, but the corresponding series diverge everywhere.

The author thanks Yu. S. Il'yashenko, V. P. Palamodov, A. N. Tyurin, A. S. Pyartli, A. D. Bryuno, and F. A. Bogomolov for useful discussions.

*All, except for a set of Lebesgue measure zero in the space of modules of one-dimensional vector bundles on elliptic curves (this space of modules has the complex dimension two).

Moscow State University. Translated from Funktsional'nyi Analiz i Ego Prilozheniya, Vol. 10, No. 4, pp. 1-12, October-December, 1976. Original article submitted April 2, 1976.

Originally publ. in: Funkts. Anal. Prilozh. 10(4), 1-12, © Russ. Acad. Sciences 1976
English transl.: Funct. Anal. Appl. 10(4), 249-259, © Consultants Bureau, New York 1976
Reprint by Springer-Verlag Berlin Heidelberg 2016 with kind permission.

§2. Resonances and Bifurcations

2.1. Nonresonant Case. We consider a holomorphic vector field v with a singular point $O \in C^n$. By $\lambda_1, \ldots, \lambda_n$ we denote the eigenvalues of a linearization of field v at point O.

Definition 2.1.1 [2]. The singular point O is of Poincaré type (of Siegel type) if the convex hull of the collection $\{\lambda_k\}$ does not contain (does contain) 0.

The numerical relation $\lambda_k = (m, \lambda), m_i \geqslant 0, \sum m_i \geqslant 2$, is called a resonance.

We consider the space C^n with coordinates $\lambda_1, \ldots, \lambda_n$.

Each resonance defines a resonance plane in this space. To a singular point of Poincaré (Siegel) type there corresponds in C^n a Poincaré domain and a Siegel domain. In a Poincaré domain the resonance planes lie discretely, while in a Siegel domain they are everywhere dense.

If $\{\lambda_k\}$ is a resonant collection of eigenvalues, then a formal change of variables exists, turning it into its linear part.

Poincaré's Theorem [10]. If $\{\lambda_k\}$ is a nonresonant collection and belongs to a Poincaré domain, then the field v is turned into a linear one by a diffeomorphism biholomorphic in a neighborhood of point O.

Siegel's Theorem [8]. For almost all collections $\{\lambda_k\}$ in a Siegel domain the field v is turned into a linear one by a diffeomorphism biholomorphic in a neighborhood of point O.

Siegel [9] also gave examples of nonresonant collections $\{\lambda_k\}$ for which there exist vector fields nonlinearizable by a biholomorphic diffeomorphism. These collections are "abnormally close to resonant ones" and form a set of Lebesgue measure zero. Bryuno [5] has constructed many examples of such kind.

2.2. Passage through Resonance between Two Eigenvalues. A vector field of general position is nonresonant. Unremovable resonances are encountered only when the field being examined depends upon parameters. Thus, the question arises on what happens to a vector field depending on a parameter when the parameter passes through a resonance value.

The simplest case of resonances in a Poincaré domain has been considered in [2]. It was ascertained that although the topology of the phase curves in a real domain is not changed when passing through resonance at a singular point of Poincaré type, a topological bifurcation takes place in a complex domain.

More precisely, let us consider a small sphere with center at a singular point of a vector field in the complex space C^2. If the singular point is of Poincaré type, then the complex phase curves intersect the sphere's surface transversally. A family of smooth real curves are formed on this surface. When passing through the resonance $\lambda_1 = n\lambda_2$ from one of the limit cycles another cycle branches off on the sphere, going around the original one n times. This branched cycle is the materialization of resonance in the Poincaré case.

In the Siegel case the situation is far more complicated. An optimistic conjecture on the materialization of resonances, made in [2], turns out to be false [6]. However, Pyartli [11] established the following results.

1) When passing through the resonance $k_1\lambda_1 + k_2\lambda_2 = 0$ in C^2, in the general case a branching takes place from a separatrix of the singular point of an invariant manifold, whose equation has in the first approximation the form

$$z_1^{k_1} z_2^{k_2} = \varepsilon,$$

where ε characterizes the deviation from resonance, while z_1 and z_2 are phase coordinates.

2) An analogous result for the same resonance has been obtained in C^n under restrictive conditions on the remaining eigenvalues.

3) For "abnormally commensurable" λ_1 and λ_2 the general case is the existence of an infinite number of invariant manifolds corresponding to the different resonances in any neighborhood of the singular point, which implies the divergence of the Poincaré series.

The special case of resonance $\lambda_1 + \lambda_2 = 0$ with pure imaginary λ was previously investigated in detail by Poincaré, Andronov [1], and Hopf [12] in a real domain in connection with the study of a mild loss of stability. Pyartli's paper is based on Hopf's method. Other proofs of the first two results of Pyartli were suggested by Bryuno [6].

2.3. Resonance between Three Eigenvalues. The resonance

$$k_1\lambda_1 + k_2\lambda_2 + k_3\lambda_3 = 0,$$

next in complexity, where the triangle with vertices λ_1, λ_2, λ_3 contains 0, also was investigated by Pyartli and Bryuno. Here too was proved a branching from the separatrix of the singular point of the invariant manifold for n = 3. Let z_1, z_2, z_3 be the phase coordinates and let ε be the deformation parameter (resonance corresponds to $\varepsilon = 0$). Then the invariant manifolds fill up, in the space with coordinates (z, ε), a holomorphic hypersurface whose equation in the first approximation has the form

$$z_1^{k_1} z_2^{k_2} z_3^{k_3} = \varepsilon$$

(in a suitable coordinate system).

Pyartli proved that for "abnormally commensurable" $(\lambda_1, \lambda_2, \lambda_3)$ forming a Siegel triangle, in the general case in any neighborhood of the equilibrium position $O \in \mathbb{C}^3$ there are infinitely many invariant manifolds of the form described, corresponding to the different resonances.

From what follows it will be seen that when a sufficiently large piece of a resonance invariant manifold is present in a neighborhood of a nonresonant point $O \in \mathbb{C}^3$ the convergences of the Poincaré–Siegel series are hindered in this region. Therefore, from the cited result of Pyartli it follows that for "abnormally commensurable" $(\lambda_1, \lambda_2, \lambda_3)$ the general case is the divergence of the series mentioned in any neighborhood of the origin.

2.4. The Discrete Time Case. Let us consider the local autodiffeomorphism $A: \mathbb{C}^n \to \mathbb{C}^n$ in the neighborhood of a fixed point O. By $\lambda_1, \ldots, \lambda_n$ we denote the eigenvalues of the linearization of A at O.

The preceding theory carries over to this case with the following changes.

Resonances: $\lambda_k = \lambda_1^{m_1} \ldots \lambda_n^{m_n}$ $(m_i \geqslant 0;\ \Sigma m_i \geqslant 2)$.

Poincaré domain: all $|\lambda_s| > 1$ or all $|\lambda_s| < 1$.

Siegel domain: $|\lambda_i| \geqslant 1$ and $|\lambda_j| \leqslant 1$ exist.

Remark 2.4.1. To a linear vector field in $\mathbb{C}^n$ there corresponds a linear transformation in $\mathbb{C}^{n-1}$ ("Poincaré's successor function"). Namely, let the field be given by the differential equation

$$\dot{z}_1 = \alpha_1 z_1, \ldots, \dot{z}_n = \alpha_n z_n,$$

where $\alpha_n \neq 0$. Let us consider the solution with initial conditions on plane $z_n = 1$ at $t = 0$. The value of the solution at $t = 2\pi i/\alpha_n$ belongs to that same plane in $\mathbb{C}^{n-1}$. The mapping $A: \mathbb{C}^{n-1} \to \mathbb{C}^{n-1}$ obtained has the eigenvalues $\lambda_s = e^{2\pi i \alpha_s/\alpha_n}$ $(s = 1, \ldots, n-1)$.

An analogous construction of the successor function exists (under a few additional assumptions) in the nonlinear case too. Therefore, the results on invariant manifolds and bifurcations for mappings imply the corresponding results for vector fields.

Besides, in the majority of cases the connection noted between equations and mappings is better used as a heuristic means for guessing the results in one area from the results existing in another; however, it is more convenient to carry out the proofs independently in both situations.

2.5. Bifurcation of Invariant Manifolds of a Diffeomorphism. Resonance between three eigenvalues of a vector field, considered in Paragraph 2.3,

$$k_1\alpha_1 + k_2\alpha_2 + k_3\alpha_3 = 0,$$

corresponds in the discrete case to a resonance of the form

$$\lambda_1^{m_1}\lambda_2^{m_2} = 1,$$

where in the Siegel domain $m_1 > 0$, $m_2 > 0$, $|\lambda_1| \neq 1 \neq |\lambda_2|$. We assume that λ_1 is greater than 1 in absolute value and λ_2 is less than 1 in absolute value.

In this case the results of Pyartli and Bryuno signify the existence of invariant manifolds filling, in the space with coordinates (ε, z_1, z_2), a surface whose equation starts with terms of the form

$$\varepsilon = z_1^{m_1} z_2^{m_2}.$$

Here ε is the deviation from resonance, while z_1 and z_2 are suitable (depending smoothly on ε) phase coordinates. For a fixed $\varepsilon \neq 0$ the invariant manifold constructed is homeomorphic to a cylinder. The linking coefficients of the direction circumference of this cylinder with coordinate axes in C^2 equal m_1 and m_2.

Let us show that the existence of a sufficiently large piece of such a resonance invariant manifold in a neighborhood of a nonresonant fixed point of a diffeomorphism hinders the linearization of the diffeomorphism in this neighborhood. Consequently, when resonance manifolds exist in any neighborhood of a fixed point the linearizing series diverge everywhere.

We remark that all mappings in C^2 with $|\lambda_1| > 1 > |\lambda_2|$ are topologically equivalent to each other. In particular, they can all be linearized and have many invariant cylinders. However, as we see, analytic invariant cylinders are a great rarity.

§3. Fundamental Construction

With a resonance invariant manifold of a mapping in C^2 we intend to connect an elliptic curve imbedded in a holomorphic surface. With this aim we introduce certain definitions.

3.1. Local Shifts. We consider the cylinder $S^1 \times \mathbf{R}$. The addition of 1 to the second coordinate is called a standard shift. Let D_0 be a domain on the cylinder, containing $S^1 \times [0, 1]$. The restriction of the standard shift on D_0 yields the diffeomorphism t: $D_0 \to D_1 = t(D_0)$. We note that the intersection $D_0 \cap D_1$ contains the circumference $S^1 \times 1$. We denote the union $D_0 \cup D_1$ by D.

Let M be a two-dimensional manifold, M_0 and M_1 be regions of it, and f: $M_0 \to M_1$ be a homeomorphism.

Definition 3.1.1. Homeomorphism f is called a local shift if there exists a homeomorphism h: $M \to D$ taking M_0 into D_0, M_1 into D_1, and f into t.

We obviously have

Proposition 1. Let M be a complex curve and f: $M_0 \to M_1$ be a holomorphic local shift. Then the pasting together of each point $z \in M_0$ and its image $f(z) \in M_1$ defines a compact complex curve homeomorphic to a torus, i.e., an elliptic curve $\Gamma = M/f$.

Let us now consider the direct product of the cylinder by the plane $(S^1 \times \mathbf{R}) \times \mathbf{R}^2$. Let T: $(S^1 \times \mathbf{R}) \times \mathbf{R}^2 \to (S^1 \times \mathbf{R}) \times \mathbf{R}^2$ be a shift by 1 along $\mathbf{R}$, E_0 be a neighborhood of $D_0 \times 0$, $E_1 = TE_0$, $E = E_0 \cup E_1$.

Let N be a real smooth four-dimensional manifold, $M \subset N$ be a two-dimensional submanifold, N_1 and N_2 be regions in $N = N_1 \cup N_2$, F: $N_1 \to N_2$ be a real homeomorphism.

Definition 3.1.2. Homeomorphism F is called a local shift of N along M if there exists a real homeomorphism H: $N \to E$ taking N_0 into E_0, N_1 into E_1, F into T, and M into D.

Proposition 2. Let N be a complex surface, $M \subset N$ be a complex curve, and F: $N_0 \to N_1$ be a holomorphic local shift. Then the pasting together of each point $z \in N_0$ and its image $F(z) \in N_1$ defines a holomorphic complex surface $\Sigma = N/F$ being a neighborhood of the elliptic curve $\Gamma = M/f$.

3.2. Construction of an Elliptic Curve on a Resonance Invariant Manifold of a Linear Transformation. Let A: $\mathbf{C}^2 \to \mathbf{C}^2$ be a linear transformation with eigenvalues λ_1, λ_2, $|\lambda_1| < 1 < |\lambda_2|$. We assume that the eigenvalues satisfy the resonance relation $\lambda_1^{m_1}\lambda_2^{m_2} = 1$, where m_1 and m_2 are relatively prime. Then the cylinder with equation $z_1^{m_1}z_2^{m_2} = 1$ (where z_1 and z_2 are the coordinates in the proper basis) is invariant relative to A.

We remark that the restriction of A on this cylinder yields a holomorphic shift. Indeed, we uniformize the curve $\lambda_1^{m_1}\lambda_2^{m_2} = 1$ in the parameter $\Lambda \neq 0$ by the formula $\lambda_1 = \Lambda^{m_2}$, $\lambda_2 = \Lambda^{-m_1}$; on the cylinder we introduce the parameter $Z \neq 0$ by the formula $z_1 = Z^{m_2}$, $z_2 = Z^{-m_1}$. Then the action of A on the cylinder takes the form

$$Z \mapsto \Lambda Z.$$

This transformation is a holomorphic shift since $|\Lambda| > 1$. The corresponding elliptic curve is $C^*/\{\Lambda\} \cong C/2\pi Z + \omega Z$, where $\Lambda = e^{i\omega}$.

We note that in the case of resonance the linear transformation has an integral one-parameter family of holomorphic invariant cylinders $z_1^{m_1} z_2^{m_2} = c\ (c \neq 0)$. The elliptic curves constructed on all these cylinders are isomorphic.

A suitable neighborhood of any such cylinder (or of a sufficiently large finite piece of it) is turned, under factorization with respect to the action of A, into a neighborhood of an elliptic curve on a complex surface. This surface is the direct product of the elliptic curve by C. Indeed, the homotheties $z \mapsto kz$ define a projection onto the elliptic curve, while the mapping $z \mapsto z_1^{m_1} z_2^{m_2}$ defines a projection onto the second factor.

In particular, the self-intersection index of the elliptic curve in the surface constructed equals 0.

3.3. Construction of an Elliptic Curve on a Resonance Invariant Manifold of a Nonlinear Transformation. Let $A(\varepsilon)$: $U \to C^2$ be a biholomorphic mapping of the region $U \subset C^2$ into C^2, depending holomorphically on a parameter ε. We assume that ε varies in a neighborhood of zero in **C** and that all the mappings $A(\varepsilon)$ leave the origin in C^2 in place.

By λ_1 and λ_2 we denote the eigenvalues of the linearization of mapping A(0) at O. We assume that $|\lambda_1| < 1 < |\lambda_2|$ and that $\lambda_1^{m_1} = \lambda_2^{m_2}$, where m_1 and m_2 are relatively prime.

For the family A of general position, when passing through resonance when $\varepsilon = 0$, an invariant holomorphic resonance cylinder branches off from a separatrix of the fixed point (see Paragraph 2.5). We fix a sufficiently small ε and we consider the restriction of $A(\varepsilon)$ on this cylinder. It can be verified that $A(\varepsilon)$ induces a local holomorphic shift on the corresponding part of the cylinder [this follows from: 1) The cylinder has in the first approximation the equation $z_1^{m_1} z_2^{m_2} = c(\varepsilon)$; 2) $A(\varepsilon)$ is close to the linearization of A(0) at O; 3) the linearization of A(0) at O acts on the cylinder $z_1^{m_1} z_2^{m_2} = c$ as a local shift (see Paragraph 3.2)].

Thus, for sufficiently small $|\varepsilon|$ the mapping $A(\varepsilon)$ defines an elliptic curve $\Gamma(\varepsilon)$ imbedded in a surface $\Sigma(\varepsilon)$. In the homologies of curve $\Gamma(\varepsilon)$ there is a distinguished circumference (the image of the cylinder's direction circumference). We can represent the curve $\Gamma(\varepsilon)$ as

$$\Gamma(\varepsilon) \approx C/2\pi Z + \omega(\varepsilon) Z,$$

where 2π corresponds to the distinguished circumference. The function $\omega(\varepsilon)$ has the limit ω_0 as $\varepsilon \to 0$. From the formulas in Paragraph 3.2 it follows that $\lambda_1 = e^{i\omega_0 m_2}$, $\lambda_2 = e^{-i\omega_0 m_1}$.

The self-intersection index of curve $\Gamma(\varepsilon)$ in surface $\Sigma(\varepsilon)$ equals zero (see Paragraph 3.2). Therefore, topologically $\Sigma(\varepsilon)$ is the direct product of $\Gamma(\varepsilon)$ by a disk. However, analytically $\Sigma(\varepsilon)$ is not necessarily a direct product. Furthermore:

1) $\Sigma(\varepsilon)$ may not be a holomorphic bundle over $\Gamma(\varepsilon)$; a neighborhood of $\Gamma(\varepsilon)$ in $\Sigma(\varepsilon)$ may not admit of holomorphic mappings onto $\Gamma(\varepsilon)$, identical on $\Gamma(\varepsilon)$.

It will be thus, e.g., when together with $\Gamma(\varepsilon)$ there is in $\Sigma(\varepsilon)$ a family of elliptic curves with various values of module ω

2) $\Gamma(\varepsilon)$ may not admit of deformations in $\Sigma(\varepsilon)$ other than shifts of $\Gamma(\varepsilon)$ with respect to itself.

This will be so, for instance, when the normal bundle of $\Gamma(\varepsilon)$ in $\Sigma(\varepsilon)$ is analytically nontrivial.

Positive results on the structure of $\Sigma(\varepsilon)$ are presented in §4.

3.4. Nonlinearizability of a Mapping in a Region Containing a Resonance Cylinder.

THEOREM 1. If an elliptic curve $\Gamma(\varepsilon)$ is undeformable in its own neighborhood in $\Sigma(\varepsilon)$, then the mapping A is not linearizable by a biholomorphic change of variables in a neighborhood of point O containing a part of the holomorphic invariant cylinder, needed for the construction of $\Gamma(\varepsilon)$.

Indeed, a holomorphic invariant cylinder of a linear mapping can always be deformed by means of a small homothety.

Thus, we obtain an estimate from above of the radius of convergence of the Poincaré–Siegel series in terms of a holomorphic invariant cylinder with a nontrivial normal bundle of the corresponding elliptic curve.

Another proof of the nonlinearizability was pointed out to me by Yu. S. Il'yashenko to whom I had communicated the preceding results.

THEOREM 2 (Il'yashenko). If a linear mapping has a holomorphic invariant cylinder whose direction circumference has the linking coefficients m_1 and m_2 with the proper axes, while the corresponding elliptic curve is

$$\mathbf{C}/2\pi\mathbf{Z} + \omega\mathbf{Z}$$

(moreover, 2π corresponds to the cylinder directrix), then the eigenvalues equal $\lambda_1 = e^{i\omega m_2}$, $\lambda_2 = e^{-i\omega m_1}$, and, consequently, there holds the resonance $\lambda_1^{m_1}\lambda_2^{m_2} = 1$.

Proof. Let z_1 and z_2 be proper coordinates. The differential forms dz_k/z_k on the cylinder are holomorphic and invariant relative to the mapping; consequently, they define holomorphic forms on the elliptic curve.

Let us compute the integrals of these forms with respect to the generators of the homology group of a torus. One of the generators corresponds to the direction circumference γ on the cylinder. For it

$$\oint dz_1/z_1 = 2\pi i m_2, \quad \oint dz_2/z_2 = -2\pi i m_1,$$

since the linking coefficient of γ with the axis $z_1 = 0$ equals m_2 and with the axis $z_2 = 0$ equals m_1. The second generator corresponds to a segment δ connecting the point z with its image Az on the cylinder's surface. For it

$$\int dz_1/z_1 = \ln \lambda_1, \quad \int dz_2/z_2 = \ln \lambda_2$$

(these relations define branches of logarithms). But all holomorphic forms on an elliptic curve are proportional. Thus,

$$\frac{\omega}{2\pi} = \frac{\ln \lambda_1}{2\pi i m_2} = -\frac{\ln \lambda_2}{2\pi i m_1},$$

QED.

COROLLARY 3.4.1. Let A be a diffeomorphism with a fixed point O and let $\lambda_{1,2}$ be the eigenvalues. Let us assume that in some neighborhood U of the fixed point there is a holomorphic cylinder invariant relative to A and let ω be the module of the corresponding elliptic curve. Then, if λ and ω are not connected by the relation in Theorem 2 (where m_1 and m_2 are linking coefficients with the separatrices of the singular point), then in region U diffeomorphism A is not analytically equivalent to a linear one.

3.5. Divergence of the Poincaré Series. From the results obtained follows

THEOREM 3. If in an arbitrarily small neighborhood of a nonresonant fixed point O of a local diffeomorphism in $\mathbf{C}^2$ there are holomorphic invariant cylinders on which the diffeomorphism acts as a local shift, then this diffeomorphism is analytically not equivalent to a linear one in any neighborhood of the fixed point O, and, consequently, the Poincaré series diverge everywhere.

Pyartli established that such an accumulation of invariant cylinders to the singular point is the general case for mappings whose eigenvalues are "abnormally well approximated by resonance ones." Consequently, for such eigenvalues the divergence of the Poincaré series everywhere also is the general case.

The results of this section carry over easily to vector fields in $\mathbf{C}^3$ close to a singular point of Siegel type. An elliptic curve is the materialization of the resonance $m_1\lambda_1 + m_2\lambda_2 + m_3\lambda_3 = 0$. The points of this curve are the phase curves of the field, lying on the invariant resonance surface

$$z_1^{m_1} z_2^{m_2} z_3^{m_3} = c\varepsilon + \ldots$$

§4. Structure of a Neighborhood of an Elliptic Curve

4.1. Preliminary Normal Form. We consider the elliptic curve

$$\Gamma = \{\varphi \in \mathbf{C} \bmod 2\pi \mathbf{Z} + \omega \mathbf{Z}\},$$

where $\omega = \sigma + i\tau$ ($\tau > 0$). In order to construct the imbedding of Γ into a holomorphic surface Σ, we consider the biholomorphic mapping

$$\begin{cases} r \\ \varphi \end{cases} \mapsto \begin{cases} r\lambda e^{ip\varphi}(1 + ra(r, \varphi)), \\ \varphi + \omega + rb(r, \varphi). \end{cases}$$

Here p is an integer, λ is a complex number not equal to 0, a and b are holomorphic functions defined in the product of the strip $|\mathrm{Im}\,\varphi| < \rho$ by the disk $|r| < \varkappa$ and periodic with period 2π in φ. By pasting together each point (r, φ) and its image, we obtain the neighborhood Σ of the elliptic curve Γ from a suitable sufficiently small neighborhood of the plane of the complex variable φ.

Proposition 4.1.1. Each pair (Γ, Σ), where Γ is an elliptic curve imbedded in a holomorphic surface Σ, is locally (close to Γ) biholomorphically equivalent to one of the pairs constructed above.

This proposition stems from the following facts a) and b).

a) Every holomorphic imbedding of a cylinder into a holomorphic surface is locally biholomorphically equivalent to an imbedding into a direct product of a cylinder by a complex straight line.

(This can be proved by a sequential normalization of jets of higher and higher order. V. P. Palamodov explained to me that this is a standard consequence of the fact that a ring is a Stein manifold. Il'yashenko gave a direct proof.)

b) All holomorphic one-dimensional vector bundles over elliptic curves are analytically equivalent to bundles resulting from mappings

$$\begin{cases} r \\ \varphi \end{cases} \mapsto \begin{cases} \lambda e^{ip\varphi} r \\ \varphi + \omega. \end{cases}$$

(A. N. Tyurin communicated to me that this result is due to N. H. Abel.)

Using a), we obtain Σ by a pasting together of the points

$$\begin{cases} r \\ \varphi \end{cases} \mapsto \begin{cases} rA(r, \varphi), \\ \varphi + \omega + rB(r, \varphi). \end{cases}$$

We apply b) to a normal bundle of Γ in Σ. We conclude that after a suitable change of coordinates, $A(0, \varphi)$ takes the form $\lambda e^{ip\varphi}$, which proves the proposition stated above.

The self-intersection index of curve Γ in Σ equals $-p$ if $\mathrm{Im}\,\omega > 0$. Consequently, a neighborhood of an elliptic curve with a zero self-intersection index in the complex plane is obtained by pasting together

$$\begin{cases} r \\ \varphi \end{cases} \mapsto \begin{cases} r\lambda(1 + ra(r, \varphi)), \\ \varphi + \omega + rb(r, \varphi). \end{cases}$$

4.2. Formal Normal Form.

Definition 4.2.1. A pair of numbers (λ, ω) is called a resonance pair if $\lambda^n e^{ik\omega} = 1$ for some integers n and k together not equal to 0.

We note that resonance pairs form an everywhere-dense set in the space of all pairs of complex numbers (λ, ω). The one-dimensional holomorphic vector bundle corresponding to a resonance pair (λ, ω) is trivial over some finite cyclic covering of the original elliptic curve.

Definition 4.2.2. The "mapping"

$$f\begin{pmatrix} r \\ \varphi \end{pmatrix} = \begin{pmatrix} rA(r, \varphi) \\ \varphi + \omega + rB(r, \varphi) \end{pmatrix},$$

where A and B are formal power series in r with coefficients analytic on the real φ axis and 2π-periodic in φ, $A(0, \varphi) \neq 0$, is called a formal pasting together.

The "mapping"

$$g\begin{pmatrix} r \\ \varphi \end{pmatrix} = \begin{pmatrix} rC(r, \varphi) \\ \varphi + rD(r, \varphi) \end{pmatrix},$$

where C and D are formal power series in r with coefficients analytic in the strip $0 \leqslant \operatorname{Im} \varphi \leqslant \operatorname{Im} \omega$ and 2π-periodic in φ, $C(0, \varphi) \neq 0$, is called a formal change of variables.

The formal change of variables g acts on the formal pasting together f by the formula $f \mapsto g \circ f \circ g^{-1}$ (the right-hand side is determined by a natural substitution of formal series and is itself a formal pasting together).

THEOREM 4.2.1. If (λ, ω) is a nonresonance pair, then every formal pasting together

$$\begin{cases} r \\ \varphi \end{cases} \mapsto \begin{cases} r\lambda(1 + ra(r, \varphi)), \\ \varphi + \omega + rb(r, \varphi) \end{cases}$$

is reduced by a formal change of variables to the linear normal form

$$\begin{cases} r \\ \varphi \end{cases} \mapsto \begin{cases} \lambda r, \\ \varphi + \omega. \end{cases}$$

LEMMA 4.2.1. Consider the equation in u:

$$\lambda^n u(\varphi + \omega) - u(\varphi) = v(\varphi),$$

where v is a 2π-periodic function analytic in the strip $\alpha \leqslant \operatorname{Im} \varphi \leqslant \beta$. If $\tau = \operatorname{Im} \omega > 0$, $\lambda \neq 0$, and $\lambda^n e^{ik\omega} \neq 1$ for all integers k, then the equation has a 2π-periodic solution u analytic in the strip $\alpha < \operatorname{Im} \varphi < \beta + \tau$.

Proof of Lemma 4.2.1. Let $v(\varphi) = \Sigma v_k e^{ik\varphi}$, $u(\varphi) = \Sigma u_k e^{ik\varphi}$. Then $u_k = \frac{v_k}{\lambda^n e^{ik\omega} - 1}$. As $k \to +\infty$, v_k decreases as $e^{k\alpha}$ but $e^{ik\omega}$ tends to zero. Hence, u_k decreases as $e^{k\alpha}$. As $k \to -\infty$, v_k decreases as $e^{-|k|\beta}$ but $e^{ik\omega}$ grows as $e^{|k|\tau}$ $(\tau = \operatorname{Im} \omega > 0)$. Consequently, u_k decreases as $e^{-|k|(\beta+\tau)}$. Hence, follows the convergence of the Fourier series for u when $\alpha < \operatorname{Im} \varphi < \beta + \tau$.

Proof of Theorem 4.2.1. Let $ra = r^n a_n(\varphi) + \ldots$, $rb = r^n b_n(\varphi) + \ldots$, where the dots denote terms of order higher than n. We make a formal change of variables, in which $C(r, \varphi) = 1 + r^n C_n(\varphi)$, $rD(r, \varphi) = r^n D_n(\varphi)$. A direct substitution convinces us that the coefficients of r^n in ra and in rb after the change have the form

$$\tilde{a}_n(\varphi) = a_n(\varphi) + \lambda^n C_n(\varphi + \omega) - C_n(\varphi),$$
$$\tilde{b}_n(\varphi) = b_n(\varphi) + \lambda^n D_n(\varphi + \omega) - D_n(\omega).$$

We find $C_n(\varphi)$ and $D_n(\varphi)$ from the equations $\tilde{a}_n(\varphi) = 0$, $\tilde{b}_n(\varphi) = 0$. By Lemma 4.2.1 these equations have solutions analytic in the strip $0 \leqslant \operatorname{Im} \varphi \leqslant \tau$. We have constructed a formal change of variables, after which the powers of the lower-order terms in ra and rb are increased.

Repeating this construction for n = 1, 2, . . ., we obtain a formal change of variables after which ra and rb vanish completely.

4.3. Analytic Normal Form

Definition 4.3.1. A pair of complex numbers (λ, ω), where $\operatorname{Im} \omega \neq 0$, and $\lambda \neq 0$, is said to be normal if constants $c > 0$ and $\nu > 0$ exist such that

$$|\lambda^n e^{ik\omega} - 1| > c(|n| + |k|)^{-\nu}$$

for all integers k and n $(n \neq 0)$.

It is easy to prove

Proposition 4.3.1. Nonnormal pairs form an everywhere-dense set of Lebesgue measure 0.

THEOREM 4.3.1. If (λ, ω) is a normal pair, then every holomorphic pasting together

$$\begin{cases} r \\ \varphi \end{cases} \mapsto \begin{cases} r\lambda(1 + ra(r, \varphi)), \\ \varphi + \omega + rb(r, \varphi) \end{cases}$$

is reducible to the linear normal form

$$\begin{cases} r \\ \varphi \end{cases} \mapsto \begin{cases} r\lambda, \\ \varphi + \omega \end{cases}$$

by a holomorphic change of variables.

Let us translate this theorem into the language of the imbeddings of an elliptic curve.

Definition 4.3.2. A holomorphic vector bundle ξ is said to be stiff if for every imbedding with the normal bundle ξ of the base into a complex manifold a sufficiently small neighborhood of the imbedded base is biholomorphically equivalent to a neighborhood of the null section of bundle ξ.

From the Proposition 4.3.1 and the Theorem 4.3.1 stated above follows

COROLLARY 4.3.1. Almost all (in the Lebesgue measure sense) one-dimensional vector bundles of degree 0 over an elliptic curve are stiff.

The proof of Theorem 4.3.1 is carried out by Newton's method, standard after Kolmogorov's papers (see [3, 4], e.g.). The author thanks Il'yashenko and Pyartli who derived the necessary estimates; they will be published elsewhere (for the more general situation when r and φ are multidimensional).

4.4. Materialization of Resonances. Let us now assume that a normal bundle of the elliptic curve Γ in surface Σ is a resonance normal bundle, i.e., in the notation introduced above,

$$\lambda^n e^{ik\omega} = 1.$$

for some $n > 0$ and for an integer k. Such a resonance is an exceptional phenomenon. But if we examine not one curve $\Gamma \subset \Sigma$, but a one-parameter family of imbeddings $\Gamma(\varepsilon) \subset \Sigma(\varepsilon)$, then the passage through resonance as ε varies is unremovable by a small stirring of the family. It turns out that under a passage through such a resonance in a general one-parameter family of imbeddings $\Gamma(\varepsilon) \subset \Sigma(\varepsilon)$ an elliptic curve, being a "materialization" of the resonance, branches off from the elliptic curve $\Gamma(0)$.

The equation of this curve (in the preceding notation) is

$$r^n e^{ik\varphi} = \varepsilon$$

[after choosing a suitable family parameter ε and after a suitable change of coordinates (r, φ), depending on ε; we take it that resonance corresponds to $\varepsilon = 0$ and that lower-order resonances are absent: $\lambda^m e^{il\varphi} \neq 1$ for $0 < m < n$].

We present here the derivation of the equation of the branched curve at the level of formal series. By arguing as in Paragraph 4.2, we reduce the pasting together to the form

$$\begin{cases} r \\ \varphi \end{cases} \mapsto \begin{cases} r\lambda\,(1 + \alpha\varepsilon + aw + A), \\ \varphi + \omega + \beta\varepsilon + bw + B, \end{cases}$$

where α, β, a, b are constants, $w = r^n e^{ik\varphi}$, A and B are series in powers of ε and w, starting with terms of degree 2. Such a substitution takes w into $w(1 + \gamma\varepsilon + cw + C)$, where $\gamma = n\alpha + ik\beta$, $c = na + ikb$, C are terms of second and higher degrees in ε and w.

The equation $\gamma\varepsilon + cw + C = 0$ defines the branched curve. For a family of general position $\gamma \neq 0$, $c \neq 0$. After a suitable change of coordinates ε and r this equation takes the form $r^n e^{ik\varphi} = \varepsilon$.

The convergence is investigated in just the same way as in the cited papers of Pyartli and Bryuno.

Remark 4.4.1. It is not difficult to verify that the condition $\lambda^n e^{ik\omega} = 1$ signifies precisely that the normal bundle is analytically trivial over some finitely sheeted cyclic covering of the elliptic curve.

Topologically all the bundles examined are trivial. In particular, the elliptic curve branched under resonance can be projected (nonholomorphically) onto the curve $\Gamma(\varepsilon)$ "along the r-direction." This projection is a topological finitely sheeted cyclic covering of a torus. It is the very covering over which the normal bundle becomes trivial at the instant of passing through resonance.

When (n, k) = d > 1 (but there are no lower-order resonances, $\lambda^m e^{il\omega} = 1, 0 < m < n$), the branched curve is unconnected. In this case it consists of d components each of which n/d-sheetedly topologically covers the original torus.

4.5. Divergence of the Linearization. For certain nonresonance bundles [i.e., pairs (λ, ω)] the series in Paragraph 4.2, taking the pasting together to a normal form, diverge. For specially selected holomorphic pastings together this can be verified by the methods of Siegel [9] or Bryuno [5] (also see [13]).

The branching of the curves materializing the resonance enables us to "explain" the divergence of the linearizing pasting together of the series in the following way. We assume that (λ, ω) is a nonresonance pair, but is very close to resonance. Then in a small neighborhood of the original elliptic curve there is, in general, another elliptic curve, precisely, the curve materializing the resonance. If the pair (λ, ω) is sufficiently close to an infinite number of resonances, then in an arbitrarily small neighborhood of the original elliptic curve there is an infinite number of curves materializing different resonances and cyclically covering the original curve.

Meanwhile a normal bundle of the original curve is a nonresonance normal bundle. Nonresonance normal bundles of degree 0 do not have sections over any cyclic finitely sheeted covering of the elliptic curve. Therefore, in a normal bundle of the original elliptic curve there are no elliptic curves cyclically covering the original curve. Consequently, no neighborhood whatsoever of the original curve on the surface can be mapped biholomorphically onto a neighborhood of the zero section of the normal bundle; therefore, the series in Paragraph 4.2 diverge for pastings together of general position if the pair (λ, ω) is too well approximated by resonance pairs.

4.6. Generalizations. The results of Paragraph 4.3 admit of generalization to the case of a multidimensional r (i.e., to the case of the imbedding of the elliptic curve in a multidimensional manifold with a "purely zero" normal bundle). Our methods also permit us to give a simple proof, for the case of elliptic curves, of the well-known (Grauert) theorem on the stiffness of negative bundles [7]. Clearly, stiffness holds with probability 1 also for the sums of negative and zero bundles.

The question on the normal form of neighborhoods of curves of genus greater than 1 also deserves investigation (in this case we should examine a Schottky uniformization instead of a ring).

It is also possible to generalize to the case of a multidimensional variable φ. A special case of one such generalization is Siegel's theorem on the linearization of a mapping in a neighborhood of a fixed point.

Note as well the analogy between compact complex submanifolds of analytic manifolds and the limit cycles of differential equations; in a manner similar to how a limit cycle can vanish only if the monodromy operator has an eigenvalue 1, an elliptic curve on a surface, having a zero self-intersection index, does not vanish under small deformations of the enveloping surface if the normal bundle is analytically nontrivial. F. A. Bogomolov pointed out the following general formulation to me: A compact submanifold of a complex manifold does not vanish under a small deformation of the enveloping manifold if the one-dimensional homologies of the normal sheaf is trivial.

LITERATURE CITED

1. A. A. Andronov and E. A. Leontovich, "Certain cases of the dependence of limit cycles on a parameter," Uch. Zap. Gor'k. Univ., 3 (1939); Collected Works of A. A. Andronov [in Russian], Izd. Akad. Nauk SSSR (1956), p. 188.
2. V. I. Arnol'd, "Remarks on the singularities of root codimension in complex dynamic systems," Funkts. Anal. Prilozhen., 3, No. 1, 1-6 (1969).
3. V. I. Arnol'd, "Singularities of smooth mappings," Usp. Mat. Nauk, 23, No. 1, 3-44 (1968).
4. V. I. Arnol'd, "On the mappings of a circumference onto itself," Izv. Akad. Nauk SSSR, Ser. Matem., 25, No. 1, 21-86 (1961).
5. A. D. Bryuno, "Analytic form of differential equations," Trudy Mosk. Matem. Obshch., 25, 120-262 (1971).

6. A. D. Bryuno, "Normal form of differential equations with a small parameter," Matem. Zametki, 16, No. 3, 407-414 (1974).
7. H. Grauert, "Über Modifikationen und exzeptionelle analytische Mengen," Ann. Math., 146, 331-368 (1962).
8. C. L. Siegel, "Iteration of analytic functions," Ann. Math., 43, 607-612 (1942).
9. C. L. Siegel, "On the integrals of canonical systems," Ann. Math., 42, 806-822 (1941).
10. H. Poincaré, Thèse (1879); Oeuvres, t. 1, Paris (1928).
11. A. S. Pyartli, "Generation of complex invariant manifolds close to a singular point of a vector field depending on a parameter," Funkts. Anal. Prilozhen., 6, No. 4, 95-96 (1972).
12. E. Hopf, "Abzweigung einer periodischen Lösung von einer stationären Losung," Berich. Sächs. Akad. Wiss., Leipzig, Math. Phys. Kl., 94, No. 19, 15-25 (1942).
13. A. Ogus, "The formal Hodge filtration," Invent. Math., 31, No. 3, 193-228 (1976).

FRACTIONAL POWERS OF OPERATORS AND HAMILTONIAN SYSTEMS

I. M. Gel'fand and L. A. Dikii

In [1] it was discovered that the nonlinear Korteweg–de Vries (KdV) equation admits in some sense of an exact integration procedure. The principal bases of this procedure were clarified in [2]: It was connected with the problem of seeking differential operators of any order for the Sturm–Liouville operator $L = -d^2/dx^2 + u(x)$, which would commute in the maximally possible way with L. Usually this is called the construction of Lax's L,A-pairs.* In [3, 4] it was shown that the KdV equation is a Hamiltonian system having a complete collection of first integrals in involutions. Already in [2] it was noticed that instead of a second-order operator L we can choose higher-order operators and look for those which pair with them. An algorithm for this was proposed in [5]. This was done in the modern way in [6] on the basis of a development of the technique in [7]. As a result, a system of equations was constructed generalizing the KdV equation and the complete integrability was proved of the corresponding stationary time-independent equations (also see survey [8]). As far as we know, the Hamiltonian mechanics of these systems, analogous to that for the case of second-order operators L, has not been constructed anywhere.

In the present article we shall show that by a sequential application of the technique suggested in the authors' previous articles [9, 10] we can construct a theory of generalized systems of the KdV type, including the Hamiltonian structure.

1. Ring of Polynomials of $u_k(x)$, $u_k'(x)$, ... By A we denote the ring of polynomials of several functions $u_k(x)$ and their derivatives of any order. The algebra and the variational calculus in such a ring of one function were presented in detail in [9]. Here we list briefly the information needed for the case of any finite number of functions. Differentiations or "vector fields" $\sum b_{k,i}\frac{\partial}{\partial u_k^{(i)}}$, $b_{k,i}\in A$ act in the ring. The collection of differentiations is named TA. There is one preferred differentiation $\frac{d}{dx}=\sum u_k^{(i+1)}\frac{\partial}{\partial u_k^{(i)}}$. Let dA/dx be the set of elements of A, representable in the form df/dx, $f\in A$. We set $\tilde{A} = A/(dA/dx)$. The elements of $\tilde{A}$ are called functionals [if we examine some boundary conditions on $u_k(x)$, making it possible to talk about the integrals $\int f\,dx$, e.g., the condition of dying out at $\pm\infty$ or of periodicity, etc., then a one-to-one correspondence exists between the equivalence

*In what follows we shall call them P,L-pairs since, firstly, the letter A will be firmly occupied, and, secondly, it honors P. Lax.

Moscow State University. Translated from Funktsional'nyi Analiz i Ego Prilozheniya, Vol. 10, No. 4, pp. 13-29, October-December, 1976. Original article submitted June 22, 1976.

LOSS OF STABILITY OF SELF-OSCILLATIONS CLOSE TO RESONANCE AND VERSAL DEFORMATIONS OF EQUIVARIANT VECTOR FIELDS

V. I. Arnol'd

UDC 517.9+513.83+513.88

A connection exists between 1) problems on the loss of stability of a periodic motion and 2) problems on the bifurcations of a phase portrait invariant relative to a finite rotation group of a plane.

More precisely, an equivariant vector field on a plane is obtained as a result of averaging along a Seifert foliation formed by fast-motion trajectories under resonance.

In this paper we examine the problem on equivariant fields and their topologically versal deformations in the functional space of all equivariant fields. These results yield a first approximation for the stability loss problem.

1. AVERAGING IN A SEIFERT FOLIATION

A partitioning of the variety $\mathbf{C} \times S^1$ into the integral curves of the equation $dz/dt = ip/q$ is called a Seifert foliation of type (p, q). Here $z \in \mathbf{C}$, $t \bmod 2\pi \in S^1$, and p and q are primes.

Let us consider a q-sheeted covering of the space of a Seifert foliation. The foliation induced on it is diffeomorphic to the Seifert foliation of type $(p, 1)$, i.e., to the trivial bundle with base $\mathbf{C}$ and with the fiber a circle.

Let v be a vector field in the variety $\mathbf{C} \times S^1$ of the Seifert foliation of type (p, q). A vector field on base $\mathbf{C}$ of the q-sheeted covering obtained by an averaging along the fibers of a covering bundle from a field on a covering space, induced by field v, is called the averaging of v along the Seifert foliation.

Proposition 1. A field averaged along a Seifert foliation of type (p, q) is invariant to a rotation of $\mathbf{C}$ by an angle $2\pi/q$.

Proof. A Seifert foliation can be obtained from the cylinder $\mathbf{C} \times I$ by pasting the bases together after a rotation by the angle $2\pi p/q$. In this model the averaging is reduced to the averaging over the q segments. A rotation of $\mathbf{C}$ by the angle $2\pi/q$ leads only to a permutation of the segments, whence follows the equivariance of the averaged field.

From the proof of Proposition 1 it follows that to study the loss of stability of cycles close to resonances it is necessary to investigate the loss of stability of the positions of equilibrium of the equivariant fields.

2. PRINCIPAL EQUIVARIANT FIELDS

Let F be a vector field on the plane of a complex variable z. We treat F as a complex-valued (not necessarily holomorphic) function on $\mathbf{C}$. The Taylor series of this function at zero can be written in the form $\sum F_{k,l} z^k \bar{z}^l$. The following is obvious.

Proposition 2. Let the field F go into itself under a rotation of plane $\mathbf{C}$ through an angle $2\pi/q$. Then the coefficients $F_{k,l}$ are nonzero only when $k - l$ is congruent to 1 modulo q.

COROLLARY 1. Differential equations invariant relative to rotations through an angle $2\pi/q$ have the following form:

$$\dot{z} = zA(|z|^2) + B\bar{z}^{q-1} + O(|z|^{q+1}) \quad (q > 2). \tag{1}$$

Moscow State University. Translated from Funktsional'nyi Analiz i Ego Prilozheniya, Vol. 11, No. 2, pp. 1-10, April-June, 1977. Original article submitted December 6, 1976.

Originally publ. in: Funkts. Anal. Prilozh. 11(2), 1-10, © Russ. Acad. Sciences 1977
English transl.: Funct. Anal. Appl. 11(2), 85-92, © Consultants Bureau, New York 1977
Reprint by Springer-Verlag Berlin Heidelberg 2016 with kind permission.

Definition 1. An Eq. (1) in which the term $O(|z|^{q+1})$ has been dropped is called a principal equation invariant relative to a rotation through the angle $2\pi/q$. The right-hand side of a principal equation is called the principal q-equivariant field.

Example 1. Principal equations invariant relative to rotation groups of third and fourth order have the respective forms

$$\dot{z} = \varepsilon z + Az|z|^2 + B\bar{z}^2, \quad \dot{z} = \varepsilon z + Az|z|^2 + B\bar{z}^3. \tag{2}$$

To formulate the results of the investigations of surgeries on phase portraits of equivariant vector fields dependent on parameters, it is convenient to introduce the following definitions.

3. EQUIVARIANT VERSAL DEFORMATIONS

Let us consider a family v_λ of vector fields invariant relative to the action of a group G on the phase space and dependent on the parameter λ belonging to a neighborhood of the point O of space $\mathbf{R}^k$ (called the base of the family). The number of parameters in the family is called the dimension of the base. The germ of the family at the point $\lambda = 0$ is called the equivariant deformation of field v_0.

Definition 2. An equivariant deformation v_λ is called an equivariantly topologically orbitally versal (versal, for short) deformation of field v_0 if for any other equivariant deformation w_μ of the same field there exists a continuous mapping φ of the bases of the deformation and a family, depending continuously on μ, of homeomorphisms h_μ of the phase space, commuting with the action of G, such that h_μ takes the phase curves of field w_μ into the phase curves of field $v_{\varphi(\mu)}$ with the motion direction preserved.

In other words, an equivariant deformation is versal if every other equivariant deformation is topologically orbitally equivalent to the deformation induced by the versal deformation.

Analogous definitions can be given for the germs of vector fields, as well as for deformations in the class of fields with special properties (e.g., having a fixed linear part at a singular point). We now go on to the construction of versal deformations.

4. PRINCIPAL DEFORMATIONS

We consider a vector field on a plane, invariant relative to a rotation through the angle $2\pi/q$, $q > 2$.

Definition 3. A field is called singular if the linear part of the field at zero equals 0. A two-parameter family of fields $v_\varepsilon = \varepsilon z + v_0$, where the parameters are the real and the imaginary parts of the complex number ε, is called the principal deformation of the q-equivariant principal singular field v_0 ($q > 2$).

Example 2. In the cases $q = 3$ and $q = 4$ the principal deformations are yielded by Eqs. (2) in which ε is treated as a parameter and the complex coefficients A and B are fixed.

Remark 1. The objects defined earlier arise in the study of the loss of stability of a cycle as a pair of (complex-conjugate) multipliers passes through the unit circle. Systems with such a transition form a hypersurface in the functional space of all systems. This hypersurface is one of three hypersurfaces bounding the stability domain. The other two hypersurfaces correspond to the passage of one multiplier through the unit circle at the point 1 and at the point -1, respectively.

The boundary of the hypersurface corresponding to the passage of a complex pair of multipliers consists of two parts (of two surfaces of codimension two in the functional space of all systems). One of these surfaces of codimension 2 corresponds to the pair of multipliers, equal to 1, forming a Jordan cell of second order, while the other, to the same cell with two eigenvalues -1.

The study of stability loss in a neighborhood of these boundary surfaces of codimension two leads to the investigation of bifurcations in two-parameter families of vector fields on a plane, having a linear part nilpotent to a Jordan cell of second order and symmetric relative to rotations through an angle $2\pi/q$, $q = 2$ (for the multipliers -1) or $q = 1$ (for the multipliers 1). In order to include the consideration of these cases into the general scheme, it is convenient to give the following definitions.

Definition 4. A field, invariant relative to a rotation of the plane through an angle $2\pi/q$, $q = 1$ or 2, is called singular if its linear part at zero is nilpotent to a second-order Jordan cell.

In other words, when $q = 1$ or $q = 2$ a singular field is that field whose linear part is the field of the phase velocity of the equation $\ddot{x} = 0$ phase plane $(x, y = \dot{x})$. It is easy to prove the following proposition.

Proposition 3. A singular field, invariant relative to the rotation of the plane through an angle $2\pi/q$, $q = 1$ or $q = 2$, is led by a diffeomorphism commuting with rotation to the field of the phase velocities of the equations

$$\ddot{x} = ax^3 + bx^2y + O(|x|^5, |y|^5) \quad (q = 2),$$
$$\ddot{x} = ax^2 + bxy + O(|x|^3, |y|^3) \quad (q = 1).$$

Definition 5. The equations

$$\ddot{x} = ax^3 + bx^2y \ (q = 2), \quad \ddot{x} = ax^2 + bxy \ (q = 1)$$

and the vector fields giving them on the phase plane are called the principal singular equations and fields, respectively, for $q = 2$ and $q = 1$.

Definition 6. The deformation consisting in the adding on the right-hand side of a second order the terms $\alpha x + \beta y$ ($q = 2$) and $\alpha + \beta x$ ($q = 1$) is called the principal deformation of a principal singular field for $q = 2$ and $q = 1$, respectively.

Definition 7. A resonance (p, q) is said to be strong if $q \leq 4$.

We list the principal deformations of q-equivariant fields in cases of strong resonance:

$$\dot{z} = \varepsilon z + Az|z|^2 + B\bar{z}^3, \quad q = 4, \qquad \ddot{x} = \alpha x + \beta y + ax^3 + bx^2y, \ q = 2,$$
$$\dot{z} = \varepsilon z + Az|z|^2 + B\bar{z}^2, \quad q = 3, \qquad \ddot{x} = \alpha + \beta x + ax^2 + bxy, \quad q = 1.$$

Here the variables z, ε, A, and B are complex and x, y, α, β, a, and b are real; we have denoted the deformation parameters by Greek letters; $y = \dot{x}$.

5. VERSALITY OF PRINCIPAL DEFORMATIONS

THEOREM 1. For each q all the principal singular fields can be separated into degenerate and nondegenerate ones so that:

1) the degenerate fields form the union of a finite number of subvarieties in the space of principal singular fields;

2) the nondegenerate fields form the union of a finite number of open connected domains;

3) the principal deformation of the germ of a nondegenerate field at zero is versal;

4) the principal deformations of the germs of nondegenerate fields at each component of connectivity are topologically equivalent.

The word theorem is used here in the sense it usually has in the works by Thom, Smale, et al. To be precise, Theorem 1 has not been proved.

Except for the case $q = 4$, the nondegeneracy conditions can be written out explicitly:

$$a \neq 0, \ b \neq 0 \ \text{for} \ q = 1, 2; \quad \operatorname{Re} A(0) \neq 0, \quad B \neq 0 \ \text{for} \ q = 3.$$

For $q = 4$ at least the conditions

$$|A|^2 \neq |B|^2, \quad |\operatorname{Re} A| \neq |B|, \quad |\operatorname{Im} A| \neq (|B|^2 + \operatorname{Re}^2 A)/|\sqrt{|B|^2 - \operatorname{Re}^2 A}$$

should be added onto the previous ones. From Theorem 1 it is easy to derive the following corollary.

COROLLARY 2. In the functional space of two-parameter families of vector fields invariant relative to the group of rotations through angles that are multiples of $2\pi/q$, an open everywhere-dense set is formed by the families for which the fields are singular only for individual isolated values of the parameters and in the neighborhood of these values the family is topologically equivalent to a versal deformation of a nondegenerate principal singular field.

In other words, let the eigenvalues of the linearization of a vector field on a plane, invariant relative to rotations through angles that are multiples of $2\pi/q$, equal zero. Let us consider a field of general position with the properties mentioned. From it we form a two-parameter deformation of general position in the class of fields with the same symmetry.

We assert that a neighborhood (independent of the parameters) of point O exists such that the deformation constructed can be led by a homeomorphism with the same symmetry, depending continuously on the parameters, to the normal form indicated in Sec. 4. (More precisely, the homeomorphisms lead the phase portraits

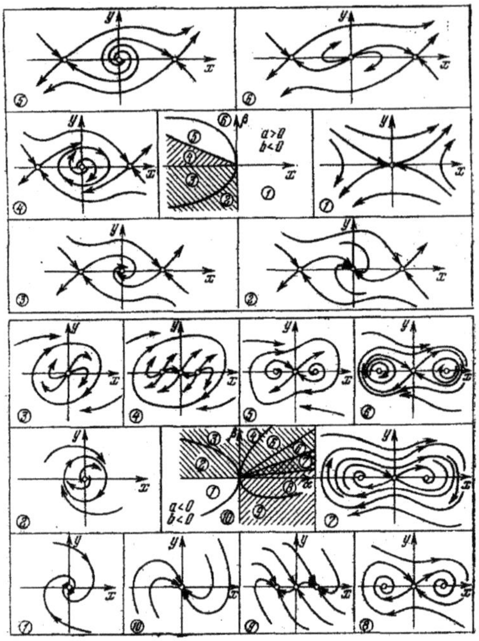

Fig. 1

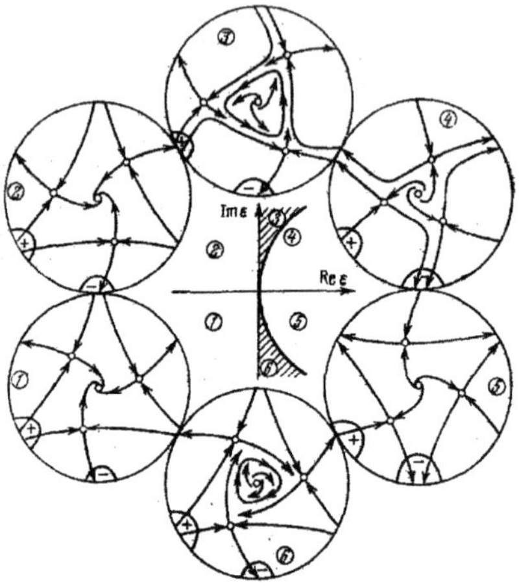

Fig. 2

of the corresponding system to a normal form.)

Thus, Theorem 1 formulated earlier reduces the description of all bifurcations to the description of bifurcations in principal deformations of nondegenerate singular fields.

6. DESCRIPTION OF THE BIFURCATIONS

In case q = 1, Theorem 1 formulated earlier was proved in 1971 by Bogdanov (see [1-3] wherein the bifurcations are described).

In case q = 2 we can achieve b < 0 by time changes. The bifurcation diagram (the "dial-plate") on the plane (α, β) and the surgery on the phase portrait for the cases $a > 0$ and $a < 0$ are shown in Fig. 1.

In case q = 3 we can achieve Re A < 0 by time changes. The bifurcation diagram and the surgeries are shown in Fig. 2. Figure 3 shows certain variants of the surgeries for q = 4. When q ≥ 5 we can achieve Re A < 0 by time changes; the bifurcation diagram and the surgeries (Fig. 4) are virtually known. We note that the zone of existence of fixed points approach the imaginary axis of ε by a narrow tongue whose boundaries have a common tangent:

$$\operatorname{Im} \varepsilon \approx P(\operatorname{Re} \varepsilon) \pm Q\,|\operatorname{Re} \varepsilon|^{(q-2)/2}, \quad q \geqslant 5.$$

The proofs of Theorem 1 and of the assertions made earlier are simple in case q ≥ 5, are contained in [2, 3] in case q = 1, are unknown in case q = 4, and are outlined later in cases q = 2, 3.

7. CASE OF SYMMETRY OF THIRD ORDER

1°. Let A = 0; in this case the system is obtained by some Hamiltonian rotation of a field. More precisely, we consider the equilateral triangle formed by the singular points (the saddles). The sides of this triangle define three linear inhomogeneous functions. The product of these three functions is the Hamiltonian function needed.

The sign of the derivative of this function relative to our system $\dot{z} = \varepsilon z + B\bar{z}^2$ is determined by the sign of the real part of parameter ε. This enables us to use the function indicated as a Lyapunov function. Thus, when A = 0 the principal deformation can be used without difficulty.

Fig. 3

2°. In the general case, by the changes $t = T/|\varepsilon|$ and $z = Z|\varepsilon|$ we bring the system to the form: $\dot{Z} = EZ + \overline{Z}^2 + |\varepsilon| AZ|Z|^2$, $E = \varepsilon/|\varepsilon|$.

In the region in which $|Z|$ is small in comparison with $1/\varepsilon$, the third term can be looked upon as a small perturbation. In the region where the argument E is not close to $\pm\pi/2$, this perturbation does not change the picture obtained in 1°. However, if the real part of E is small in comparison with the complex part, then the system can be treated as a slightly perturbed Hamiltonian system $Z = \pm iZ + \overline{Z}^2$.

The unperturbed Hamiltonian function H was described in 1°.

3°. Let us compute the rate of change of H along the solutions of the system. Integrating along the level line $H = h$, we obtain the condition for the generation of a cycle from this line in the form $\oint \mu r^2 + \lambda r^4 d\varphi = 0$, where $\mu = -\sigma/\tau$, $\lambda = a\tau$, $\varepsilon = \sigma + i_\tau$, $A = -a + ib$; r and φ are polar coordinates on the plane.

By ρ we denote the radius of inertia of the domain bounded by the elliptic curve $H = h$. Then the condition for the generation of a cycle precisely from the line $H = h$ as $|\varepsilon|$ passes through zero can be written in the form $\sigma = \rho^2\tau^2 a$.

The most feasible value of ρ corresponds to the triangle formed by the separatrices of the saddle. The function $\rho(h)$ is monotone.

4°. The subsequent justification of the versality of the principal family and of the form of the bifurcation diagrams and phase portraits is carried out as in [2, 3].

8. CASE OF SYMMETRY OF SECOND ORDER

1°. By extensions of x and by changes of time and also of the parameters the family can be brought to the form

$$\dot{x} = \alpha x + 2\beta y + a x^3 + b x^2 y, \quad a = \pm 1, b = -2.$$

We investigate this family with parameters (α, β).

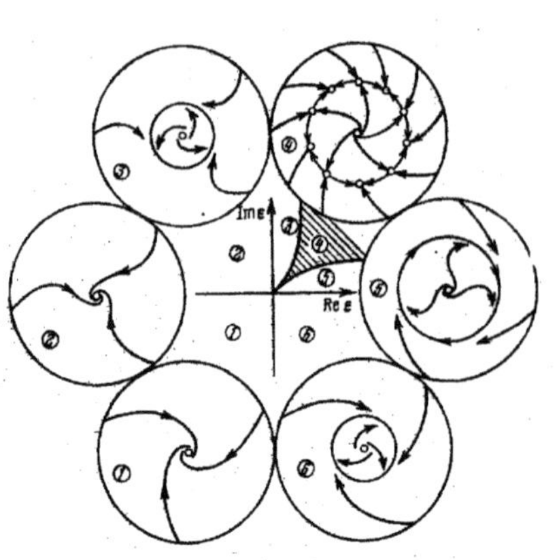

Fig. 4

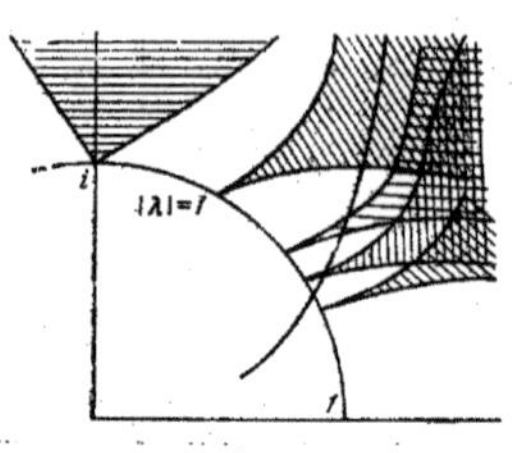

Fig. 5

2°. If $|\beta| \ll \sqrt{\alpha}$, then we make the changes $x = \sqrt{|\alpha|/|a|}x'$ and $t = t'/\sqrt{|\alpha|}$, reducing $|\alpha|$ and $|a|$ to 1. We obtain an almost-Hamiltonian system. The Hamiltonian function has the form $H = y^2/2 - \operatorname{sgn}\alpha\, x^2/2 - \operatorname{sgn} a x^4/4$. The dissipative terms have the form $2\beta' y + b' x^2 y$, $\beta' = \beta\sqrt{|\alpha|}$, $b' = \sqrt{|\alpha|}b/|a|$.

3°. By integrating the rate of change of H along the level line H = h, we obtain the condition for the generation of a cycle from this very line as α and $\beta = u|\alpha|$ pass through zero: $u = r^2$, where

$$r^2 = \iint x^2\,dx\,dy \Big/ \iint dx\,dy$$

is the square of the radius of inertia relative to the y axis of the domain bounded by line H = h.

4°. If $|\alpha| \ll |\beta|$, then we make the change $x = \lambda x'$ and $t = \varkappa t'$, taking $|a|$ to 1 and $|b|$ to $|\beta|$: $\lambda = \sqrt{|\beta/b|}$, $\varkappa = \sqrt{|b/\beta|}$. We obtain $\alpha' = |b/\beta|\alpha$, $b' = \sqrt{|\beta/b|}$, $\beta' = \sqrt{|b/\beta|}\beta$, $a' = a$;

$$\ddot{x} = ax^3 + \beta' 2y(1 \pm x^2) + \alpha' x.$$

Here both parameters $\beta' \sim \sqrt{|\beta|}$ and $\alpha' \sim \alpha/\beta$ are small. With $\beta' = 0$ we obtain the Hamiltonian system

$$H = \frac{y^2}{2} - \frac{\alpha' x^2}{2} - \frac{ax^4}{4}.$$

Subsequent investigation is carried out in the usual fashion.

9. ZEROS OF ELLIPTIC INTEGRALS

It was shown earlier that the investigation of the behavior of the cycles in out families is reduced to the solving of special cases of the following "weakened 16th problem of Hilbert":

Let H be a real n-th-degree polynomial and P be a real m-th-degree polynomial in variables x and y. How many zeros can the function

$$I(h) = \iint_{H \leqslant h} P\,dx\,dy$$

have?

For the investigation of symmetries of second order we need the case $P = \mu + \lambda x^2$ and $H = y^2/2 - \alpha x^2/2 - ax^4/4$. When $P = \mu + \lambda x^2$ the question reduces to the investigation of the monotonicity of the function r(h), where r is the radius of inertia relative to the y axis of the domain bounded by the cycle.

LEMMA 2. The function r(h) on the intervals between the critical values of H goes into itself as shown

TABLE 1

Values of α and a	$-1,+1$	$+1,-1$	$+1,-1$	$-1,-1$
Interval h	0,1/4	−1/4,0	$0,+\infty$	$0,+\infty$
Behavior of r	↑	↓	↓↑	↑

in Table 1 [in the third case r(h) decreases at first and then increases].

Il'yashenko proved this lemma without requiring calculations.

10. THE SUCCESSOR FUNCTION

The applications of our constructs to the investigation of stability loss of a cycle are based on the following.

LEMMA 3. We consider the map f: $(R^2, O) \to (R^2, O)$, having a fixed point O with the eigenvalues $e^{\pm 2\pi ip/q}$ (and with a Jordan cell of second order if q = 1 or q = 2). Then for any N the iteration f^q can be represented in a sufficiently small neighborhood of point O in the form of a sum $f^q = g + h$, where $h(z) = O(|z|^N)$, while g is a transformation of the phase flow of the vector field invariant relative to the finite cyclic group of diffeomorphisms γ.

We note, in particular, that g commutes with rotation through an angle $2\pi/q$. The map f^q itself, in general, does not commute with the action of any finite group whatever and is not included in the flow. Lemma 3 shows however that at the level of formal series f^q is included in the flow and commutes with a finite group.

The proof of Lemma 3 is carried out by the usual scheme of constructing the Poincaré–Dulac–Birkhoff forms.

LEMMA 4. We consider a deformation f_λ of the map $f_0 = f$, satisfying the hypotheses of Lemma 3. Then for any N the iteration f_λ^q can be represented in a sufficiently small neighborhood of the point O in the form of a sum $f_\lambda^q = g_\lambda + h_\lambda$, where $h_\lambda(z) = O(|z|^N)$, while g_λ is a transformation of the phase flow of a vector field v_λ invariant relative to a finite cyclic group of diffeomorphisms γ_λ. (Here f_λ, g_λ, h_λ, v_λ, and γ_λ depend smoothly on parameter λ varying in a neighborhood of zero.)

The proof is based on the fact that the reduction to normal form of the terms of degree no higher than N, with the resonance terms left as they are, is effected smoothly dependent on the parameters of the diffeomorphism.

By combining Lemma 4 with the description of the bifurcations of the phase flows from the preceding sections, we obtain information on the loss of stability of the fixed point O of map f (or of the periodic motion for which f is the successor function).

Remark 2. A family of differential equations in a neighborhood of a (p, q)-resonance periodic solution in the space of the q-sheeted covering also can be led directly to a normal form [4]. The application of the standard Poincaré–Dulac–Birkhoff method in this situation leads the family of vector fields, $2\pi q$-periodic in time, to the sum of a q-symmetric field not depending on time and of a remainder $O(|z|^N)$.

11. DISCUSSION

1°. To translate the results obtained into the language of bifurcations of periodic solutions, it is necessary to replace the fixed points found on the plane by closed trajectories in space, which are separatrices of the points, by the attracting and repelling surfaces of these closed trajectories which are limit cycles on the plane, and by invariant tori (cf. [5, 6]). The only essential difference will be in the surgeries on the separatrices: whereas on the plane the separatrices under bifurcations instantly pass one through the other, in space this process is stretched out, with the formation of the usual homoclinic (or heteroclinic) pictures.

Invariant tori in space collapse earlier than a cycle on a plane becomes a loop of separatrices; however, all these purely three-dimensional effects are weak (take place on terms of arbitrarily high degree in the normal forms) in comparison with the two-dimensional ones considered earlier.

2°. The consideration of stability loss as a two-parameter, and not a one-parameter, phenomenon enables us to easily undestand certain circumstances which otherwise seem to be astonishing.

Let us consider a two-parameter family in which the multiplier itself is taken as a parameter. On the plane of parameter values we draw the region of existence of periodic solutions subtended with q turns along the fundamental solution, having made p turns across it. This region goes onto the unit circle at the point $e^{2\pi ip/q}$ as a narrow (when q > 4) tongue (its width at a distance σ from the circle is of the order of $\sigma^{(q-2)/2}$; see Sec. 6). Therefore, a curve of general position of the plane of the multiplier intersects an infinite number of tongues close to the unit circle (Fig. 5).

Consequently, in a general one-parameter family in which the cycle loses stability without a strong resonance, long-period cycles get born and die in infinite number close to the instant of loss of stability.

A proof of this fact, independently of whether there is a (weak) resonance at the instant of stability loss, has recently been obtained by Kozyakin [7].

3°. By examining curves of general location of the bifurcation diagrams of strong resonances, presented in Sec. 6, we can describe sequences of surgeries that are universal but seem to be nonlocal from the one-parameter point of view.

For example, in case q = 2 one of the possibilities is the following sequence of events: a stable cycle mildly loses stability with the formation of a torus on which an overwinding is rapidly formed so that the form of the meridian of the torus approximates to a figure eight; on passing to the center of the figure eight (where the unstable cycle is located) the attracting set, remaining close to the torus with the meridian almost tightened down to a figure eight, collapses close to the homoclinic separatrix. In this case the phase trajectory twists around now one, then the other half of the collapsed torus, jumping from one side to the other randomly. This description is similar to the phenomenon observed in a numerical experiment by Gertsenshtein and Shmidt [8].

(It seems that a part of the results of this article was announced by F. Takens in a 1974 preprint unavailable in Moscow.)

LITERATURE CITED

1. V. I. Arnol'd, "Lectures on bifurcations and versal families," Usp. Mat. Nauk, 27, No. 5, 119-185 (1972).
2. R. I. Bogdanov, "Versal deformations of a singular point of a vector field on a plane in the case of zero eigenvalues," Proceedings of the I. G. Petrovskii Seminar [in Russian], No. 2, Moscow State Univ. (1976), pp. 37-65.
3. R. I. Bogdanov, "Bifurcations of a limit cycle of one family of vector fields on a plane," Proceedings of the I. G. Petrovskii Seminar [in Russian], No. 2, Moscow State Univ. (1976), pp. 23-35.
4. A. D. Bryuno, "Instability in a Hamiltonian system and the distribution of asteroids," Mat. Sb., 83, 213-312 (1970).
5. R. J. Sacker, "On invariant surfaces and bifurcation of periodic solutions of ordinary differential equations," Report IMMNYU 333, New York Univ. (1964).
6. R. J. Sacker, "A new approach to the perturbation theory of invariant surfaces," Commun. Pure Appl. Math., 18, No. 4, 717-732 (1965).
7. V. S. Kozyakin, "Subfurcation of periodic oscillations," Dokl. Akad. Nauk SSSR, 232, No. 1, 25-27 (1977).
8. S. Ya. Gertsenshtein and V. M. Shmidt, "Nonlinear development and interaction of perturbations of finite amplitude under convective instability of a rotating flat layer," Dokl. Akad. Nauk SSSR, 225, No. 1, 59-62 (1975).

Some Problems in the Theory of Differential Equations

V.I. Arnold

(Published in the collection "Unsolved Problems in Mechanics and Applied Mathematics", Moscow Univ. Press, Moscow, 1977, 3–9)

Many problems discussed below are of general character. However, at the end I am formulating several independent isolated problems.

1 Intrinsically unstable limiting regimes

During the last 10–15 years in the qualitative theory of ordinary differential equations one observed substantial progress comparable with the rapid development of this theory in the works of Poincaré. It is well known that Poincaré's limit cycles had no applications for several decades. However, when demands of radio engineering required a mathematical theory of nonlinear auto-oscillations, it turned out that the necessary toolbox has been already supplied by Poincaré in a complete form.

The modern achievements of the qualitative theory are now in the same position that the theory of limit cycles was in before the advance of radio engineering: the mathematical toolbox has been already developed and is waiting for its applications. Here, I am going to briefly discuss what the aforementioned achievements are and what kind of applications they might have.

It is well known how important some simplest trajectories, equilibria and limit cycles, are for the study of trajectories of a dynamical system in the phase space. The simplest equilibria and limit cycles are structurally stable in the sense that under a small change of the differential equation they do not disappear but only slightly deform.

All "generic" systems with a two-dimensional phase space possess a similar property of structural stability (as was proven by Andronov, Pontryagin, De Baggis, and Peixoto).

Originally publ. in: Unsolved Problems of Mechanics and Applied Mathematics, 3-9, © Moscow Univ. Press, Moscow, 1977
Translated by Boris Khesin and Andrey Khesin.
Springer-Verlag Berlin Heidelberg 2016 with kind permission.

There was the following view on the general properties of trajectories of differential equations with a higher-dimensional phase space, which prevailed since the time of Poincaré until the beginning of the sixties. It was assumed that here one can also single out a finite number of sufficiently simple motions (equilibria and cycles) such that under a small change of the vector field the whole picture of phase curves does not change qualitatively, but only slightly deforms. If the system in question exhibits a more complicated motion, such a motion could be destroyed by a small perturbation of the system, so that it splits into simple motions and a structurally stable system is obtained.

More precisely, this point of view was expressed by the so-called structural stability conjecture (claiming that structurally stable systems are dense) and by the conjecture on the organization of structurally stable systems (finiteness of the number of cycles, etc.). Both conjectures were proven in the two-dimensional case. However, as became clear in the beginning of the sixties in the works of Smale, Anosov, and others, in higher dimensions both of these conjectures are false.

First, the second conjecture was disproven. Namely, structurally stable systems with a compact phase space that have an infinite (and even everywhere dense) set of closed orbits were constructed.

Systems with exponential divergence of trajectories (S. Smale proposed to call them Anosov systems) are important examples of complicated structurally stable systems. Examples of systems with exponential divergence of trajectories were already known at the time of Poincaré (the principle of "minor causes may yield serious consequences"). For instance, the inertial motion of a material point over a smooth surface of negative curvature possesses the property of exponential divergence of trajectories (J. Hadamard). Ya.G. Sinaĭ showed that the simplest two-dimensional systems of statistical mechanics also have similar properties: in a hard ball system in a planar box, a small change in the initial conditions implies an exponential divergence of trajectories after a few collisions. Lately L.A. Bunimovich proved exponential divergence of billiard trajectories for a large variety of billiards.

The trajectories of the above-mentioned systems are rather tangled in the phase space due to their property of exponential instability. Some of them are closed and, furthermore, in a vicinity of every point of the phase space there is a closed (very long) trajectory. Other trajectories, on the contrary, densely fill the space, returning many times to any domain.

And now it turned out that all this complicated and complex picture with rapidly diverging from each other trajectories is structurally stable, i.e., it is preserved under any small perturbation of the vector field or differential equation affording the trajectories (on each surface of constant energy).

In "physical" terms the above conclusions mean that in oscillating systems with many degrees of freedom there could occur such limiting regimes that attract all motions in their vicinity, but themselves possess intrinsic instability. Namely, the "phases" of the oscillations of the limiting regime depend heavily on the initial conditions so that a small change in the initial conditions leads to an exponentially fast divergence of the phases of the unperturbed and perturbed motions.

However, such intrinsically unstable regimes are preserved under any small change of the system parameters, similarly to the persistence of limit cycles.

Now the problem on intrinsically unstable limiting regimes can be formulated as follows: find mechanical (or physical, chemical, etc.) phenomena described by systems with exponential divergence of trajectories and with intrinsically unstable regimes attracting nearby motions.

The conjectural solution of this problem discussed since the beginning of the sixties is related to the turbulence problem. In this case the role of the phase space is played by the infinite-dimensional functional space of the initial conditions (i.e., the velocity fields) of a viscid incompressible fluid. The dynamical system is presumably defined by the Navier–Stokes equations (presumably, in view of the absence of the existence and uniqueness theorems). It is assumed that for any finite Reynolds number, in the phase space there is a finite-dimensional attractor (since the viscosity kills higher harmonics), so that the limiting regime of the fluid motion is described by a finite number of parameters. As is well known, for a large viscosity this limiting set is a point (a stationary flow). As the viscosity decreases (i.e., the Reynolds number increases), the stationary solution becomes unstable, and, furthermore, a secondary periodic flow can arise (i.e., the attracting set becomes a cycle).

One expects that for still larger Reynolds numbers the attractor will have higher dimension, and a complicated intrinsically exponentially unstable motion will be able to occur on it. The corresponding fluid motion would be neither periodic nor almost periodic (as there arises a continuous spectrum), so that the observer could call it a turbulent motion.

Validity of the conjecture described above might be partially confirmed by a computer experiment with Galerkin approximations for the Navier–Stokes equation.

However, making such an experiment poses several interesting computational problems.

2 Computational aspects of the qualitative and ergodic theories

In the qualitative theory of differential equations and in the ergodic theory, many notions have been developed to describe the behavior of dynamical systems: invariant manifolds, invariant measures, ergodicity, mixing, spectrum, entropy. However, for an experimental computer study of the system, many straightforward questions like "find all invariant measures", or "find all closed trajectories", or "describe the spectrum" lead to unfeasible calculations, whose complexity is not matched by the value of the information obtained.

For instance, finding an unstable periodic trajectory or precise tracing of a certain separatrix, which is necessary for a complete description of the topological picture of trajectories, can be very laborious. On the other hand, the behavior details of the separatrix under consideration might not affect the properties of the vast majority of trajectories.

Similarly, it is practically difficult to distinguish between ergodic and non-ergodic systems, between slowly mixing and non-mixing ones, a discrete spectrum with too many frequencies is almost undistinguishable from a continuous spectrum, etc.

Therefore, in order to make a computer carry out a qualitative or ergodic study of a dynamical system, one should learn how to pose realistic questions, remote from the ones usually considered in abstract classification theorems, and one should not request a too precise answer.

Here are the high-priority questions in this type of research:

1) Teach the computer to recognize whether the trajectory entered a vicinity of an attracting invariant set;

2) If so, teach the computer to determine the dimension of this set (0 for an equilibrium, 1 for a cycle, etc.) and, possibly, its topology;

3) Teach the computer to calculate ergodic characteristics of the motion in this set; first of all, to recognize if there is exponential instability of trajectories on the limiting set under consideration (i.e., if the entropy is positive).

The ability to respond to these questions would allow one to study experimentally the limiting regimes of multi-dimensional dynamical systems.

3 The averaging method

The theory of non-conservative perturbations of multi-frequency systems is still practically devoid of rigorous results. The standard averaging method is as follows: to determine the slow evolution of trajectories, the perturbation is averaged over the phases of the fast oscillations. The "evolution" or "secular" equations thus obtained are expected to describe the mean drift, under perturbations, of the quantities preserved for the fast motion. However, the justification of this very helpful method involves significant difficulties. The reason is that during the evolution of a trajectory, the frequencies of the fast motions are (slowly) changing themselves, and at some moments a resonance (commensurability of the frequencies) occurs. In the latter case, for certain relations between the phases, a jam at the resonance may happen, i.e., capturing of the trajectory by a resonant motion having a fewer number of frequencies. To describe the evolution of such trajectories captured by a resonance, the averaging method over the set of all phases is clearly inapplicable.

The problem of estimating the proportion of trajectories captured by resonances, as well as determining a possible long-time behavior of such trajectories, has not yet been solved even for the simplest case of two-frequency systems

$$\begin{aligned} \dot{\varphi}_1 &= \omega_1(I) + \varepsilon f_1(I,\varphi), \\ \dot{\varphi}_2 &= \omega_2(I) + \varepsilon f_2(I,\varphi), \\ \dot{I} &= \qquad \varepsilon g(I,\varphi). \end{aligned}$$

During the last years in the theory of conservative perturbations of Hamiltonian systems, there were obtained exponentially small upper bounds for the evolution rate of the slow variables (N.N. Nekhoroshev). However, the lower bounds are only obtained in separate examples. Another open question is to find the probability distribution for the quantity measured how far the slow variables go from their initial values under small conservative perturbations of a generic integrable Hamiltonian system.

4 The theory of stability of equilibria

Many efforts have been made to find stability criteria for equilibria in the case where some of the roots of the characteristic equation are purely imaginary. However, currently it is proven that there do not exist algebraic criteria of this type (A.D. Bruno, Yu.S. Il′yashenko).

The following question on the algorithmic solvability of the stability problem remains open: is there an algorithm that allows one to decide between stability and instability of an equilibrium of a system of differential equations whose right-hand sides are polynomials with rational coefficients?

A close question on the existence of a limit cycle for a polynomial vector field on the plane is also probably algorithmically unsolvable.

Here a model problem is whether the period of an Abelian integral (the form under integration and the curve are given by polynomials with rational coefficients) is positive.

An even simpler model is a problem on the positivity of a sum of quasi-polynomials (i.e., products of exponents and polynomials). However here there is an algorithm: this follows from an affirmative solution of the Hilbert problem about transcendental numbers (as was explained to me by Yu.V. Matiyasevich).

I will also mention here the classical problem on the inversion of the Lagrange–Dirichlet theorem: prove instability of an equilibrium of the system $\ddot{x} = -\dfrac{\partial U}{\partial x}$ with an analytic potential energy in the case where the equilibrium is not at a local minimum.

The proof of instability published by N.G. Chetaev is erroneous; in the case of two degrees of freedom there also exists a proof by V.P. Palamodov, but it remains unpublished.

5 Estimating the number of limit cycles

One of the questions of Hilbert's 16th problem is how many limit cycles the system

$$\dot{x} = P(x,y), \qquad \dot{y} = Q(x,y)$$

can possess, if P and Q are polynomials of degree 2.

There are examples with 3 cycles, but whether the number of cycles can be larger than 3 is still unknown. I.G. Petrovskiĭ's deep ideas on the connections of this problem with the topology of the foliation of the complex projective plane defined by the complex integral curves, unfortunately, have not yet been realized.

6 Fixed points of symplectic diffeomorphisms

The theory of periodic orbits of Hamiltonian equations leads to interesting topological questions, as is well known since Poincaré's times.

In the following problem we discuss a generalization of the so called "last geometric theorem" of Poincaré, which was proven by Birkhoff.

Let $A : \mathbb{T}^2 \to \mathbb{T}^2$ be the mapping of a two-dimensional torus to itself determined by a periodic planar vector field f by the formula

$$A(x) = x + f(x).$$

Assume that 1) the mapping A preserves the areas (i.e., $\det(DA(x)/Dx) \equiv 1$) and that 2) A does not move the center of mass of the torus (i.e. the mean value of f is equal to 0).

Is it true that under these assumptions the mapping A has at least one fixed point?

Conjecturally, the minimal possible number of fixed points is 4 (if one counts the algebraic number of the points, taking into account their multiplicities) and, in particular, at least 3 geometrically distinct points.

More generally, instead of a torus one can consider a compact symplectic manifold (i.e. a manifold with a fixed nondegenerate closed differential 2-form on it). A diffeomorphism of a symplectic manifold is said to be symplectic, if it preserves this 2-form. A symplectic diffeomorphism is said to be homologous to the identity if it belongs to the commutator subgroup of the group of symplectic diffeomorphisms homotopic to the identity. A symplectic diffeomorphism homologous to the identity can be obtained from the identity diffeomorphism as the time t flow of a non-autonomous system of canonical differential equations with a univalued (but time-dependent) Hamiltonian.

Presumably, the number of fixed points of a symplectic diffeomorphism homologous to the identity is bounded below by the number of critical points of smooth functions on the manifold under consideration. (One counts the critical points algebraically or geometrically depending on whether one counts algebraically or geometrically the fixed points.)

However, this is proven only for the cases of an annulus (by Poincaré and Birkhoff) and a two-dimensional sphere: as was recently pointed out by A.I. Shnirel′man and proven by N.A. Nikishin, every area-preserving diffeomorphism of a two-dimensional sphere to itself admits at least two geometrically distinct fixed points.

Bifurcations of Discrete Dynamical Systems

V.I. Arnold and A.P. Shapiro

(Preprint: Far East Scientific Center of the Academy of Sciences of the USSR, Institute of Automation and Control Processes, Vladivostok, 1977, 24 pp.)[1]

Abstract We consider the conditions of appearance of stable cycles in discrete dynamical systems. We study only typical cases, i.e., investigate only those features, which do not change under small variations of parameters. The study uses the techniques of formal power series for which one proves the Poincaré and Siegel theorems on rectifying the mappings by means of a nonlinear change of coordinates. Bifurcations of one-parameter families of mappings are examined in detail. Examples of bifurcation curves are given. We prove that generically the multiplicity of the arising and disappearing cycles is equal to two. We also outline a method for finding the regions in the parameter space corresponding to the presence of cycles.

The Poincaré theorem

Consider a mapping

$$f : \mathbb{R}^n \to \mathbb{R}^n.$$

[1] The present work is a lecture course given by V.I. Arnold at the 7th Summer Mathematical School in the city of Vladivostok, August 1975. The notes of these lectures were written down and prepared for publication by A.P. Shapiro.

The results described below deal with periodicity properties of iterations of smooth mappings. They are known as a folklore to some mathematicians specializing in dynamical systems, but, to the best of the authors' knowledge, there is no systematic exposition of this topic in Russian. On the other hand, the properties of periodic discrete trajectories are not always analogous to those of periodic solutions of systems of differential equations, and it is worth treating them separately.

Qualitative study of periodic discrete trajectories may also be useful for exploring the appearance of self-oscillations in dynamical systems, as well as for examining the origin of the appearance of cycles in biological populations.

Originally publ. as: Preprint. Far Eastern Scientific Center, USSR Acad. Sciences, Inst. of Automation and Control Processes, Vladivostok, 24 pp., © Far Eastern Scientific Center, USSR Acad. Sciences, Inst. of Automation and Control Processes, Vladivostok, 1977
Translated by Boris Khesin and Andrey Khesin.
Springer-Verlag Berlin Heidelberg 2016 with kind permission.

Definition. A point $x \in \mathbb{R}^n$ is called a *fixed point* of the mapping f if $x = f(x)$.

Define $x_k := f(x_{k-1})$. A fixed point is *locally asymptotically stable* if there is $\varepsilon > 0$ such that whenever $\rho(x_1, x) < \varepsilon$ one has $x_k \to x$ as $k \to \infty$. Similarly, one can introduce the notion of global asymptotic stability and Lyapunov stability.

Theorem. *If $f : \mathbb{R}^n \to \mathbb{R}^n$ has a fixed point $\bar{x}$ and $\left|\operatorname{Spec} f_x'(\bar{x})\right| < 1$, then the point $\bar{x}$ is stable. If at least one eigenvalue lies outside of the unit disk, the fixed point is unstable.*

Proof is straightforward by using estimates for the Euclidean norm $\|\bar{x} - x_k\|$. □

One can also consider the case where one or several eigenvalues lie on the unit circle. The more degenerate the case that is being dealt with, the more difficulties arise. On the other hand, "in general position" degenerate cases do not occur: if one encounters a degeneration, one can get rid of it by an arbitrarily small perturbation.

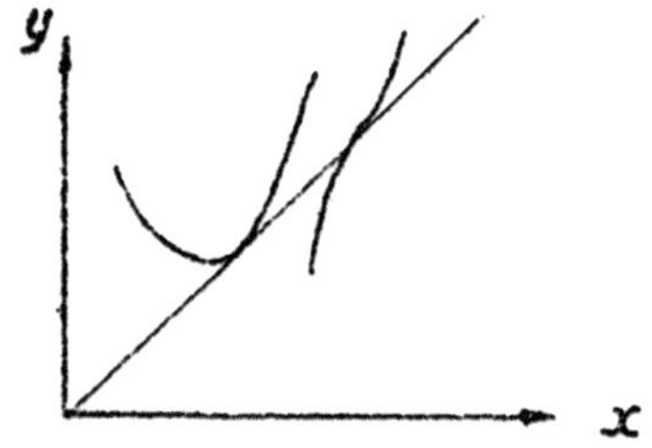

A degenerate case

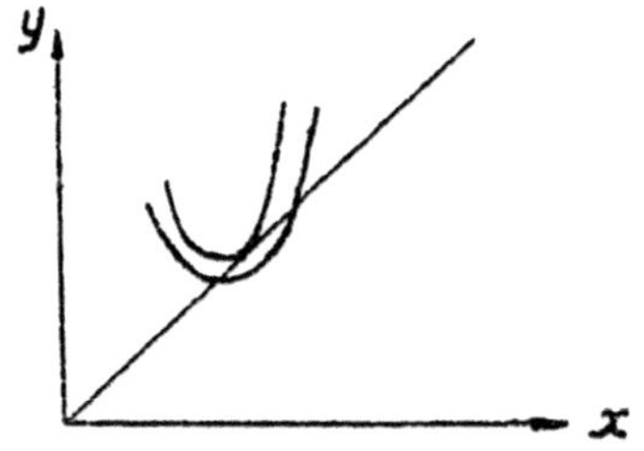

Removing the singularity

The situation changes if one deals with a family of mappings depending on a parameter. Consider a family $x \mapsto F(x, \varepsilon)$, $F(x, 0) = f(x)$. If f has a singularity, the singularity can move to a different function of the same family under a perturbation.

Example. Consider the family $x \mapsto x + x^2 + \varepsilon$. For $\varepsilon = 0$, the fixed point $\bar{x} = 0$ has a singularity, since $f'(\bar{x}) = f'(0) = 1$. Assume that a perturbation is given by adding a small term $\alpha(x)$ whose derivative is small as well.

After the perturbation the family is

$$x \mapsto x + x^2 + \varepsilon + \alpha(x),$$

while its derivative is $F'(x,\varepsilon) = 1 + 2x + \alpha'(x)$. The sign of the quantity $2x + \alpha'(x)$ for sufficiently large $|x|$ coincides with the sign of x, and hence $2x + \alpha'(x)$ vanishes somewhere in a neighborhood of the origin. Denote by x_1 the point where $2x_1 + \alpha'(x_1) = 0$, i.e., $F'(x_1, \varepsilon) = 1$. Thus the singularity moves to the mapping corresponding to $\varepsilon = -x_1^2 - \alpha(x_1)$.

In the functional space of all mappings, the mappings that have a singularity (i.e. the mappings whose derivative has an eigenvalue on the unit circle) form a hypersurface, see the Figure. If a family intersects this hypersurface, any nearby family intersects it as well.

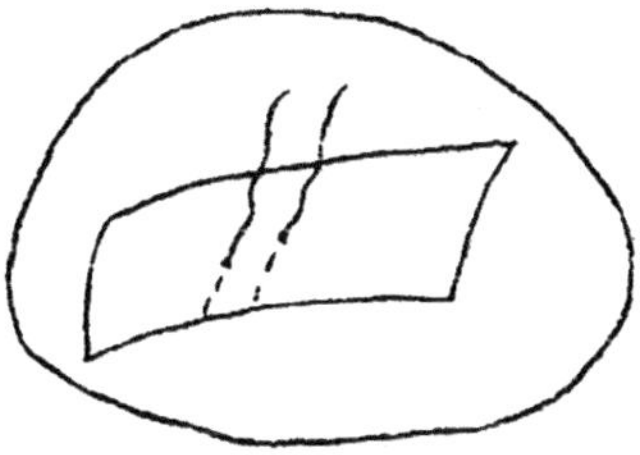

The hypersurface of mappings with singularities

This hypersurface has codimension one. It has a complicated structure and also contains strata of higher codimensions, corresponding to degenerations of higher orders. A one-parameter family can intersect those strata, but by means of a small perturbation one can avoid intersections with strata of codimensions greater than one.

Structure of the space in a neighborhood of a fixed point

Consider a mapping

$$f : \mathbb{C}^n \to \mathbb{C}^n$$

with a fixed point at the origin, $f(0) = 0$.

Problem. Make this mapping as simple as possible by means of an appropriate coordinate change.

Let $f(x)$ be given by a formal power series

$$f(x) = \Lambda x + (Ax, x) + \dots$$

Here Λx is its linear part.

We say that f has a *resonance* if the eigenvalues $\lambda_1,\dots,\lambda_n$ of the matrix Λ satisfy the relation

$$\lambda_1^{m_1}\dots\lambda_n^{m_n}=\lambda_i,$$

where $m_i\geq 0$ are integers and $\sum m_i\geq 2$.

Theorem (Poincaré). *If f does not have a resonance, there is a change of coordinates such that in the new coordinates the mapping becomes linear.*

Proof. Let $y=\Lambda x+F(x)$. Introduce the coordinate change

$$X=x+H(x),\qquad Y=y+H(y),$$

where F and H are formal power series. It is easy to check that

$$Y=\Lambda X+\Phi(X).$$

Rewrite this equation in the old coordinates:

$$\Lambda x+F(x)+H\big(\Lambda x+F(x)\big)=\Lambda\big(x+H(x)\big)+\Phi\big(x+H(x)\big).$$

Hence

$$F(x)+H\big(\Lambda x+F(x)\big)-\Lambda H(x)=\Phi\big(x+H(x)\big).$$

Let $k>1$ be the lowest degree of the terms entering F. It turns out that one can choose H in such a way that all its terms of orders less than k vanish, while the corresponding Φ has all terms of orders less than or equal to k vanishing.

Indeed, let F_k and H_k be homogeneous terms of degree k of the mappings F and H, and let $\mathfrak{M}$ be the ideal of series without the constant term. Then, if the aforementioned H exists, one has

$$F_k(x)+H_k(\Lambda x)-\Lambda H_k(x)=0\mod\mathfrak{M}^{k+1}.$$

Set

$$L(H)=\Lambda H-H\Lambda.$$

Consider this linear operator in the space of polynomials of degree k.

Lemma. *The eigenvectors of L are the monomials $x^m=x_1^{m_1}\dots x_n^{m_n}$ multiplied by the basis unit vectors ℓ_i. The corresponding eigenvalue is equal to $\lambda_i-\lambda^m$.*

Proof of Lemma is a straightforward computation. The completeness of the system of the eigenvectors is evident. □

Since there are no resonances, the operator L is invertible, i.e., the equation $L(H)=F_k$ can be solved, and H turns out to be as required. Furthermore,

$$h^{m,i}=\frac{f^{m,i}}{\lambda_i-\lambda^m},$$

where $h^{m,i}$ and $f^{m,i}$ are the corresponding coefficients of H and F.

Now, by similar reasoning, we make a change of coordinates which gets rid of the terms of order $k+1$ and so on by induction. The composition of all the changes makes sense and gives a formal solution. This completes the proof of the theorem. □

It is interesting to find out when the obtained series are convergent.

Theorem (Poincaré). *If the absolute values of all eigenvalues of* Λ *are less than* 1 (*or all greater than* 1), *then there exists a convergent change of coordinates.*

Proof. Indeed, recall that $h^{m,i} = f^{m,i}/(\lambda_i - \lambda^m)$. Let all λ_i lie outside of the unit circle. Then λ^m is large for large $|m|$ and the terms of the corresponding series are bounded above by those of a convergent series. If $|\lambda_i| < 1$, then by the coordinate change $w = 1/z$ one reduces the problem to the one above. □

Theorem (Siegel). *For almost all* Λ *the mapping* f *is linearizable.*

The resonance points form hypersurfaces. The whole space of parameters is divided into two domains. In the Poincaré domain the resonance hypersurfaces are discrete, while in the Siegel domain they form an everywhere dense set.

Now we are going to examine mappings that admit a resonance. While studying families one cannot get rid of a resonance by a small perturbation. The resonance condition is

$$\lambda^m = \lambda_i,$$

where $m_j \geq 0$ and $\sum m_j \geq 2$.

From this equality it follows that

$$\lambda^s = 1,$$

where $s = m - \ell_i$ is called the *resonance vector*. We have $\sum s_j \geq 1$ and $s_j \geq 0$ with a possible exception of s_i which can be equal to -1.

The set of vectors s satisfying the condition $\lambda^s = 1$ form a group, which is a subgroup of $\mathbb{Z}^n$ (the n-dimensional group of integers). These vectors form a lattice. If the vector space spanned by $\{s\}$ is $\mathbb{R}^\nu$, this is called a resonance of multiplicity ν. The most important case is that of $\nu = 1$, a simple resonance.

Theorem. *If* f *has a resonance, one can introduce new coordinates in which*

$$Y = \Lambda X + \Phi(X),$$

where Φ *contains only resonant terms:*

$$Y_j = X_j \left(\lambda_j + \sum_k \Phi^{kj} X^k \right) \qquad \textit{where} \quad \lambda^k = 1.$$

Proof is similar to the proof of the Poincaré theorem. Let $y = \Lambda x + F_2(x)$. If $\lambda^m \neq \lambda_j$, take

$$h^{m,j} = \frac{f^{m,j}}{\lambda_j - \lambda^m}.$$

Otherwise, set $h^{m,j} = 0$. If one makes the coordinate change, then

$$Y = \Lambda X + R_2(X) + F_3(X),$$

where R_2 contains only resonant terms. By a similar consideration, after $k-2$ steps one obtains

$$y = \Lambda x + R(x) + F_k(x), \tag{1}$$

where R stands for resonant terms of degrees from 2 to $k-1$.

Let us show that an appropriate coordinate change kills in (1) all non-resonant terms of order k, that is $Y = \Lambda X + R(X) + \Phi_k(X) \mod \mathfrak{M}^{k+1}$. We have:

$$\begin{aligned} &\Lambda x + R(x) + F_k(x) + H_k\big(\Lambda x + R(x) + F_k(x)\big) \\ &= \Lambda\left(x + H_k(x)\right) + R\big(x + H_k(x)\big) + \Phi_k\left(x + H_k(x)\right). \end{aligned}$$

Note that

$$\begin{aligned} R\big(x + H_k(x)\big) &= R(x) \mod \mathfrak{M}^{k+1}, \\ \Phi_k\left(x + H_k(x)\right) &= \Phi_k(x) \mod \mathfrak{M}^{k+1}, \\ H_k\big(\Lambda x + R(x) + F_k(x)\big) &= H_k(\Lambda x) \mod \mathfrak{M}^{k+1}. \end{aligned}$$

Therefore what we need is equivalent to

$$F_k(x) + H_k(\Lambda x) - \Lambda H_k(x) = \Phi_k(x) \mod \mathfrak{M}^{k+1}.$$

As above one can choose H_k so that the non-resonant terms will be killed. □

While one can employ the series obtained, they are generally divergent. Usually one makes use of a finite number of terms, among which there are resonant ones.

Consider the simplest case of $n = 1$.

1. The resonance $\lambda^2 = 1$. After applying the transformation one obtains

$$y = x(-1 + \varphi_2 x^2 + \varphi_4 x^4 + \ldots).$$

2. The resonance $\lambda^s = 1$. Here after applying the transformation one obtains

$$y = x(\lambda + \varphi_s x^s + \varphi_{2s} x^{2s} + \ldots).$$

Apply the above procedure to study a family of functions $y = f(x, \varepsilon)$. Let $f(0,0) = 0$, i.e., 0 is a fixed point of the mapping, and $f'_x(0,0) = -1$, i.e., $\lambda = -1$.

Consider a fixed point $x = x(\varepsilon)$, a root of the equation $x = f(x,\varepsilon)$. To apply the implicit function theorem, in addition to differentiability of $f(x,\varepsilon)$ one needs the condition $f'_x(0,0) \neq 1$, which is satisfied in our case. Hence there exists a fixed point $x_0 = x_0(\varepsilon)$, at least for sufficiently small values of ε. Introduce the argument change $x \rightsquigarrow x - x_0(\varepsilon)$. In the new coordinates the point 0 is fixed for all ε, so that one can set $f(0,\varepsilon) \equiv 0$.

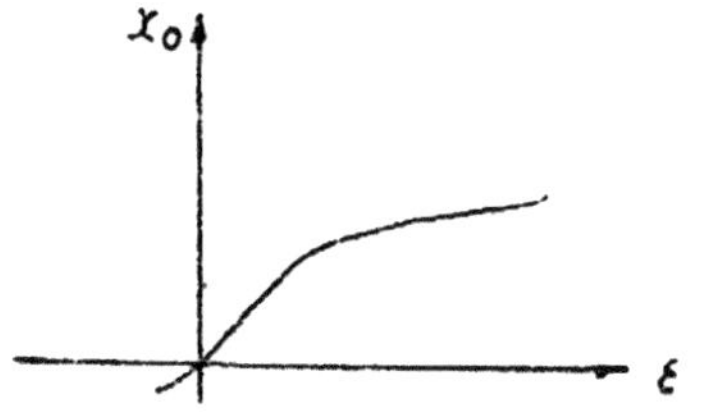

A fixed point as a function of the parameter

We have

$$y = \Lambda(\varepsilon)x + f_2(\varepsilon)x^2 + f_3(\varepsilon)x^3 + \dots$$

Here $\Lambda(\varepsilon)$, $f_2(\varepsilon)$, $f_3(\varepsilon)$, ... are smooth functions of ε,

$$\Lambda(0) = -1, \qquad \left.\frac{d\Lambda}{d\varepsilon}\right|_0 \neq 0. \tag{2}$$

One can use $\Lambda(\varepsilon) + 1$ as a new parameter. Under the conditions (2) there is a smooth one-to-one correspondence between the new parameter and ε in a sufficiently small neighborhood of the origin. For simplicity we will keep the original notation ε for the new parameter. We have

$$y = (-1 + \varepsilon)x + f_2(\varepsilon)x^2 + \dots$$

For $\varepsilon \neq 0$ the linearization theorem can be applied, however for different ε the regions of the argument, where linearization is possible, differ. As $|\varepsilon|$ decreases, the region decreases as well, see the Figure. One can study the mappings separately for each value of ε, but then the most interesting phenomena will not be observed.

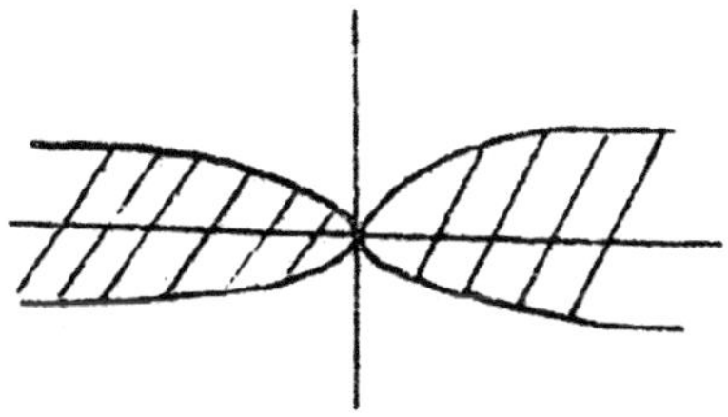

The region of applicability of the linearization theorem

For $\varepsilon \neq 0$ one can linearize the mapping, but the coefficients arising in the coordinate change will have a pole at $\varepsilon = 0$. Therefore for $\varepsilon \neq 0$ it is natural to keep the terms that become resonant.

We pick, as usual, the coefficients of the coordinate change as

$$h^{m,j} = \frac{f^{m,j}}{\lambda^m - \lambda_j} \qquad \text{if} \quad \lambda^m(0) \neq \lambda_j(0),$$
$$h^{m,j} = 0 \qquad \text{if} \quad \lambda^m(0) = \lambda_j(0).$$

In our case of $\lambda_j = -1$ we obtain

$$y = x(\Lambda + \Phi_2 x^2 + \Phi_4 x^4 + \ldots),$$

where $\Lambda = -1 + \varepsilon$.

Let us study the question on cycles in this mapping. We look for cycles of length two. We have

$$y_1 = x(\Lambda + \Phi_2 x^2 + \ldots),$$
$$y_2 = y_1(\Lambda + \Phi_2 y_1^2 + \ldots),$$

or

$$y_2 = x(\Lambda + \Phi_2 x^2 + \ldots)\Big(\Lambda + \Phi_2 x^2(\Lambda + \Phi_2 x^2 + \ldots)^2 + \ldots\Big).$$

This implies that

$$y_2 = \Lambda^2 x + C_3 x^3 + \ldots,$$

where $C_3 = \Phi_2 \Lambda + \Lambda^3 \Phi_2$ depends smoothly on ε, and

$$C_3(0) = -2\Phi_2(0).$$

Generically ("generally speaking") $\Phi_2(0) \neq 0$, i.e., $C_3(0) \neq 0$. The fixed points of the second iteration of the mapping in question satisfy the equation

$$x = \Lambda^2 x + C_3 x^3 + F_5(x, \varepsilon).$$

Getting rid of the trivial fixed point $x = 0$ we obtain

$$1 = 1 - 2\varepsilon + \varepsilon^2 + C_3 x^2 + G_4(x, \varepsilon),$$

or

$$2\varepsilon - \varepsilon^2 = C_3 x^2 + G_4(x, \varepsilon).$$

Let us keep only the principal terms in the equation. For small x and ε we have at the fixed point

$$\varepsilon \approx C_3 \frac{x^2}{2}.$$

The relation between ε and the fixed points is depicted in the two Figures below.

For $C_3 > 0$, when ε passes through 0, a cycle of length two appears with oscillations of order $\sqrt{\varepsilon}$. For $C_3 < 0$ the cycle that existed for negative ε disappears.

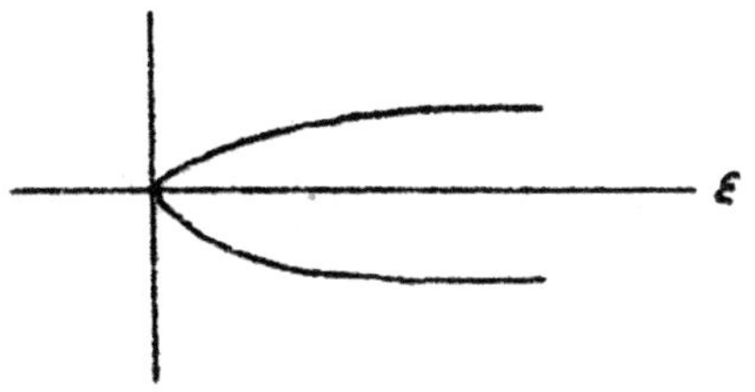

The relation between the coordinates of the fixed points and ε for $C_3 > 0$

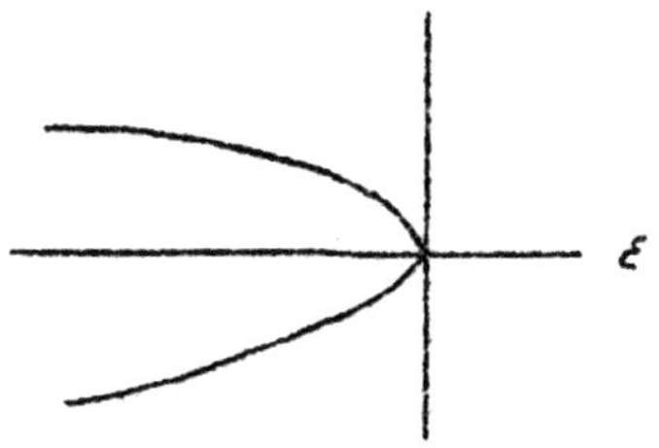

The relation between the coordinates of the fixed points and ε for $C_3 < 0$

Now we study stability of the fixed point and the cycle. Note that

$$f'(0) = \Lambda = -1 + \varepsilon.$$

According to the criterion mentioned above, the point $x = 0$ is stable for $\varepsilon > 0$ and unstable for $\varepsilon < 0$. (One considers only small values of ε.)

To investigate the cycle stability we differentiate the second iteration:

$$f_2'(x,\varepsilon) = \Lambda^2 + 3C_3x^2 + \dots$$

For the cycle elements one has

$$C_3x^2 = 2\varepsilon + \dots$$

Therefore,

$$f_2'(x,\varepsilon) = 1 + 4\varepsilon + O(\varepsilon^2).$$

If $\varepsilon > 0$ then the derivative (the eigenvalue of f_2) is greater than 1 and the cycle is unstable. For $\varepsilon < 0$ the cycle is stable. Stability of the fixed point and the cycle is shown in the Figures.

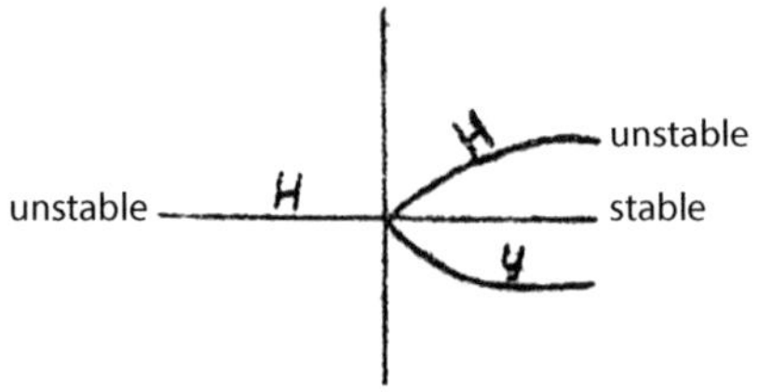

Stability of the cycles of lengths 1 and 2 for $C_3 > 0$

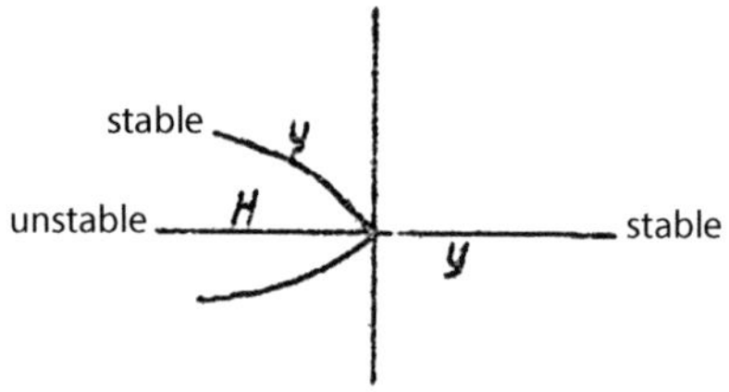

Stability of the cycles of lengths 1 and 2 for $C_3 < 0$

Note that for a given value of ε only one cycle is stable.

For the Ricker function $f = Axe^{-x}$ the coefficient C_3 is negative, while an increase in A corresponds to a decrease in ε. In our notations after a change of variables $f = (x + \ln A)e^{-x} - \ln A$ and $\varepsilon = 2 - \ln A$.

Bifurcations of cycles of length 1 *(fixed points)*

Consider a mapping $y = f(x)$. Its fixed points are solutions of the equation

$$f(x) = x. \tag{3}$$

The set of solutions of this equation can be an arbitrary closed set. Generically, this equation has only isolated solutions. More precisely, the set of functions f for which equation (3) has only isolated solutions is an open and everywhere dense subset of the space of continuous functions. By means of a small perturbation of the mapping f, one can always ensure that at the fixed points x_0 of f the corresponding differential $f'_x(x_0)$ does not have an eigenvalue equal to 1, i.e.,

$$\det(f'_x - E) \neq 0.$$

The situation changes if instead of a mapping $f(x)$ one considers a one-parameter family $y = F(x, \varepsilon)$ with $F(x, 0) = f(x)$. For each ε the number of fixed points is finite, but in the space (x, ε) they form a curve or several curves. If this curve intersects the set of mappings for which $\lambda = 1$, one cannot get rid of such an intersection by means of a small perturbation of the family. Self-intersections and other singularities can be eliminated by a small variation of the family (see the Figures). A typical family does not have such fixed points.

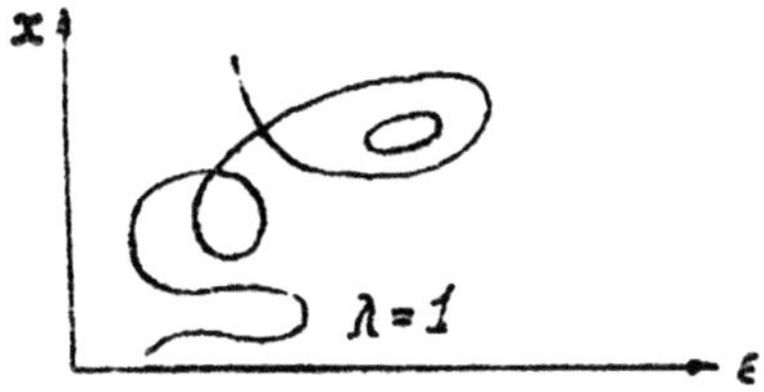

Fixed points of a family $F(x, \varepsilon)$

Fixed points of a deformed family $F(x, \varepsilon)$

Definition. A fixed point is said to be *unramified* if in its neighborhood the curve $F(x, \varepsilon) - x = 0$ can be given in the form $x = x(\varepsilon)$, where $x(\varepsilon)$ is a continuous function.

Theorem. *Given a point* $x(\varepsilon)$, *if* $F_x'(x, \varepsilon)$ *does not have an eigenvalue equal to* 1, *then the point* $x(\varepsilon)$ *is unramified.*

Proof follows from the assumption of the theorem and the implicit function theorem. □

Thus, the vanishing of the determinant,

$$\det\big(F_x'(x, \varepsilon) - E\big) = 0,$$

is a necessary condition of ramification.

Definition. The points that are not unramified are called the *points of ramification or bifurcation*. The curve $x = F(x, \varepsilon)$ is called the *bifurcation curve*.

Remark. At the bifurcation points the curve $x = x(\varepsilon)$ has the tangent directed along the eigenvector of the mapping F_x', corresponding to the eigenvalue equal to 1.

The bifurcation curve can have a tangency with its tangent of orders 3 or 4, but this would be an exception.

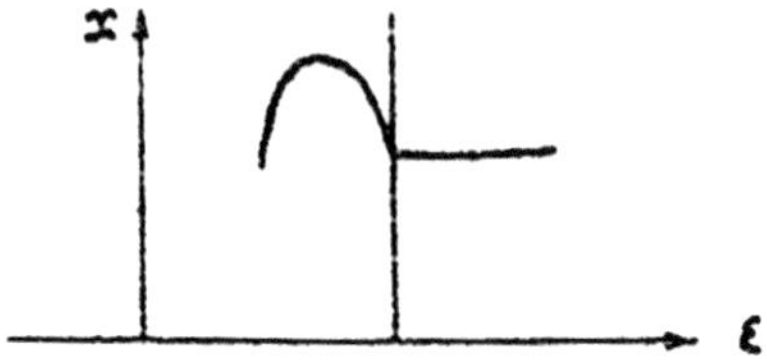

The tangency of the bifurcation curve

Definition. A bifurcation point is said to be *nondegenerate* if the second derivative at this point does not vanish.

Remark. There exist bifurcation points in which two fixed points collide, while in other bifurcation points two fixed points may be born.

Theorem. *In the functional space, typical families yield smooth curves with nondegenerate bifurcation points. (The set of such curves is open and everywhere dense.)*

Generically, for each value of ε there would be no more than one bifurcation point for each bifurcation value.

Now consider a family of mappings $F(x,\varepsilon_1,\varepsilon_2)$ depending on two parameters. Typically for each pair of values $(\varepsilon_1,\varepsilon_2)$ the set of fixed points is discrete. Therefore the bifurcation surface has dimension 2 in the $(n+2)$-dimensional space.

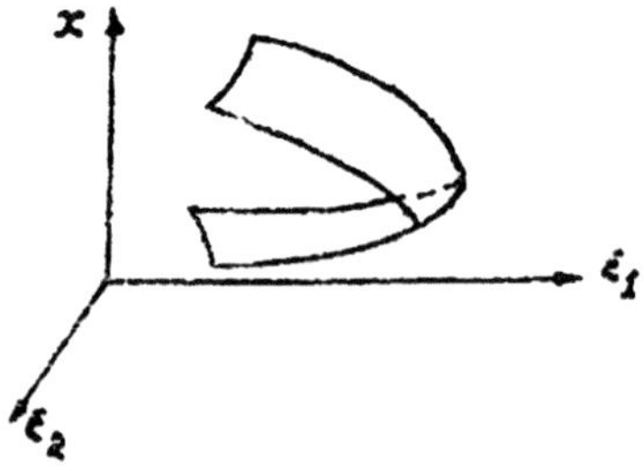

A bifurcation surface depending on two parameters

Generically this surface is smooth. At the points where the determinant $\det(F'_x - E)$ vanishes, the kernel of the operator contains vectors orthogonal to the plane $(\varepsilon_1,\varepsilon_2)$. The set of such points is a smooth curve. At some points of this curve, the kernel of the operator can be tangent to the curve itself.

Example. $F(x, \varepsilon_1, \varepsilon_2) = x^3 - (\varepsilon_2 - 1)x - \varepsilon_1$. Fixed points satisfy the equation

$$x^3 - \varepsilon_2 x - \varepsilon_1 = 0.$$

Singular curves (folds) correspond to the equation $\det(F'_x - E) = 0$, i.e. $3x^2 = \varepsilon_2$. Project this fold to the plane $(\varepsilon_1, \varepsilon_2)$. We obtain the semicubic parabola

$$\varepsilon_1 = a\varepsilon_2^{3/2}, \qquad a = \pm\frac{2}{3^{3/2}}.$$

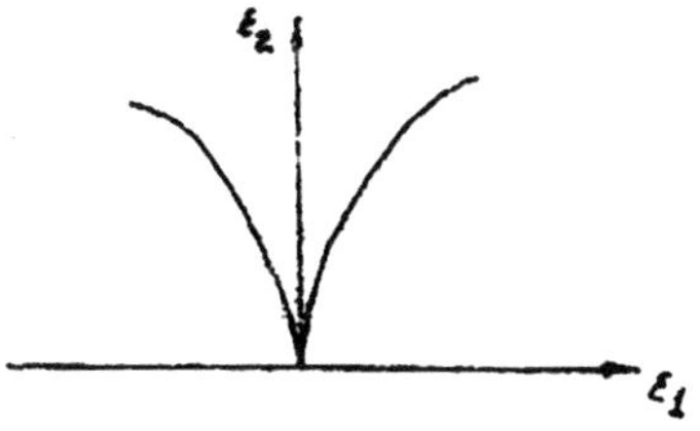

The projection of the fold to the parameter plane

Inside the domain bounded by this curve (i.e. inside the cusp) each point of the plane $(\varepsilon_1, \varepsilon_2)$ corresponds to three fixed points. For instance, for $\varepsilon_1 = 0$ and positive ε_2 the fixed points are $x_1 = 0$, $x_2 = \sqrt{\varepsilon_2}$, and $x_3 = -\sqrt{\varepsilon_2}$. These solutions extend to the semicubic parabola. Parameter values outside of this domain correspond to a single fixed point. For instance, for $\varepsilon_2 = 0$ one has the fixed point x satisfying the equation $x^3 = \varepsilon_1$ which determines a cubic parabola on the plane $\varepsilon_2 = 0$. The whole bifurcation surface looks as follows:

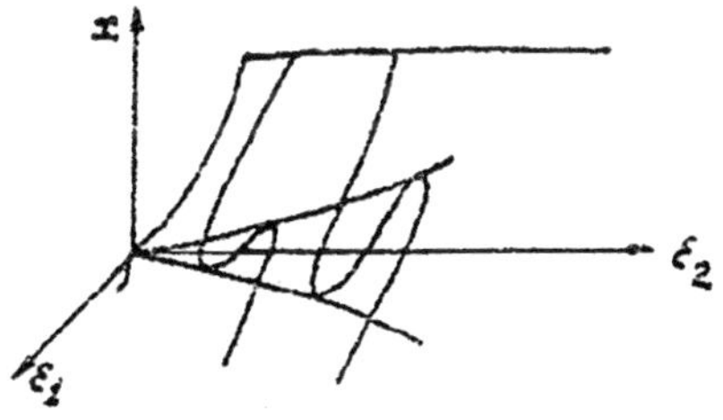

Bifurcation surface depending on two parameters

Transversality theorems

A mapping $f : \mathbb{R}^m \to \mathbb{R}^n$ induces the mapping of the corresponding tangent spaces

$$T_x f : T_x\mathbb{R}^m \to T_{f(x)}\mathbb{R}^n. \quad (4)$$

Definition. If the mapping (4) is not surjective for a given point x, i.e., its image is not the whole space $\mathbb{R}^n$, then x is called a *critical point* and $f(x)$ is a *critical value*.

Lemma (Sard[2]). *The measure of critical values in $\mathbb{R}^n$ is equal to zero.*

Proof will be outlined for $m = n = 1$. By the definition of a critical point one can choose such a small neighborhood of this point that its image will be much smaller than the neighborhood itself. By using Borel's lemma, one can cover the critical values in such a way that the sum of the lengths of the covering sets will be much less than the length of the segment. The remaining estimates are evident. □

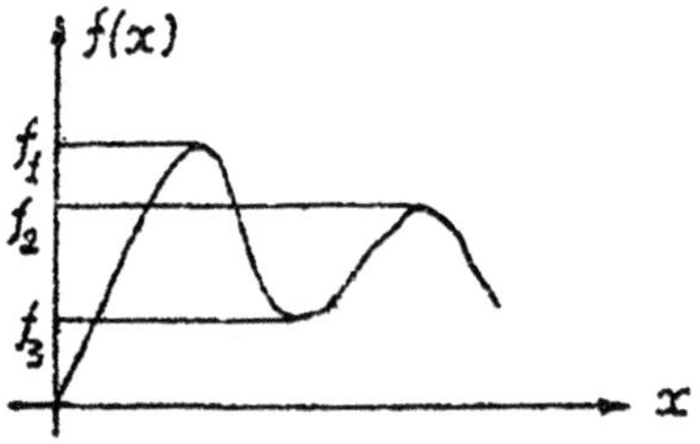

Critical values of a mapping $\mathbb{R}^1 \to \mathbb{R}^1$

Consider a manifold M, its regular submanifold Σ, and a compact manifold K. Let $f : K \to M$ be a smooth mapping.

Definition. The mapping f at a point $x \in K$ is said to be *transversal* to $\Sigma \subset M$ if either $f(x) \notin \Sigma$ or

$$(T_x f)T_x K + T_{f(x)}\Sigma = T_{f(x)}M.$$

Recall that two vector subspaces L_1 and L_2 of a vector space L are said to be transversal if $L_1 + L_2 = L$. Thus, the mapping f is transversal to Σ at a point x if the image of the tangent space at $x \in K$ and the tangent space at $f(x) \in \Sigma$ are transversal.

Definition. The mapping f is said to be *transversal* if it is transversal at each point $x \in K$.

---

[2] A. Sard was a student of M. Morse.

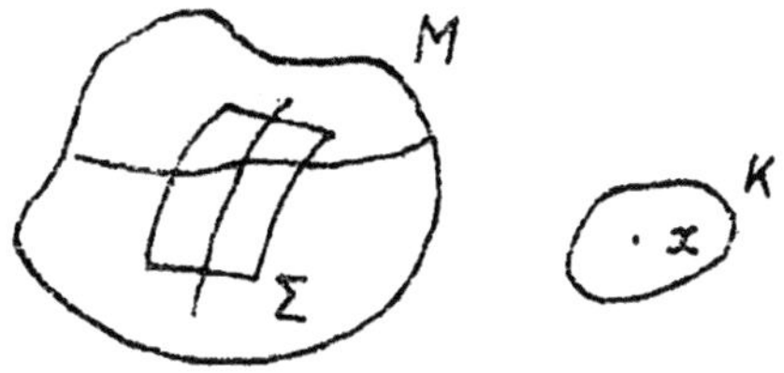

Transversal mapping

The Weak Transversality Theorem. *Let K be a compact manifold and Σ a closed submanifold of M. Then the mappings transversal to Σ form an open and everywhere dense set.*

Proof. Transversality is preserved under small perturbations. Non-transversality disappears under an arbitrarily small deformation of the mapping. □

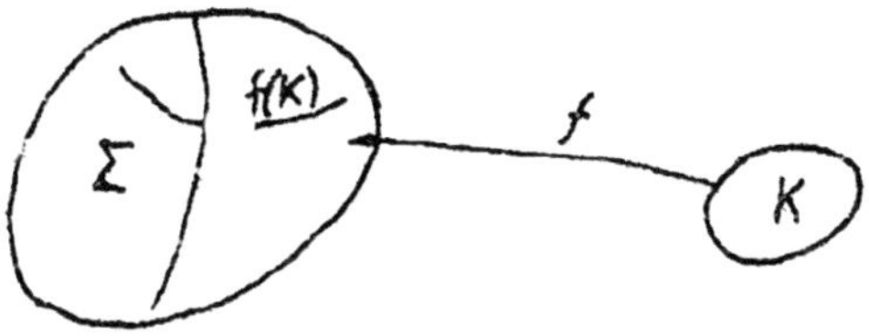

Concerning the Weak Transversality Theorem

Definition. Let M and N be smooth manifolds. Consider a mapping $f: M \to N$. The terms of the Taylor series of f at a point $x \in M$ up to the kth order inclusively are called the *k-jet* of f at the point x. The set of all k-jets turns out to be a manifold and is called the *manifold of k-jets*. We will denote it by $J^k(M,N)$.

For instance, the manifold $J^1(\mathbb{R}^2,\mathbb{R})$ is five-dimensional, while $J^2(\mathbb{R}^2,\mathbb{R})$ is eight-dimensional.

Consider $\Sigma \subset J^k(M,N)$, a submanifold of the manifold of k-jets. Let $f: M \to N$ and let $\hat{f}$ be the *k-jet extension* of f, i.e., $\hat{f}(x) = (j^k f)(x)$ where the operator j^k means taking the Taylor kth order polynomial of the function.

The Strong Transversality Theorem. *Among smooth mappings, the set of those mappings f for which $\hat{f}$ is transversal to Σ is open and everywhere dense.*

Consider the following particular case: $M = \mathbb{R}$, $N = \mathbb{R}$, and $k = 1$. Then $J^1(\mathbb{R},\mathbb{R}) = \{(x,y,p)\}$. Let Σ be the set of critical points. It is clear that

$$\Sigma = \{(x,y,p) \mid p = 0\},$$
$$\hat{f}(x) = (x, f(x), f'(x)).$$

The mapping $\partial/\partial x$ takes this vector to

$$\left(1, f'(x), f''(x)\right).$$

The transversality conditions are as follows:

$$1)\ f'(x) \neq 0,$$
$$2)\ f'(x) = 0, \quad f''(x) \neq 0.$$

Evidently, these are either non-critical points x or critical nondegenerate ones.

Let $M = \mathbb{R}^m$ and $N = \mathbb{R}^n$. In this case $\dim J^1(M,N) = m + n + mn$. The space $J^1(M,N)$ contains a submanifold of mappings of rank r:

$$J^1 \supset \Sigma = \left\{ j^1 f(x) \,\middle|\, \operatorname{rank}\left(\frac{\partial f}{\partial x}\right) = r \right\}.$$

Denote by Σ_r the submanifold of matrices of rank r. Then for $m = n$ one has

$$\operatorname{codim} \Sigma_r = (n-r)^2.$$

In particular, codim $\Sigma_n = 0$, codim $\Sigma_{n-1} = 1$, codim $\Sigma_{n-2} = 4$. By a small perturbation a one-parameter family of matrices can be made generic, i.e. such that all matrices of the family have rank no less than $n-1$. Two- and three-parameter families of matrices possess the same property. Families depending on four parameters may contain irremovable elements of rank $n-2$.

Consider a mapping $x \mapsto f(x)$. Assume that its linear part has an eigenvalue equal to 1. The corresponding eigensubspace is $\ker\left(f'(x) - E\right)$. If $f(x)$ belongs to a one-, two-, or three-parameter family, then, generally speaking, this space is one-dimensional.

Lemma. *The set*

$$\Sigma_r = \left\{ j^1 f(x) \in J^1 \,\middle|\, \operatorname{rank}\left(\frac{\partial f}{\partial x}\right) = r \right\}$$

is a submanifold of J^1 *of codimension* $(m-r)(n-r)$.

Proof. One has $\hom(\mathbb{R}^m, \mathbb{R}^n) \cong \mathbb{R}^{mn}$. Consider an operator of rank r and choose bases of the spaces in such a way that the matrix of the operator has the form

$$\begin{pmatrix} E & 0 \\ 0 & 0 \end{pmatrix}.$$

The codimension is equal to the number of independent equations determining a given submanifold. In our case one has to equate to zero all bordering (i.e. containing the given submatrix) minors of order $r+1$. There are $(m-r)(n-r)$ such minors. If one puts ε in the entry (p,q), where $p > r$, $q > r$, then the corresponding minor is equal to $\varepsilon + O(\varepsilon^2)$, while the corresponding differential is $e_{pq} \cdot \varepsilon + O(\varepsilon^2)$, where e_{pq} is the element of the natural basis in the matrix space. Such differentials for different p and q are linearly independent. This proves the lemma. □

Exercise. Find the codimension of the manifold of symmetric matrices of rank r.

The set of matrices whose rank is less than n is not a manifold but a stratified manifold.

Definition. A union of manifolds of various dimensions is a *stratified manifold*, while the "participating" manifolds are *strata*, if:

1) the intersection of different strata is empty;
2) the boundary of each stratum belongs to the union of strata of smaller dimensions;
3) a manifold transversal to a smaller stratum is also transversal to a larger stratum in a neighborhood of the latter.

Theorem. *If the conditions* 1), 2), 3) *are satisfied, then by a small perturbation of a mapping f one can achieve its transversality to each stratum of a stratified manifold.*

Proof. Indeed, start with strata of the lowest dimension and perturb the mapping f in such a way that it will be transversal to those strata. Thanks to 3), in neighborhoods of the strata of the lowest dimension, the mapping will also be transversal to all the higher-dimensional strata that contain the strata of the smallest dimension in their closure. Then, outside of those neighborhoods, one achieves transversality to strata of the next dimension, etc. □

Bifurcations of fixed points

Definition. A family $F(x,\varepsilon)$ is called a *deformation* of a mapping $f(x)$ if $F(x,0) = f(x)$.

Let $f(0) = 0$. We pose the following

Problem: Study the cycles of the mapping $F(x,\varepsilon)$, or, equivalently, study the equation

$$F^{(n)}(x,\varepsilon) = x.$$

Let us study possible bifurcations of a fixed point.

Theorem. *For an open and everywhere dense set of families, the multiplicity of the cycle appearing or disappearing at a fixed point is equal to 2. One can get rid of any other cycle by a small perturbation.*

Proof. Note that the Jacobi matrix of the nth iteration of the mapping F at a fixed point equals the nth power of the Jacobi matrix of F itself. Consequently, if λ is an eigenvalue of $F'_x(0,0)$, then λ^n is an eigenvalue of $F'^{(n)}_x(0,0)$. The fixed point is bifurcating if at least one of the eigenvalues is a root of unity, $\lambda^n = 1$.

Introduce the set of $m \times m$ matrices possessing an eigenvalue equal to a primitive root of unity of degree n:

$$\Sigma_n = \left\{A \in \operatorname{End} \mathbb{R}^m \mid \ker(A^n - E) \neq \{0\}, \quad \ker(A^s - E) = \{0\} \text{ for } 0 < s < n\right\}.$$

We claim that $\operatorname{codim} \Sigma_1 = \operatorname{codim} \Sigma_2 = 1$ and $\operatorname{codim} \Sigma_n = 2$ for $n \geq 3$. Indeed, $\lambda = 1$ for $n = 1$ and $\lambda = -1$ for $n = 2$. So, in both the cases, λ is real. The matrices with such a real eigenvalue λ satisfy a single equation $\det(A - \lambda E) = 0$. For a complex λ the latter equation is equivalent to two real ones. If $n \geq 3$ then the primitive roots of unity are complex, hence $\operatorname{codim} \Sigma_n = 2$ for $n \geq 3$.

Therefore, a one-parameter family in the space of matrices, mappings, or jets can be slightly deformed to avoid intersections with Σ_n for $n \geq 3$. The intersection with Σ_1 means the presence of a fixed point, while the intersection with Σ_2 means the presence of a cycle of length 2. The theorem is proven. □

Appearance of cycles

Let us study in more detail how the birth of a cycle of length 2 takes place ($n = 2$, $\lambda = -1$). According to the normal form theorem, the mapping F after an appropriate change of coordinates can be written in the form

$$F : \begin{pmatrix} x_1 \\ \vdots \\ x_m \end{pmatrix} \mapsto \begin{pmatrix} x_1\left[(-1+\varepsilon) + \Phi^{2,1} x_1^2 + \ldots\right] \\ \vdots \\ x_m\left[\lambda_m(\varepsilon) + \Phi^{2,m} x_1^2 + \ldots\right] \end{pmatrix}.$$

Neglect the terms $O(x_1^4)$. Let F_1 be the mapping given by the first row of F and let π be the projection of an m-dimensional vector to its first coordinate. The system admits a factorization. The mapping F_1 determines either the birth of a cycle of length 2 at the value $\varepsilon = 0$, or the death of such a cycle.

$$\begin{array}{ccc} \mathbb{R}^m & \xrightarrow{F} & \mathbb{R}^m \\ \pi \downarrow & & \downarrow \pi \\ \mathbb{R} & \xrightarrow{F_1} & \mathbb{R} \end{array}$$

Definitions. 1. Two deformations $F(x,\varepsilon)$ and $\Phi(x,\varepsilon)$ are said to be *equivalent*, if one of them can be obtained from the other by a coordinate change $H(x,\varepsilon)$ depending smoothly on both the coordinate and the parameter, i.e.

$$H\big(F(x,\varepsilon),\varepsilon\big) = \Phi\big(H(x,\varepsilon),\varepsilon\big). \tag{5}$$

2. Two deformations are *topologically equivalent*, if there exists a homeomorphism H continuously depending on ε for which (5) is satisfied.

3. Two deformations are said to be *equivalent with respect to a bifurcation of an equilibrium*, if there is a differentiable change of coordinates $H(x,\varepsilon)$, for which the equilibria of the first deformation are taken to the equilibria of the second one.

4. Let $y = F(x,\varepsilon)$ be a deformation. The space $\{\varepsilon\}$ is the *base* of this deformation. If $\varepsilon = \varphi(\lambda)$ then $y = F\big(x,\varphi(\lambda)\big)$ is called an *induced deformation*.

5. A deformation is said to be *versal*, if any other deformation is equivalent to a deformation induced from this one.

Example. The mapping $x \mapsto -x + x^3$ has a topologically versal deformation $x \mapsto (-1+\varepsilon)x + x^3$, which we studied above.

Theorem. *In the functional space of all families $F(x,\varepsilon)$, an open and everywhere dense set is formed by those families whose bifurcation curves are smooth, although may intersect. Upon exclusion of such intersection points, the rest consists of smooth curves.*

We studied above the bifurcation of an equilibrium into a cycle of length 2. Let us look what may happen with a cycle. It can appear or it can disappear, but its multiplicity cannot change arbitrarily for typical families. Actually, cycles of any order 2^ν can arise:

Theorem. *In a generic one-parameter family, a cycle of length $2p$ can arise from a cycle of length p.*

Proof follows from the fact that F^p is generic. It remains to cite the corresponding result for a fixed point. □

Loss of stability

Assume that while passing through the value $\varepsilon = 0$, a stable fixed point becomes unstable, and set $\lambda(\varepsilon)$ to be an eigenvalue of the matrix $F'_x(0,\varepsilon)$. For a stable fixed point one has the inequality $|\lambda(\varepsilon)| < 1$. When the stability is being lost, one of the eigenvalues crosses the unit circle. If this happens through $\lambda = -1$, a bifurcation occurs: a stable cycle of length 2 arises. The amplitude of oscillations is of order $\sqrt{\varepsilon}$. For small ε, the points of the cycle are close to the fixed point. This situation is called a *soft loss of stability.*

While crossing through $\lambda = -1$, a *hard loss of stability* can occur as well. In that case a stable cycle arises from a stable fixed point. As ε approaches 0, the basin of attraction of the fixed point shrinks to zero.

It remains to consider the case of a complex λ crossing the unit circle. By a small perturbation, one can always ensure λ not to be a root of unity, $\lambda^n \neq 1$. Therefore we consider a generic one-parameter family of mappings $F(x,\varepsilon)$ such that λ_1, one of the eigenvalues of $F'_x(0,0)$, lies on the unit circle and is not a root of unity. Then the other eigenvalue $\lambda_2 = \overline{\lambda_1}$ also lies on the unit circle and is not a root of unity. Therefore one encounters a resonance, $\lambda_1\lambda_2 = 1$.

Typically, the speed of crossing the circle is non-zero,

$$\frac{d|\lambda|}{d\varepsilon} \neq 0.$$

Generically, $\lambda_1\lambda_2 = 1$ is the only resonance.

Choose the first two coordinates to be complex conjugates: $z_2 = \overline{z_1}$. Then after an appropriate change of coordinates, F takes the following form:

$$F : \begin{pmatrix} z_1 \\ z_2 \\ \vdots \end{pmatrix} \mapsto \begin{pmatrix} z_1\left(\lambda_1 + a_1 z_1 z_2 + a_2 z_1^2 z_2^2 + \ldots\right) \\ z_2\left(\lambda_2 + \overline{a_1} z_1 z_2 + \overline{a_2} z_1^2 z_2^2 + \ldots\right) \\ \vdots \end{pmatrix}.$$

This means that the plane (z_1, z_2) is mapped to itself. Due to the condition $|\lambda_k| < 1$ for $k \geq 3$, the rest of the space contracts to this (z_1, z_2)-plane. Therefore, an analysis of the two-dimensional case allows one to study the higher-dimensional setting.

The series determining the mapping F are divergent, but we are going to consider only a finite number of terms and to take the corrections into account. If one confines oneself to one term only, then the mapping

$$F : \begin{pmatrix} z_1 \\ z_2 \end{pmatrix} \mapsto \begin{pmatrix} \lambda_1 z_1 \\ \lambda_2 z_2 \end{pmatrix}$$

is a rotation by the angle $\arg\lambda_1$. The actual mapping F is a rotation by an angle depending on $\tau = z_1 z_2 = |z_1^2|$. Indeed, the product of the coordinates of the image is

$$\begin{aligned} z_1' z_2' &= z_1 z_2 (\lambda_1 + a_1 z_1 z_2 + \ldots)(\lambda_2 + \overline{a_1} z_1 z_2 + \ldots) \\ &= \tau\Big(1 + (\lambda_1 \overline{a_1} + \overline{\lambda_1} a_1)\tau + \ldots\Big). \end{aligned}$$

The circle $|z_1|^2 = \tau$ is mapped to a circle, but of different radius. Introduce the polar coordinates

$$z_1 z_2 = \tau, \quad \arg z_1 = \varphi;$$

$$\ln z_1' = \ln z_1 + \ln\lambda_1 + \ln\left(1 + \frac{a_1}{\lambda_1}\tau + \ldots\right).$$

Let $\lambda_1 = \lambda_1(\varepsilon)$. Then

$$\begin{pmatrix} \tau \\ \varphi \end{pmatrix} \mapsto \begin{pmatrix} \tau(|\lambda_1|^2 + A\tau + \ldots) \\ \varphi + \arg\lambda_1 + (*) \end{pmatrix},$$

where $(*)$ is a series in τ with real coefficients, i.e.,

$$\begin{pmatrix} \tau \\ \varphi \end{pmatrix} \mapsto \begin{pmatrix} \tau(|\lambda_1|^2 + A\tau + \ldots) \\ \varphi + \arg\lambda_1 + B\tau \ldots \end{pmatrix}. \tag{6}$$

Note that the transformation of the radii does not depend on the angles. On the other hand, the transformation of the angles does depend on the radius. Since the value $|\lambda|^2$ is close to 1 for small ε, one can take

$$\varepsilon = |\lambda|^2 - 1$$

in lieu of ε. Drop in (6) all terms starting with the third order in τ. Then the value $\tau = -\varepsilon/A$ will be critical for τ. For this value, the new squared radius τ' coincides with τ. The critical value is positive provided that $\varepsilon A < 0$. Assume that $\varepsilon > 0$ and $A < 0$. Then all circles inside the critical one are mapped to circles of bigger radius, while those outside of the critical circle are mapped to circles of smaller radius. One encounters an attracting invariant set, the critical circle.

The angle is changing by $(\arg\lambda_1 + \text{correction})$. If this change is commensurable with π then a cycle appears. If one takes into account all terms of the series, the conclusion about the existence of an invariant curve is still valid. The invariant curve is close to a circle.

The latter statement holds not only for λ with an argument incommensurable with π, but also for most commensurable ones. In fact, one only has to exclude the roots of unity of degrees 3 and 4.

Example. Examine the mapping $\varphi \mapsto \varphi + \alpha + k \sin \varphi$.

If the mean rotation number $\mu(\alpha, k)$ is equal to $2\pi p/q$, then cycles arise and half of them are stable, while the other half are unstable. The regions of cycle formation are everywhere dense on the real axis.

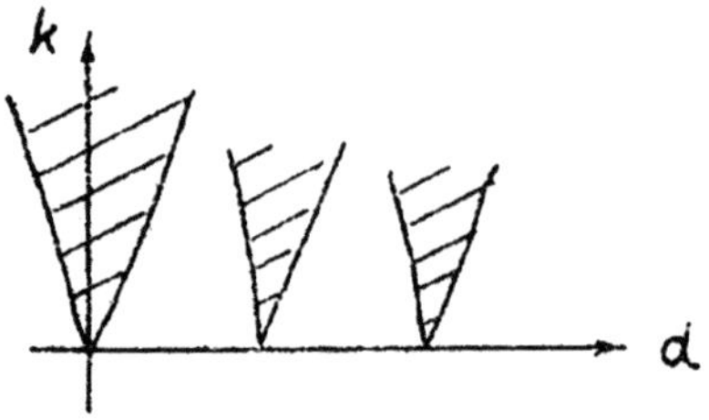

Regions of cycle formation

Consider the regions of cycle formation in the λ-space. From each root of unity there emanates a tongue in which the mapping has a cycle of the corresponding multiplicity. Namely, if this is a point $\exp(2\pi i p/q)$, where $\gcd(p,q) = 1$, then the region corresponds to cycles of multiplicity q, half of them being stable. If λ moves outside of the circle, the curve will intersect infinitely many tongues, i.e., it contains infinitely many cyclic regions.

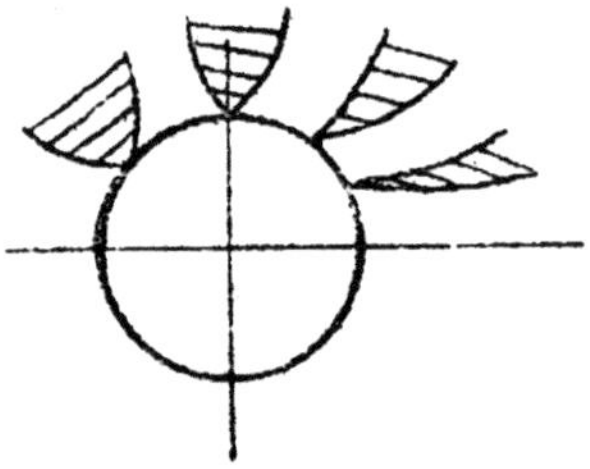

The regions of cycle appearance in the λ-space

Exercises.

1. Study the structure of the tongues in a vicinity of a given root of unity.
2. How many parameters does a versal family of deformations contain as λ crosses the unit circle?

В МОСКОВСКОМ МАТЕМАТИЧЕСКОМ ОБЩЕСТВЕ

ЗАСЕДАНИЯ МОСКОВСКОГО МАТЕМАТИЧЕСКОГО ОБЩЕСТВА

Заседание 11 октября 1977 г.

1. В. И. Арнольд «Индекс особой точки векторного поля, неравенство Петровского — Олейник и смешанные структуры Ходжа».

Неизвестно, какое максимальное значение может иметь индекс особой точки вещественного векторного поля, компоненты которого — однородные многочлены данной степени. Недавно Г. Левин, Д. Эйзенбуд и Г. Химшиашвили нашли алгебраическую формулу для индекса особой точки: индекс равен сигнатуре естественно определяемой квадратичной формы на локальном кольце особенности. Эта формула позволяет оценивать индекс сверху. В 1949 г. И. Г. Петровский и О. А. Олейник указали неравенство, ограничивающее топологические инварианты вещественных алгебраических гиперповерхностей. Оказывается, эти неравенства следуют из оценки сверху модуля индекса особой точки векторного поля. Чтобы оценить сверху модуль индекса особой точки поля, компоненты которого — многочлены степени k от p переменных, рассмотрим куб $(0, k+1)^p$ и проведем через его центр гиперплоскость, перпендикулярную главной диагонали. Индекс оценивается сверху числом целых точек из этой гиперплоскости, лежащих строго внутри куба. В случае, когда размерность p меньше 4, указанная оценка точна, что следует из результатов И. Г. Петровского (1938). В докладе рассказано о перечисленных результатах, а также об их обобщениях, связанных с диаграммами Ньютона и с недавними результатами и гипотезами Ж. Стинбринка о вычислении смешанных структур Ходжа в гомологиях локального многообразия уровня голоморфной функции. По-видимому, правильную оценку индекса дает среднее число Ходжа на пространстве инвариантных циклов.

Заседание 18 октября 1977 г.

1. А. Е. Туманов, Г. М. Хенкин «Голоморфные инварианты вещественных гиперповерхностей комплексного пространства и их приложения».

До недавнего времени считалось, что в многомерном комплексном анализе нет естественного аналога теоремы Римана о конформном отображении односвязной области на круг. Ситуация изменилась после яркой работы Мозера (1973), в которой, в развитие идей Пуанкаре, указываются все канонические гиперповерхности в комплексном пространстве, на которые (локально) голоморфно отображаются произвольные гиперповерхности, и работы Фефермана (1974), показавшей, что результаты Мозера дают классификацию не только гиперповерхностей, но и областей, ограниченных этими гиперповерхностями. Как показал Чёрн (1974), инварианты Мозера тесно связаны с формами связности и кривизны для частично комплексных структур на гиперповерхностях, введенных Э. Картаном еще в 1932 г. Указанные результаты уже имеют значительное количество приложений (Александер, Бернс, Вебстер, Дидрих, Наруки, Пинчук, Уэллс, Феферман, Фримен, Шнайдер и др.). Наиболее интересные из этих приложений также изложены в докладе.

INDEX OF A SINGULAR POINT OF A VECTOR FIELD, THE PETROVSKII – OLEINIK INEQUALITY, AND MIXED HODGE STRUCTURES

V. I. Arnol'd

UDC 513.60+517.9

There is an interesting connection between the topology of real algebraic manifolds, on the one hand, and the local theory of singularities of smooth mappings, on the other [2]. This connection can be used in both directions. On the one hand, contemporary successes in the theory of singularities give an upper bound on the index of a singular point of a vector field, which reduces to a new, entirely simple proof of the Petrovskii–Oleinik inequality [13] in real algebraic geometry; here both Petrovskii–Oleinik inequalities (relating to the cases of even-dimensional and odd-dimensional hypersurfaces) are united into one.

On the other hand, from results of real algebraic geometry, the sharpness of the bound indicated above for homogeneous polynomial vector fields on the plane and in space follows. In these cases, the greatest possible modulus of the index is attained by gradient vector fields (whether this is true in spaces of more than three dimensions and whether the bound in question is sharp in the multidimensional case is unknown).

The topology of real algebraic manifolds also gives models for the geometry of real singularities. The usual scheme of argument in real algebraic geometry (cf. [6, 10]) is this: in order to estimate some invariant of real topological type, one looks for an appropriate invariant of a complex object majorizing the first. Invariants of complex objects are constant for almost all terms of a complex irreducible family (since degenerate cases correspond to complex hypersurfaces in the space of parameters of the family, and the complement of such hypersurfaces is connected). Hence, the invariants of a complex object can be computed in terms of discrete data of the problem (the degree, the Newton diagram, etc.). Thus, estimating invariants of real topological type splits into two problems: the search for a majorizing invariant of a complex object and its calculation from discrete data.

We consider from this point of view the problem of the structure of smooth functions in a neighborhood of a complicated critical point. Among the series of invariants of real topological type (such as Morse numbers, the number of points of different indices for all real Morsivizations, etc.) the Poincaré index of the gradient of the function is singled out. The problem arises of estimating this index by means of discrete data (the Newton diagram, etc.).

On the other hand, the complex geometry of a singularity reduces to the consideration of mixed Hodge structures [8]. Mixed Hodge structures reflect the structure of a singularity by nonsingular projective manifolds of various dimensions. To each singularity is associated a collection of integral Hodge numbers $h^{p,q}$; following Steenbrink [15], we denote by $h_1^{p,q}$ the Hodge number of the space of cycles on which all characteristic values of monodromy are equal to one.

Numerous examples lead to the following conjecture:

$$|\mathrm{ind}| \leqslant h_1^{n/2,\, n/2}, \qquad (*)$$

where ind on the left-hand side means the index of a singular point of the gradient of a function of an even number n of variables.

In order to obtain from this an estimate for the index of a singular point of the gradient of a function of an odd number of variables, it is necessary to append a superfluous variable and to add its square to the function.

In the case when the function considered is a nondegenerate homogeneous polynomial, the right-hand side is explicitly computable and the conjecture (*) turns right into the unified formulation of the Petrovskii–Oleinik inequality for even- and odd-dimensional hypersurfaces.

Moscow State University. Translated from Funktsional'nyi Analiz i Ego Prilozheniya, Vol. 12, No. 1, pp. 1-14, January-March, 1978. Original article submitted September 6, 1977.

0016-2663/78/1201-0001 $07.50 ©1978 Plenum Publishing Corporation

Originally publ. in: Funkts. Anal. Prilozh. 12(1), 1-14, © Russ. Acad. Sciences 1978
English transl.: Funct. Anal. Appl. 12(1), 1-12, © Consultants Bureau, New York 1978
Reprint by Springer-Verlag Berlin Heidelberg 2016 with kind permission.

Thus, (*) is a direct generalization of the Petrovskii–Oleinik inequalities to the case of singularities of smooth functions.

As far as I know, the calculation of the Hodge numbers by means of the Newton diagram has been carried out at the present time only for functions of two variables. In the general case one has only the conjecture of Steenbrink [16]. Thus, a definitive estimate of the index by means of the Newton diagram depends on two unproven conjectures, and the proof of this estimate would support each of these conjectures.

Hodge numbers (really not mixed, but of pure Hodge structures) were first applied in the topology of real algebraic manifolds in papers of the school of Rokhlin [17, 18], to whom the author gratefully dedicates this paper.

1. Estimate of the Index of a Singular Point of a Vector Field

We consider a vector field in n-dimensional real space, whose components are homogeneous polynomials of fixed degree $m-1$ (e.g., the gradient of a homogeneous polynomial of degree m). Let us assume that O is a singular point of the field of finite multiplicity (i.e., isolated in the complex domain).

Definition. By the Petrovskii number we mean the number of integral points strictly interior to the cube $(0, m)^n$, lying in the hyperplane passing through the center of the cube and perpendicular to the principal diagonal of the cube. Notation:

$$\Pi_n(m) = (\# k = (k_1, \ldots, k_n),\ 0 < k_s < m,\ \sum k_s = mn/2).$$

THEOREM. The index of a singular point of finite multiplicity of a vector field in $\mathbf{R}^n$, whose components are homogeneous polynomials of degree $m-1$, does not exceed in modulus the Petrovskii number $\Pi_n(m)$.

Proof. The estimate is based on the explicit formula for the index of a singular point of a (not necessarily homogeneous) vector field, discovered by Khimshiashvili [19] and Eisenbud and Levine [9].

We denote by f_s the components of the field considered and by $Q_f = \mathbf{R}[[x_1, \ldots, x_n]]/(f_1, \ldots, f_n)$ the local $\mathbf{R}$-algebra of mapping $\mathbf{R}^n \to \mathbf{R}^n$, defined by the functions f_s.

We denote by $\mu = \dim_{\mathbf{R}} Q_f$ the multiplicity of the singular point, and we consider in the $\mathbf{R}$-algebra Q_f an ideal I such that $I^2 = 0$; let us assume that I is one of the maximal elements among the ideals with this property.

From the cited formula for the index, it follows that

$$|\operatorname{ind}| = \mu - 2 \dim_{\mathbf{R}} I.$$

Now let f_s be a homogeneous polynomial of degree $m-1$. In this case, μ is equal to the number of integral points strictly interior to the cube $(0, m)^n$; as $\mathbf{R}$-generators of the local ring, one can take the cosets of monomials; the number of monomials of each fixed degree d is equal here to the number of integral points strictly interior to the cube on the corresponding plane perpendicular to the principal diagonal (cf. [1]):

$$v_d = (\# k = (k_1, \ldots, k_n),\ 0 < k_s < m,\ \sum k_s = d + n).$$

We consider the $\mathbf{R}$-linear subspace of Q_f spanned by the classes of monomials whose degree is greater than half the maximal possible $(d > n(m-2)/2)$. This linear space J is an ideal in Q_f, and its square is equal to zero. Hence

$$|\operatorname{ind}| \leqslant \mu - 2 \dim_{\mathbf{R}} J.$$

The right-hand side of the last inequality is equal to the number of basis monomials of middle degree, i.e., equal to $\Pi_n(m)$, and the theorem is proved.

Remark 1. The theorem remains in force for singular points of infinite multiplicity which are real-isolated (the proof is by considerations of continuity).

Remark 2. A. G. Khovanskii, to whom I communicated the proof above, remarked that from the theorem follows an upper bound by the number $\Pi_n(m)$ for the sum of the indices of singular points of a vector field in $\mathbf{R}^n$, the components of which are inhomogeneous polynomials of degree $m-1$ in general position. He also gave his proof of the bound on the sum of the indices and its generalization, which is intermediate between that given above and the proof of [13]; here it appeared that the argument of Petrovskii contained many elements of the proof of the index formula from [9, 19].

2. Left-Hand Side of the Petrovskii – Oleinik Inequalities

The left-hand side of the original Petrovskii–Oleinik inequalities for projective hypersurfaces of even dimension is the modulus of the reduced (lowered by one) Euler characteristic of the hypersurface, and for hypersurfaces of odd dimension and even degree it is the modulus of the difference of the Euler characteristics of the parts of projective space corresponding to positive and negative values of the polynomial (cf. [12, 13, 17, and 18]).

THEOREM. The left-hand side of the Petrovskii–Oleinik inequalities in both cases indicated above is the modulus of the index of the gradient of the homogeneous polynomial defining the hypersurface.

Proof. We use the formulas expressing the index of the gradient of a (not necessarily homogeneous) function f at a critical point of finite multiplicity by means of the reduced Euler characteristic of the local level manifolds of the function.

Let B be a sufficiently small ball in $\mathbf{R}^n$ with center at the singular point O of the function f with critical value 0. For sufficiently small $|\varepsilon| \neq 0$, the set $V_\varepsilon = (f^{-1}(\varepsilon) \cap B)$ is a manifold with boundary; it is called a local level manifold of the function f at the critical point O. We denote by $\tilde{\chi}_+$ (respectively, $\tilde{\chi}_-$) the reduced (lowered by one) Euler characteristic of the manifold V_ε for $\varepsilon > 0$ (respectively, $\varepsilon < 0$). It is easy to prove (cf., e.g., [19]).

LEMMA. The index of a singular point of the gradient of a function f is connected with the reduced Euler characteristic of its local level manifolds by the relations

$$\text{ind} = -\tilde{\chi}_- = \begin{cases} -\tilde{\chi}_+, & \text{if } n \text{ is even}, \\ \tilde{\chi}_+, & \text{if } n \text{ is odd}. \end{cases}$$

The proof of the lemma is obtained from a direct application of Morse theory. Actually, according to the latter, the index of a singular point of a gradient is equal to the relative Euler characteristic of the set where $f \leq \varepsilon$ modulo the set where $f \leq -\varepsilon$, and consequently, of $f \leq 0$ modulo $f < 0$. But the set where $f = 0$ is locally contractible to the singular point. Hence $\text{ind} = -\tilde{\chi}_-$. This is the first assertion of the lemma. The second follows from the first, since if n is even, then changing the sign of f does not change the index, and if it is odd then it changes its sign.

We turn now to the case when f is a homogeneous polynomial in $\mathbf{R}^n$. We consider the sphere S^{n-1}.

The equation $f = 0$ cuts out a hypersurface on the sphere, which we denote by $\hat{A}$; we denote by $\hat{B}_+$ the part of the sphere where $f \geq 0$, and by $\hat{B}_-$ the part where $f \leq 0$. It is easy to see that $\hat{B}_+$ is diffeomorphic to V_ε, and $\hat{B}_-$ is diffeomorphic to $V_{-\varepsilon}$ (for $\varepsilon > 0$).

It follows from the additivity of the Euler characteristic that

$$\chi(S^{n-1}) = \chi(\hat{B}_+) + \chi(\hat{B}_-) - \chi(\hat{A}). \tag{1}$$

We denote by A the hypersurface in $\mathbf{RP}^{n-1}$ defined by the equation $f = 0$. The manifold $\hat{A}$ is then a two-sheeted, unramified covering of A. If the degree of f is even, then the manifolds with boundary $\hat{B}_+$ and $\hat{B}_-$ are also two-sheeted, unramified coverings of the manifolds with boundary B_+ and B_-, lying in $\mathbf{RP}^{n-1}$ and defined by the conditions $f \geq 0$ and $f \leq 0$, respectively.

Let us assume that the number of variables n is even. Then the quantities appearing in (1) have the following values: $\chi(S^{n-1}) = 0$, $\chi(\hat{B}_+) = \tilde{\chi}_+ + 1 = 1 - \text{ind} = \chi(\hat{B}_-)$, $\chi(\hat{A}) = 2\chi(A)$. Consequently, it follows from (1) that

$$\text{ind} = 1 - \chi(A). \tag{2}$$

Let us assume that the number of variables n is odd, and the degree of f is even. Then the quantities appearing in (1) have the following values: $\chi(S^{n-1}) = 2$, $\chi(\hat{B}_+) = 2\chi(B_+) = 1 + \text{ind}$, $\chi(\hat{B}_-) = 2\chi(B_-) = 1 - \text{ind}$, $\chi(\hat{A}) = 0$. Consequently, it follows from (1) that

$$\text{ind} = \chi(B_+) - \chi(B_-) = 1 - 2\chi(B_-). \tag{3}$$

The theorem is proved.

3. Right-Hand Sides of the Petrovskii – Oleinik Inequalities

The right-hand sides of the original Petrovskii–Oleinik inequalities for the case of a polynomial of degree m in n homogeneous variables (i.e., for a hypersurface of degree m in $\mathbf{RP}^{n-1}$) have the following form (cf. [3, 12, 13]). We consider the cube $(0, m)^{n-1}$. Through the centers of all faces of dimensions $n-2$, $n-4$, $n-6, \ldots$ we draw hyperplanes perpendicular to the main diagonal. These hyperplanes cut the cube into an odd number of slices. The right-hand side of the Petrovskii–Oleinik inequality is the number of integral points strictly interior to the middle such slice, i.e.,

$$N_n(m) = \left(\#, l = (l_1, \ldots, l_{n-1}) \colon 0 < l_s < m, \frac{n-1}{2} m < \sum l_s < \frac{nm}{2}\right).$$

THEOREM. The right-hand side of the Petrovskii–Oleinik inequality coincides with the Petrovskii number $\Pi_n(m)$, defined in Sec. 1:

$$N_n(m) = \Pi_n(m).$$

Proof. We project the cube $(0, m)^n$ onto the cube $(0, m)^{n-1}$ along the last coordinate axis. This projection establishes a one-to-one correspondence between the integral points strictly interior to the large cube, lying on a hyperplane passing through the center of the cube and normal to the principal diagonal, and the integral points strictly interior to the smaller cube lying inside the middle slice described above: $(n-1)m/2 < \Sigma l_s < nm/2 \Leftrightarrow 0 < l_n < m$, where $l_n = (nm/2) - \Sigma l_s$. The theorem is proved.

We now give another interpretation of the right-hand sides of the Petrovskii–Oleinik inequalities.

THEOREM. The right-hand side of the Petrovskii–Oleinik inequalities for a hypersurface defined by a homogeneous polynomial f is equal to the Hodge number $h^{n/2, n/2}$ of the mixed Hodge structure on a nonsingular local level manifold of the polynomial f if the number of variables n is even, and to the Hodge number $h^{(n+1)/2, (n+1)/2}$ of the mixed Hodge structure on a nonsingular local level manifold of the polynomial $f(x) + y^2$ if the number of variables n of the polynomial f is odd and the degree is even.

Proof. We use Steenbrink's equation [15], which expresses the number $h^{p,q}$ of the mixed Hodge structure of a quasihomogeneous singularity in terms of the generators of the local algebra of the gradient of the map.

Let f be a quasihomogeneous, nondegenerate complex polynomial in the variables $x_1, \ldots, x_n$ with weights $\alpha_1, \ldots, \alpha_n$, i.e., $f(t^{\alpha_1}x_1, \ldots, t^{\alpha_n}x_n) = tf(x)$ and the critical point O of the polynomial f be complex isolated. We consider the local algebra

$$Q_f = \mathbf{C}\,[[x_1, \ldots, x_n]]/(\partial f/\partial x_1, \ldots, \partial f/\partial x_n).$$

Let $\{x^k\}$ be μ monomials defining a C-basis in Q_f. We consider the μ differential forms $\omega_k = x^k dx_1 \wedge \cdots \wedge dx_n$. We consider their quasihomogeneous degrees

$$l_k = (k+1, \alpha) = (k_1+1)\alpha_1 + \ldots + (k_n+1)\alpha_n.$$

The characteristic numbers of the monodromy operator are expressed in terms of the numbers l_k by the formula $\lambda_k = e^{2\pi i l_k}$.

The equation $l_k = q$ (where q is an integer) defines a hyperplane in the space of exponents k. Such hyperplanes divide a collection of μ exponents k into n parts between hyperplanes and $n-1$ additional parts landing on hyperplanes. Thus, we get a partition of the number μ into $n + (n-1)$ parts. The numbers of exponents in these parts are the Hodge numbers.

More precisely, the Hodge numbers $h^{p,q}$ form a matrix $(0 \le p, q \le n-1)$, in which the nonzero numbers fill out only two slanting lines: the numbers of basis monomials occurring strictly between hyperplanes lie on the auxiliary diagonal

$$h^{p,q} = (\# k \colon q < l_k < q+1) \quad \text{for } p+q = n-1; \tag{4}$$

the numbers of basis monomials landing on a hyperplane lie on the following slanting line:

$$h^{p,q} = (\# k \colon q = l_k) \quad \text{for } p+q = n. \tag{5}$$

We apply Steenbrink's equations (4) and (5) to the case of a homogeneous polynomial of degree m and an odd number n of variables. In this case $\alpha_s = 1/m$, and the number of basis monomials of any degree is the same as for the polynomial $f = x_1^m + \ldots + x_n^m$. For this polynomial the points $k+1$ are all the integral points of the cube $(0, m)^n$, and the plane $(k+1, \alpha) = q$ is the hyperplane passing through a vertex of the cube. Thus, the Hodge numbers are the numbers of integral points in slices between hyperplanes perpendicular to the main diagonal of the cube and passing through a vertex of the cube (for $p+q = n-1$), and on these hyperplanes (for $p+q = n$).

In particular, $h^{n/2,n/2} = \Pi_n(m)$ if n is even. The first assertion of the theorem is proved.

Now let n be odd. We consider the quasihomogeneous polynomial $f(x_1, \ldots, x_n) + x_0^2$. We have $\alpha_1 = \ldots = \alpha_n = 1/m$, $\alpha_0 = 1/2$. The degrees of the basis monomials are the same as for the polynomial $x_1^m + \ldots + x_n^m + x_0^2$. For this polynomial the points k + 1 are all integral points of the n-dimensional cube $(0, m)^n \times 1$ in (n + 1)-dimensional space. The planes $(k + 1, \alpha) = q$ cut hyperplanes out of the planes of this n-dimensional cube, dividing it into parts; the integral points interior to and on the boundary of the slices formed give the Hodge numbers according to (4) and (5).

Let $(\varkappa_1, \ldots, \varkappa_n, 1)$ be the coordinates of an indicated integral point. Then from (4) and (5) we have

$$h^{p,q} = (\# \varkappa \in (0, m)^n\colon m(q - 1/2) < \textstyle\sum \varkappa_i < m(q + 1/2)) \text{ for } p + q = n,$$

$$h^{p,q} = (\# \varkappa \in (0, m)^n\colon m(q - 1/2) = \textstyle\sum \varkappa_i) \qquad \text{for } p + q = n + 1.$$

In particular,

$$h^{(n+1)/2,\,(n+1)/2} = (\# \varkappa \in (0, m)^n\colon \textstyle\sum \varkappa_i = mn/2) = \Pi_n(m).$$

The theorem is proved.

Remark. In the quasihomogeneous case the characteristic numbers of monodromy on spaces corresponding to $h^{p,q}$ from the auxiliary diagonal $(p + q = n - 1)$ are different from 1, and on spaces corresponding to the next line $(p + q = n)$ are equal to 1. Hence, the Hodge numbers $h^{n/2,n/2}$ and $h^{(n+1)/2,(n+1)/2}$ which take part in the theorem proved are equal in the quasihomogeneous case to the special Hodge numbers

$$h^{n/2,\,n/2} = h_1^{n/2,\,n/2}, \quad h^{(n+1)/2,\,(n+1)/2} = h_1^{(n+1)/2,\,(n+1)/2}$$

(corresponding to the characteristic number 1 of the monodromy operator).

4. Petrovskii – Oleinik Inequalities

We consider a real projective hypersurface A of degree m, defined by a homogeneous polynomial f of n variables.

The Petrovskii–Oleinik inequalities consist of the following:

$$|\chi(A) - 1| \leqslant N_n(m), \quad \text{if } n \text{ is even};$$

$$|\chi(B_+) - \chi(B_-)| \leqslant N_n(m), \quad \text{if } n \text{ is odd and } m \text{ is even}$$

Here χ is the Euler characteristic, B_+ and B_- are the parts of $\mathbf{R}P^{n-1}$ defined by the conditions $f \geq 0$ and $f \leq 0$, respectively, the number $N_n(m)$ is defined in Sec. 3.

THEOREM. The Petrovskii–Oleinik inequality, in the case of an even as well as of an odd number of variables n, has the form

$$|\operatorname{ind}| \leqslant \Pi_n(m),$$

where ind is the index of the singular point O of the gradient of the homogeneous polynomial of degree m defining the hypersurface under consideration.

Proof. The left-hand sides of both Petrovskii–Oleinik inequalities coincide with |ind| according to Sec. 2, and the right-hand sides coincide with $\Pi_n(m)$ according to Sec. 3. The theorem is proved.

COROLLARY 1. The Petrovskii–Oleinik inequalities follow from the bound on the index of a singular point of a vector field cited in Sec. 1.

COROLLARY 2. The inequality of Sec. 1 for the index of a singular point of a vector field in $\mathbf{R}^n$, whose components are homogeneous polynomials of degree m – 1, is sharp for $n \leq 3$; the equality holds for gradient fields.

Proof. For $n \leq 3$ the Petrovskii–Oleinik inequalities are sharp.

Remark. For n = 2 it is easy to define an extremal field explicitly as $(\operatorname{grad} \operatorname{Re} z^m)$. For n = 3 the construction of an extremal field is not so simple. Rejection of gradientness of the field facilitates the problem of construction of examples. D. B. Fuks proved sharpness for m = 3 for any n: $f = (x_1 + \ldots + x_n)^3 - x_1^3 - \ldots - x_n^3$.

5

5. Case of Functions of Two Variables

We consider real functions of two variables with fixed Newton diagram Γ. What is the largest value which the index of a singular point O of the gradient of such a function can have?

THEOREM. The modulus of the index of a singular point O of finite multiplicity of the gradient of a function of two variables with fixed Newton diagram does not exceed the number of interior integral points on the diagram.

Example. For homogeneous functions of degree m, the number of interior integral points is equal to $m-1$ according to the bound of Sec. 1 and the example of Sec. 4.

Proof. It is not hard to verify that the index is equal to one minus the number of real branches of the zeroth level set.

The number of real branches does not exceed the number of complex branches. The number of complex branches does not exceed the number of segments into which integral points divide the compact faces of the Newton diagram (and is equal to this number for almost all functions with the given diagram). The theorem follows directly from the cited equalities and inequalities.

Remark. The number of complex branches without units can be evaluated as the dimension of the image of the homology of the boundary of a complex local nonsingular level manifold under inclusion of the boundary in the level manifold (k holes on a Riemann surface generate a subgroup of rank $k-1$ because of the relation: the sum of all holes is homologically null).

One can verify that this number of components of the boundary of a nonsingular level manifold lowered by one is the Hodge number $h_1^{1,1}$ of the mixed Hodge structure in the cohomology of this level manifold.

From the theorem proved follows:

COROLLARY. The index of the gradient of a function of two variables at a critical point of multiplicity μ does not exceed in modulus $C\sqrt{\mu}$, where the constant C is independent of the function.

The proof is based on the fact that the number of integral points on the diagram is bounded above by the quantity 2r, where (r, r) is a point of intersection of the diagram with the bisector of the coordinate angle, and μ is bounded below by a quantity of order r^2 (since monomials corresponding to integral points situated below the diagram and at a distance greater than 1 from it define C-independent elements in the local algebra).

Remark. In the case of n variables there is an analogous bound on the modulus of the index by the quantity $\mu^{(n-1)n}$, both for gradients and for general fields with a singular point of multiplicity μ [9].

Remark. An estimate of the index of a singular point by means of the Newton diagram for a nongradient vector field on the plane can be obtained from the recent work of F. S. Berezovskii (Pushchino, 1977).

6. Hodge Numbers and Newton Diagram

6.1. Newton Diagram. The Newton diagram of a power series is introduced in [1, 11] and is defined as follows. In the space of exponents of terms of the series one marks exponents of monomials appearing in the series with nonzero coefficients. For each of the marked exponents one constructs an octant consisting of points each of whose coordinates is greater than or equal to the corresponding coordinate of the distinguished point. The Newton polyhedron is the convex hull of the union of the octants so obtained. The Newton diagram is made up of the compact faces of the Newton polyhedron.

A Newton diagram is called suitable if it intersects all the coordinate axes. A suitable Newton diagram defines the Newton filtration in the ring of power series: a series has filtration d [belongs to A (d), $d \in \mathbf{Q}$], if its Newton polyhedron lies in the polyhedron obtained from the Newton polyhedron of the original diagram by a dilation of d times with center at zero.

6.2. Poincaré Series of a Face. With each face σ of a Newton diagram is connected the ring A_σ which is the ring of power series, the exponents of the nonzero terms of which lie in the cone with directrix σ and vertex O. It is convenient to define the dimension of the face $d(\sigma)$ as the dimension of this cone and to introduce a formal zero-dimensional face O with ring A_O, consisting of the constants. The Newton filtration defines a filtration in the ring A_σ. We denote by PA_σ the Poincaré series defined by this filtration:

$$P_{A_\sigma}(t) = \sum_{\alpha \in \mathbf{Q}} p_{A_\sigma}(\alpha)\, t^\alpha,$$

[$p_{A_\sigma}(\alpha)$ is the number of monomials of quasidegree α in A_σ, i.e., the dimension of the factor space $A_\sigma(\alpha)/A_\sigma(>\alpha)$].

Example. The Newton diagram of the series $x_1^m + \ldots + x_n^m$ is the simplex $m_s \geq 0$, $\Sigma m_s = m$. The faces σ of dimension n correspond to the ring $A_\sigma = A$. The Poincaré series is

$$P_{A_\sigma}(t) = (1 - t^{1/m})^{-n}.$$

This is the generating function for the number of integral points on planes perpendicular to the diagonal of an octant.

6.3. Interior Poincaré Series of a Face. It is convenient to introduce another "interior Poincaré series"

$$Q_{A_\sigma}(t) = \sum_{\alpha \in \mathbf{Q}} q_{A_\sigma}(\alpha) t^\alpha$$

[$q_{A_\sigma}(\alpha)$ is the number of monomials of quasidegree α in A_σ, whose exponents lie in the interior of the cone of the face σ, i.e., do not lie in cones of proper subfaces]. We denote by $\tau \leq \sigma$ the relation "τ is a subface of σ." The series P and Q can be expressed in terms of one another:

$$P_{A_\sigma} = \sum_{\tau \leqslant \sigma} Q_{A_\tau}, \quad Q_{A_\sigma} = \sum_{\tau \leqslant \sigma} (-1)^{d(\sigma)-d(\tau)} P_{A_\tau}$$

(in the second case one uses the Euler characteristic of the star of the face τ; we note that $0 \leq \sigma$, $P_{A_0} = Q_{A_0} = 1$).

Example. In the example considered earlier

$$Q_{A_\sigma}(t) = t^{n/m}(1 - t^{1/m})^{-n}.$$

The Poincaré series P_{A_σ} (which means also Q_{A_σ}) in the general case is explicitly computed easily by subdividing σ into simplices and using the cited example.

6.4. Local Ring of a Face. We consider a series f with fixed Newton diagram. By the principal part of the series is meant the sum of all terms of the series whose exponents belong to the Newton diagram.

Each face σ distinguishes the σ-principal part of the series, f_σ being the sum of all terms of the principal part of the series f, whose exponents belong to the face σ. The local ring of the face σ (for the series f) is defined as the factor ring

$$B_\sigma = A_\sigma/(F_{\sigma,1}, \ldots, F_{\sigma,n}), \text{ where } F_{\sigma,s} = x_s \partial f_\sigma / \partial x_s.$$

The series f is called Γ-nondegenerate (for the Newton diagram Γ) if all the rings B_σ are finite-dimensional (cf. Sec. 1.20 of Kushnirenko's paper [11]).

6.5. Poincaré Polynomials of a Face and Interiorness. The Newton filtration defines a filtration in the ring B_σ; we denote its Poincaré polynomial by

$$P_{B_\sigma}(t) = \sum_{\alpha \in \mathbf{Q}} p_{B_\sigma}(\alpha) t^\alpha$$

[$p_{B_\sigma}(\alpha)$ is the number of terms of degree α in the collection of monomials from A_σ, whose classes form a basis in the linear space B_σ].

Of fundamental value in what follows is the "interior Poincaré polynomial"

$$Q_{B_\sigma}(t) = \sum_{\alpha \in \mathbf{Q}} q_{B_\sigma}(\alpha) t^\alpha$$

[$q_{B_\sigma}(\alpha)$ is the number of monomials of the basis indicated above, whose exponents lie strictly interior to the cone of the face σ].

We assume as before that $P_{B_0} = Q_{B_0} = 1$; the expressions for P and Q in terms of one another are just as in 6.3:

$$P_{B_\sigma} = \sum_{\tau \leqslant \sigma} Q_{B_\tau}, \quad Q_{B_\sigma} = \sum_{\tau \leqslant \sigma} (-1)^{d(\sigma)-d(\tau)} P_{B_\tau}.$$

Example. Let $f(z) = z_1^m + \ldots + z_n^m$, σ be the unique face of dimension $d(\sigma) = n$ (a simplex). Then Q_{B_σ} is the generating polynomial of the number of integral points strictly interior to the cube $(0, m)^n$ on planes perpendicular to the principal diagonal. More precisely,

$$P_{B_\sigma}(t) = \left(\frac{1-t}{1-t^{1/m}}\right)^n, \quad Q_{B_\sigma}(t) = \left(\frac{t^{1/m}-t}{1-t^{1/m}}\right)^n.$$

Remark. In general, P_B can always be expressed in terms of P_A:

$$P_{B_\sigma}(t) = (1-t)^{d(\sigma)} P_{A_\sigma}(t).$$

(This follows from the acyclicity of the Koszul complex constructed on the elements $F_{\sigma,s}$ in A_σ, cf. [11].) In particular, p_{B_σ} and q_{B_σ} are independent of just which Γ-nondegenerate function was used to construct the ring B_σ.

6.6. Combinatorial Polynomial of a Face. To each face τ of the Newton diagram in n-dimensional space we associate a polynomial R_τ in a variable s, defined in the following way:

$$R_\tau(s) = \sum_{\sigma > \tau} (-1)^{n-k(\sigma)} (s-1)^{k(\sigma)-d(\sigma)},$$

where $k(\sigma)$ is the dimension of the minimal coordinate subspace containing σ.

The polynomial R_τ is called the combinatorial polynomial of the face τ.

Example 1. Let τ be an interior face of the Newton diagram, i.e., $k(\tau) = n$. Then the combinatorial polynomial has the form

$$R_\tau(s) = \alpha_0 + (s-1)\alpha_1 + (s-1)^2\alpha_2 + \ldots + (s-1)^\nu,$$

where α_0 is the number of faces of highest dimension adjacent to τ, α_1 is that for the next dimension, etc., $\nu = n - d(\tau) = \operatorname{codim} \tau$.

Example 2. Let $\tau = O$. Then $R_\tau(1) = \beta_0 - \beta_1 + \beta_2 - \ldots \pm 1$, where β_0 is the number of faces of highest dimension in the whole diagram, β_1 is the number of highest dimensional faces of the traces of the diagram in the coordinate planes, etc.

6.7. Generating Polynomial of the Newton Diagram. This polynomial of two arguments s and t is defined as

$$P(s,t) = \sum_\tau R_\tau(s)\, Q_\tau(s,t),$$

where R is the combinatorial polynomial of the face and Q is expressed in terms of the interior Poincaré polynomial by the equation

$$Q_\tau(s,t) = (s/t)^{d(\tau)/2} Q_{B_\tau}(t).$$

The generating polynomial of the Newton diagram can be written as a finite sum

$$P(s,t) = \sum p_{\alpha,\beta}\, s^\alpha t^\beta$$

(2α are integers, β are rational).

Steenbrink's conjecture, formulated later in this article, expresses the Hodge numbers in terms of the coefficients of this polynomial.

6.8. Characteristic Pairs. The characteristic pairs of monodromy are defined by Steenbrink [16] and are in one-to-one correspondence with its characteristic numbers. They are pairs of the form (l, w), where l is a rational number, and w is an integer. To each pair corresponds a characteristic number $\lambda = e^{2\pi i l}$ of the action of monodromy on the space $H^{p,q}$ of the mixed Hodge structure, where q is the integral part of l and w is $p+q$ for $\lambda \neq 1$ and $p+q-1$ for $\lambda = 1$.

It is clear that the collection of all μ of characteristic pairs determines all the Hodge numbers $h^{p,q}$ and $h_1^{p,q}$ and the characteristic numbers of monodromy on each of the spaces $H^{p,q}$.

Example. Let the number of arguments n of the function f be even. Then the Hodge number $h_1^{n/2,n/2}$ is equal to the number of characteristic pairs of the form $(n/2, n-1)$.

Example. Let f be a quasihomogeneous, nondegenerate function of n variables. Then, as was proved by Steenbrink [14], the characteristic pairs have the form $(l, n-1)$, where l is the quasidegree of the differential form $x^m dx_1 \wedge \cdots \wedge dx_n$, $\{x^m\}$ is the system of monomial generators of the local ring.

6.9. Steenbrink's Conjecture. Let f be a Γ-nondegenerate function of n variables with suitable Newton diagram Γ. We denote by $C(l, w)$ the number of characteristic pairs (l, w). Steenbrink's conjecture consists of the following expression for the number C in terms of the coefficients of the generating polynomial of the simplicial Newton diagram:

$$C(n-(\alpha+\beta),\ 2n-1-2\alpha)=p_{\alpha,\beta}. \qquad (**)$$

Example. $C(n/2, n-1) = p_{n/2,0}$.

Remark. The conjecture formulated in Steenbrink's preprint [16] differs from that cited here (probably due to misprints).

COROLLARY. From the conjectures (*) and (**) follows the following upper bound for the index of the gradient of a real, Γ-nondegenerate function of n variables in terms of the Newton diagram:

$$|\operatorname{ind}|\leqslant p_{n/2,0}(\Gamma). \qquad (***)$$

Here n can be even as well as odd (addition of the square of a superfluous variable increases n by one, l by 1/2, w by one; here the exponents α in the generating polynomial are increased by 1/2, and β are unchanged).

6.10. Examples. The conjectures (*), (**), and (***) are confirmed by many examples, in particular, they are verified for quasihomogeneous functions, for functions of two variables, or functions obtained from them by adding squares of additional variables. A more striking example is furnished by the function

$$f(x, y, z)=-x^2y^2z^2+x^a+y^b+z^c, \quad a, b, c>6.$$

It is easy to compute the index of the gradient, representing it by the zeroth level surface of the function f. If (a, b, c) are odd and relatively prime, then the local zeroth level set is homeomorphic to a union of two planes, and if one of them is even, then three. Correspondingly |ind| = 1 in the first case and |ind| = 2 in the second.

Calculating from (**) shows that $p_{3/2,0} = 1$ in the first case and $p_{3/2,0} = 2$ in the second. This confirms (***).

Calculations of Hodge numbers for the function of four variables obtained from f by adding a square of a new variable give, according to (**) for relatively prime a, b, and c

$$h_1^{1,1}=1,\ h_1^{2,2}=1,\ h_1^{3,3}=1 \quad (a, b, c \text{ odd}),$$
$$h_1^{1,1}=1,\ h_1^{2,2}=2,\ h_1^{3,3}=1 \ (a \text{ even},\ b \text{ and } c \text{ odd}).$$

This example bears on the conjecture (*).

7. Remarks

7.1. Combinatorial polynomials of simplicial Newton diagrams have special properties.

More generally we consider a simplicial decomposition of a sphere in $\mathbf{R}^{\nu}$. We denote by α_0 the number of simplices of highest dimension, by α_1 the next, . . . , by $\alpha_{\nu-1}$ the number of vertices. We form the combinatorial polynomial

$$R(s)=\alpha_0+(s-1)\alpha_1+\ldots+(s-1)^{\nu-1}\alpha_{\nu-1}+(s-1)^{\nu}.$$

Example 1 of Sec. 6.6 suggests the following conjecture: the polynomial R(s) is reflexive and has positive coefficients. In particular, if ν is odd, then $R(-1) = 0$, i.e., $\alpha_0 - 2\alpha_1 + 4\alpha_2 - \ldots = 2^{\nu}$.

A. G. Khovanskii indicated to me that in some cases the formulated conjecture reduces to Serre duality and that Klee [20] and Ehlers [21] proved the reflexivity for simplicially subdivided manifolds with the Euler characteristic of the sphere. V. I. Danilov and also A. G. Kushnirenko and D. B. Fuks proved the conjecture on positivity by the method of Reisner [23], described by Stanley [22].

7.2. In the case when all the faces of the Newton diagram are simplices, the coefficients of the combinatorial polynomials of all faces are apparently nonnegative. In particular, apparently, the alternating sum of the numbers of faces of highest dimension in the coordinate planes, $\beta_0 - \beta_1 + \ldots \pm 1$, is nonnegative (Example 2 of Sec. 6.6).

In the general case, it follows from Steenbrink's conjecture (**) that all coefficients of the generating polynomial of the Newton diagram are nonnegative. Obviously they are also symmetric with respect to the point $\alpha = n/2$, $\beta = 0$.

7.3. Steenbrink's conjecture generalizes Kushnirenko's formula for the Milnor numbers [11] and Varchenko's formula for the characteristic polynomial of monodromy [5]. In our notation these formulas take the forms

$$\mu = \sum R_\tau(1)\mu(\tau), \quad \chi = \prod \chi_\tau^{R_\tau(1)},$$

where χ_T is a polynomial with roots $e^{2\pi i \alpha}$ of multiplicity $q_{B_T}(\alpha)$, and $\mu(\tau)$ is its degree; the numbers $q_{B_T}(\alpha)$ are defined in Sec. 6.4; the sum and product are taken over all faces of the Newton diagram (including zero).

In the simplicial case, the numbers $R_T(1)$, apparently, are nonnegative.

7.4. It is natural to begin the proof of the conjecture (*) with the study of complex conjugation σ in the (reduced) homology H_* of the local level set $f = \varepsilon > 0$. Let us assume for definiteness that the number of variables $n \equiv 3 \pmod 4$. Then, by analogy with the Atiyah–Singer formula for involutions [4], one may think that

$$\text{ind} = (\text{sgn}\langle \sigma\cdot, \cdot\rangle - \text{sgn}\langle \sigma'\cdot, \cdot\rangle)/2,$$

where σ' is the action of complex conjugation in homology, $f = -\varepsilon$, and the brackets are for the intersection index.

Gusein-Zade, whom I informed of this conjecture, proved it for functions $f(x, y) + z^2$, and in the general case also proved the formula

$$\text{ind} = \text{sgn}\,\text{Var}^{-1}\sigma$$

(cf. [7] for the definition of the isomorphism Var: $H^* \to H_*$, the operator $\text{Var}^{-1}\sigma$ is symmetric).

Another approach to the proof of (*) is based on the equation

$$2i = 2 - \chi,$$

connecting the index i of a singular point of a gradient vector field in four-dimensional space with the Euler characteristic χ of the boundary of the real local level manifold. The boundary **RK** of the real level manifold is the set of fixed points of the involution σ of complex conjugation on the boundary **CK** of the complex level manifold, having real dimension $2n - 3$. For $n > 1$ the dimensions of both middle homology groups of the boundary of the complex local level manifold are equal to the dimension of the homology space invariant with respect to monodromy. Thus, (*) suggests the necessity of considering mixed Hodge structures in the middle homology of the boundary of the complex level manifold.

By the Lefschetz formula (cf. [4]), $\chi(\mathbf{RK}) = \Sigma(-1)^k \text{tr}(\sigma | H_k(\mathbf{CK}))$. The manifold CK has nontrivial homology only in dimensions $0, n-2, n-1, 2n-3$. Since σ changes the orientation of CK, we have $\chi(\mathbf{RK}) = 2 - 2\,\text{tr}(\sigma | H_{n-1}(\mathbf{CK}))$, $i = \text{tr}(\sigma | H_{n-1}(\mathbf{CK}))$. The inclusion of the boundary in the local nonsingular level manifold identifies $H_{n-1}(\mathbf{CK})$ with the group of cycles invariant under monodromy. Consequently, the index i is equal to the trace of the complex conjugation σ on the invariant cycles. Since σ interchanges p and q in $H^{p,q}$, we get the estimate $|i| \le \Sigma_{1,1}^{p,p} = h_1^{n/2, n/2}$ of the index in terms of the sum of the dimensions of the middle Hodge components of the space of invariant cocycles (the second subscript one indicates that one considers the characteristic and not the adjoint space). This, apparently, proves (*).

7.5. Resolution of singularities determines on the boundary of the local level manifold a singular structure as the union of fibrations with toroidal fibers.

Definition. By a Clemens complex structure on a closed real smooth manifold E is meant a continuous map $p: E \to B$, where the base B is a complex algebraic manifold which is a union of a finite number of transversely intersecting closed smooth projective manifolds B_i, and where p satisfies condition *Cl*, which is formulated below.

By a stratum of depth k we mean any nonempty intersection of k of the manifolds B_i; the corresponding open stratum is the complement in the given stratum of all strata of greater depth.

Condition *Cl*. Over each open stratum of depth k, the map p is a smooth fibration whose fiber is the k-dimensional torus $(S^1)^k$.

A real Clemens structure is defined analogously but the base is real and the fiber over a stratum of depth k is $(S^0)^k$, i.e., consists of 2^k points.

Proposition. The boundary of a complex (real) local critical level manifold of a complex (real) function of n variables at a singular point admits a complex (real) Clemens structure with $(n-2)$-dimensional base, which is the divisor cut out by the old critical level manifold on the closed algebraic manifold glued in at the singular point under the resolution of singularities.

7.6. We consider an arbitrary manifold equipped with a real Clemens structure.

Proposition. The Euler characteristic of a manifold equipped with a real Clemens structure is equal to the sum of the Euler characteristics of all closed strata of the base of odd depth.

It is clear that this proposition is only interesting in the case of an even-dimensional manifold and that it gives its Euler characteristic as the sum of the characteristics of all closed even-dimensional strata of the base.

Combining this proposition with the Atiyah–Singer representation of the Euler characteristic of each stratum as the signature of the involution of complex conjugation in the cohomology of the complex stratum, we get the following corollary.

COROLLARY. The Euler characteristic of the boundary of a real local level manifold of a real function of an even number of variables at a singular point does not exceed the sum of the middle Hodge numbers $h^{p,p}$ of all even-dimensional strata of the trace cut out of the complex critical level manifold of the function by the closed complex algebraic manifold glued in by resolution.

Remark. For complex Clemens structures and singularities with even-dimensional base, the role of the sum of the Euler characteristics is played by the sum of the signatures. This sum is independent of the resolution and is equal to $1 - \Sigma(-1)^q h^{p,q}$ (A. G. Khovanskii), and is also equal to $\operatorname{sgn} \mathrm{Var}^{-1}$ on $\mathrm{Ker}\,(T-1)$ for $f + z^2$.

7.7. The Newton diagram determines a partial resolution of the singularity. A. G. Kushnirenko proposed carrying the constructions of 7.5–7.6 over to this case in the hope of verifying (***), omitting (**) and (*).

According to Sec. 6, the right-hand side of (***) has the form of the sum, over all faces τ of the Newton polyhedron, of products $q_\tau r_\tau$, where $q_\tau = q_{B_\tau}(d/2)$ can be called the number of interior generators of middle degree in the local ring of the face, and r_τ can be called the combinatorial coefficient of the face; this is the coefficient of $s^{(n-d)/2}$ in the combinatorial polynomial R_τ of the face τ of dimension $d = d(\tau)$.

Proposition. Let the diagram dual to the Newton diagram of a function of an even number of variables be simplicial. Then the combinatorial coefficient of each interior face of even dimension is equal to one, and of odd dimension to zero.

LITERATURE CITED

1. V. I. Arnol'd, "Normal forms of functions in the neighborhood of degenerate critical points," Usp. Mat. Nauk, 29, No. 2, 11–49 (1974).
2. V. I. Arnol'd, "Some unsolved problems of the theory of singularities. The theory of cubic formulas and the application of functional analysis to problems of mathematical physics," Tr. Sem. S. L. Soboleva, No. 1, Novosibirsk (1976).
3. V. I. Arnol'd, "Modern development of the work of I. G. Petrovskii in the topology of real algebraic manifolds," Usp. Mat. Nauk, 32, No. 3, 215–216 (1977).
4. M. F. Atiyah and I. M. Singer, "The index of elliptic operators. III," Usp. Mat. Nauk, 26, No. 1, 127–182 (1969).
5. A. N. Varchenko, "The characteristic polynomial of the monodromy and the Newton diagram," Invent. Math., 37, 253–262 (1976).
6. A. Hurwitz, "Über Riemannsche Flächen mit gegebenen Verzweigungspunkten," Math. Ann., 39, 1–61 (1891).
7. S. M. Gusein-Zade, "Monodromy groups of isolated singularities of hypersurfaces," Usp. Mat. Nauk, 32, No. 2, 23–65 (1977).
8. P. Deligne, "Poids dans la cohomologie des variétés algébriques," in: Proceedings of the International Congress of Math., Vol. I, Vancouver (1974), pp. 79–85.
9. D. Eisenbud and H. Levine, "The topological degree of a finite C$^\infty$-map germ," Ann. Math., 106, No. 1, 19–38 (1977).
10. F. Klein, Riemannsche Flächen (lit.) Vorlesungen, Vols. I, II, Göttingen (1892, new printing 1906).
11. A. G. Kuchnirenko, "Polyhedres de Newton et nombres de Milnor," Invent. Math., 32, No. 1, 1–32 (1976).
12. I. G. Petrovskii, "On the topology of real plane algebraic curves," Ann. Math., 39, 187–209 (1938).
13. I. G. Petrovskii and O. A. Oleinik, "On the topology of real algebraic surfaces," Izv. Akad. Nauk SSSR, Ser. Mat., 13, 389–402 (1949).

14. J. H. M. Steenbrink, "Intersection form for quasihomogeneous singularities," Report 75-09, Amsterdam Univ. (1975).
15. J. H. M. Steenbrink, "Mixed Hodge structure on the vanishing cohomology," Report 76-06, Amsterdam Univ. (1976).
16. J. H. M. Steenbrink, "Appendix to [15]," Amsterdam Univ. (1977).
17. V. M. Kharlamov, "A generalization of Petrovskii's inequality," Funkts. Anal. Prilozhen., 8, No. 2, 50-56 (1974).
18. V. M. Kharlamov, "A generalization of Petrovskii's inequality. II," Funkts. Anal. Prilozhen., 9, No. 3, 93-94 (1975).
19. G. N. Khimshiashvili, "On the local degree of a smooth map," Soobshch. Akad. Nauk GruzSSR, 85, No. 2, 309-311 (1977).
20. V. Klee, "A combinatorial analog of Poincaré duality theorem," Can. J. Math., 16, 517-531 (1964).
21. F. Ehlers, "Eine Klasse Komplexer Mannigfaltigheiten und die Auflösung einiger isolierten Singularitäten," Math. Ann., 218, 127-156 (1975).
22. R. P. Stanley, "The upper bound conjecture and Cohen-Macaulay rings," Stud. Appl. Math., 54, No. 2, 135-142 (1975).
23. G. A. Reisner, "Cohen-Macaulay quotients of polynomial rings," Ph. D. Thesis, Minn. Univ. (1974).

BEST LINEAR AND NONLINEAR APPROXIMATIONS FOR SMOOTH FUNCTIONS

B. E. Klots

UDC 517.5

1. We begin with definitions. In the Banach space X, suppose given sets K and N. By the deviation of K from N, we mean the quantity

$$\delta(K, N; X) = \sup_{x\in K} \inf_{y\in N} \| x - y \|. \tag{1}$$

Along with deviation (1) one considers the linear deviation of K from N:

$$\Delta(K, N; X) = \inf_{AK\subset N} \sup_{x\in K} \| x - Ax \|. \tag{2}$$

The infimum here is taken over all continuous linear operators $A: X \to X$ such that $AK \subset N$. It is clear that

$$\delta(K, N; X) \leqslant \Delta(K, N; X). \tag{3}$$

By $L_p = L_p([0, 2\pi])$, we shall, as usual, denote the space of real 2π-periodic summable functions

$$\| x \|_p^p = \frac{1}{2\pi}\int_0^{2\pi} |x(t)|^p dt, \quad 1 \leqslant p \leqslant \infty; \quad \| x \|_\infty = \text{ess sup} | x(t) |;$$

C is the space of all continuous periodic functions; $M = C^*$ is the space of regular Borel measures on the circle, realized as the segment $[0, 2\pi]$ with the ends identified; S is the space of 2π-periodic generalized functions; $\mathcal{T}_N$ is the subspace of trigonometric polynomials of degree not greater than N; $\mathcal{T} = \bigcup \mathcal{T}_N$.

Let $0 < n$ be an even number. We introduce the operators

$$\begin{gathered} I^n x(t) = (\Psi_n * x)(t) = \frac{1}{2\pi}\int_0^{2\pi} \Psi_n(t - s) x(s)\, ds, \\ \Psi_n(t) = i^{-n} + \sum_{k\neq 0} (ik)^{-n} e^{ikt}. \end{gathered} \tag{4}$$

We note that $\operatorname{Ker} I^n = 0$. The operator $(I^n)^{-1}$ will be denoted by D^n. It is easy to see that D^n differs from the n-th derivative by a linear functional,

G. M. Krzhizhanovskii Energy Institute. Translated from Funktsional'nyi Analiz i Ego Prilozheniya, Vol. 12, No. 1, pp. 15-24, January-March, 1978. Original article submitted March 26, 1976; revision submitted April 25, 1977.

 0016-2663/78/1201-0012$07.50

Russian Math. Surveys 33:5 (1978), 99–116 Uspekhi Mat. Nauk 33:5 (1978), 91–105

CRITICAL POINTS OF FUNCTIONS ON A MANIFOLD WITH BOUNDARY, THE SIMPLE LIE GROUPS B_k, C_k, AND F_4 AND SINGULARITIES OF EVOLUTES

V. I. Arnol'd

Contents

Introduction

In the investigation of degenerate critical points of functions, a useful instrument are the simple Lie groups whose Dynkin diagrams contain only ordinary edges, that is, the groups A_k, D_k, and E_k (see [1] and [9]). It turns out that the groups with double edges, that is, B_k, C_k, and F_4, are connected in a similar way with the theory of critical points of functions on a manifold with boundary.

The bifurcation diagrams of B_2 and B_3 appeared unexpectedly in the investigation of the geometry of families of evolutes of plane curves in general position. The classification given below of simple critical points of functions on a manifold with boundary is the result of an attempt to understand this observation.

Going over to Dynkin diagrams with edges of different lengths leads to a distinctive variant of the Picard–Lefschetz theory in which, along with vanishing cycles, "vanishing half-cycles" are involved. The example of simple singularities suggests in the general situation a new, rather unusual, variant of relative monodromy on the space of vanishing cycles and half-cycles. In particular, for the relative case we have succeeded in defining a form of intersections, and consequently, elliptic, parabolic, and hyperbolic singularities. Below we show that elliptic singularities on a manifold with boundary are precisely the simple singularities. We have found three one-parameter families of quasihomogeneous parabolic singularities (which apparently exhaust all the parabolic singularities). We also classify all unimodel singularities of functions of two variables.

The author wishes to thank V. I. Borzenko, V. V. Goryunov, V. I. Guts, V. I. Matov, D. B. Fuks, S. M. Husein-Zade, and V. P. Palamodov for useful discussions.

§1. Classification of simple critical points of functions on the boundary of a manifold

In what follows, a manifold with boundary is understood to be a smooth (real or complex) manifold with a fixed smooth hypersurface. Two functions on a manifold with boundary are called *equivalent* if one goes over into the other under a diffeomorphism of the manifold that takes the boundary into itself.

On the boundary we consider a distinguished point O. The group of germs of diffeomorphisms of a manifold with boundary at a distinguished point that keep the boundary fixed acts on the spaces of germs and jets of functions at the distinguished point for which this is a critical point with critical value zero.

DEFINITION. The *modality* of a critical point of a function with critical value zero is defined as the smallest number m such that a sufficiently small neighbourhood of the jet of this function in the space of jets of arbitrarily high order intersects at most finitely many of at most m-parameter families of orbits of the action indicated above.

A critical point is called *simple* if its modality is zero.

THEOREM 1. *The simple critical points of functions on the boundary of a manifold with boundary are exhausted, to within equivalence, by the following list of germs of functions $f(x, y)$ at the point $x = y = 0$ of the boundary $x = 0$:*

$$B_k = \pm x^k \pm y^2, \qquad C_k = xy \pm y^k, \qquad F_4 = \pm x^2 + y^3; \qquad k \geqslant 2.$$

Here equivalence of functions of different numbers of variables is understood to be stable equivalence, that is, equivalence of a function of a large number of variables to the sum of a function of a smaller number of variables and a non-degenerate quadratic form in the missing variables.

REMARK 1. The set of non-simple critical points has codimension 3 in the space of functions with critical point O and critical value 0.

REMARK 2. The codimensions of the sets of simple critical points of the types B_k, C_k, and F_k are equal to $k-2$.

REMARK 3. The singularities B_2 and C_2 are equivalent.

The proof of Theorem 1 is carried out in §8.

§2. Versal deformations and bifurcation diagrams

We carry over to the case of a manifold with boundary the basic ideas of the theory of critical points of functions (see [2] and [5]).

A versal deformation of the germ of a smooth function at a distinguished point of the boundary is understood henceforth to be a versal deformation for the group of diffeomorphisms that take the boundary into itself. For the germ at zero of a function $f(x_1, \ldots, x_n)$ at the point $x = 0$ of the boundary $x_1 = 0$ we can take as such a deformation one of the form

$$F(x, \lambda) = f(x) + \lambda_1 e_1 + \ldots + \lambda_\mu e_\mu,$$

where the numbers λ_k are the parameters of the deformation and the e_k are monomials in x that give a basis of the local ring

$$Q_{f\,|\,x_1} = \mathbf{C}\,[[x_1, \ldots, x_n]]/(x_1 f_1, f_2, \ldots, f_n), \qquad f_k = \partial f/\partial x_k$$

(for definiteness we consider the complex case). The space $\mathbf{C}^\mu$ with coordinates $\lambda_1, \ldots, \lambda_\mu$ is called the *base* of the deformation F.

We define the *multiplicity* of a critical point by the formula

$$\mu\,(f|x_1) = \dim_{\mathbf{C}} Q_{f\,|\,x_1}.$$

Critical points of infinite multiplicity form a set of infinite codimension. In what follows we assume that the multiplicities of the critical points under consideration are finite.

The non-singular local level manifolds and the bifurcation diagram of zeros for a critical point of finite multiplicity of a holomorphic function on a manifold with boundary $x_1 = 0$ are defined in the following way.

We fix a versal deformation F and choose a sufficiently small ball $B_\rho = \{x \in \mathbf{C}^n : |x| \leqslant \rho\}$. Depending on ρ we choose a sufficiently small δ and consider for a λ in the ball $|\lambda| \leqslant \delta$ the local level set $V_\lambda = \{x \in B_\rho : F(x, \lambda) = 0\}$.

DEFINITION. The local level set V_λ is called *non-singular* if 1) 0 is not a critical value for $F(\cdot, \lambda)$, 2) the manifold V_λ is transversal to the boundary (that is, the hyperplane $x_1 = 0$).

The set of singular values λ is denoted by

$$\Sigma_{f\,|\,x_1} = \{\lambda \in \mathbf{C}^\mu : |\lambda| \leqslant \delta, \quad V_\lambda \text{ singular}\}.$$

The germ of the hypersurface Σ at zero is called the *bifurcation diagram of zeros* for f. It is determined uniquely to within local biholomorphic diffeomorphisms of the base of the deformation at zero (and does not depend on the choice of F nor on ρ and δ).

REMARK. The bifurcation diagram of zeros of the germ of a function on a manifold with boundary consists of two hypersurfaces, corresponding to the two conditions of non-singularity 1) and 2): the level manifold may be non-smooth or non-transversal to the boundary.

The roots for the groups B_k, C_k, and F_4 (and the mirrors corresponding to them) are also of two unmixed types. This suggests the results presented below.

THEOREM 2. *The bifurcation diagram of zeros for a simple critical point B_μ, C_μ, or F_μ, embedded in $\mathbf{C}^\mu$, is biholomorphically equivalent to the manifold of irregular orbits of the corresponding group generated by reflections that acts on the complexification of Euclidean space.*

The complement of the bifurcation diagram is a space $K(\pi, 1)$, where π is Brieskorn's braid group [12], constructed with respect to the corresponding group generated by reflections.

PROOF. The versal deformations for B_μ (and $n = 1$) and for C_μ (and $n = 2$) can be taken in the form

$$B_\mu : F = x^\mu + \lambda_1 x^{\mu-1} + \ldots + \lambda_\mu,$$
$$C_\mu : F = xy + y^\mu + \lambda_1 y^{\mu-1} + \ldots + \lambda_\mu.$$

A local level set is smooth if the polynomial with the coefficients $\lambda_1, \ldots, \lambda_\mu$ has no multiple roots. It is transversal to the hyperplane $x = 0$ if zero is not a root of the polynomial.

The action of the group generated by reflections of B_μ (or C_μ) on the complexification $\mathbf{C}^\mu$ of Euclidean space reduces to arbitrary changes of sign of the coordinates z_k and permutations of them. The space of orbits is identified with the manifold $\mathbf{C}^\mu$ of polynomials with roots z_k^2. The space of irregular orbits (that is, the image of the mirrors $z_k = 0$, $z_i = \pm z_j$) under the factorization mapping $\mathbf{C}^\mu \to \mathbf{C}^\mu$, $z \mapsto \lambda$, is a space of polynomials with multiple ($z_i^2 = z_j^2$) or zero ($z_k = 0$) roots.

Information on the homotopy groups of the complement is now obtained in the usual way of [8] and [12]. In the cases B_μ and C_μ the theorem is proved.

In the case F_4 the proof proceeds by means of the theorems of Brieskorn [11] and Deligne [15] (see also a paper of Slodowy [19]).

REMARK. Versal deformations of quasihomogeneous singularities (and all the normal forms of simple singularities mentioned above are quasihomogeneous) can be taken to be quasihomogeneous. In the case of simple singularities the weights of the variables λ_k turn out to be positive; they are connected with the indices p_k of the corresponding Weyl groups and with their Coxeter numbers N just as in the cases A, D, and E ([2]).

More precisely, let $f(x_1, x_2)$ be a quasihomogeneous simple function on a manifold with boundary $x_1 = 0$ of degree 1 for weights $\deg x_k = \alpha_k$.

Then the Coxeter number N is defined by the formula

$$2/N = \alpha_1 + 2\alpha_2 - 1,$$

and the weights d_k of the basis monomials e_k are given by

$$Nd_k = p_k - 1 \qquad (k = 1, \ldots, \mu).$$

The numerical values are:

Type	α_1	α_2	N	e_k	$N \deg e_k$
B_μ	$1/\mu$	$1/2$	2μ	$1, x, \ldots, x^{\mu-1}$	$0, 2, \ldots, 2\mu-2$
C_μ	$(\mu-1)/\mu$	$1/\mu$	2μ	$1, y, \ldots, y^{\mu-1}$	$0, 2, \ldots, 2\mu-2$
F_4	$1/2$	$1/3$	12	$1, x, y, xy$	$0, 4, 6, 10$

It is easy to verify that the quasihomogeneous functions with critical point on a manifold with boundary $x_1 = 0$ for which the weights of all the λ_k are positive (that is, for which all the $d_k < 1$), are exhausted by the simple singularities B_μ, C_μ and F_4 (see [3]).

§3. Vanishing cycles and half-cycles

The bifurcation diagrams of the singularities B_μ and C_μ are the same, and the preceding arguments do not explain which of the series of singularities is connected with the root system of B_μ and which with that of C_μ. The distinction between short and long roots becomes apparent only in an attempt to associate (co)homology classes with the roots.

The fundamental group of the complement of the bifurcation diagram acts on the various homology groups that can be defined by means of a local nonsingular level manifold V_λ and boundary $x_1 = 0$. We denote by V'_λ the intersection of V_λ with the hyperplane $x_1 = 0$ of $\mathbf{C}^n$.

THEOREM 3. *The factor-space V_λ/V'_λ has the homotopy type of a bouquet of μ spheres of dimension $n-1$.*

We use the following notation:

$$\mu = \mu(f|x_1) = \dim_{\mathbf{C}} \mathbf{C}[x_1, \ldots, x_n]]/(x_1 f_1, f_2, \ldots, f_n);$$

$$\mu_1 = \mu(f) = \dim_{\mathbf{C}} \mathbf{C}[[x_1, \ldots, x_n]]/(f_1, \ldots, f_n);$$

$$\mu_0 = \mu(f|x_1 = 0) = \dim_{\mathbf{C}} \mathbf{C}[x_2, \ldots, x_n]]/(f_2, \ldots, f_n(0, x_2, \ldots, x_n)).$$

Thus, V_λ is homotopy equivalent to a bouquet of μ_1 of dimension $n-1$ (vanishing cycles) and its submanifold V'_λ is homotopy equivalent to a bouquet of μ_0 spheres of dimension $n-2$ (for $n = 2$, V'_λ reduces to $\mu_0 + 1$ points). It is easy to verify that $\mu = \mu_0 + \mu_1$ and that the μ spheres in Theorem 1 can be obtained from the μ_1 spheres representing the vanishing cycles of f and the μ_0 discs that cover the μ_0 vanishing spheres of the function $f|x_1 = 0$ inside the complement $V_\lambda - V'_\lambda$.

EXAMPLE. Let $n = 2$. Then (V_λ, V'_λ) is a Riemann surface with Betti number $b_1 = \mu_1$ and $\mu_0 + 1$ distinguished points on it. A basis of the one-dimensional relative cycles for (V_λ, V'_λ) can be chosen from the μ_1 circles that emanate from a distinguished point and determine vanishing cycles on V_λ and the μ_0 segments that join this distinguished point to the others. We can choose the system of circles and segments so that V_λ / V'_λ shrinks to its image in V_λ / V'_λ.

We call these generators vanishing cycles (for circles or spheres) and *vanishing half-cycles* (for segments or discs).

Furthermore, we can carry over to the relative case the idea of a distinguished basis of vanishing cycles. To this end we consider a line $\mathbf{C}^1$ in general position in the base of a versal deformation of f near the point 0 of the boundary $x_1 = 0$. This line intersects the bifurcation diagram of the zeros of $\Sigma(f \mid x_1)$ in μ points, of which μ_0 are on the intersection with the part of the bifurcation diagram where transversality is lost, and μ_1 on that where V_λ loses smoothness. To points of the second kind there correspond ordinary vanishing cycles. To points of the first kind there correspond singularities in a neighbourhood of which V_λ is given by an equation of the form $x_1 + x_2^2 + \ldots + x_n^2 = \varepsilon$. As $\varepsilon \to 0$, the cycle $\operatorname{Im} x = 0, x_2^2 + \ldots + x_n^2 = \varepsilon$, $x_1 = 0$ vanishes on V'_λ; a covering disc on V_λ is given by the equation $x_1 = \varepsilon - x_2^2 - \ldots - x_n^2$, $\operatorname{Im} x = 0$, $x_2^2 + \ldots + x_n^2 < \varepsilon$; this disc also determines the vanishing half-cycle. The basis of these μ vanishing cycles and half-cycles is also called *distinguished.*

The proofs of the theorem and the assertions are little different from those of the analogous assertions in the absolute case (see, for example, [17], [13], and [14]).

§4. The form of intersections

Although $\dim_{\mathbf{C}} H_{n-1}(V_\lambda, V'_\lambda) = \mu$, the most natural results are obtained by considering the action of the fundamental group of the complement of the bifurcation diagram not on this space, but on another space of the same dimension: the space of relative cycles with torsion coefficients. This new space is defined by means of a two-sheeted covering, and the introduction of it is a decisive factor in the present article (similar to the introduction of a two-sheeted covering of a plane, ramified along a curve, which is a decisive factor in [4] and subsequent articles on the geometry of real algebraic curves).

Let f be the germ of a holomorphic function of $x_1, \ldots, x_n$ at the point 0 of a manifold with boundary $x_1 = 0$. We consider a two-sheeted ramified covering $\varphi\colon \mathbf{C}^n \to \mathbf{C}^n$ with ramification on the boundary $x_1 = 0$. We denote by $\hat{f}$ the function f lifted to the covering, this is, $\hat{f} = \varphi^* f$. In other words,

$$\hat{f}(z; x_2, \ldots, x_n) = f(z^2; x_2, \ldots, x_n).$$

The local non-singular level manifold $\hat{V}\colon \hat{f} = \varepsilon$ doubly covers the local non-singular level manifold $V\colon f = \varepsilon$ with ramification along $V^{\cdot}\colon f = \varepsilon, x_1 = 0$. The mapping $z \mapsto -z$ determines a holomorphic involution of $\hat{V}$.

DEFINITION. The *form of intersections* of a singularity f at a point of the boundary $x_1 = 0$ is defined as the pair consisting of: 1) the space H^- of *anti-invariant* homology classes under the action of the involution on $H_{n-1}(\hat{V}, \mathbf{Z})$, and 2) the restriction to this space of the anti-invariant cycles H^- of the form of intersections of the covering manifold $\hat{V}$.

THEOREM 4. *The manifold $\hat{V}$ is homotopy equivalent to the bouquet of $\hat{\mu} = \mu_0 + 2\mu_1$ spheres of middle dimension (equal to $n-1$).*

The involution $z \mapsto -z$ of the pair $(\hat{V}, \hat{V}')$ is homotopy equivalent to an involution of the bouquet of the $\hat{\mu}$ spheres under which $2\mu_1$ of the spheres intersect in pairs and on μ_0 of the spheres the involution is the reflection in the equator.

For the μ_0 anti-invariant spheres and the μ_1 pairs of symmetric spheres we can take the geometric complete inverse images of the vanishing cycle and half-cycles of §3, having oriented these circuits in a suitable way.

EXAMPLE. For $n = 2$, $\hat{V}$ is a two-sheeted ramified covering of V with ramification at points of V'. The segment joining two points of V' on V is lifted to $\hat{V}$ as a circle. This circle changes orientation under the involution.

Theorem 4 is proved by the standard arguments (see [13] and [14]), like Theorem 3.

REMARK. To the homologies described above there correspond in analysis integrals of the form

$$\int \frac{\alpha(x)}{\sqrt{x_1}} \frac{dx_1 \wedge \ldots \wedge dx_n}{df}$$

over cycles of H^-.

§5. The monodromy group

The fundamental group of the complement of the bifurcation diagram $\Sigma(f\,|\,x_1)$ acts on the space defined in §4, preserving the form of intersections. We describe the system of generators of the monodromy group defined in this way.

DEFINITION. The homology class of the inverse image of a vanishing half-cycle in H^- is called a *short cycle*, and an element of H^- that is the difference of two orthogonal symmetric vanishing cycles of $\hat{V}$ is called a *long cycle.*

A *distinguished basis* of H^- is defined as a basis consisting of μ_0 short and μ_1 long cycles, corresponding to the distinguished basis of the relative homology in $H_{n-1}(V, V')$ constructed in §3 from the μ_0 vanishing half-cycles and μ_1 cycles corresponding to the points of intersection of a line in general position with the bifurcation diagram $\Sigma(f\,|\,x_1)$.

Let V be the non-singular level manifold corresponding to a point $* \in \mathbf{C}^1 \setminus \Sigma$, and p_i the points of intersection of $\mathbf{C}^1$ with Σ. A distinguished basis $\{e_i\}$ in H^- is constructed with respect to the system of paths α_i that do not intersect outside $*$ and lead from $*$ in p_i to $\mathbf{C}^1$. (The cycle e_i vanishes along the path α_i.) We consider the system of generators γ_i of the fundamental group $\pi_1(\mathbf{C}^1 \setminus \Sigma, *)$ that corresponds to this system of paths (the loop γ_i goes along α_i to p_i, goes round p_i in the positive direction, and returns to the old path).

We denote by T_γ the monodromy transformation $H^- \to H^-$ corresponding to $\gamma \in \pi_1$.

THEOREM 5. *The generator γ_i acts on H^- according to the generalized Picard–Lefschetz formulae:* $T_{\gamma_i}(x) = x + (-1)^{n(n+1)/2}(x, e_i)e_i/2$ *if e_i is a long cycle;* $T_{\gamma_i}(x) = x + (-1)^{n(n+1)/2}(x, e_i)\, e_i$ *if e_i is a short cycle. Here, as before, n is the number of arguments of the functions.*

The proof is the same as for the usual formulae (see [14]); I am grateful to S. M. Husein-Zade for verifying this.

§6. Elliptic singularities

The Dynkin diagrams $A_{2\mu-1}$, $D_{\mu+1}$, and E_6 have non-trivial involutions. An involution of the diagram induces an involution of the corresponding Euclidean space which preserves the scalar product, the set of simple roots, and the lattice generated by them.

We consider a subspace of our Euclidean space that is fixed under the involution. In this subspace the lattice of integral combinations of the simple roots cuts out a sublattice, which consists of all integral combinations of the simple roots in which the roots transposed by the involution have equal coefficients. It is, therefore, generated by those simple roots that are fixed under the involution, and by the sums of transposed roots and their images. The following lemma is well known and easy to verify.

LEMMA. *A subspace of the space of the diagram $A_{2\mu-1}$, $D_{\mu+1}$, or E_6 that is fixed under the involution and is endowed with the scalar product, the lattice, and the basis of this lattice described above, is none other than the Euclidean space of one of the diagrams B_μ, C_μ, or F_4, respectively, endowed with the standard system of simple roots and the lattice of their integral combinations.*

REMARK. We normalize the system of roots so that the scalar square of each short root is 2 and that of each long root is 4 (this coincides with the usual normalization in the cases C_μ; in the remaining cases our squares are twice as large as usual).

We now turn to singularities of functions of n variables on a manifold with boundary. We assume that $n \equiv 3 \bmod 4$ (this can be achieved without leaving the stable equivalence class).

DEFINITION. A singularity is called *elliptic* (respectively, *parabolic* or *hyperbolic*) if its quadratic form is negative definite (respectively, semidefinite, or has 1 positive square).

THEOREM 6. 1) *Elliptic singularities are precisely the simple singularities B_μ, C_μ, and F_4.*

2) *We can obtain the form of intersections of an elliptic singularity from the corresponding Dynkin diagram by considering in the Euclidean space the system of simple roots (short roots with square 2 and long roots with square 4); the form is defined on the lattice generated by the roots, and differs from the scalar product of the surrounding Euclidean space only in sign.*

The action of the fundamental group of the complement of the bifurcation diagram on H^- is the natural representation of the Brieskorn braid group on the corresponding Weyl group.

3) *For simple singularities f of the types B_μ, C_μ, and F_4 the function $\hat{f}$ also has a simple singularity of the types $A_{2\mu-1}$, $\hat{D}_{\mu+1}$, and E_6, respectively. Distinguished bases for $\hat{f}$ in $H_{n-1}(\hat{V})$ and for f in $H^-_{n-1}(\hat{V})$ can be chosen so that they are obtained from one another by the construction described in the lemma. The role of the involution of the Dynkin diagram is played by the composition of the involution of the covering (induced by the change of sign of z; see* §4) *and the involution -1 on $H_{n-1}(\hat{V})$.*

Theorem 6, 1) is proved in the same way as the analogous theorem of Tyurina [21] for the usual simple singularities; we need only calculate the quadratic forms of the first adjoining non-simple singularities. These singularities are considered in §§7–9.

2) and 3) are verified directly; this can be done very quickly by the method of Husein-Zade [14], which is easily adapted to the relative case (or the even case with respect to z).

§7. Classification of singularities

In this section we make a start on the classification of critical points of holomorphic functions $f(x_1, x_2, \ldots, x_n)$ with respect to diffeomorphisms that preserve the hyperplane $x_1 = 0$ and the critical point O; the critical value is also assumed to be zero. The classification is to within stable equivalence. For $n = 2$ the variables (x_1, x_2) are denoted by (x, y). Thus, the equation of the boundary in this case is $x = 0$, and y is the coordinate along the boundary.

LIST OF NORMAL FORMS

$$\text{Series } F\colon\quad \begin{array}{ccccccc} F_{k,0} & \leftarrow & F_{k,1} & \leftarrow & F_{k,2} & \leftarrow & \cdots \\ & & \uparrow & & & & \\ & & F_{6k+2} & \leftarrow & F_{6k+3} & \leftarrow & F_{6k+4} \leftarrow (F_{k+1,0}). \end{array}$$

Notation	Normal form	Restrictions	μ	cod
F_{6k+2}	$x^{3k+1}+y^3+\mathbf{a}x^{2k+1}y$	$k\geqslant 1$	$6k+2$	$5k$
F_{6k+3}	$x^{2k+1}y+y^3+\mathbf{a}x^{k+3}y^2$	$k\geqslant 1$	$6k+3$	$5k+1$
F_{6k+4}	$x^{3k+2}+y^3+\mathbf{a}x^{2k+2}y$	$k\geqslant 1$	$6k+4$	$5k+2$
$F_{k,0}$	$x^{3k}+bx^ky^2+y^3+\mathbf{c}x^{2k+1}y$	$k\geqslant 1$, $4b^3+27\neq 0$	$6k$	$5k-2$
$F_{k,p}$	$\mathbf{a}x^{3k+p}+x^ky^2+y^3$	$k\geqslant 1$, $a_0\neq 0$	$6k+p$	$5k-2+p$

Here $\underset{\sim}{a} = a_0 + \ldots + a_{k-1}^k x^{k-1}$ and $\underset{\sim}{c} = c_0 + \ldots + c_{k-2}x^{k-2}$ for $k \geqslant 2$; for $k = 1$, $\underset{\sim}{c} = 0$. Here and later, cod denotes the codimension of the class in the space of functions with critical point 0 and critical value zero.

We say that a function f *adjoins* a class of functions X if in any small neighbourhood of the jet of f in the space of jets of arbitrarily high order there are jets of functions of the class X.

THEOREM 7. *Every function with critical point of corank <2 on the* *boundary is either stably equivalent to B_μ, C_μ, or F_4, or to one of the functions of the series F or K, or it adjoins one of the classes K_0^*, K_0^{**}, or K_0^{***}.*

Every function with a critical point of corank $\geqslant 2$ on the boundary is either stably equivalent to one of the functions $f(x, y, z)$ of the class L_6 at the point 0 of the boundary $x = 0$

$$L_6 = y^3 + z^3 + axz + xy, \quad a^3 \neq 1, \quad \mu = 6, \quad \text{cod} = 3,$$

or it adjoins L_6.

Series K.

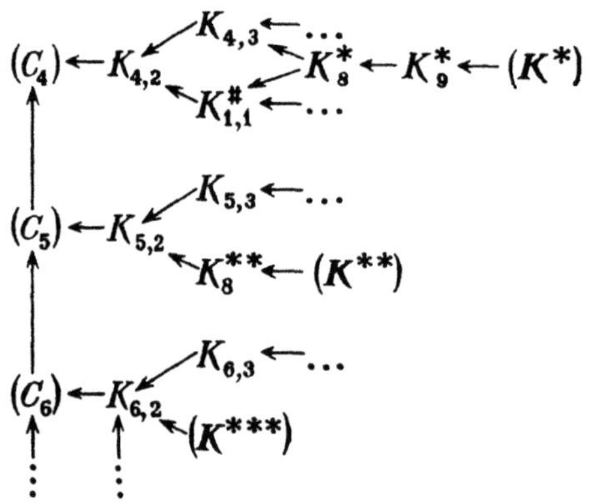

Notation	Normal form	Restrictions	μ	cod
$K_{4,2}$	$y^4+axy^2+x^2$	$a^2 \neq 4$	6	3
$K_{4,q}$	$y^4+xy^2+ax^q$	$a \neq 0,\ q > 2$	$q+4$	$q+1$
$K_{p,q}$	$y^p+xy^2+ax^q$	$a \neq 0,\ p > 4,\ q \geqslant 2$	$p+q$	$p+q-3$
$K^{\#}_{1,2p-3}$	$(y^2+x)^2+ax^py$	$a \neq 0$	$2p+3$	$2p$
$K^{\#}_{1,2p-4}$	$(y^2+x)^2+ax^p$	$a \neq 0$	$2p+2$	$2p-1$
K_8^*	$y^4+x^2y+ax^3$	—	8	5
K_9^*	$y^4+x^3+ax^2y^2$	—	9	6
K_0^*	$y^4+ax^2y^2+bx^3y+x^4+cx^3y^2$	$\Delta(a, b) \neq 0$	12	7
K_8^{**}	$y^5+x^2+axy^3$	—	8	5
K_0^{**}	$y^5+x^2y+(a+by)xy^3$	$a^2 \neq 4$	10	6
K_0^{***}	$y^6+(a+by)xy^3+x^2$	$a^2 \neq 4$	10	6

The parameters a, b, and c in all our normal forms are moduli, that is, a singularity equivalent to the given one occurs by changing these parameters finitely many times.

The proof of this theorem is sketched in §8.

§8. The determinant of singularities

We shall use the abridged notation of [5], the only difference being that all the equivalences preserve the boundary ($x = 0$). In the determinant, capital letters denote the stable equivalence classes defined above (for the definition of B_μ, C_μ and F_4, see §1). It is implied that each function depends on as many variables as are explicitly mentioned: $n = 2$ in Theorems **2–41**, and $n = 3$ in Theorems **43** and **44**. By F we always mean a function with critical point 0 and critical value zero; $\approx$ means quasihomogeneous equivalence of quasijets; $j_{(x^m)}f$ means the quasijet of f consisting of monomials whose exponents lie not above the hyperplane determined by the exponents of the m monomials in parentheses; $\Rightarrow$ means "implies", and $|\Rightarrow$ means "see".

THE DETERMINANT.

1. $\mu(f) < \infty \Rightarrow$ one of four:

the corank of the restriction of f to the boundary equal to 0 $|\Longrightarrow$ **2**,
1 $|\Longrightarrow$ **3**,
2 $|\Longrightarrow$ **42**,
>2 $|\Longrightarrow$ **46**.

2. The corank of the restriction of f to the boundary equal to 0 $\Rightarrow f \in B_k$, $k \geqslant 2$.

3. The corank of the restriction of f to the boundary equal to 1 $\Rightarrow$ the restriction belongs to A_{p-1}, $p > 2$.

4. $f|_{x=0} \in A_{p-1}$, $p > 2 \Longrightarrow f \sim y^p + x\varphi(x, y)$, $\varphi(0, 0) = 0$.

5. $f = y^p + x\varphi(x, y)$, $\varphi(0, 0) = 0$, $p > 2 \Longrightarrow$ one of five:

$p = 3$ $|\Longrightarrow$ **6**,
$p = 4$ $|\Longrightarrow$ **16**,
$p = 5$ $|\Longrightarrow$ **27**,
$p = 6$ $|\Longrightarrow$ **33**,
$p > 6$ $|\Longrightarrow$ **38**.

Series F.

6. $j_{y^3, x}f = y^3 \Longrightarrow$ one of three:

$j_{y^3, xy}f \approx y^3 + xy$ $|\Longrightarrow$ **7**,
$j_{y^3, x^2}f \approx y^3 + x^2$ $|\Longrightarrow$ **8**,
$j_{y^3, x^2}f \approx 0$ $|\Longrightarrow$ **13_1**.

7. $j_{y^3, xy}f = y^3 + xy \Longrightarrow f \in C_3$.

8. $j_{y^3, x^2}f = y^3 + x^2 \Longrightarrow f \in F_4$.

In Theorems **9–15**, $k \geqslant 1$.

9_k. $j_{y^3, x^{3k}}f = y^3 \Longrightarrow$ one of four:

$j_{y^3, x^{3k+1}}f \approx y^3 + x^{3k+1}$ $|\Longrightarrow$ **10_k**,
$j_{y^3, x^{2k+1}y}f \approx y^3 + x^{2k+1}y$ $|\Longrightarrow$ **11_k**,
$j_{y^3, x^{3k+2}}f \approx y^3 + x^{3k+2}$ $|\Longrightarrow$ **12_k**,
$j_{y^3, x^{3k+2}}f \approx y^3$ $|\Longrightarrow$ **13_{k+1}**.

10$_k$. $j_{y^3, x^{3k+1}} f = y^3 + x^{3k+1} \Longrightarrow f \in F_{6k+2}$.

11$_k$. $j_{y^3, x^{2k+1}y} f = y^3 + x^{2k+1}y \Longrightarrow f \in F_{6k+3}$.

12$_k$. $j_{y^3, x^{3k+2}} f = y^3 + x^{3k+2} \Longrightarrow f \in F_{6k+4}$.

13$_k$. $j_{y^3, x^{3k-1}} f \approx y^3 \Longrightarrow$ one of three:

$$j_{y^3, x^{3k}} f \approx y^3 + bx^k y^2 + x^{3k},\ 4b^3 + 27 \neq 0 \models\!\!\Longrightarrow 14_k,$$

$$j_{y^3, x^{3k}} f \approx y^3 + x^k y^2 \models\!\!\Longrightarrow 15_k,$$

$$j_{y^3, x^{3k}} f \approx y^3 \models\!\!\Longrightarrow 9_k.$$

14$_k$. $j_{y^3, x^{3k}} f = y^3 + bx^k y^2 + x^{3k},\ 4b^3 + 27 \neq 0 \Longrightarrow f \in F_{k,0}$.

15$_k$. $j_{y^3, x^k y^2} f = y^3 + x^k y^2 \Longrightarrow f \in F_{k,p},\quad p > 0$.

Series K_4.

16. $j_{y^4, xy} f \approx y^4 + xy \Longrightarrow f \in C_4$.

17. $j_{y^4, xy} f = y^4 \Longrightarrow$ one of four:

$$j_{y^4, x^2} f \approx y^4 + axy^2 + x^2,\ a^2 \neq 4 \models\!\!\Longrightarrow 18,$$

$$j_{y^4, x^2} f \approx y^4 + xy^2 \models\!\!\Longrightarrow 19,$$

$$j_{y^4, x^2} f \approx (y^2 + x)^2 \models\!\!\Longrightarrow 20,$$

$$j_{y^4, x^2} f \approx y^4 \models\!\!\Longrightarrow 21.$$

18. $j_{y^4, x^2} f = y^4 + axy^2 + x^2,\ a^2 \neq 4 \Longrightarrow f \in K_{4,2}$.

19. $j_{y^4, x^2} f = y^4 + xy^2 \Longrightarrow f \in K_{4,q},\ q > 2$.

20. $j_{y^4, x^2} f = (y^2 + x)^2 \Longrightarrow f \in K^{\#}_{1,q},\ q \geqslant 1$.

21. $j_{y^4, x^2} f = y^4 \Longrightarrow$ one of three:

$$j_{y^4, x^2 y} f \approx y^4 + x^2 y \models\!\!\Longrightarrow 22,$$

$$j_{y^4, x^3} f \approx y^4 + x^3 \models\!\!\Longrightarrow 23,$$

$$j_{y^4, x^3} f = y^4 \models\!\!\Longrightarrow 24.$$

22. $j_{y^4, x^2 y} f = y^4 + x^2 y \Longrightarrow f \in K^*_8$.

23. $j_{y^4, x^3} f = y^4 + x^3 \Longrightarrow f \in K^*_9$.

24. $j_{y^4, x^3} f = y^4 \Longrightarrow$ one of two:

$$j_{y^4, x^4} f \approx y^4 + ax^2y^2 + bx^3 y + y^4,\ \Delta(a, b) \neq 0 \models\!\!\Longrightarrow 25,$$

f adjoins the preceding class $| \Rightarrow$ **26**.

25. $j_{y^4, x^4} f = y^4 + ax^2y^2 + bx^3 y + x^4,\ \Delta(a, b) \neq 0 \Longrightarrow f \in K^*_0$.

26. f adjoins K^*_0 and $f \in K^*$.

Series K_5.

27. $j_{y^5, xy} f \approx y^5 + xy \Longrightarrow f \in C_5$.

28. $j_{y^5, xy} f = y^5 \Longrightarrow$ one of four:

$$j_{y^5, xy^2} f \approx y^5 + xy^2 \models\!\!\Longrightarrow 29,$$

$$j_{y^5, x^2} f \approx y^5 + x^2 \models\!\!\Longrightarrow 30,$$

$$j_{y^5, x^2 y} f \approx y^5 + x^2 y + axy^3,\quad a^2 \neq 4 \models\!\!\Longrightarrow 31,$$

f adjoins the preceding class $| \Rightarrow$ **32**.

29. $j_{y^5,\, xy^2} f = y^5 + xy^2 \Longrightarrow f \in K_{5,\,q},\ q \geqslant 2.$

30. $j_{y^5,\, x^2} f = y^5 + x^2 \Longrightarrow f \in K_8^{**}.$

31. $j_{y^5,\, x^2y} f = y^5 + x^2y + axy^3,\ a^2 \neq 4 \Longrightarrow f \in K_0^{**}.$

32. f adjoins K_0^{**} and $f \in K^{**}$.

Series K_6.

33. $j_{y^6,\, xy} f \approx y^6 + xy \Longrightarrow f \in C_6.$

34. $j_{y^6,\, xy} f = y^6 \Longrightarrow$ one of three:

$j_{y^6,\, xy^2} f \approx y^6 + xy^2 \mid\!\Longrightarrow$ **35**,

$j_{y^6,\, x^2} f \approx y^6 + axy^3 + x^2,\quad a^2 \neq 4 \mid\!\Longrightarrow$ **36**,

f adjoins the preceding class $\mid\Rightarrow$ **37**.

35. $j_{y^6,\, xy^2} f = y^6 + xy^2 \Longrightarrow f \in K_{6,\,q},\ q \geqslant 2.$

36. $j_{y^6,\, x^2} f = y^6 + axy^3 + x^2,\ a^2 \neq 4 \Longrightarrow f \in K_0^{***}.$

37. f adjoins K_0^{***}, $f \in K^{***}$.

Series K_p, $p > 6$.

38$_p$. $j_{y^p,\, xy} f \approx y^p + xy \Longrightarrow f \in C_p.$

39$_p$. $j_{y^p,\, xy} f = y^p \Longrightarrow$ one of two:

$j_{y^p,\, xy^2} f \approx y^p + xy^2 \mid\!\Longrightarrow$ **40**$_p$,

$j_{y^p,\, xy^2} f \approx y^p \mid\!\Longrightarrow$ **41**$_p$.

40$_p$. $j_{y^p,\, xy^2} f = y^p + xy^2 \Longrightarrow f \in K_{p,\,q},\ q \geqslant 2.$

41$_p$. $j_{y^p,\, xy^2} f = y^p \Longrightarrow f$ adjoins K_0^{***}.

The corank of the restriction to the boundary greater than 1.

42. The corank of the restriction of f to the boundary equal to 2 $\Rightarrow$ one of two:

the restriction to the boundary equivalent to $y^3 + z^3$ $\mid\Rightarrow$ **43**,

f adjoins the preceding class $\mid\Rightarrow$ **45**.

43. $f(0,\, y,\, z) = y^3 + z^3 \Longrightarrow$ one of two:

$j_{y^3,\, z^3,\, xy} f \approx y^3 + z^3 + axz + xy,\ a^3 \neq 1 \mid\!\Longrightarrow$ **44**,

f adjoins the preceding class $\mid\Rightarrow$ **45**.

44. $j_{y^3,\, z^3,\, xy} f = y^3 + z^3 + axz + xy,\ a^3 \neq 1 \Longrightarrow f \in L_6.$

45. f adjoins L_6, $f \in L^*$, cod $L^* = 5$.

46. The corank of the restriction to the boundary greater than 2 $\Rightarrow f$ adjoins L_6.

The theorem of §7 is an obvious consequence of Theorems **1–46**.

These theorems are proved by standard methods (see [3], [5], [6], and [7]), namely:

Number of theorem	Method of proof
2, 3, 4, 7, 8, 10, 11, 12, 13, 14, 16, 17, 18, 22, 23, 24, 25, 27, 30, 31, 33, 36, 38, 39, 42, 43, 44	(Semi)quasihomogeneous normal form
2, 3, 6, 9, 21, 28, 34, 42, 43	Rotation of the Newton ruler
15, 19, 20, 29, 35, 40,	Spectral sequence
1, 5, 26, 32, 37, 41, 45, 46	Tautology

§9. Some corollaries

Many corollaries follow from the theorems given above, of which we mention here only a few.

COROLLARY. 1. *Each singularity, except* B_μ, C_μ, F_4, *adjoins either* $F_{1,0}$ *or* $K_{4,2}$ *or* L_6.

Since singularities of these three classes are unimodal, Theorem 1 of §1 follows.

By the method of Husein-Zade it is easy to verify that singularities of these three classes are parabolic (two zero squares). This proves Theorem 6 of §6.

Husein-Zade has also calculated the signatures of a number of other singularities of functions of two variables in the list of §7. The singularities $F_{1,k}$, $K_{p,q}$, and $K^{\#}_{1,q}$ are hyperbolic (one zero and one positive square). The singularities $F_{2,0}$, K_8^*, K_9^*, K_0^*, K_8^{**}, K_0^{**}, K_0^{***}, F_8, F_9, and F_{10} have two positive squares.

COROLLARY 2. *The parabolic singularities of functions of two variables are precisely* $F_{1,0}$ *and* $K_{4,2}$.

Apparently all parabolic singularities are exhausted by the three families $F_{1,0}$, $K_{4,2}$, and L_6.

COROLLARY 3. *The hyperbolic singularities of functions of two variables are precisely* $F_{1,k}$ *with* $k > 0$, $K_{p,q}$ *with* $(p, q) > (4, 2)$, *and* $K^{\#}_{1,q}$.

COROLLARY 4. *All simple and unimodal singularities of functions of two variables are contained in the following diagram*:

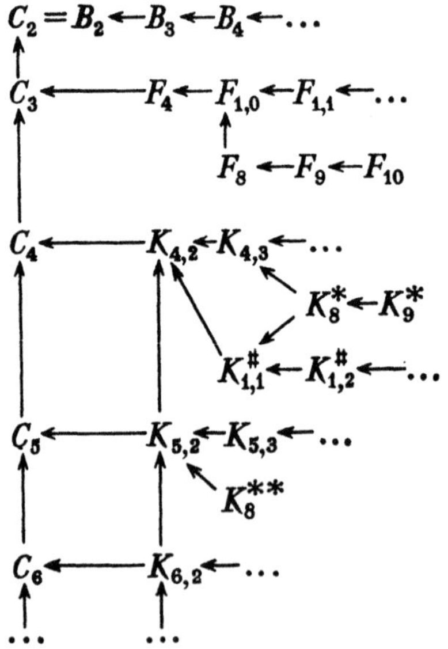

The classification of critical points of functions on a manifold with boundary is equivalent to that of critical points of functions that are even with respect to one of the coordinates (see [9]). All our results can, therefore, be interpreted as a classification of equivariant critical points of functions for the simplest action of a group of order 2. The initial part of this classification has been rediscovered several times by various authors (see, for example, [9], [18], [20], and [22]). Guts has even found a complete list of simple singularities of symmetric functions, but the main point – the connection between critical points of functions on a manifold with boundary and the simple groups B_μ, C_μ, and F_4 – has apparently gone unnoticed.

The only missing simple Lie group G_2 arises naturally in the classification of simple functions that are invariant under the action of the symmetric group of permutations of the coordinates (that is, under the classification of simple symmetric functions). The classification problem corresponding to G_2 is related to the group of permutations of three elements and can also be interpreted as the problem of classifying functions on a plane to within diffeomorphisms that preserve a semicubical parabola (see [9]).

§ 10. Singularities of a family of evolutes

Let γ be a plane curve, parametrized by the natural parameter s (arc length), that is, a smooth map $\gamma\colon \mathbf{R} \to \mathbf{R}^2$ of a line in the Euclidean plane with $|d\gamma/ds| = 1$.

The *family of evolutes* of γ is defined as the family of images of the lines $s + t = \text{const}$ in the plane with the coordinates (s, t) under the mapping

$$f: \mathbf{R}^2 \to \mathbf{R}^2, \quad f(s, t) = \gamma(s) + t\gamma'(s)$$

(s is the coordinate of the point of contact of the string with the curve, and t is the length of the free part of the string).

The family of evolutes of a plane curve arises as the family of level lines of time in the problem of the shortest path in a plane with an obstruction bounded by the curve.

THEOREM A. *In a neighbourhood of a point of non-zero curvature on a curve the family of evolutes is diffeomorphic to the family of plane semi-cubical parabolas $y = (x - c)^{2/3}$ in a neighbourhood of the axis $y = 0$.*

The proof is based on the following construction. We consider the mapping $\hat{f}\colon \mathbf{R}^2 \to PT^*\mathbf{R}^2$ that associates with the pair (s, t) a line element of the tangent to the evolute at $f(s, t)$. It is easy to verify that $\hat{f}$ is an immersion.

By contracting the domain of definition of f, we may suppose that the image of $\hat{f}$ is a smooth surface M in the three-dimensional space of line elements. We may also carry over the curves $s + t = \text{const}$ to M by means of $\hat{f}$; it remains to investigate the restriction to M of the projection $p\colon PT^*\mathbf{R}^2 \to \mathbf{R}^2$.

It is easy to compute that this restriction (and consequently the composition $f = p \circ \hat{f}$) has a fold whose image is the curve under investigation and whose inverse image is the set of its normal elements. Moreover, curves of the family $s + t = \text{const}$ touch the kernel of the derivative of our mapping. This guarantees the reduction to the indicated normal form (see [10]) under some non-degeneracy condition, which is satisfied because the curvature is non-zero.

We now suppose that the curvature of the curve at some point vanishes, but its first derivative is non-zero; we then talk of an ordinary point of inflexion (example: $y = x^3$).

DEFINITION. We call a line element in the plane *singular* with respect to γ if it either lies on the surface $M = \hat{f}(\mathbf{R}^2)$ described above or is applied at a point of $\gamma(\mathbf{R})$.

THEOREM B. *The set of singular elements in a neighbourhood of an element of the normal to the curve at an ordinary point of inflexion is locally diffeomorphic to the bifurcation diagram B_3 in $\mathbf{R}^3$* (we have in mind diffeomorphisms of the ambient three-dimensional manifolds).

In a neighbourhood of an element of the normal at a point of non-zero curvature the set of singular elements is locally diffeomorphic to the direct product of the bifurcation diagram B_2 in $\mathbf{R}^2$ and a line.

PROOF. We fix Cartesian orthonormal coordinates in which the curve is given by the equation $y = x^3 + \ldots$. Through the point of application of each line element of the plane we draw the line perpendicular to it and consider the points of intersection of this line with our curve. For elements close to an element of the normal at the point of inflexion, there are three such points of intersection (generally speaking, complex and not necessarily distinct) close to $x = 0$. Let (x_1, x_2, x_3) be the abscissae of the points of intersection, and x_0 the abscissa of the point of application of the element. We associate with our element the polynomial of degree 3 with the roots $\xi_i = x_i - x_0$. It is easy to verify that this defines a local diffeomorphism $PT^*\mathbf{R}^2 \to \mathbf{R}^3$. The surface M goes over to the set of polynomials with multiple roots, and the surface of elements applied to the curve to the set of polynomials with zero roots. These two sets also constitute the bifurcation diagram B_3.

In the non-analytic case these arguments are placed by a reference to the Malgrange division theorem [16].

REMARK 1. This theorem makes it possible to split the investigation of singularities of the family of evolutes into two problems: the investigation of singularities of the family of curves on M and that of the additional singularities that arise by the projection of M onto $\mathbf{R}^2$.

The first problem is solved by applying the theorem of [9] on the normal form of a function to the space of the bifurcation diagram.

REMARK 2. My attention was drawn to the problem of the singularities of a family of evolutes by V. I. Borzenko. Having drawn the surface M, I recognized in it part of the bifurcation diagram B_3 studied by V. V. Goryunov. An analysis of the causes of this coincidence also led to the results of §§1–9.

References

[1] V. I. Arnol'd, Normal forms of functions near degenerate critical points, the Weyl groups A_k, D_k, E_k, and Lagrangian singularities, Funktsional. Anal. i Prilozhen. **6**:4 (1972), 3–25. MR **50** # 8595.
= Functional Anal. Appl. **6** (1972), 254–272.

[2] ———, Remarks on the stationary phase method and the Coxeter numbers, Uspekhi Mat. Nauk **28**:5 (1973), 17–44. MR **53** # 1635.
= Russian Math. Surveys **28**:5 (1973), 19–48.

[3] ———, Normal forms of functions in a neighbourhood of degenerate critical points, Uspekhi Mat. Nauk **29**:2 (1974), 11–49.
= Russian Math. Surveys **29**:2 (1974), 10–50.

[4] ———, The situation of ovals of real plane algebraic curves, the involutions of four-dimensional smooth manifolds, and the arithmetic of integral quadratic forms, Funktsional. Anal. i Prilozhen. **5**:3 (1971), 1–9. MR **44** # 3999.
= Functional Anal. Appl. **5** (1971), 169–176.

[5] ———, Critical points of smooth functions and their normal forms, Uspekhi Mat. Nauk **30**:5 (1975), 3–65. MR **54** # 8701.
= Russian Math. Surveys **30**:5 (1975), 1–75.

[6] ———, A spectral sequence for the reduction of functions to normal forms, in the collection "Problems of mechanics and mathematical physics", Nauka, Moscow 1976, 7–20.

[7] ———, A spectral sequence for the reduction of functions to normal forms, Funktsional. Anal. i Prilozhen. **9**:3 (1975), 81–82. MR **52** # 4332.
= Functional Anal. Appl. **9** (1975), 251–253.

[8] ———, Braids of algebraic functions, and cohomology of swallow tails, Uspekhi Mat. Nauk **23**:4 (1968), 247–248. MR **38** # 156.

[9] ———, Wave front evolution and equivariant Morse lemma. Comm. Pure Appl. Math. **29** (1976), 557–582 (in English). MR **55** # 9148.

[10] ———, *Dopolnitel'nye glavy teorii obyknovennykh differentsial'nykh uravnenii* (Additional chapters of the theory of ordinary differential equations), Nauka, Moscow 1978, §4.

[11] E. V. Brieskorn, Singular elements of semisimple algebraic groups, Proc. Internat. Congress Mathematicians, Nice 1970, vol. 2, Gauthier–Villars, Paris 1971, 279–284.

[12] ———, Sur les groupes de tresses (d'après V.I. Arnold), Sém. N. Bourbaki **24** (1971/72), exp. 401.

[13] ———, Die Monodromie der isolierten Singularitäten von Hyperflächen, Manuscripta Math. **2** (1970), 103–161. MR **42** # 2509.

[14] S. M. Husein-Zade, Monodromy groups of isolated singularities of hypersurfaces, Uspekhi Mat. Nauk **32**:2 (1977), 23–65.
= Russian Math. Surveys **32**:2 (1977, 23–69.

[15] P. Deligne, Les immeubles des groupes de tresses généralisés, Invent. Math. **17** (1972), 273–302. MR **54** # 10659.

[16] B. Malgrange, Ideals of differentiable functions, Tata Inst. Fund. Research, Bombay; Oxford Univ. Press, London 1967. MR **35** # 3445.
Translation: *Idealy differentsiruemykh funktsii*, Mir, Moscow 1968.

[17] J. Milnor, Singular points of complex hypersurfaces, Annals of Math. Studies, **61**, Princeton Univ. Press, Princeton, N.J., 1968. MR **39** # 969.
Translation: *Osobye tochki kompleksnykh giperpoverkhnostei*, Mir, Moscow 1971.

[18] T. Poston and I. N. Stewart, Catastrophe theory and its applications, Pitman, London 1978.
[19] P. Slodowy, Einfache Singularitäten und einfache algebraische Gruppen, Regensburger Math. Schriften **2**, Univ. Regensburg 1978.
[20] ———, Einige Bemerkungen zur Entfaltung symmetrischer Funktionen, Math. Z. **158** (1978), 157–170.
[21] G. N. Tyurina, The topological properties of isolated singularities of complex spaces of codimension 1, Izv. Akad. Nauk SSSR Ser. Mat. **32** (1968), 605–620. MR **37** # 3053. = Math. USSR-Izv. **2** (1968), 557–571.
[22] G. Wassermann, Classification of singularities with compact Abelian symmetry, Regensberger Math. Schriften **1**, Univ. Regensburg 1977.

Received by the Editors 5 June 1978

Translated by E. J. F. Primrose

Uspekhi Mat Nauk 34:2 (1979), 3–38 Russian Math. Surveys, 34:2 (1979), 1–42

INDICES OF SINGULAR POINTS OF 1-FORMS ON A MANIFOLD WITH BOUNDARY, CONVOLUTION OF INVARIANTS OF REFLECTION GROUPS, AND SINGULAR PROJECTIONS OF SMOOTH SURFACES

V. I. Arnol'd

Contents

Introduction

The notion of the index of a singular point of a vector field does not generalize directly to the case of manifolds with boundary. However, it turns out that a generalization to the case of boundary singularities becomes possible if vector fields are replaced by differential 1-forms. For a 1-form on a manifold with boundary we define singular points and their indices so that the sum of the indices of all (interior and boundary) singular points of any 1-form with isolated singularities on a compact manifold with boundary is equal to the Euler characteristic of the manifold. For boundary singularities we define several integral invariants (indices) and we establish the equivalence of their geometric definitions with algebraic definitions in terms of the signatures of quadratic forms on local rings.

In the special case when the 1-form is the differential of a function and the boundary singularity of the function is simple, the resulting index may also be described as the signature of a quadratic form, constructed using the special operation of "convolving" the invariants of the corresponding reflection group. Moreover, the family of quadratic forms on a local ring that arises through the operation of multiplication and the family of quadratic forms on the dual space that arises through the operation of convolution, are dual to each other in a certain sense. One of the results of studying this duality is the assignment of a Lie algebra whose dimension is equal to the dimension of the space to each irreducible reflection group in Euclidean space: the corresponding Lie group is formed by the linearizations of those diffeomorphisms of the orbit manifold of the reflection group that map the set of non-regular orbits into itself.

This technique leads to the solution of a number of problems on normal forms of various geometric objects. In particular, we consider the projection of smooth surfaces from 3-space onto the plane. The singularities of projections in general position were already classified in a classical paper of Whitney. However, by choosing the direction of projection in a special way we can add more complicated singularities. There turn out to be ten singularities such that for surfaces in general position all the singularities of the projections with respect to all the directions are equivalent to these ten.

Precise definitions and statements of these and other results and conjectures are contained in §§1–7 (indices), §§8–9 (convolutions) and §10 (projections).

The author thanks V. V. Goryunov, V. I. Danilov, V. M. Zakalyukin, A. G. Kushnirenko, O. V. Lyashko, O. A. Platonova, A. G. Khovanskii, A. N. Shoshitaishvili, M. A. Shubin, and D. B. Fuks for useful discussions.

§1. Singularities and their indices

We consider a 1-form on a manifold with boundary. A boundary point is called a *boundary singularity* if this form vanishes on the whole tangent plane to the boundary at the point.

We assume that this singularity is isolated, and we define its index. We fix a non-degenerate equation for the boundary, that is, a function f that vanishes on the boundary, has no critical points on the boundary, and is positive inside the manifold.

DEFINITION. The *index* i_+ of a boundary singularity is the integer indicating the number of points on a small hemisphere with centre at the singularity at which points the form is positively proportional to the differential of the equation of the boundary.

Here the points on the hemisphere are included with their orientations. More precisely, we consider the cosphere bundle of the manifold restricted to this hemisphere. Our 1-form ω and the differential df of the equation of the boundary determine a pair of sections of this bundle. The number i_+ is the intersection index of these sections. (The orientations are chosen so that the index of the singularity at 0 for the form $\omega = x_1 dx_1 + \ldots + x_n dx_n$ is 1.)

EXAMPLE. We consider a function of two variables $F(x_1, x_2)$ on the half-plane $x_1 \geqslant 0$. The 1-form dF has as boundary singularities the critical points of the restriction of F to the boundary. The part of the half-plane where the values of F are less than that at the critical point approaches the critical point through several sectors. It is not hard to show that $i_+ = 1 - k_+$, where k_+ is the number of sectors of the set of smaller values in the half-plane.

REMARK 1. The index i_+ does not depend on the non-degenerate function f in its definition. In fact, any two such functions can be joined by a homotopy in the class of non-degenerate functions. The section generated by df on the boundary of our hemisphere does not vary under homotopy, so that its intersection index with the section generated by ω does not vary.

REMARK 2. The index of a 1-form ω can be defined for any transversally oriented hypersurface with boundary lying in the domain of regularity of f, under the assumptions that the boundary of the hypersurface lies on the boundary of the manifold in question and that the hypersurface does not pass through the singularities of the form (neither the interior nor the boundary ones).

With this aim we consider the cosphere bundle over the domain of regularity of f, excluding from the base small neighbourhoods around the singularities of ω. This is an oriented manifold of dimension $2n - 1$ (if the original manifold M is n-dimensional). The form df defines a submanifold of dimension n, and the restriction of ω to the hypersurface in question defines a submanifold of dimension $n - 1$. A (local) orientation of M defines (local) orientations of both these submanifolds. Under a change of the local orientation of M, both these induced orientations vary. Therefore, the intersection index is well-defined; it is also the index of the form along a transversally oriented hypersurface.

If the hypersurface is the boundary of a domain in which f is regular, then the index of ω along a hypersurface transversally oriented along an outward normal is equal to the sum of the indices i_+ of the boundary singularities of the form in this domain and the usual indices of the interior singular points of the form in the domain. (This follows from standard properties of the intersection index.)

The detailed description of the homology group on which the index is defined is left to the reader.

REMARK 3. On a Riemannian manifold with boundary one can define the boundary singularities of a vector field as the points where the field is orthogonal to the boundary. The index i_+ of a boundary singularity of the field indicates the number of points (including signs) on a small hemisphere with centre at the singular point at which the field is directed along the normal to the boundary. However, in contrast to the index of a boundary singularity of a form, the index of a boundary singularity of a field is not a diffeomorphism-invariant: it depends on the Riemannian metric.

§2. A theorem on the sum of indices

We consider a 1-form on a compact manifold with boundary and assume that it has only isolated (both interior and boundary) singularities.

THEOREM. *The sum of the indices of all singularities of* 1*-forms on a compact manifold with boundary (that is, the sum of the ordinary indices for the interior singular points and the indices* i_+ *for the boundary singularities) is equal to the Euler characteristic of the manifold.*

PROOF. We consider the double $\widetilde{M}$ of M. By definition,

$$\widetilde{M} = \{(x, z) \in M \times \mathbf{R}\colon z^2 = f(x)\},$$

where $f = 0$ is a non-degenerate equation for the boundary ∂M. $\widetilde{M}$ is a closed manifold.

The natural projection $\pi\colon \widetilde{M} \to M$ allows us to lift ω to $\widetilde{M}$: we obtain a 1-form $\widetilde{\omega} = \pi^*\omega$ on $\widetilde{M}$. Every interior singular point of ω on M corresponds to a pair of singular points of $\widetilde{\omega}$ on $\widetilde{M}$, each of them having the same index. The remaining singular points of $\widetilde{\omega}$ are precisely the points $(x, 0)$, where x is a boundary singularity of ω. We denote by $\widetilde{i}$ the index of this singularity for $\widetilde{\omega}$ and by i_0 the index of the restriction of ω to ∂M at x.

LEMMA. $\widetilde{i} = 2i_+ - i_0$.

This lemma will be proved in §5.

END OF THE PROOF OF THE THEOREM. The sum $\widetilde{I}$ of the indices of all the singular points of $\widetilde{\omega}$ on $\widetilde{M}$ is equal to the Euler characteristic $\chi(\widetilde{M})$, and the sum I_0 of the indices of all the singular points of the restriction of ω to ∂M is equal to $\chi(\partial M)$. We denote by I_+ the sum of the indices of all the boundary singularities of ω, and by I the sum of the indices of all its interior singularities. By the lemma,

$$\widetilde{I} = 2I + 2I_+ - I_0.$$

But $\chi(\widetilde{M}) = 2\chi(M) - \chi(\partial M)$. Comparing this with the preceding formula we obtain $I + I_+ = \chi(M)$, as was required.[1]

The proof given later for the lemma proceeds in the following way. With a boundary singularity of a form we associate a singularity of a vector field tangent to the boundary (§4). For such fields we define a number of indices and establish relations between them (§3). Next we verify that the corresponding indices for forms and fields are equal (§§4–5), and we derive a relation between the indices of forms from the corresponding relation for fields.

[1] The generalized Lefschetz theorem is proved similarly: the alternating sum of the traces of the automorphisms of the homology groups under a diffeomorphism of a compact manifold with boundary to itself is equal to the sum of the ordinary indices of the interior fixed points and the indices i_+ of the boundary fixed points defined above.

§3. Indices of singular points of adapted vector fields

A vector field on a manifold with boundary is said to be *adapted* if the vector of the field at every boundary point is tangent to the boundary. Several numbers (indices) are associated with a boundary singular point of an adapted vector field.

In Euclidean n-space $\mathbf{R}^n$ with the coordinates $(x_1, \ldots, x_n)$ we consider the half-space $x_1 \geqslant 0$. Let v be an adapted field with an isolated singular point O. We denote by S the unit sphere $S^{n-1} = \{x: |x| = 1\}$, and for sufficiently small $\varepsilon > 0$ we construct the usual spherical mapping $f\colon S \to S$, $f(x) = v(\varepsilon x)/|v(\varepsilon x)|$. If v is adapted, then f takes an equator of the sphere $(x_1 = 0)$ into itself. This equator, which we denote by S_0, divides S into a right and a left hemisphere ($x_1 \geqslant 0$ and $x_1 \leqslant 0$), which we denote by S^+ and S^-, respectively. We denote by σ the involution of reflection in the equator (changing the sign of x_1).

DEFINITION. The following four numbers are the *partial indices* of a singular point of an adapted field:

$$\left.\begin{aligned} i_+ &= \deg[(S^+/S_0) \to (S/S^-)] && \text{(right-right index)} \\ i_- &= \deg[(S^-/S_0) \to (S/S^+)] && \text{(left-left index)} \end{aligned}\right\} \text{direct,}$$

$$\left.\begin{aligned} j_+ &= \deg[(S^+/S_0) \to (S/S^+)] && \text{(right-left index)} \\ j_- &= \deg[(S^-/S_0) \to (S/S^-)] && \text{(left-right index)} \end{aligned}\right\} \text{crossed.}$$

The three numbers below are called the *complete indices*:

$$\begin{aligned} i &= \deg[(f \mid S^+) \vee (\sigma f \mid S^-)] && \text{(complete relative index),} \\ i_1 &= \deg(S \to S) && \text{(usual index),} \\ i_0 &= \deg(S_0 \to S_0) && \text{(restricted index).} \end{aligned}$$

The two numbers below are called the *covering indices*:

$$\tilde{i}_+ = \deg[(f \mid S^+) \vee (\sigma f \sigma \mid S^-)], \quad \tilde{i}_- = \deg[(f \mid S^-) \vee (\sigma f \sigma \mid S^+)].$$

Here the arrows denote the mappings of spheres that are induced by the spherical mapping f of v.

THEOREM. *The following relations hold between the indices of an adapted vector field*:

$$\begin{aligned} i_+ + i_- &= i_1 + i_0, \\ i_+ - i_- &= i, \\ 2i_+ &= \tilde{i}_+ + i_0, \\ 2i_- &= \tilde{i}_- + i_0. \end{aligned}$$

LEMMA A. *The complete indices can be expressed by the partial indexes*: $i = i_+ - i_- = j_+ - j_-$, $i_1 = i_+ + j_- = i_- + j_+$, $i_0 = i_+ - j_+ = i_- - j_-$.

PROOF. The formulae for i_1 and for i follow from the definition of degree as the number of inverse images of a point of S^+ (first formulae) or of S^- (second formulae). The formulae for i_0 are obtained from the following commutative diagrams induced by $f \mid S^+$:

$$\begin{array}{ccccc} S_0 \to & S^+ \to & S^+/S_0 & 0 \to \pi_{n-1}(S^+) \to \pi_{n-1}(S^+, S_0) \xrightarrow{\alpha} \pi_{n-2}(S_0) \to 0 \\ \downarrow & \downarrow_{f|S^+} & \downarrow & \qquad \downarrow \qquad\qquad \beta\downarrow \qquad\quad \delta \quad \downarrow\gamma \\ S_0 \to & S \to & S/S_0, & 0 \to \pi_{n-1}(S) \to \pi_{n-1}(S, S_0) \to \pi_{n-2}(S_0) \to 0. \end{array}$$

Under the usual orientation for spheroids ($\partial[S^+] = [S_0] = -\partial[S^-]$) we have:

$$\alpha([S^+, S_0]) = [S_0], \quad \gamma([S_0]) = i_0[S_0],$$
$$\beta([S^+, S_0]) = (i_+[S^+, S_0] + j_+[S^-, S_0]),$$
$$\delta([S^+, S_0]) = [S_0], \quad \delta([S^-, S_0]) = -[S_0].$$

It follows from the commutativity that $i_0 = i_+ - j_+$, as required.

LEMMA B. *The covering indices can be expressed by the partial indexes*:

$$\tilde{i}_+ = i_+ + j_+, \quad \tilde{i}_- = i_- + j_-.$$

PROOF. Considering the inverse images of a point of S^+ under the mapping $(f \mid S^+) \vee (\sigma f \sigma \mid S^-)$, we obtain i_+ inverse images in S^+ and j_+ inverse images in S^-; similarly for $\tilde{i}_-$.

The theorem now follows from the formulae of Lemmas A and B.

COROLLARY. *All nine indices can be expressed in terms of the three complete indexes*:

$$\tilde{i}_+ = i_1 + i, \quad \tilde{i}_+ + \tilde{i}_- = 2i_1, \quad 2i_+ = i_1 + i_0 + i,$$
$$\tilde{i}_- = i_1 - i, \quad \tilde{i}_+ - \tilde{i}_- = 2i, \quad 2i_- = i_1 + i_0 - i.$$

§4. Connection between the indices of fields and forms

In Euclidean space $\mathbf{R}^n$ with the coordinates $(x_1, \ldots, x_n)$ we consider the right and left half-spaces $\mathbf{R}^n_+$ $(x_1 \geqslant 0)$ and $\mathbf{R}^n_-$ $(x_1 \leqslant 0)$. Let $\Omega = \omega_1 dx_1 + \ldots + \omega_n dx_n$ be a differential form in $\mathbf{R}^n$ with an isolated boundary singularity at the origin. This singularity can be regarded as a boundary singularity for both half-spaces, so that right and left indices arise; for them we introduce the notation

$$i_+(\Omega, \mathbf{R}^n_+) = i_+(\Omega) \qquad \text{(right index of the form)},$$
$$i_+(\Omega, \mathbf{R}^n_-) = i_-(\Omega) \qquad \text{(left index of the form)}.$$

LEMMA A. *The right and left indices of Ω are the same as those of the form*

$$\Omega_1 = |x_1| \omega_1 dx_1 + \omega_2 dx_2 + \ldots + \omega_n dx_n.$$

PROOF. We construct a family of forms $(0 \leqslant \lambda \leqslant 1)$:

$$\Omega_\lambda = [1 - \lambda(1 - |x_1|)]\omega_1 dx_1 + \omega_2 dx_2 + \ldots + \omega_n dx_n.$$

The singular points of the forms Ω_λ (both interior and boundary) do not depend on λ. Moreover, the condition of proportionality of Ω_λ with dx_1 has the form $\omega_2 = \ldots = \omega_n = 0$, hence, does not depend on λ.

The coefficient of proportionality $[1 - \lambda(1 - |x_1|)]\omega_1$ does not change sign as λ varies (if $x_1 \neq 0$), that is, its sign is the same as that of ω_1. Therefore, the indices of Ω_λ along a small right (left) hemisphere with centre at the origin do not change as λ varies from 0 to 1, that is, they are the same for Ω and for Ω_1, as required.

LEMMA B. *The right and left indices of* Ω_1 *are equal to the indices* i_+ *and* i_-, *as defined in* §3, *for the adapted field*

$$W = |x_1| \omega_1 \frac{\partial}{\partial x_1} + \omega_2 \frac{\partial}{\partial x_2} + \ldots + \omega_n \frac{\partial}{\partial x_n},$$

that is, $i_+(\Omega_1) = i_+(W), i_-(\Omega_1) = i_-(W)$.

PROOF. Under the spherical mapping of W the number of (oriented) inverse images in the right hemisphere of the pole of the right hemisphere S^+ is equal to the number of points on a small right hemisphere with centre at zero at which the form Ω_1 is positively proportional to dx, and similarly for the left hemisphere, as required.

We introduce three more vector fields

$$V = x_1\omega_1 \frac{\partial}{\partial x_1} + \omega_2 \frac{\partial}{\partial x_2} + \ldots + \omega_n \frac{\partial}{\partial x_n},$$
$$V_1 = \omega_1 \frac{\partial}{\partial x_1} + \omega_2 \frac{\partial}{\partial x_2} + \ldots + \omega_n \frac{\partial}{\partial x_n},$$
$$V_0 = \omega_2 \frac{\partial}{\partial x_2} + \ldots + \omega_n \frac{\partial}{\partial x_n} \Big| \mathbf{R}^{n-1} \quad (x_1 = 0).$$

LEMMA C. *The complete indices of zero of the adapted field* W *is equal to the usual indices at zero of the constructed fields, namely,*

$$i(W) = \text{ind } V, \quad i_1(W) = \text{ind } V_1, \quad i_0(W) = \text{ind } V_0.$$

PROOF. For i_0 this is the definition, for i_1 the proof is as in Lemma A. The formula for i follows from the fact that V is the same as W on S^+ and as σW on S^-.

THEOREM. *The right and left indices* i_+ *and* i_- *of* Ω *at zero are connected with the indices* i, i_1, *and* i_0 *of the singular points of* V, V_1, *and* V_0 *by the relations*

$$i_+(\Omega) + i_-(\Omega) = i_1 + i_0,$$
$$i_+(\Omega) - i_-(\Omega) = i.$$

PROOF. Lemmas A, B, and C express all five of these indices in terms of the corresponding indices of W, and the theorem in §3 contains the required relations between them.

§5. Covering indices

We regard the double $\widetilde{\mathbf{R}}^n_+$ of the right half-space $x_1 \geqslant 0$ as the space with coordinates $(z, x_2, \ldots, x_n)$ equipped with the projection $\pi_+ : \widetilde{\mathbf{R}}^n_+ \to \mathbf{R}^n_+$, taking the point $(z, x_2, \ldots, x_n)$ into $(z^2, x_2, \ldots, x_n)$. The form Ω is pulled back to

$$\pi_+^*\Omega = 2z\tilde{\omega}_1\, dz + \tilde{\omega}_2\, dx_2 + \ldots + \tilde{\omega}_n\, dx_n, \quad \tilde{\omega} = \pi_+^*\omega.$$

LEMMA. *The index of the singular point zero of the form* $\pi_+^*\Omega$ *on the double* $\widetilde{\mathbf{R}}^n_+$ *is equal to the covering index* $\tilde{i}_+$ (*defined in* §3) *of the adapted field*

$$W = |x_1| \omega_1 \frac{\partial}{\partial x_1} + \omega_2 \frac{\partial}{\partial x_2} + \ldots + \omega_n \frac{\partial}{\partial x_n}.$$

PROOF. In calculating the index we can replace a small sphere in the double by the surface $z^4 + x_2^2 + \ldots + x_n^2 = \varepsilon^2$, which is the inverse image of the small sphere $|x| = \varepsilon$. We consider the spherical mapping of the field

$$\widetilde{V}_+ = 2z\widetilde{\omega}_1 \frac{\partial}{\partial z} + \widetilde{\omega}_2 \frac{\partial}{\partial x_2} + \ldots + \widetilde{\omega}_n \frac{\partial}{\partial x_n}.$$

The inverse image of the centre $(\partial/\partial z)$ of the right hemisphere under this spherical mapping is obtained from the points at which $\widetilde{\omega}_2 = \ldots = \widetilde{\omega}_n = 0$, $z\widetilde{\omega}_1 > 0$. The inverse images in the right half-space $(z > 0)$ correspond to the inverse image S^+ of the pole of the right hemisphere S^+, under the spherical mapping f of W. The inverse images of the left half-space $(z < 0)$ in S^+ under f correspond to the pole of the left hemisphere S^-. This relation (completed by a check of the orientations) leads to the formula

$$\operatorname{ind} \widetilde{V}_+ = \deg [(f \mid S^+) \vee (\sigma f \sigma \mid S^-)] = \widetilde{i}_+(W),$$

as was required.

In just the same way we can prove the analogous formulae for the double of the left half-space: for $\pi_-(z, x_2, \ldots, x_n) = (-z^2, x_2, \ldots, x_n)$

$$\pi_-^*(\Omega) = -2z\widetilde{\omega}_1\, dz + \widetilde{\omega}_2\, dx_2 + \ldots + \widetilde{\omega}_n\, dx_n, \qquad \widetilde{\omega} = \pi_-^*\omega,$$

$$\widetilde{V}_- = -2z\widetilde{\omega}_1 \frac{\partial}{\partial z} + \widetilde{\omega}_2 \frac{\partial}{\partial x_2} + \ldots + \widetilde{\omega}_n \frac{\partial}{\partial x_n}, \qquad \operatorname{ind} \widetilde{V}_- = \widetilde{i}_-(W).$$

THEOREM. *The indices $\widetilde{i}_+$ and $\widetilde{i}_-$ of the singular points of the lift of Ω to the doubles of the left and right half-spaces are associated with the left, right, and restricted indices i_+, i_-, and i_0 of the form by the relations*

$$\widetilde{i}_+ = 2i_+ - i_0, \qquad \widetilde{i}_- = 2i_- - i_0.$$

PROOF. The preceding lemma and the lemmas of §4 identify the covering and the other indices of Ω with the corresponding indices of the adapted field W, and for them the assertion was already proved in §3.

These theorems imply that all the indices of Ω are equal to the corresponding indices of W and satisfy the same relations. We give to each of the indices of the form the name of the corresponding index of the field. For example:

$i(\Omega) = i_+(\Omega) - i_-(\Omega)$ is the complete relative index,

$i_1(\Omega)$ is the usual index,

$i_0(\Omega)$ is the index of the restriction of the form to the boundary,

$\widetilde{i}_+(\Omega)$ is the index of the form lifted to the double of the right half-space.

REMARK. In [6] Khovanskii considers a vector field P in $\mathbf{R}^n$ whose components are homogeneous polynomials $P_1, \ldots, P_n$ of degrees $m_1, \ldots, m_n$, satisfying a non-degeneracy condition that is valid almost everywhere. In $\mathbf{R}^{n-1} = \{x\colon x_1 = 1\}$ he constructs a vector field with the components

$$p_k(x_2, \ldots, x_n) = P_k(1, x_2, \ldots, x_n) \qquad (k = 2, \ldots, n).$$

The sum of the indices of all the singular points of this field at which $p_1 > 0$ is denoted by ind^+, and those at which $p_1 < 0$ by ind^-. Khovanskii found best possible estimates for ind^+, for ind^-, for $\text{ind}^+ - \text{ind}^-$, and for $\text{ind}^+ + \text{ind}^-$ in terms of the number of lattice points in certain polyhedra. He also noted that in the case

$$m_1 + \ldots + m_n \equiv n \bmod 2,$$

the index of the singular point O of the field P in $\mathbf{R}^n$ is equal to the characteristic $\text{ind}^+ - \text{ind}^-$ and he raised the question of the meaning of the characteristic $\text{ind}^+ - \text{ind}^-$ in terms of the singular point O of P for a different parity of the sum of the exponents. We can now answer this question. The following proposition is easy to prove.

PROPOSITION. *In the notation introduced above,* $\text{ind}^+ = i_+$, $\text{ind}^- = j_+$, *and the Kronecker characteristic* $\text{ind}^+ - \text{ind}^-$ *is equal to the covering index* $\tilde{i}_+$ *of the form* $\Sigma P_k dx_k$ *at* O *on the manifold with boundary* $x_1 \geqslant 0$.

The covering index $\tilde{i}_+$ is also equal to the sum $i_1 + i$, by the formula of the corollary to Lemma B in §3. For a field with homogeneous components, depending on the parity of the sum $m_1 + \ldots + m_n$, one of the terms i_1 or i vanishes (this is easily obtained from a comparison of the vectors of the field $(P_1, \ldots, P_n)$ and $(x_1 P_1, P_2, \ldots, P_n)$, respectively, at antipodal points). The sums of the degrees for both fields are different from 1. Thus, the *characteristic* $\text{ind}^+ - \text{ind}^-$, depending *on the parity of the sum of the degrees, is equal either to the usual index* i_1 *of the singular point* O *of the field* $P = (P_1, \ldots, P_n)$ *(the case* $m_1 + \ldots + m_n \equiv n \bmod 2$), *or to our complete index* $i(P)$, *which is equal to the usual index of* O *of the field* $(x_1 P_1, P_2, \ldots, P_n)$ *(the case* $m_1 + \ldots + m_n \equiv n + 1 \bmod 2$).

§6. Algebraic definitions

Let

$$\Omega = \omega_1\, dx_1 + \ldots + \omega_n\, dx_n$$

be a 1-form with an isolated boundary singularity O in the half-space $\mathbf{R}^n_+$ $(x_1 \geqslant 0)$.

We consider the following three local algebras over $\mathbf{R}$:

$$Q = \mathbf{R}\,[[x_1, x_2, \ldots, x_n]]/(x_1\omega_1, \omega_2, \ldots, \omega_n),$$
$$Q_1 = \mathbf{R}\,[[x_1, x_2, \ldots, x_n]]/(\omega_1, \omega_2, \ldots, \omega_n),$$
$$\widetilde{Q} = \mathbf{R}\,[[z, x_2, \ldots, x_n]]/(z\widetilde{\omega}_1, \widetilde{\omega}_2, \ldots, \widetilde{\omega}_n).$$

Here $\widetilde{\omega}(z, x_2, \ldots, x_n) = \omega(z^2, x_2, \ldots, x_n)$.

We assume that the dimensions μ, μ_1, and $\widetilde{\mu}$ of these algebras over $\mathbf{R}$ are finite and consider the corresponding Jacobians

$$J = \partial(x_1\omega_1, \omega_2, \ldots, \omega_n)/\partial(x_1, x_2, \ldots, x_n),$$
$$J_1 = \partial(\omega_1, \omega_2, \ldots, \omega_n)/\partial(x_1, x_2, \ldots, x_n),$$
$$\widetilde{\widetilde{J}} = \partial(z\widetilde{\omega}_1, \widetilde{\omega}_2, \ldots, \widetilde{\omega}_n)/\partial(z, x_2, \ldots, x_n).$$

Let

$$q: Q \to \mathbf{R}, \quad q_1: Q_1 \to \mathbf{R} \text{ and } \widetilde{\widetilde{q}}: \widetilde{Q} \to \mathbf{R}$$

be any linear forms taking positive values on the corresponding Jacobians. The multiplication in the local algebra then defines a *non-degenerate* symmetric bilinear form:

$$\varphi(a, b) = q(ab) \quad \text{for } a, b \in Q,$$
$$\varphi_1(a, b) = q_1(ab) \quad \text{for } a, b \in Q_1,$$
$$\widetilde{\widetilde{\varphi}}(a, b) = \widetilde{\widetilde{q}}(ab) \quad \text{for } a, b \in \widetilde{Q},$$

respectively.

THEOREM. *The signatures of these three forms are equal, respectively, to the complete relative index i, the usual index i_1, and the covering index $\widetilde{i}_+$ of the boundary singularity of Ω at zero.*

COROLLARY. *The three signatures defined above are connected by the relation $\widetilde{i}_+ = i_1 + i$ (the third signature is equal to the sum of the first two).*

PROOF OF THE THEOREM. By a theorem of Eisenbud–Levine [3] and Khimshiashvili [5], these signatures are equal to the usual indices of singular points of the corresponding fields V, V_1, and $\widetilde{V}$. According to §§4 and 5, these usual indices are equal to the corresponding indices i, i_1, and $\widetilde{i}_+$ of the boundary singularity.

Although the corollary follows immediately from the theorem and formulae in §3, we give here a direct algebraic proof.

PROOF OF THE COROLLARY. A) *The linear space $\widetilde{Q}$ splits into the direct sum $\widetilde{Q} = \widetilde{Q}_{\text{even}} + \widetilde{Q}_{\text{odd}}$ of even and odd constituents* mod z.

For we split $\mathbf{R}[[z, x_2, \ldots, x_n]]$ into the direct sum of elements that are odd and even mod z. The generators of the ideal $(z\widetilde{\omega}_1, \widetilde{\omega}_2, \ldots, \widetilde{\omega}_n)$ are purely even $(\widetilde{\omega}_2, \ldots, \widetilde{\omega}_n)$ or purely odd $(z\widetilde{\omega}_1)$. Therefore, the ideal splits into the direct sum of the odd and even parts. Consequently, the factor-algebra

$$\widetilde{Q} = \mathbf{R}[[z, x_2, \ldots, x_n]]/(z\widetilde{\omega}_1, \widetilde{\omega}_2, \ldots, \widetilde{\omega}_n)$$

also splits into a direct sum.

B) *The space $\widetilde{Q}_{\text{even}}$ is a subalgebra of $\widetilde{Q}$ and is naturally isomorphic to Q.*

In fact, the even elements mod z of the ideal $(z\widetilde{\omega}_1, \widetilde{\omega}_2, \ldots, \widetilde{\omega}_n)$ in $\mathbf{R}[[z, x_2, \ldots, x_n]]$ belong to the ideal $(x_1\omega_1, \omega_2, \ldots, \omega_n)$ in $\mathbf{R}[[x_1, x_2, \ldots, x_n]]$ (we identify the latter algebra of even elements mod z in $\mathbf{R}[[z, x_2, \ldots, x_n]]$, assuming that $x_1 = z^2$).

C) *The mapping $\mathbf{R}[[x_1, \ldots, x_n]] \to \mathbf{R}[[z, x_2, \ldots, x_n]]$ that associates with $a(x)$ the element $z\widetilde{a}(x)$ defines an isomorphism of linear spaces $Q_1 \to \widetilde{Q}_{\text{odd}}$.*

For the elements of the ideal $(\omega_1, \ldots, \omega_n)$ in $\mathbf{R}[[x_1, \ldots, x_n]]$ are mapped to elements of the ideal

$$(z\tilde{\omega}_1, \ldots, z\tilde{\omega}_n) \subset (z\tilde{\omega}_1, \tilde{\omega}_2, \ldots, \tilde{\omega}_n),$$

so that multiplication by z gives a linear mapping $Q_1 \to \tilde{Q}$. The image of this mapping is exactly $\tilde{Q}_{\text{odd}}$. We claim that this mapping is an isomorphism. Suppose that the image of a in $\tilde{Q}$ is zero, that is,

$$z\tilde{a}(x) = \alpha_1 z\tilde{\omega}_1 + \alpha_2\tilde{\omega}_2 + \ldots + \alpha_n\tilde{\omega}_n$$

$(\alpha_k \in \mathbf{R}[[z, x_2, \ldots, x_n]])$. Then

$$z(\tilde{a} - \alpha_1\tilde{\omega}_1) = \alpha_2\tilde{\omega}_2 + \ldots + \alpha_n\tilde{\omega}_n,$$

that is, $z(\tilde{a} - \alpha_1\tilde{\omega}_1)$ belongs to the ideal $(\tilde{\omega}_2, \ldots, \tilde{\omega}_n)$ in $\mathbf{R}[[z, x_2, \ldots, x_n]]$. But z is not a zero-divisor in

$$\mathbf{R}[[z, x_2, \ldots, x_n]]/(\tilde{\omega}_2, \ldots, \tilde{\omega}_n),$$

because the sequence $(z, \tilde{\omega}_2, \ldots, \tilde{\omega}_n)$ is regular, by the Cohen–Macaulay theorem, $\tilde{Q}$ being finite-dimensional. Hence, we have the expansion

$$\tilde{a} - \alpha_1\tilde{\omega}_1 = \beta_2\tilde{\omega}_2 + \ldots + \beta_n\tilde{\omega}_n$$

with coefficients $\beta \in \mathbf{R}[[z, x_2, \ldots, x_n]]$. Replacing each coefficient α_1, β_k by its even part, we find that

$$a = \gamma_1\omega_1 + \ldots + \gamma_n\omega_n$$

with coefficients $\gamma_k \in \mathbf{R}[[x_1, \ldots, x_n]]$. Consequently, the class of a in Q_1 is zero, and our mapping is an isomorphism.

D) *The mapping*

$$\mathbf{R}[[x_1, \ldots, x_n]] \to \mathbf{R}[[x_1, \ldots, x_n]]$$

taking $a(x)$ to $x_1a(x)$ defines an isomorphism of the linear space Q_1 with the linear space of the ideal (x_1) in Q.

The elements of the ideal $(\omega_1, \ldots, \omega_n)$ go into elements of the ideal $(x_1\omega_1, \omega_2, \ldots, \omega_n)$, so that the mapping is defined. It is clear that it is "onto"; we verify that the kernel is (0). If

$$x_1a(x) = \alpha_1x_1\omega_1 + \alpha_2\omega_2 + \ldots + \alpha_n\omega_n,$$

then $x_1(a - \alpha_1\omega_1)$ belongs to the ideal $(\omega_2, \ldots, \omega_n)$ in $\mathbf{R}[[x_1, \ldots, x_n]]$. But the sequence $(x_1, \omega_2, \ldots, \omega_n)$ is regular by sense of the Cohen–Macaulay theorem (since Q is finite-dimensional). Consequently, there is an expansion

$$a - \alpha_1\omega_1 = \beta_2\omega_2 + \ldots + \beta_n\omega_n,$$

that is, the class of a in Q_1 is zero.

E) *The Jacobians corresponding to the rings Q, Q_1, and $\tilde{Q}$ are equal to*

$$J = \omega_1 J_2 + x_1 J_1, \quad J_1, \quad \textit{and} \quad \tilde{\tilde{J}} = \tilde{\omega}_1\tilde{J}_2 + 2z^2\tilde{J}_1,$$

where

$$J_k = \det\,(\partial\omega_i/\partial x_j), \qquad i, j \geqslant k.$$

F) LEMMA. *The classes* $\omega_1 J_2$ *and* $x_1 J_1$ *in* Q *are proportional, and*

$$\omega_1 J_2 / x_1 J_1 = \mu_0/\mu_1,$$

where

$$\mu_0 = \dim_{\mathbf{R}} \mathbf{R}[[x_1, x_2, \ldots, x_n]]/(x_1, \omega_2, \ldots, \omega_n).$$

REMARK. More generally, *we consider the rings*

$$Q_{ab} = \mathbf{R}[[x]]/(ab, \omega_2, \ldots, \omega_n),\ Q_a = \mathbf{R}[[x]]/(a, \omega_2, \ldots, \omega_n),$$
$$\text{and}\quad Q_b = \mathbf{R}[[x]]/(b, \omega_2, \ldots, \omega_n),$$

of finite dimensions μ_{ab}, μ_a, *and* μ_b, *respectively, with the Jacobians*

$$J(ab) = aJ(b) + bJ(a),$$
$$J(a) = \partial(a, \omega_2, \ldots, \omega_n)/\partial(x_1, \ldots, x_n),$$
$$J(b) = \partial(b, \omega_2, \ldots, \omega_n)/\partial(x_1, \ldots, x_n).$$

Then the classes of both the terms of the Jacobian $J(ab)$ *in* Q_{ab} *are proportional, and*

$$[aJ(b)]/[bJ(a)] = \mu_b/\mu_a.$$

PROOF. The line spanned by the Jacobian is the annihilator of the maximal ideal. Let us consider one of the terms, say $[aJ(b)]$. We claim that it belongs to the annihilator of the maximal ideal. The products of $aJ(b)$ by the elements of the maximal ideal of Q_{ab} are generated by the elements $x_i aJ(b)$ in $\mathbf{R}[[x]]$. But $x_i J(b)$ lies in the ideal $(b, \omega_2, \ldots, \omega_n)$. Consequently, $x_i aJ(b)$ lies in the ideal $(ab, \omega_2, \ldots, \omega_n)$, that is, $[aJ(b)]$ belongs to the annihilator. It remains to compute the coefficient of proportionality. We assume that the various ω are holomorphic (the general case reduces easily to this one). We use the standard formula for the multiplicity of a holomorphic mapping: for small ρ

$$\left(\frac{1}{2\pi i}\right)^n \oint_{|f_k|=\rho_k} \frac{J\prod dx}{\prod f_k} = \mu,$$

where

$$J = \partial(f_1, \ldots, f_n)/\partial(x_1, \ldots, x_n),$$
$$\mu = \dim_{\mathbf{C}} \mathbf{C}[[x_1, \ldots, x_n]]/(f_1, \ldots, f_n).$$

Applying this formula in the case

$$f_1 = ab,\ f_2 = \omega_2, \ldots, f_n = \omega_n,$$

we find

$$(2\pi i)^n \mu_{ab} = \oint_{|f_k|=\rho_k} \frac{J(b)\,dx}{b\omega_2 \ldots \omega_n} + \oint_{|f_k|=\rho_k} \frac{J(a)\,dx}{a\omega_2 \ldots \omega_n}.$$

In the first integral we can replace the chain $|ab| = \rho_1$, $|\omega_k| = \rho_k$ by a homologous chain in the complement of the zeros of the denominator, namely,

$|b| = \rho_1$, $|\omega_k| = \rho_k$, and in the second by $|a| = \rho_1$, $|\omega_k| = \rho_k$. Again using the multiplicity formula we find

$$\oint_{|f_k|=\rho_k} \frac{aJ(b)\,dx}{ab\omega_2 \dots \omega_n} = (2\pi i)^n \mu_b, \quad \oint_{|f_k|=\rho_k} \frac{bJ(a)\,dx}{ab\omega_2 \dots \omega_n} = (2\pi i)^n \mu_a.$$

The mapping

$$\alpha \mapsto \left(\frac{1}{2\pi i}\right)^n \oint_{|f_k|=\rho_k} \frac{\alpha\, dx}{\prod f_k}$$

gives a linear form on the local algebra

$$Q = \mathbf{C}[[x_1, \dots, x_n]]/(f_1, \dots, f_n).$$

We have shown above that the values of this form on the elements $[J(ab)]$, $[aJ(b)]$, and $[bJ(a)]$ proportional in Q are, respectively, equal to μ, μ_b, and μ_a, as was required.

G) *The choice of the quadratic forms on Q_1, on Q, and on $\widetilde{Q}$.* Let $q_1 : Q_1 \to \mathbf{R}$ be a linear form with positive values on J_1. We denote by $\pi: \widetilde{Q} \to Q$ the projection of $\widetilde{Q}$ onto $\widetilde{Q}_{\text{even}} \approx Q$, which annihilates $\widetilde{Q}_{\text{odd}}$. Let $P: Q \to x_1 Q$ be any projection onto the ideal in Q generated by x_1. Let $\rho: Q_1 \to x_1 Q$ be the isomorphism of part D). We define linear forms $q: Q \to \mathbf{R}$ and $\tilde{\tilde{q}} : \widetilde{Q} \to \mathbf{R}$ by

$$q = q_1 \circ \rho^{-1} \circ P, \quad \tilde{\tilde{q}} = q_1 \circ \rho^{-1} \circ P \circ \pi.$$

We consider the corresponding symmetric bilinear forms on Q_1, Q, and $\widetilde{Q}$: $\varphi_1(a, b) = q_1(ab)$ for a, b in Q_1, $\varphi(a, b) = q(ab)$ for a, b in Q, and $\tilde{\tilde{\varphi}}(a, b) = \tilde{\tilde{q}}(ab)$ for a, b in $\widetilde{Q}$. We note that

$$q_1([J_1]) > 0, \quad q([J]) > 0, \quad \tilde{\tilde{q}}([\tilde{\tilde{J}}]) > 0.$$

For it follows from Lemma F that

$$q([J]) = \frac{\mu_0 + \mu_1}{\mu_1} q_1([J_1]), \quad \tilde{\tilde{q}}([\tilde{\tilde{J}}]) = \frac{\mu_0 + 2\mu_1}{\mu_1} q_1([J_1]).$$

H) *The subspaces $\widetilde{Q}_{\text{even}}$ and $\widetilde{Q}_{\text{odd}}$ are $\tilde{\tilde{\varphi}}$-orthogonal.*

For, $\pi(a_{\text{even}}, b_{\text{odd}}) = 0$.

I) *The restriction of $\tilde{\tilde{\varphi}}$ to $\widetilde{Q}_{\text{even}}$ is φ (under the identification of Q and $\widetilde{Q}_{\text{even}}$ in part* B)),

For $\widetilde{Q}_{\text{even}}$ is a subalgebra of $\widetilde{Q}$, and $\tilde{\tilde{q}} = q \circ \pi$.

J) *The restriction of $\tilde{\tilde{\varphi}}$ to $\widetilde{Q}_{\text{odd}}$ is φ_1 (under the identification of Q_1 and $\widetilde{Q}_{\text{odd}}$ in part* C)),

For using C), G), and D) in succession, we find that

$$\tilde{\tilde{\varphi}}(z\tilde{a}, z\tilde{b}) = q_1 \circ \rho^{-1} \circ P \circ \pi (z^2 \tilde{a}\tilde{b}) = q_1 \circ \rho^{-1} \circ P(x_1 ab) = \\ = q_1 \circ \rho^{-1}(x_1 ab) = q_1(ab) = \varphi_1(a, b).$$

K) *It follows from* H), I), *and* J) *that the signature of* $\widetilde{\widetilde{\varphi}}$ *is equal to the sum of the signatures of* φ *and* φ_1, *as required.*

§7. Indices of critical points of functions on the boundary of a manifold

By the indices of a critical point of a function we mean the corresponding indices of its differential. Let f: $(\mathbf{R}^n, O) \to (\mathbf{R}, 0)$ be a smooth function with an isolated boundary singularity O on the half-space $\mathbf{R}^n_+$ $(x_1 > 0)$. We fix a sufficiently small ρ-neighbourhood of O and then we choose a sufficiently small $\varepsilon > 0$. We consider the manifold with boundary

$$M_\varepsilon = \{x\colon f(x) = -\varepsilon, \ |x| \leqslant \rho\}.$$

We divide it into the parts in the half-spaces $\mathbf{R}^n_+$ $(x_1 \geqslant 0)$ and $\mathbf{R}^n_-$ $(x_1 \leqslant 0)$ and in the hypersurface $\mathbf{R}^{n-1}$ $(x_1 = 0)$:

$$M_\varepsilon = M^+ \cup M^-, \ M^+ \cap M^- = M^0.$$

Let $\widetilde{M}_+$ and $\widetilde{M}_-$ denote the corresponding doubles.

THEOREM. *The indices of the boundary singularity of the differential of f are connected with the Euler characteristics of the corresponding level sets by the relations*

$$i_+ = 1 - \chi(M^+), \qquad i_- = 1 - \chi(M^-),$$
$$\tilde{i}_+ = 1 - \chi(\widetilde{M}^+), \qquad \tilde{i}_- = 1 - \chi(\widetilde{M}^-),$$
$$i_1 = 1 - \chi(M_\varepsilon), \qquad i_0 = 1 - \chi(M^0).$$

PROOF. The last four formulae are obvious (see, for example, [5]), and the first two follow from them by the relations between the indices (§3) and between the Euler characteristics

$$\chi(\widetilde{M}_+) = 2\chi(M^+) - \chi(M^0).$$

REMARK. Actually, the definitions of the indices i_+ and i_- were suggested by these formulae on the one hand, and by the equality of the signatures of the two families of quadratic forms connected with the so-called simple singularities on the other. These two families of quadratic forms and the connections between them are discussed in §§8 and 9.

§8. Indices of simple boundary singularities

The simplest boundary singularities of functions are connected with the simple Lie algebras A_μ, B_μ, C_μ, D_μ, E_6, E_7, E_8, F_4 (only G_2 is not included):

A_μ: $f = x_1 \pm x_2^{\mu+1} + \ldots,$
B_μ: $f = \pm x_1^\mu + \ldots,$
C_μ: $f = x_1 x_2 \pm x_2^\mu + \ldots,$
D_μ: $f = x_1 + x_2^2 x_3 \pm x_3^{\mu-1} + \ldots,$
E_6: $f = x_1 + x_2^3 \pm x_3^4 + \ldots,$
E_7: $f = x_1 + x_2^3 + x_2 x_3^3 + \ldots,$
E_8: $f = x_1 + x_2^3 + x_3^5 + \ldots,$
F_4: $f = \pm x_1^2 + x_2^3 + \ldots$

Here the dots denote a non-degenerate quadratic form in the remaining variables (that is, $\pm x_k^2 \pm \ldots \pm x_n^2$, where $k = 2$ for B, 3 for A, C, and F, and 4 for D and E); the boundary equation is $x_1 = 0$.

Apart from the quadratic forms on the local algebras

$$Q = \mathbf{R}[[x_1, \ldots, x_n]]/(x_1 \partial f/\partial x_1, \partial f/\partial x_2, \ldots, \partial f/\partial x_n),$$

which were described in §6, there are also quadratic forms on the dual spaces that are connected with the simple singularities; the signatures of these forms also turn out to be equal (in modulus) to the indices $i_+ - i_-$ of the correspond- boundary singularities. These new quadratic forms are defined immediately for all reflection groups.

Let G be an finite irreducible reflection group in the Euclidean space $\mathbf{R}^\mu$. The G also acts on the complexified space $\mathbf{C}^\mu$. The orbit *variety* $B = \mathbf{C}^\mu/G$ is also diffeomorphic to $\mathbf{C}^\mu$. A coordinate system in B is given by a basis of the ring of symmetric (G-invariant) functions (polynomials) in $\mathbf{C}^\mu$. The natural projection π: $\mathbf{C}^\mu \to B$ takes all the reflecting planes into some (in general, reducible) hypersurface $\Sigma \subset B$. This hypersurface is called the *variety of non-regular orbits* (or *discriminant variety*).

We denote by $\mathcal{O}$ the algebra of germs of holomorphic functions on B at O. We define a bilinear operation of "*convolution of invariants*" Φ: $\mathcal{O} \times \mathcal{O} \to \mathcal{O}$ by the relation

$$\Phi(a, b) = \pi_* \langle \nabla \pi^* a, \nabla \pi^* b \rangle,$$

where $\langle \, , \rangle$ and ∇ are the Euclidean scalar product and gradient.

Explicit formulae are given in [1] for the convolution Φ in the case of the series A.

We denote by T the tangent space to B at zero and by T^* the dual space. The linearization of the convolution leads to the operation of "*linearized convolution*" φ: $T^* \times T^* \to T^*$, which is defined by the relation

$$\varphi(\alpha, \beta) = d(\Phi(a, b)),$$

where $\alpha = da$ and $\beta = db$ (d is the differential of a function at zero).

Let q be a point of T. We define a symmetric bilinear form on T^* by

$$\psi_q(\alpha, \beta) = \varphi(\alpha, \beta) | q.$$

THEOREM A. *If q does not belong to the tangent plane to the discriminant* *variety at zero, then the form* ψ_q *is non-degenerate.*

Before we prove Theorem A we state it in other terms.

DEFINITION. A *vector field with potential* $a \in \mathcal{O}$ is a field

$$v_a = \pi_* \nabla \pi^* a$$

on B. Obviously, the fields v_a can be expressed in terms of the convolution operation:

$$v_a b = \Phi(a, b).$$

THEOREM B. *Let* $a_1, \ldots, a_\mu$ *be a coordinate system on* B. *Then the vectors* $v_{a_1}, \ldots, v_{a_\mu}$ *are linearly independent at every point of* $B \setminus \Sigma$, *and their determinant has a first-order zero on* Σ.

DEFINITION. A linear field with potential $\alpha \in T^*$ is a field on T:

w_α = linear part of v_a at O, where α is the linear part of a at O.

The fields w_α can, obviously, be expressed in terms of the linearized convolution operation: $w_\alpha \beta = \varphi(\alpha, \beta)$.

THEOREM C. *Let* $\alpha_1, \ldots, \alpha_\mu$ *be a basis of* T^*. *Then the vectors* $w_{\alpha_1}, \ldots, w_{\alpha_\mu}$ *are linearly independent at each point of* T *that does not belong to the tangent plane* τ *to the discriminant variety* Σ, *and the determinant of the components of these vectors has a zero of order* μ *on* τ.

PROOF OF THEOREM B. The differentials of the functions a_i at each point of B are linearly independent. Hence, the differentials of $\pi^* a_i$ are independent at each point of $\mathbf{C}^\mu$ that does not lie on a reflecting hyperplane. Consequently, the vectors $\nabla \pi^* a_i$ are independent outside such a hyperplane. Therefore, the vectors $v_{a_i} = \pi_* \nabla \pi^* a_i$ are independent at each point of B outside the discriminant variety Σ.

We consider a *generic* point of the discriminant variety. We choose local coordinates $\Lambda_1, \ldots, \Lambda_\mu$ in a neighbourhood of this point such that the equation of the discriminant variety takes the form $\Lambda_1 = 0$ (and so that all the coordinates of this point are zero).

We consider one of the inverse images of this point under the mapping

$$\pi \colon \mathbf{C}^\mu \to B.$$

The functions

$$z_2 = \pi^* \Lambda_2, \ldots, z_\mu = \pi^* \Lambda_\mu \text{ and } z_1 = \sqrt{(\pi^* \Lambda_1)}$$

form a system of local coordinates in a neighbourhood of this point in $\mathbf{C}^\mu$; here the equation of the reflecting hyperplane has the form $z_1 = 0$, and the reflection corresponding to it changes the sign of the coordinate z_1 without changing the values of the remaining coordinates of the point being reflected.

We compute the Jacobian J of the components of the fields v_{Λ_i} at a point with a small non-zero Λ_1-coordinate. We have

$$J = (\rho d\Lambda_1 \wedge \cdots \wedge d\Lambda_\mu)|(v_1 \wedge \cdots \wedge v_\mu) =$$
$$= \rho \det(\partial\Lambda/\partial z)[(dz_1 \wedge \cdots \wedge dz_\mu)|(\nabla\pi^*\Lambda_1 \wedge \cdots \wedge \nabla\pi^*\Lambda_\mu)],$$

where $\rho \neq 0$. Note that $\det(\partial\Lambda/\partial z) = 2z_1$. Further, the vectors $\nabla\pi^*\Lambda_k$ are obtained from $\Sigma(\partial\Lambda_k/\partial z_i)\partial/\partial z_i$ by a non-degenerate linear transformation whose determinant g also does not vanish on the reflecting hyperplane (for $z_1 = 0$). Therefore,

$$J = \rho g 2z_1[(dz_1 \wedge \cdots \wedge dz_\mu)|(2z_1\partial/\partial z_1 \wedge \partial/\partial z_2 \wedge \cdots \wedge \partial/\partial z_\mu)] =$$
$$= 4\rho g z_1^2 = 4\rho g\Lambda_1.$$

Hence, the determinant of the components of the fields v_{Λ_i} has a first-order zero on Σ. For any coordinate system $a_1, \ldots, a_\mu$ the v_{a_i} can be expressed linearly and non-degenerately in terms of the v_{Λ_i}; therefore, the determinant of their components has a first-order zero at the generic point of Σ.

PROOF OF THEOREM C. Let $a_1, \ldots, a_\mu$ be a coordinate system such that $da_i = \alpha_i$. We consider the Jacobian J formed from the components of the v_{a_i}. This function on B has a first-order zero on Σ (Theorem B). Hence, the algebraic-geometric multiplicity of the singular point O (the order of the zero of J on a general line passing through O) is equal to μ. But Σ has a tangent plane at zero. Consequently, the initial form of J has the shape $c\lambda^\mu$, where $\lambda = 0$ is the equation of the tangent plane to Σ and $c \neq 0$.

Now we note that the determinant of the components of the v_{a_i} differs from the determinant of the components of the vectors w_{α_i} at the corresponding point by terms of order $\mu + 1$ (since w_{α_i} is the linearization of v_{a_i} at zero). Hence, the last determinant is *equal* to $c\lambda^\mu$, as required.

PROOF OF THEOREM A. The fact that the form ψ_q is non-degenerate indicates that the defining form of the linear operator

$$\Psi_q\colon T^* \to T,\ \Psi_q(\alpha)|\beta = \varphi(\alpha, \beta)|q, \quad \alpha, \beta \in T^*,$$

is non-degenerate. But the operator ψ_q takes α to the value of w_α at q (see the definition of the w_α). Hence, ψ_q being non-degenerate is equivalent to the linear independence of the system of values of the basis fields w_{α_i} at q. Thus, Theorem A follows from Theorem C, which we have just proved.

We return to the simple boundary singularity of a real function on the manifold with boundary $x_1 = 0$. The complexification of the corresponding local ring

$$Q = \mathbf{R}[[x_1, \ldots, x_n]]/(x_1\partial f/\partial x_1, \partial f/\partial x_2, \ldots, \partial f/\partial x_n)$$

is identified with the tangent space T at O to the variety B of orbits of the corresponding reflection group. For real q the form ψ_q is real on Q^*. By Theorem A it is non-degenerate if q is an invertible element of Q. Consequently, the signature of this form does not change under a continuous variation of q, and the modulus of the signature also does not depend at all on the choice of the invertible element $q \in Q$.

THEOREM D. *The signature of the form* ψ_q *defined above is equal in modulus to the index* $i = i_+ - i_-$ *of the corresponding boundary singularity.*

PROOF. Computing the indices i for all real forms of boundary singularities, we obtain

Type	A_{2k}	A_{2k+1}	B_{2k}	B_{2k+1}	C_{2k}	C_{2k+1}	D^+_{2k}	D^-_{2k}	D_{2k+1}	E_6	E_7	E_8	F_4
$\lvert i \rvert$	1	0	0	1	0	1	0	2	1	0	1	0	0

Here the sign in $D^\pm_\mu$ corresponds to the sign in the normal form $x_2^2 x_3 \pm x_3^{\mu-1}$.

Calculating the signatures of the ψ_q in these cases, we obtain the same answers. For in the cases A_μ, B_μ, C_μ, and D_μ it is fairly easy to compute the linearized convolution operation φ (see the remarks below), and the computations of the signature of ψ_q can be done explicitly. In the cases E_μ and F_4 the moduli of the signatures can be found without calculating φ, but at the expense of considerations of quasi-homogeneity (this is also applicable in the cases A_μ, B_μ, C_μ, and D_{2k+1}).

We carry out this computation completely for the case E_6. The normal form is $x_1 + x_2^3 \pm x_3^4$. The local algebra Q is the same as for the usual (non-boundary) singularity E_6 of the function $x^3 \pm y^4$.

Q is generated (over $\mathbf{R}$) by the basis $e_1 = 1$, $e_2 = y$, $e_3 = x$, $e_4 = y^2$, $e_5 = xy$, $e_6 = xy^2$. The corresponding coordinates $\lambda_1, \ldots, \lambda_6$ in Q form a basis of Q^*.

These elements of Q^* determine the basis invariants of the corresponding reflection group. Their degrees are, respectively, equal to 12, 9, 8, 6, 5, and 2. We equip the ring of polynomials in λ with a "quasi-homogeneous grading", in which $\deg \lambda_1 = 12$, $\deg \lambda_2 = 9, \ldots, \deg \lambda_6 = 2$. If a is a homogeneous invariant of degree k, and b is of degree l, then $\langle \nabla a, \nabla b \rangle$ is a homogeneous invariant of degree $k + l - 2$. This allows us to control the action of the convolution operation Φ on the grading.

We form a matrix from the functions $\Phi(\lambda_p, \lambda_q) = \pi_*\langle \pi^*\lambda_p, \pi^*\lambda_q \rangle$ on Q. Its elements are (quasi-)homogeneous polynomials in λ whose degrees are given by the following table:

	λ_6	λ_5	λ_4	λ_3	λ_2	λ_1
λ_6	2	5	6	8	9	12
λ_5	5	8	9	11	12	
λ_4	6	9	10	12		
λ_3	8	11	12			
λ_2	9	12				
λ_1	12					

(the unoccupied spaces are filled with numbers larger than 12). From this table it is clear that the functions in the lower right (unoccupied) corner of the matrix of Φ are decomposable invariants (they lie in the square of the maximal ideal of $\mathcal{O}$); they do not contribute to the linearized convolution φ.

For the calculation of the signature of ψ_q we take as q the element e_1. Each element $\varphi(\lambda_p, \lambda_q)$ is a linear combination of the functions λ_r, and its value on e_1 is equal to the coefficient of λ_1 in this linear combination. But the degree of λ_1 is equal to 12, which is the largest of all the numbers in the left upper corner of our matrix.

Therefore, $\psi_{e_1}(\lambda_p, \lambda_q) \neq 0$ only for dual p and q (such that $p + q = 7$). Consequently, the quadratic form on Q^* corresponding to ψ_{e_1} reduces to $\xi_1\xi_6 + \xi_2\xi_5 + \xi_3\xi_4$ and hence, has signature zero.

The same argument allows us to compute the modulus of the signature in all the cases except D_{2k} (this is the only case when the set of degrees contains a repetition). We give here only the result of a direct computation of the linearized convolution operation φ for D_μ.

The normal form is $x_1 + x_2^2 \pm x^{\mu-1}$. The local algebra Q is the same as for the usual (non-boundary) singularity $D_\mu^\pm$ of the function $x^2y \pm y^{\mu-1}$. Q is generated (over $\mathbf{R}$) by the basis $e_1 = y^{\mu-2}, \ldots, e_{\mu-1} = 1; e_\mu = x$. The corresponding coordinates $\lambda_1, \ldots, \lambda_\mu$ in Q form a basis of Q^*.

Let $(z_1, \ldots, z_\mu)$ be coordinates in $\mathbf{R}^\mu$, where the reflection group D_μ acts by permuting the coordinates and changing the signs of an even number of coordinates. For the μ basis invariants we can take, for example,

$$s_2 = \sum z_i^2, \quad s_4 = \sum z_i^4, \quad \ldots, \quad s_{2\mu-2} = \sum z_i^{2\mu-2}, \quad P_\mu = z_1 \ldots z_\mu.$$

We compute the linearized convolution operation φ on the basis invariants. We denote the field $\partial/\partial z_i$ by ∂_i. In succession we find

$$\nabla s_{2k} = 2k \sum z_i^{2k-1}\partial_i, \quad \langle \nabla s_{2k}, \nabla s_{2l}\rangle = 4kl s_{2k+2l-2},$$
$$\nabla P_\mu = P z_1^{-1}\partial_1 + \ldots + P z_\mu^{-1}\partial_\mu, \quad \langle \nabla s_{2k}, \nabla P_\mu\rangle = 2k \sum z_i^{2k-2} P_\mu,$$
$$\langle \nabla P_\mu, \nabla P_\mu\rangle = \sum (P_\mu/z_i)^2 = \sigma_{\mu-1}(z^2) = (-1)^\mu(\mu-1)^{-1}s_{2\mu-2} + \ldots$$

Here the dots denote decomposable invariants (contained in the square of the ideal generated by $s_2, \ldots, s_{2\mu-2}, P_\mu$). We note that $\langle \nabla s_{2k}, \nabla s_{2l}\rangle$ is also decomposable for $2k + 2l > 2\mu$, and that $\langle \nabla s_{2k}, \nabla P_\mu\rangle$ is decomposable for $k \neq 1$; in the last case $\langle \nabla s_2, \nabla P_\mu\rangle = 2\mu P_\mu$. Finally, the linearized convolution φ is given by the table

	s_2	s_4	$\ldots$	$s_{2\mu-2}$	P_μ
s_2	$4s_2$	$8s_4$	$\ldots$	$4(\mu-1)s_{2\mu-2}$	$2\mu P_\mu$
s_4	$8s_4$		$\cdot$	0	0
$\vdots$	$\vdots$		$\cdot$	$\vdots$	$\vdots$
$s_{2\mu-2}$	$4(\mu-1)s_{2\mu-2}$	0	$\ldots$	0	0
P_μ	$2\mu P_\mu$	0	$\ldots$	0	$\dfrac{(-1)^\mu s_{2\mu-2}}{\mu-1}$.

The isomorphism of the complexification of Q with the tangent space T to the orbit space of D_μ on $\mathbf{C}^\mu$ at O is constructed in the following way. We consider the function $x^2y \pm y^{\mu-1} + \Sigma\lambda_k e_k$. A point $\Sigma\lambda_k e_k \in \mathbf{C}Q$ is singular if zero is a critical value of this function of x and y. The set of all singular points is called the *discriminant variety. Under the mapping of* $\mathbf{C}Q$ *into* B *the discriminant variety is mapped to the variety of non-regular orbits.*

Computing the singular points $\Sigma\lambda_k e_k$ we obtain

$$2xy + \lambda_\mu = 0, \quad x^2 + p'(y) = 0, \quad x^2y + p(y) + \lambda_\mu x = 0,$$

where

$$p(y) = \pm y^{\mu-1} + \lambda_1 y^{\mu-2} + \ldots + \lambda_{\mu-1}.$$

Eliminating x from the first equation, we conclude that the point $\Sigma\lambda_k e_k$ is singular if and only if the polynomial $4yp(y) - \lambda_\mu^2$ has a multiple root.

On the other hand, the variety of non-regular orbits Σ in B can be described in the following way. Let $\sigma_1, \ldots, \sigma_\mu$ be the fundamental symmetric functions of $z_1^2, \ldots, z_\mu^2$. The invariants $\sigma_1, \ldots, \sigma_{\mu-1}, P_\mu$ can be taken as coordinates on B. We form the equation

$$Z^\mu - \sigma_1 Z^{\mu-1} + \ldots + (-1)^{\mu-1}\sigma_{\mu-1}Z + (-1)^\mu P_\mu^2 = 0.$$

The point with the coordinates $(\sigma_1, \ldots, \sigma_{\mu-1}, P_\mu)$ belongs to the variety of non-regular orbits if and only if this equation has a multiple root (since the reflecting planes are given by the equations $z_i^2 = z_j^2$, that is, $Z_i = Z_j$).

Comparing the coefficients of this polynomial with the coefficients of the polynomial

$$4yp - \lambda_\mu^2 = \pm 4y^\mu + 4\lambda_1 y^{\mu-1} + \ldots + 4\lambda_{\mu-1}y - \lambda_\mu^2,$$

we find

$$\sigma_k = \pm(-1)^k\lambda_k \quad (k = 1, \ldots, \mu - 1); \; P_\mu = \lambda_\mu \sqrt{(\mp(-1)^\mu/2)}.$$

But by Newton's formula,

$$k\sigma_k = (-1)^{k-1}s_{2k} + \text{decomposable elements.}$$

Consequently, up to decomposable elements,

$$s_{2k} = \mp k\lambda_k, \quad 2P_\mu = \lambda_\mu \sqrt{(\mp(-1)^\mu)}.$$

These relations allow us to calculate the linearized convolution operation φ in the basis $\lambda_1, \ldots, \lambda_\mu$. The resulting table for $\varphi(\lambda_k, \lambda_l)$ has the form

	λ_1	λ_2	$\ldots$	$\lambda_{\mu-1}$	λ_μ
λ_1	$\mp 4\lambda_1$	$\mp 8\lambda_2$	$\ldots$	$\mp 4(\mu-1)\lambda_{\mu-1}$	$\mp 2\mu\lambda_\mu$
λ_2	$\mp 8\lambda_2$		$\cdot$	0	0
$\vdots$	$\vdots$	$\cdot$	$\cdot$	$\vdots$	$\vdots$
$\lambda_{\mu-1}$	$\mp 4(\mu-1)\lambda_{\mu-1}$	0	$\ldots$	0	0
λ_μ	$\mp 2\mu\lambda_\mu$	0	$\ldots$	0	$4\lambda_{\mu-1}$

To compute the signature of the form ψ_q we take $q = e_{\mu-1}$ (that is, q is a unit of the ring). Then we obtain a quadratic form ψ_q of the kind $\mp(\xi_1\xi_{\mu-1} + \xi_2\xi_{\mu-2} + \ldots) + \xi_\mu^2$. If μ is odd, then this form has signature 1. But if μ is even, the signature of the form in parentheses is equal to 1, so that

$$\operatorname{sgn} \psi_q(D_{2k}^+) = 0, \quad \operatorname{sgn} \psi_q(D_{2k}^-) = 2,$$

as required.

§9. Duality

From the theorems proved above it follows that the (moduli of the) signatures of the quadratic forms are equal, of which one form is defined on the local algebra Q and the other on the dual space Q^*. This suggests that these forms are themselves related. Such a connection in fact exists. It is one of the manifestations of the astonishing duality between the multiplication $Q \times Q \to Q$ and the linearized convolution operation φ: $T^* \times T^* \to T^*$. This duality seems surprising to me because it is verified separately for each singularity, while the general reasons for it are not clear.

Let $\alpha \in Q^*$ be a general element of the dual space to Q. We define a bilinear form f_α on Q by the relation $f_\alpha(u, v) = \alpha \mid (uv)$. From the multiplication in Q we obtain a family of bilinear forms on Q, parametrized by the elements of Q^*

Precisely the same construction transforms the linearized convolution operation φ: $Q^* \times Q^* \to Q^*$ (defined in §8) into a family of bilinear forms on Q^*, parametrized by the elements of Q.

In §8 these forms were denoted by $\psi_q (q \in Q)$ (strictly speaking, they also depend on the choice of identification of the space of the local algebra Q with the tangent space T to the orbit space of a reflection group; we discuss the influence of this identification below in Remark 3).

We consider a non-degenerate quadratic form on Q. It is given by a non-degenerate operator $Q \to Q^*$. The inverse operator gives a non-degenerate form on Q^*. This form is said to be *dual* to the original one (or the *Legendre transform* of the original form).

PROPOSITION 1. *The choices of the forms* $\{f_\alpha\}$ *and* $\{\psi_q\}$ *are dual in the sense that for each non-degenerate form* f_α *the dual is one of the* ψ_q, *and vice versa.*

This (surprising) proposition is proved by direct calculations in the cases A_μ, B_μ, C_μ, and D_μ; apparently, it also holds in general.[1]

The question arises of the relation between the parameters α and q of dual forms. This relation is non-linear, but can be expressed linearly in terms of the inversion operation in Q. More accurately, the linearized displacement φ can be expressed in terms of multiplication in Q in the following way.

Let D be a quasi-homogeneous derivation in the graded local algebra Q, $D = \Sigma d_k \lambda_k \partial/\partial\lambda_k$, where $d_k = \deg e_k$ (we are concerned with a grading in which the function f from which the local algebra

[1] **Translator's note: In a letter dated 18 November 1979 the author states that A. B. Guivental has proved this conjecture.**

$$Q = \mathbf{C}[[x]]/(x_1 f_1,\ f_2,\ \ldots,\ f_n),$$

is constructed is homogeneous of degree 1). We define a **C**-linear operator $R = E - D\colon Q \to Q$. We denote by $M_p\colon Q \to Q$ the operator of multiplication by $p \in Q$, so that $M_p q = pq$. The dual operator $M_p^*\colon Q^* \to Q^*$ gives, for each $\alpha \in Q^*$, a linear mapping $N_\alpha\colon Q \to Q^*$ by the formula

$$N_\alpha p = M_p^* \alpha \qquad \text{for every} \qquad p \in Q.$$

For an α in general position the mapping N_α is invertible.

Let G be an irreducible reflection group in $\mathbf{C}^\mu$, B the orbit variety, T the tangent space to B at O, T^* the dual space, and $\varphi\colon T^* \times T^* \to T^*$ the linearized convolution operation, defined by the Euclidean structure in $\mathbf{C}^\mu$ (see §8).

PROPOSITION 2. *Let α be an element of Q^* in general position (so that N_α is an invertible operator). Then the operation $\varphi_\alpha\colon Q^* \times Q^* \to Q^*$, given by the relation*

$$\varphi_\alpha(a,\ b) = R^* N_\alpha (N_\alpha^{-1} a \cdot N_\alpha^{-1} b),$$

goes over into the linearized convolution operation $\varphi\colon T^* \times T^* \to T^*$ under some isomorphism of the linear spaces T^* and Q^*.

This (surprising) assertion is proved in the cases, A, B, C, and D by a direct calculation if α is taken to be a basis element (the coefficient of a basis monomial of highest degree, proportional to the Jacobian). But apparently it also holds in the general case,[1] and for any (general) α. In particular, the operations φ_α with different α are equivalent (they go over into each other under some linear mapping of Q^* into itself), which is not at all obvious from the formula for φ_α.

PROOF OF PROPOSITION 2 FOR THE SINGULARITIES A_μ, B_μ, C_μ, D_μ AND A SPECIAL α. Since in the cases B_μ and C_μ both the reflection groups and the graded local rings are isomorphic, the case C_μ is not considered further. The versal deformations of the singularities A_μ, B_μ, and D_μ are chosen in the following way:

$$A_\mu\colon\ x^{\mu+1} + \lambda_1 x^{\mu-1} + \ldots + \lambda_\mu \cdot 1,$$
$$B_\mu\colon\ x^{\mu} + \lambda_1 x^{\mu-1} + \ldots + \lambda_\mu \cdot 1 \ \text{(boundary: } x = 0\text{)},$$
$$D_\mu\colon\ x^2 y \pm y^{\mu-1} + \lambda_1 y^{\mu-2} + \ldots + \lambda_{\mu-1} \cdot 1 + \lambda_\mu x.$$

We denote by $e_0, e_1, \ldots, e_{\mu-1}$ the basis elements of Q over **C** (or over **R**) corresponding to the monomials in these formulae. In the dual space Q^* we choose a basis $f_0, f_1, \ldots, f_{\mu-1}$. These bases are connected with the monomials of the versal deformation and the coefficients λ_k for these monomials in the following way:

	$e_0 \ldots$	$e_{\mu-2}$	$e_{\mu-1}$	$f_0 \ldots$	$f_{\mu-2}$	$f_{\mu-1}$
A_μ, B_μ	$1 \ldots$	$x^{\mu-2}$	$x^{\mu-1}$	$\lambda_1 \ldots$	$\lambda_{\mu-1}$	λ_μ
D_μ	$1 \ldots$	$y^{\mu-2}$	x	$\lambda_1 \ldots$	$\lambda_{\mu-1}$	λ_μ

[1] This has also been proved since by A. B. Guivental (letter from author, 18 November 1979). (Transl.)

The multiplication tables in Q in these bases have the following form:

A_μ, B_μ	e_0	$e_1 \ldots e_{\mu-1}$
e_0	e_0	$e_1 \ldots e_{\mu-1}$
e_1	e_1	$\cdot^{\cdot^{\cdot}}$ 0
$\vdots$	$\vdots$	$\cdot^{\cdot^{\cdot}}$ $\vdots$
$e_{\mu-1}$	$e_{\mu-1}$	$0 \ldots 0$

D_μ	e_0	$e_1 \ldots e_{\mu-2}$	$e_{\mu-1}$
e_0	e_0	$e_1 \ldots e_{\mu-2}$	$e_{\mu-1}$
e_1	e_1	$\cdot^{\cdot^{\cdot}}$ 0	0
$\vdots$	$\vdots$	$\cdot^{\cdot^{\cdot}}$ $\vdots$	$\vdots$
$e_{\mu-2}$	$e_{\mu-2}$	$0 \ldots 0$	0
$e_{\mu-1}$	$e_{\mu-1}$	$0 \ldots 0$	$\mp(\mu-1)e_{\mu-2}$

In all cases we take $\alpha = f_0$. It is not hard to check that the operator $N_\alpha : Q^* \to Q^*$ then acts on the above bases according to the formula $N_\alpha e_k = f_k$ in all cases except one, namely, for D_μ the last basis element is transformed into $N_\alpha e_{\mu-1} = \mp(\mu - 1)f_{\mu-1}$.

The gradings in Q are defined by the conditions

$$\deg x^{\mu+1} = 1 \quad (\text{for } A_\mu)$$
$$\deg x^{\mu} = 1 \quad (\text{for } B_\mu),$$
$$\deg(x^2y,\ y^{\mu-1}) = 1 \quad (\text{for } D_\mu);$$

consequently, $\deg e_k = k/(\mu+1)$ for A_μ, $\deg e_k = 2k/2\mu$ for B_μ; $\deg e_k = 2k/(2\mu-2)$ for D_μ when $k < \mu - 1$, $\deg e_{\mu-1} = (\mu-2)/(2\mu-2)$. In place of R it is convenient to consider the proportional operator hR, where $h(A_\mu) = \mu + 1$, $h(B_\mu) = 2\mu$, $h(D_\mu) = 2\mu - 2$. The eigenvectors of the operator hR^* are f_k, and the corresponding eigenvalues are given, in accordance with the $\deg e_k$ found above, by the table

Type	f_0	f_1	$\ldots$	$f_{\mu-2}$	$f_{\mu-1}$
A_μ	2	3	$\ldots$	μ	$\mu+1$
B_μ	2	4	$\ldots$	$2\mu-2$	2μ
D_μ	2	4	$\ldots$	$2\mu-2$	μ

Finally, the operations

$$h\varphi_\alpha = hR^* N_\alpha (N_\alpha^{-1} \cdot N_\alpha^{-1} \cdot)$$

are given by the tables

A_μ	f_0	$f_1 \ldots f_{\mu-1}$
f_0	$2f_0$	$3f_1 \ldots (\mu+1)f_{\mu-1}$
f_1	$3f_1$	$\cdot^{\cdot^{\cdot}}$ 0
$\vdots$		$\cdot^{\cdot^{\cdot}}$ $\vdots$
$f_{\mu-1}$	$(\mu+1)f_{\mu-1}$	$0 \ldots 0$

B_μ	f_0	$f_1 \ldots f_{\mu-1}$
f_0	$2f_0$	$4f_1 \ldots 2\mu f_{\mu-1}$
f_1	$4f_1$	$\cdot^{\cdot^{\cdot}}$ 0
$\vdots$	$\vdots$	$\cdot^{\cdot^{\cdot}}$ $\vdots$
$f_{\mu-1}$	$2\mu f_{\mu-1}$	$0 \ldots 0$

D_μ	f_0	$f_1 \ldots f_{\mu-2}$	$f_{\mu-1}$
f_0	$2f_0$	$4f_1 \ldots (2\mu-2)f_{\mu-2}$	$\mu f_{\mu-1}$
f_1	$4f_1$	$\cdot^{\cdot^{\cdot}}$ 0	0
$\vdots$	$\vdots$	$\cdot^{\cdot^{\cdot}}$ $\vdots$	$\vdots$
$f_{\mu-2}$	$(2\mu-2)f_{\mu-2}$	$0 \ldots 0$	0
$f_{\mu-1}$	$\mu f_{\mu-1}$	$0 \ldots 0$	$\mp 2f_{\mu-2}$

The linearized convolution operation φ is computed in §8 for the case D_μ. In the remaining cases the computations are similar but simpler. In the case A_μ the group acts by permutations on the roots of the equation $x^{\mu+1} + f_0 x^{\mu-1} + \ldots + f_{\mu-1} = 0$; in the case B_μ by permutations and changes of sign of the roots of the equation $z^{2\mu} + f_0 z^{2\mu-2} + \ldots + f_{\mu-1} = 0$. The calculation of $\langle \nabla f_k, \nabla f_l \rangle$ modulo decomposable elements is easily carried out by means of Newton's formula $\sigma_k = (-1)^{k-1} s_k / k$ mod (decomposable elements). Finally, the operations φ are given by the tables

A_μ	f_0	f_1	$\cdots$	$f_{\mu-1}$
f_0	$-2f_0$	$-3f_1$	$\cdots$	$-(\mu+1)f_{\mu-1}$
f_1	$-3f_1$		$\cdot$	0
$\vdots$	$\vdots$		$\ddots$	$\vdots$
$f_{\mu-1}$	$-(\mu+1)f_{\mu-1}$	0	$\cdots$	0

B_μ	f_0	f_1	$\cdots$	$f_{\mu-1}$
f_0	$-4f_0$	$-8f_1$	$\cdots$	$-4\mu f_{\mu-1}$
f_1	$-8f_1$		$\cdot$	0
$\vdots$	$\vdots$		$\ddots$	$\vdots$
$f_{\mu-1}$	$-4\mu f_{\mu-1}$	0	$\cdots$	0

D_μ	f_0	f_1	$\cdots$	$f_{\mu-2}$	$f_{\mu-1}$
f_0	$\mp 4f_0$	$\mp 8f_1$	$\cdots$	$\mp 4(\mu-1)f_{\mu-2}$	$\mp 2\mu f_{\mu-1}$
f_1	$\mp 8f_1$		$\cdot$	0	0
$\vdots$	$\vdots$		$\ddots$	$\vdots$	$\vdots$
$f_{\mu-2}$	$\mp 4(\mu-1)f_{\mu-2}$	0	$\cdots$	0	0
$f_{\mu-1}$	$\mp 2\mu f_{\mu-1}$	0	$\cdots$	0	$4f_{\mu-2}$

Comparing these with the preceding tables, we obtain in all cases

$$\varphi = ch\varphi_\alpha,$$

where $c(A_\mu) = -1$, $c(B_\mu) = -2$, and $c(D_\mu^\pm) = \mp 2$.

Upon multiplication by a constant φ goes into an equivalent operation, because

$$\varkappa\varphi(a, b) = \varkappa^{-1}\varphi(\varkappa a, \varkappa b).$$

Consequently, the operations φ and φ_α are equivalent, as required.

REMARK 1. Proposition 1 follows from Proposition 2. For it is not difficult to calculate that N_γ is an operator of the form f_γ on Q. With an element $\gamma \in Q^*$ we associate a $q \in Q$ such that $\gamma = N_\alpha (Rq)^{-1}$. Then N_γ^{-1} is an operator of the form $\psi_{a,q} = \varphi_\alpha \mid q$ on Q^*. Thus, a form of the family $\{f_\gamma\}$ on Q corresponding to the parameter value γ and a form of the family $\{\psi_{\alpha,q}\}$ on Q^* corresponding to the parameter value q are dual to each other, provided that the element Rq is linearly independent of q, the element N_γ^{-1} of Q is linearly independent of, and these two elements are inverses of each other.

REMARK 2. Apart from the operation φ_α we can consider more general operations $\varphi_{\alpha,p}$ (respectively, $\varphi_{\alpha,p,q}$), in which instead of R we take $R_p = M_p R$ (respectively, $R_{p,q} = M_p R M_q$; here p and q are elements of Q).

However, there are identities linking these new operations with $\varphi_\alpha = \varphi_{\alpha,1,1}$:

1) $\varphi_{M_\pi^* \alpha, \pi p, q} = \varphi_{\alpha, p, q}$,

2) $\varphi_{\alpha, \pi^{-2} p, \pi q} \sim \varphi_{\alpha, p, q}$.

In particular,

$$\varphi_{\alpha, p} = \varphi_{M_{p-1}^* \alpha}, \quad \varphi_{\alpha, p, q} = \varphi_{M_{q-2p-1}^* \alpha}.$$

The calculations leading to these identities are based on the following easily proved identities:

$$N_{N_\alpha p}q = N_\alpha(pq), \quad N_\alpha M_q = M_q^* N_\alpha, \quad DM_q = M_q D + M_{D[q]} .$$

REMARK 3. The fact that instead of a single operation φ on T we obtain a whole family φ_α of operations on Q^* is not surprising, since the identification of the spaces T and Q is not canonical. For when we identify the base of the versal deformation with the variety B of orbits of a reflection group, we watch only that the discriminant variety is identified with the variety of non-regular orbits.[1] The group of diffeomorphisms $B \to B$ that preserve the variety of non-regular orbits acts on the space of such identifications.

The group indicated above can also be defined as the group of germs of diffeomorphisms of B at O, lifted by the mapping π: $\mathbf{C}^\mu \to B$ (see [1]). Thus, it is naturally isomorphic to the group of equivariant (germs of) diffeomorphisms in $\mathbf{C}^\mu$.

The corresponding Lie algebra consists of germs of vector fields tangent to the variety of non-regular orbits. This algebra is a free module with μ generators over the ring $\mathcal{O}$ of germs of holomorphic functions on B at O. As generators we can take the fields $v_{\alpha_i} = \pi_* \nabla \pi^* a_i$, where the potentials a_i form a system of local coordinates on B in a neighbourhood of zero.

From this it is clear that on the space of identifications of T with Q it is natural to consider the action of the group of linear parts of π-liftable diffeomorphisms of T at O. The corresponding Lie algebra is of dimension μ and consists of the linearizations of the fields v_a at zero. These fields were denoted above by w_β $(\beta \in T^*)$. Thus, with each reflection group in $\mathbf{C}^\mu$ there is associated a μ-dimensional Lie algebra, the Lie algebra of the fields w_β. In terms of the linearized convolution operation φ in T^* these fields can be expressed as follows: for any γ of T^* we have $w_\beta \gamma = \varphi(\beta, \gamma)$.

From Proposition 2 we can extract a description of these fields on T in terms of the algebra Q. We fix an element α from Q^* that determine the operation φ_α, and we fix an identification of T with Q that takes φ_α to φ. It is easy to prove the following proposition without using special properties of Q.

PROPOSITION 3. *Let D be a derivation of the commutative algebra Q. For any $p \in Q$ we consider* **C***-linear operators W_p taking Q into Q that are given by the formula $W_p = M_p R$, where M_p is the operator of multiplication by p and $R = E - D$. The linear operators W_p corresponding to all the possible $p \in Q$ form a μ-dimensional Lie algebra.*

REMARK 4. The operators

$$W_{p,q} = M_p R M_q \ (p, q \in Q).$$

also form a Lie algebra. This algebra can also be described as the Lie algebra of operators of the form

[1] The corresponding restriction is not imposed on the isomorphism of T^* and Q^* in Proposition 2, but it holds for a suitable normalization of α.

$$M_u + M_v D(u,\ v \in Q).$$

PROPOSITION 4. *If Q is the local algebra of a simple singularity and D is the quasi-homogeneous derivation introduced above, then under the identification of Q with T indicated above the linear vector fields on Q corresponding to the linear operators W_p of Proposition 3 are taken to the linear fields w_β on T defined above. The parameters p (of Q) and β (of Q^*) are related linearly: $\beta = N_\alpha p$.*

PROOF. We have to verify that $w_\beta \gamma \equiv (W_p)^* \gamma$. According to Propositions 1 and 3, this means that

$$R^* N_\alpha (N_\alpha^{-1} N_\alpha p \cdot N_\alpha^{-1} \gamma) \equiv R^* M_p^* \gamma,$$

that is, that

$$N_\alpha (p \cdot N_\alpha^{-1} \gamma) \equiv M_p^* \gamma.$$

But by the definition of N_α we have

$$N_\alpha (p \cdot N_\alpha^{-1} \gamma) \equiv M_p^* M_{N_\alpha^{-1}\gamma}^* \gamma \equiv M_p^* N_\alpha N_\alpha^{-1} \gamma,$$

which proves the required identity.

PROPOSITION 5. *The operations $\varphi_\alpha : Q^* \times Q^* \to Q^*$ corresponding to different α are mapped to each other by linear mappings of Q^* into itself.*

PROOF. 1°. The isomorphism $N_\alpha : Q \to Q^*$ allows us to transfer the operation φ_α from Q^* to Q. We define an operation $\Phi_\alpha : Q \times Q \to Q$ by the relation

$$\Phi_\alpha(p,\ q) = N_\alpha^{-1}\ \varphi_\alpha(N_\alpha p,\ N_\alpha q).$$

By the definition of φ_α,

$$\Phi_\alpha(p,\ q) = S_\alpha(pq),$$

where $S_\alpha = N_\alpha^{-1} R^* N_\alpha$.

2°. Let $\alpha_0 \in Q^*$ be a form equal to 1 on the Jacobian J and to 0 on all monomials of smaller degree. Then, as is not hard to check,

$$N_{\alpha_0}^{-1}\ R^* N_{\alpha_0} = D + (2E/K),$$

where K is the Coxeter number defined by

$$\deg J = 1 - (2/K).$$

3°. We denote the operator $S_{\alpha_0} = D + (2E/K)$ simply by S. If $\alpha = N_{\alpha_0} a$, then $N_{\alpha_0}^{-1} N_\alpha : Q \to Q$ is multiplication by a. Therefore, $S_\alpha = M_a^{-1} S M_a$, where a is the class in Q of a function that is non-zero at the origin.

4°. To prove that the operations Φ_α and Φ_{α_0} are similar, it suffices for any function a with $a(0) = 1$ to construct an automorphism h in the algebra Q and a function u with $u(0) = 1$ for which

$$M_a^{-1} S M_a = M_u^{-1}\ (h^{-1} S h) M_u^2.$$

5°. We look for a (semi)quasi-homogeneous formal diffeomorphism h

inducing an automorphism of Q such that

$$h^{-1}Dh = \lambda D, \tag{1}$$

where λ is a (formal) function. To solve the equation in 4° it suffices to find a (formal) solution h, λ, u of the system (1)–(2), where

$$(2/K) + a^{-1}Da = (2/K)u + u^{-1}\lambda Du^2. \tag{2}$$

6°. The equation (2) is equivalent to the system

$$(2/K) + b = (2/K)u + u^{-1}\lambda D[u^2], \tag{3}$$

$$D = \lambda u D, \tag{4}$$

where $b = a^{-1}D[a]$ is a known function.

7°. The system (3)–(4) reduces to the equation

$$D[u] = u\left(\frac{1-u}{K}+\frac{b}{2}\right). \tag{5}$$

8°. The equation (5) is solved by expanding in quasi-homogeneous terms: if $u = 1 + \Sigma u_d$, where $\deg u_d = d > 0$ (that is, $Du_d = du_d$) and $b = \Sigma b_d$, $d > 0$, then

$$\cdot(K^{-1} + d)u_d = 2^{-1}b_d + M(u_{<d}), \qquad d > 0,$$

where M is a polynomial of terms of lower degree.

9°. Having found a series u as indicated above and the series $\lambda = 1/u$, we solve (1) relative to the formal diffeomorphism $h = \exp(\gamma D)$, where γ is a formal function, by the usual method of Poincaré.

REMARK 5. The proof is applicable not only to simple singularities, but to all quasi-homogeneous (usual or boundary) singularities for which the number $-1/K$ does not occur among the weights of the basis monomials.

REMARK 6. The operator taking φ_α to φ_β has the form $\exp W_p^*$, where the operator $W_p^*: Q^* \to Q^*$ is dual to $W_p = M_p R: Q \to Q$, at least for $A_\mu, \mu < 4$.[1]

REMARK 7. Proposition 4 can be used to recover the local algebra Q from the linearized convolution operation φ. For φ determines the Lie algebra of the fields w_β. These fields can be regarded as linear operators $w_\beta: T \to T$. We choose an invertible element w_{β_0} and form the operator $w_\beta w_{\beta_0}^{-1} = u_\beta$. If we began with the operation φ for a reflection group associated with a simple singularity, then these operators u_β form an algebra isomorphic to Q.

Thus, for any reflection group we obtain a linear family of operators u_β. Calculations show that these families are always commutative algebras: for the groups $I_2(n)$, H_3, and H_4 we obtain the algebras A_2, A_3, and D_4, respectively, but with unusual gradings.

REMARK 8. Proposition 4 can be used for the construction of (generalized) linear convolution operation φ from local algebra of a not necessarily simple

[1] This conjecture has also been proved by A. B. Guivental (letter from the author, 18 November 1979) (Transl.)

quasi-homogeneous singularity. It would be interesting to find out whether these operations are connected with some generalized reflection groups in Euclidean spaces.

§ 10. Some applications

The Lie algebras of liftable fields and their linearizations w_β are to the utmost extent a useful instrument for the reduction of various geometric objects to a normal form. This method was already used in [1]. Below we state some new results of applying this method.

10.1. Normal forms of curves in the swallow-tail space. We recall that B denotes the variety of orbits of a reflection group in $\mathbf{C}^\mu$. For the group A_μ of permutations of coordinates in the plane $z_0 + \ldots + z_\mu = 0$ of $\mathbf{C}^{\mu+1}$ the variety B is identified with the base of the versal deformation $z^{\mu+1} + \lambda_1 z^{\mu-1} + \ldots + \lambda_\mu$. The basis invariants $\lambda_1, \ldots, \lambda_\mu$ determine elements of the ring $\mathcal{O}$ of germs of holomorphic functions on B at O.

The differentials of these functions at O are denoted by $d\lambda_i$ and form a basis in the space T^* dual to the tangent space T to B at O. The corresponding vector fields v_{λ_i} form a free basis of the $\mathcal{O}$-module of germs of vector fields tangent to the variety of non-regular orbits Σ (the image of the reflecting hyperplane under the factorization π: $\mathbf{C}^\mu \to B$). An explicit form of the components of these fields, that is, of the convolution operation Φ, is indicated in [1]. In particular, the linearized convolution is given by the formula

$$\varphi(d\lambda_i, d\lambda_j) = (i + j) d\lambda_{i+j-1}$$

(or 0 for $i + j - 1 > \mu$). The tangent plane to the discriminant variety Σ at O is denoted by τ. In this notation the equation of τ has the form $d\lambda_\mu = 0$.

Turning to the general case of an arbitrary irreducible reflection group, we consider a curve γ: $(\mathbf{C}, 0) \to (B, O)$. We call it *transversal* to the variety of non-regular orbits Σ at O if the velocity vector $\dot{\gamma}(0)$ does not belong to τ.

THEOREM 1. *All the curves transversal to Σ at O are mapped into each other by diffeomorphisms of B that preserve Σ (so that $\gamma_1(t) \equiv h(\gamma_0(t))$):*

In particular, in the case A_μ every such curve reduces to the form $\lambda_1 = \ldots = \lambda_{\mu-1} = 0, \lambda_\mu = t$.

We consider the germ of the holomorphic vector field u at O in B. We say that *u is transversal* to Σ at zero if the field vector at O does not belong to τ.

THEOREM 2. *All the vector fields transversal to Σ at O are mapped into each other by diffeomorphisms of the base B that preserve Σ (at least in the cases A_μ, D_μ, E_μ, B_μ, C_μ, F_4, G_2, and $I_2(n)$).*

In particular, in the case A_μ every such field is equivalent to the field $\partial/\partial\lambda_\mu$.

A simple proof of Theorem 2 can probably be obtained from the following arguments. We consider a general complex phase curve of the given field. It intersects Σ in μ points. The passage times of these points are determined up to an initial reference time. Consequently, our curve corresponds uniquely to an (unordered) set of μ complex numbers with sum zero. This mapping of the (local) manifold of phase curves (which is diffeomorphic to a neighbourhood of O in $\mathbf{C}^{\mu-1}$) into the variety of sets (which can be identified with the variety $\mathbf{C}^{\mu-1}$ of the polynomials $z^\mu + a_2 z^{\mu-2} + \ldots + a_\mu$) is proper and finite-to-one (this mapping was used by O. V. Lyashko and E. Looijenga in [4] in the cases A_μ, D_μ, and E_μ, but it can also be constructed for the other reflection groups).

Apart from the given field we consider the standard "constant" field ($\partial/\partial\lambda_\mu$ in the case A_μ). The preceding construction determines for each phase curve of our field a phase line of the constant field, up to finitely many possibilities. With our phase curve we associate the nearest line. The required diffeomorphism of the ambient spaces is now constructed from the mappings of the phase curves into the phase lines that take the (complex) time on the phase curve into time on the phase line. (The origin of reference in both cases is chosen so that the sum of the times of intersection with Σ is zero.)

Apparently, Theorem 2 remains true also in the real C^∞ (not necessarily analytic) case if we allow the field to change sign.[1]

Theorem 1 follows from Theorem 2, but we give a direct proof.

PROOF OF THEOREM 1. It follows from Theorem C of §8 that each connected component of the space $T \setminus \tau$ is an orbit of the group of the linear parts of diffeomorphisms of the base that preserve Σ (in the real case there are two components, and in the complex case one). Therefore, each vector transversal to τ is carried by this linear group into any other such vector (in the real case it is perhaps necessary to permit sign changes in the vector). Thus, it suffices to prove the equivalence of curves of the form

$$\gamma_0(t) = et, \quad \gamma_1(t) = et + \delta(t),$$

where $\delta(t) = t^2\beta(t)$ and e is a vector transversal to τ. (We have identified our space B with its tangent space at zero, after fixing the basis invariants, say.)

Now we form the family of curves $\gamma_\theta(t) = \gamma_0(t) + \theta\delta(t)$, $0 \leqslant \theta \leqslant 1$, and we look for a family of (local) diffeomorphisms $g_\theta : (B, \Sigma) \to (B, \Sigma)$ from the condition

$$\gamma_0(t) + \theta\delta(t) \equiv g_\theta(\gamma_0(t)).$$

For g we obtain the equation $\partial g/\partial\theta\,|_{\varphi(t),\theta} = \delta(t)$. We then consider a vector field h, depending on the "time" θ and corresponding to the family of diffeomorphisms g_θ : by definition,

$$\partial g/\partial\theta\,|_{x,\theta} = h(g_\theta(x), \theta).$$

In particular, for $x = \gamma_0(t)$ the following relation must hold:

$$\delta(t) = h(et + \theta\delta(t), \theta),$$

from which we can find h.

We fix a basis of liftable fields v_{a_i}. We look for a liftable field h in the form of a linear combination of basis fields

[1] As we have learned during the preparation of the manuscript, O. V. Lyashko and V. M. Zakalyukin have also proved Theorem 2 and a number of generalizations of it (see [7]).

$$h(x, \theta) = \sum h_i(x, \theta) v_{a_i}(x).$$

We have

$$v_{a_i}(et + t^2\theta\beta(t)) = t[w_{\alpha_i}(e) + tu_i(t, \theta)].$$

Here $\alpha_i = da_i$ is a vector of the basis of T^* corresponding to the coordinate system $\{a_i\}$ in B; the u_i are smooth functions.

For the coefficients h_i we obtain the equation

$$t\beta(t) = \sum h_i(et + \theta\delta(t), \theta)[w_{\alpha_i}(e) + tu_i(t, \theta)].$$

By Theorem C of §8, the vectors $w_{\alpha_i}(e)$ are linearly independent and form a basis of T. For sufficiently small $|t|$ the vectors in square brackets also form a basis. Therefore, the coefficients h_i, as functions of t and θ, are uniquely determined and regular (smooth, holomorphic, . . .). For $t = 0$ these coefficients vanish for any θ. We extend the functions h_i for each θ from the curve γ_θ, where they are already defined, to a neighbourhood of the origin in B. This extension can be chosen to be regular in θ. The resulting family of vector fields h defines a one-parameter family of local liftable diffeomorphisms g_θ for all θ from 0 to 1, because $h(0, \theta) \equiv 0$. The diffeomorphism g_1 takes γ_0 into γ_1, and Theorem 1 is proved.

10.2. Normal forms of projections of surfaces in 3-space. DEFINITION. A *projection* of a surface S is a diagram $S \to \mathbf{R}^3 \to \mathbf{R}^2$, where the left arrow is an embedding of the smooth surface S and the right arrow is the linear projection. An *equivalence* of projections of surfaces is a commutative diagram.

$$\begin{array}{ccccc} S_1 & \to & \mathbf{R}^3 & \to & \mathbf{R}^2 \\ \downarrow & & \downarrow & & \downarrow \\ S_2 & \to & \mathbf{R}^3 & \to & \mathbf{R}^2, \end{array}$$

where the vertical arrows are diffeomorphisms. Similar definitions can be given for the germs (S, O).

THEOREM 3. *In the space of compact smooth hypersurfaces in* $\mathbf{R}^3$ *there is an open everywhere dense set formed by those surfaces whose projection at any point and in any direction is locally equivalent to the projection of one of the* 10 *surfaces of the following list at* $(0, 0, 0)$ *along the x-axis*:

$$z = x,\ z = x^2,\ z = x^3 + xy,\ z = x^3 \pm xy^2,\ z = x^3 + xy^3,$$
$$z = x^4 + xy,\ z = x^4 + x^2y + xy^2,\ z = x^5 + x^3y \pm xy.$$

These projections are pairwise inequivalent. (In the complex case the projections with the minus signs in the equations for the surfaces are equivalent to the projections with the plus signs, so that the list can be shortened to eight surfaces.) The singularities of the projections of the list are not removable by a small perturbation of the surface.

For a surface in general position the first normal form in the list describes the projection in any direction at almost every point, the second at the points of some curve, the third at the remaining points. The normal forms given by the fourth and sixth equations of the list describe the projection at the remaining points in the directions belonging to some curve in the space of directions. The fifth, seventh, and eighth equations give projections that occur for a general surface at the remaining points in the remaining directions.

The proof of Theorem 3 is based on the fact that we can associate with a projection a two-parameter family of "vertical" lines (inverse images in $\mathbf{R}^3$ of points from $\mathbf{R}^2$). A surface S cuts out several points on each vertical line. Thus, we obtain a two-parameter family of zero-dimensional hypersurfaces on the vertical lines.

The points of intersection of S with the verticals can be combined into several verticals. Suppose that μ points merge at the relevant point of S on the vertical (assuming also complex points; in other words, μ is the dimension over $\mathbf{R}$ of the local ring corresponding to the intersection). From the Weierstrass preparation theorem (in the real C^∞ case, the Malgrange preparation theorem) it follows that our two-parameter family of zero-dimensional hypersurfaces is induced from the universal $(\mu - 1)$-parameter family

$$x^\mu + \lambda_1 x^{\mu-2} + \ldots + \lambda_{\mu-1} = 0. \tag{1}$$

In the universal family each $\lambda \in \mathbf{R}^{\mu-1}$ (in the complex case $\mathbf{C}^{\mu-1}$) corresponds to the zero-dimensional hypersurface $\{x\}$. Thus, the equation of S has (locally) the form

$$x^\mu + \lambda_1(y, z)x^{\mu-2} + \ldots + \lambda_{\mu-1}(y, z) = 0. \tag{2}$$

We may assume that all the coordinates of the point in question (the vertical coordinate x and the horizontal ones y and z) vanish and that $\lambda(0) = 0$. Here y and z are coordinates in the plane $\mathbf{R}^2$, which is the base of our family of vertical lines.

LEMMA 1. *The projection is equivalent to the one for which the equation of S has the form*

$$z = x^\mu + \lambda_1(y)x^{\mu-2} + \ldots + \lambda_{\mu-2}(y)x.$$

PROOF. Since S is smooth, $d\lambda_{\mu-1}(0) \neq 0$. By choosing coordinates in $\mathbf{R}^2$ (that is, a diffeomorphism of the base) we can achieve that $\lambda_{\mu-1} = -z$. For $y = 0$ the equation of S gives

$$z = x^\mu + \lambda_1(0, z)x^{\mu-2} + \ldots + \lambda_{\mu-2}(0, z)x, \qquad \lambda_k(0, 0) = 0.$$

By the implicit function theorem, $z = f(x)$, where f has a zero of order μ at the origin. A local diffeomorphism of the x-axis reduces f to the form x^μ.

Now the equation of S can be regarded as a deformation of the function $z = x^\mu$ with parameter y. By the versal deformation theorem a diffeomorphism of the x-axis depending on y reduces the equation of S to the form

$$z = x^\mu + \lambda_1(y)x^{\mu-2} + \ldots + \lambda_{\mu-1}(y).$$

This in turn takes the form indicated in the lemma by choosing the function $z - \lambda_{\mu-1}(y)$ as the new z-coordinate on the base.

LEMMA 2. *A hypersurface in general position in* $\mathbf{R}^n$ *does not admit tangent lines of order* $\mu > 2n - 1$.

PROOF. In $\mathbf{R}^n$ we fix a point, a hyperplane passing through it, and a line in the hyperplane also passing through this point. We consider the space of jets of the hypersurfaces for which this plane is tangent at the distinguished point. The set of jets of surfaces having a tangent of order μ at the distinguished point (say, the x-axis) has codimension $\mu - 2$ in this jet space. For, if the equation of the surface is $z = f(x, y, \ldots)$, then the coefficients of the expansion

$$f(x, 0) = f_2 x^2 + \ldots + f_{\mu-1} x^{\mu-1} \bmod x^\mu,$$

must vanish, and there are $\mu - 2$ of them. The dimension of the variety of lines passing through a point and lying in a hyperplane is $n - 2$. Therefore, the variety of jets of surfaces admitting a tangent line of order μ at a fixed point has codimension $\mu - n$ in the variety of jets of hypersurfaces at a point with a fixed tangent plane. By the transversality theorem, by a small perturbation of an $(n-1)$-dimensional hypersurface we can get rid of all the tangents with lines of order μ if $n - 1 < \mu - n$, as required.

COROLLARY. *For surfaces in general position in* $\mathbf{R}^3$ *we have* $\mu \leqslant 5$ *for projection in any direction and at any point.*

For by Lemma 2, a surface in $\mathbf{R}^3$ does not have contact of order more than five with a line.

Now we consider the cases $\mu = 1, 2, 3, 4, 5$ in turn. In the first two cases we obtain immediately the normal forms $z = x$ and $z = x^2$ from Lemma 1.

LEMMA 3. *For surfaces in general position the singularities of the projections with* $\mu = 3$ *are equivalent to those of the projection of the surfaces* $z = x^3 \pm xy^k$, $k \leqslant 3$, *along the x-axis at the origin.*

PROOF. Lemma 1 gives $z = x^3 + \lambda(y)x$. If λ has a zero of order k at $y = 0$, then a change of the variable y reduces the equation of the surface to the equivalent form $z = x^3 \pm y^k x$. For surfaces in general position $k = 1$ occurs for the common directions of projection (Whitney's theorem), $k = 2$ for directions on some curve in the space of directions, and $k = 3$ for individual directions; $k > 3$ does not occur for S in general position.

Before we consider the cases $\mu = 4$ and $\mu = 5$, we return to the formula (2), which associates our projection $S \to \mathbf{R}^3 \to \mathbf{R}^2$ with the mapping $\lambda \colon \mathbf{R}^2 \to \mathbf{R}^{\mu-1}$ of the base of our family of vertical lines into the base of the universal family.

The universal family is given by the diagram $\Gamma \to \mathbf{R}^\mu \to \mathbf{R}^{\mu-1}$, where the left arrow is the embedding of the hypersurface given by (1) into $\mathbf{R}^\mu$ with the coordinates (x, λ), and the right arrow is the canonical projection $(x, \lambda) \mapsto (\lambda)$. The hypersurface Γ itself is diffeomorphic to $\mathbf{R}^{\mu-1}$ (we can take $(x, \lambda_1, \ldots, \lambda_{\mu-2})$ as coordinates on it). The composition of the two arrows gives the Whitney mapping $w \colon \Gamma \to \mathbf{R}^{\mu-1}$. The formula (2) gives a mapping $\lambda \colon \mathbf{R}^2 \to \mathbf{R}^{\mu-1}$, which induces a commutative diagram

$$\begin{array}{ccccc} S & \to & \mathbf{R}^3 & \to & \mathbf{R}^2 \\ \downarrow & & \downarrow & & \downarrow \lambda \\ \Gamma & \to & \mathbf{R}^\mu & \to & \mathbf{R}^{\mu-1} \end{array}$$

whose upper row defines the projection in question, while the lower row is the diagram of the universal family. We call this the *λ-diagram* of the projection.

An *equivalence* of a λ-diagram and a λ′-diagram is a commutative diagram shaped like a prism, whose horizontal faces are a λ-diagram and a λ′-diagram, and the six vertical edges are diffeomorphisms. The following result is obvious.

LEMMA 4. *If a λ-diagram and a λ′-diagram are equivalent, then the corresponding projections are also equivalent.*

It is useful to clarify in what case the mappings λ and λ' give equivalent diagrams, that is, one is taken into the other by an automorphism of the diagram of the universal family.

An *infinitesimal automorphism* of a diagram $\Gamma \to \mathbf{R}^\mu \to \mathbf{R}^{\mu-1}$ is a vector field in the ambient space $\mathbf{R}^\mu$ that is tangent to the hypersurface Γ and projects onto the base $\mathbf{R}^{\mu-1}$.

The projection of an infinitesimal automorphism onto the base is a vector field that is liftable relative to the Whitney mapping $w\colon \Gamma \to \mathbf{R}^{\mu-1}$.

LEMMA 5. *Every Whitney-liftable (that is, liftable to* Γ*) vector field on* $\mathbf{R}^{\mu-1}$ *is the projection of some infinitesimal automorphism (that is, can be lifted to* $\mathbf{R}^\mu$*).*

PROOF. Suppose that the field $v = \Sigma v_k(\lambda)\, \partial/\partial\lambda_k$ is lifted to Γ from a field which in the coordinates $(x, \lambda_1, \ldots, \lambda_{\mu-2})$ has the form

$$V = V_0(x, \lambda_1, \ldots, \lambda_{\mu-2})\partial/\partial x + \ldots .$$

The required field on $\mathbf{R}^\mu$ is

$$V_0 \partial/\partial x + \sum v_k(\lambda)\partial/\partial\lambda_k.$$

Now we pass to the problem of the classification of the diagrams $\mathbf{R}^2 \xrightarrow{\lambda} \mathbf{R}^{\mu-1} \xleftarrow{w} \Gamma$, where w is the Whitney mapping. The equivalences here are taken to be commutative diagrams

$$\begin{array}{ccccc} \mathbf{R}^2 & \xrightarrow{\lambda} & \mathbf{R}^{\mu-1} & \xleftarrow{w} & \Gamma \\ \downarrow & & \downarrow & & \downarrow \\ \mathbf{R}^2 & \xrightarrow{\lambda(t)} & \mathbf{R}^{\mu-1} & \xleftarrow{w} & \Gamma, \end{array}$$

where the vertical arrows are diffeomorphisms depending continuously on t. In other words, we must classify the mappings of the plane into the swallow-tail space relative to the group of diffeomorphisms of the ambient space that preserve the swallow-tail (more precisely, relative to the identity component of this group). Here the specific equivalence of the two mappings λ and $\lambda' = \lambda(1)$ is called a *w-equivalence.*

We consider the space Ω of germs of smooth mappings $\lambda\colon (\mathbf{R}^2, 0) \to (\mathbf{R}^{\mu-1}, 0)$.

LEMMA 6. *Let* $\mu = 4$. *Then the germ* λ *is either w-equivalent to the germ of the embedding of the plane* $\lambda_1 = 0$ *into* $\mathbf{R}^3$ *or it belongs to a certain set of co-*

dimension 1 *in* Ω. *In the last case, the germ of* λ *is either w-equivalent to the germ of the embedding of the cylinder* $\lambda_2 = \pm \lambda_1^2$ *in* $\mathbf{R}^3$ *or it belongs to a certain set of codimension* 2 *in* Ω.

PROOF. The set of germs whose rank at O is less than 2 has codimension 2 in Ω. Therefore, the image of λ can be assumed to be the smooth surface of a swallow-tail in $\mathbf{R}^3$. Now the first assertion of the lemma follows from the equivariant Morse lemma proved in [1]. The second assertion is also contained in [1], where the beginnings are given of a classification of degenerate critical points of S_4-invariant functions (in [1] we were concerned with the whole group of diffeomorphisms that preserve the swallow-tail, but only the identity component was used in the proof).

The λ-diagrams corresponding to the two cylinders $\lambda_2 = \pm \lambda_1^2$ are equivalent. The corresponding planes and the cylinder of projection are equivalent to the projection of the surfaces $z = x^4 + xy$ and $z = x^4 + x^2y + xy^2$ along the x-axis at O. From Lemmas 5 and 6 it follows that for surfaces of general position in $\mathbf{R}^3$ the projections with $\mu = 4$ are locally equivalent to those described above. (The case of a plane occurs only for some curve in the space of projection directions, and the case of the cylinder only for isolated directions).

LEMMA 7. *Let* $\mu = 5$. *Then the mapping* λ: $(\mathbf{R}^2, O) \to (\mathbf{R}^{\mu-1}, O)$ *is either w-equivalent to the embedding of one of the two planes* $\lambda_2 = 0$, $\lambda_1 = \pm \lambda_3$ *into* $\mathbf{R}^4$, *or it belongs to a certain set of codimension* 1 *in* Ω.

PROOF. The image of λ may be assumed to be a surface in $\mathbf{R}^4$ that is smooth at zero (otherwise λ belongs to a set of codimension 3 in Ω). We consider the tangent plane to this surface at zero. The group $\{\exp W\}$, where the W are the linear operators of the fields w_β in §8 for the case of A_4, acts on the variety of two-dimensional subspaces in $T = \mathbf{R}^4$.

We introduce the notation $\partial_k = \partial/\partial\lambda_k$ ($k = 1, \ldots, 4$). The basis fields w_k are then given by the formulae

$$\begin{aligned} w_1 &= 2\lambda_1\partial_1 + 3\lambda_2\partial_2 + 4\lambda_3\partial_3 + 5\lambda_4\partial_4, \\ w_2 &= 3\lambda_2\partial_1 + 4\lambda_3\partial_2 + 5\lambda_4\partial_3, \\ w_3 &= 4\lambda_3\partial_1 + 5\lambda_4\partial_2, \\ w_4 &= 5\lambda_4\partial_1. \end{aligned}$$

LEMMA 8. *The orbit of a plane* $\mathbf{R}^2$ *in general position in* $\mathbf{R}^4$ *under the action of the group* $\{\exp W\}$ *contains one of the two planes generated by the vectors* ∂_4 *and* $\partial_1 \pm \partial_3$; *the union of the remaining orbits is a proper subvariety of the space* $G_{4,2}$ *of planes in* $\mathbf{R}^4$.

PROOF. It is easy to prove the following proposition.

PROPOSITION 1. *The orbit of the basis vector* ∂_k *under the action of* $\{\exp W\}$ *is the half-space* $\lambda_k > 0$ *of the plane* $\lambda_{k+1} = \lambda_{k+2} = \ldots = 0$.

A plane $\mathbf{R}^2$ in general position in $\mathbf{R}^4$ intersects the orbit of ∂_4 (that is, the half-space $\lambda_4 > 0$). Therefore, the action of the group can reduce this plane to the plane π generated by the vectors ∂_4 and $e = a\partial_1 + b\partial_2 + c\partial_3$.

A further reduction will be carried out so that the plane always contains ∂_4.

PROPOSITION 2. *Suppose that a smooth one-parameter family of linear transformations taking π into planes containing ∂_4 is given. Then the velocity of the motion at ∂_4 belongs to π.*

For let us denote the positions of the vectors ∂_4 and e at the moment ε by $d(\varepsilon) = \partial_4 + \alpha\varepsilon + \dots$ and $e(\varepsilon) = e + \beta\varepsilon + \dots$ The coefficients of the expansion of ∂_4 in the basis $d(\varepsilon)$ and $e(\varepsilon)$ of the plane $\pi(\varepsilon)$ are denoted by $p(\varepsilon) = 1 + \gamma\varepsilon + \dots$ and $q(\varepsilon) = \delta\varepsilon + \dots$ ($\partial_4 = p(\varepsilon)d(\varepsilon) + q(\varepsilon)e(\varepsilon)$). Equating the coefficients for ε, we find that $\alpha = -\gamma\partial_4 - \delta e$, as required.

PROPOSITION 3. *Conversely, if we are given a one-parameter family of linear transformations taking π into $\pi(\varepsilon)$, and if the velocity at ∂_4 at each moment of time ε belongs to $\pi(\varepsilon)$, then all the planes $\pi(\varepsilon)$ contain ∂_4.*

For in this case the velocity with which the distance of ∂_4 from $\pi(\varepsilon)$ changes is identically zero.

The values of the basis fields w at ∂_4 are $w_1 = 5\partial_4$, $w_2 = 5\partial_3$, $w_3 = 5\partial_2$, and $w_4 = 5\partial_1$. Thus, the field

$$kw_1 + K(cw_2 + bw_3 + aw_4)$$

from any real k and K defines a motion whose velocity at ∂_4 at the initial moment belongs to π. The velocity of e is then equal to

$$k[2a\partial_1 + 3b\partial_2 + 4c\partial_3)] + K[c(3b\partial_1 + 4c\partial_2) + b(4c\partial_1)].$$

According to Proposition 3, we obtain equivalent planes π (belonging to the same orbit of $\{\exp W\}$) if we transfer e with respect to the phase curve of each of the two systems

$$\dot a = 2a, \quad \dot b = 3b, \quad \dot c = 4c; \tag{3}$$

$$\dot a = 7bc, \quad \dot b = 4c^2, \quad \dot c = 0. \tag{4}$$

For π in general position $c \neq 0$. We make $c = 1$ by dilating e. Now from (4) we obtain

$$\dot a = 7b, \ \dot b = 4; \ b = b_0 + 4t, \ a = a_0 + 7b_0 t + 14t^2.$$

We make $b = 0$ by choosing t suitably. For $b = 0$ the system (3) allows us to vary $|a|/|c|$. For π in general position we have $a \neq 0$ and $c \neq 0$. By transferring along the phase curve of the first system (3) we can achieve that $|a| = |c|$. By dilating e we make $|a| = |c| = 1$. This proves Lemma 8.

PROPOSITION 4. *Each of the two planes $\lambda_2 = 0$, $\lambda_3 = \pm\lambda_1$ in $\mathbf{R}^4$ is w-stable, in the sense that the germ of every surface with this tangent plane at O is w-equivalent to the germ of the corresponding plane at O.*

PROOF. We verify the relevant infinitesimal condition. With this aim we compute the tangent space to the variety of systems of equations $f_1 = 0$, $f_2 = 0$, which yield surfaces equivalent to the given one.

This tangent space may be assumed to be the space of vectors $(\delta f_1, \delta f_2) \in A^2$, where A is a ring of functions of the class in question (say, $A = \mathbf{R}[[\lambda]]$). The system goes over into an equivalent one if we add to one equation the other one multiplied by an arbitrary (small) function. Therefore, this tangent space contains the submodule of A^2 generated by the vectors $(f_1, 0)$, $(f_2, 0)$, $(0, f_1)$, and $(0, f_2)$. The system goes over to an equivalent one also under the action of the one-parameter group of diffeomorphisms given by a w-liftable vector field in $\mathbf{R}^4$. Therefore, the tangent space also contains all the vectors of the form (vf_1, vf_2), where v is a liftable field. The vectors of this kind form a submodule of A^2 with four generators $(v_k f_1, v_k f_2)$, where $v_1, \ldots, v_4$ are liftable fields of a basis.

Now we consider the linear parts of our four generators at zero, which have the form $(w_k f_1, w_k f_2)$, where w_k are the basic linear fields described above. Substituting $f_1 = \lambda_2$, $f_2 = \lambda_3 \pm \lambda_1$, we obtain eight linear parts of the generators of the submodule of A^2 in the form

$$\begin{pmatrix}\lambda_2\\0\end{pmatrix}\begin{pmatrix}0\\\lambda_2\end{pmatrix}\begin{pmatrix}\lambda_3 \mp \lambda_1\\0\end{pmatrix}\begin{pmatrix}0\\\lambda_3 \mp \lambda_1\end{pmatrix}\begin{pmatrix}3\lambda_2\\4\lambda_3 \mp 2\lambda_1\end{pmatrix}\begin{pmatrix}4\lambda_3\\5\lambda_4 \mp 3\lambda_2\end{pmatrix}\begin{pmatrix}5\lambda_4\\\mp 4\lambda_3\end{pmatrix}\begin{pmatrix}0\\\mp 5\lambda_4\end{pmatrix}.$$

Comparing the first column with the fifth, and the result with the fourth, we see that all the $(0, \lambda_k)$ occur among the linear parts of the vectors of the submodule. Now by considering columns 3, 5, 6 and 7 we see that all the $(\lambda_k, 0)$ also occur (this result can be seen from the fact that the orbit of π in the variety of planes is open). Since all linear vector-functions of λ occur among the linear parts of the vectors of the submodule, our submodule contains all the variations $(\delta f_1, \delta f_2)$ for which the functions δf_1 and δf_2 vanish at zero (Hadamard's lemma). This proves the infinitesimal w-stability; the actual proof of w-stability is now carried out by any of the usual methods in the theory of singularities of differentiable mappings. This proves Proposition 4.

Lemma 7 follows from Lemma 8 and Proposition 4.

Theorem 3 follows from Lemmas 2, 3, 6 and 7.

REMARK 1. The mutual inequivalence of the 10 singularities listed in the theorem is proved by comparing their sets of critical points.

For example, the projections of the surfaces $z = x^5 + x^3 y + \varepsilon xy$ along the x-axis have curves of critical values for $\varepsilon = \pm 1$, and these are diffeomorphic to the curves $Y = t^4$, $Z = t^5 + \varepsilon t^7 + \ldots$ These two curves (with $\varepsilon = +1$ and -1) are not diffeomorphic. Hence, the corresponding two projections are not equivalent.

The irremovability is verified by changing y into $y + \alpha x$ and z into $z + \beta x$, which corresponds to a small change in the direction of projection. After these changes we can verify the independence of the equations of the system relative to α, β, y, and z, which determine that the projection belongs to the given type.

For example, the equation $z = x^5 + x^3 y + xy$ is transformed to

$z = x^5 + \alpha x^4 + yx^3 + \alpha x^2 + (y - \beta)x$. The change of variables $x = \xi - \alpha/5$ takes the equation to $\xi^5 + p\xi^3 + Q\xi^2 + R\xi + S = 0$, where $P = y + \ldots$, $Q = \alpha + \ldots$, $R = y - \beta + \ldots$, $S = -z + \ldots$ (the dots denote terms of order higher than 1 in α, β, y, z). The system of equations $P = Q = R = S = 0$ has the solution $\alpha = \beta = y = z = 0$, which satisfies the conditions of the implicit function theorem. This proves the transversality of the mapping of the parameter spaces (α, β, y, z) into the variety of jets of projections to the subvariety of jets with $\mu = 5$, and hence, the irremovability of this class of singularities; on the other hand, it was proved above that the singularities of this class reduce to the indicated normal form.

REMARK 2. The above classification of mappings of the plane into the base of the universal family can also be carried out for $\mu = 3$ modulo w-equivalences. The answer has the following form: in two-parameter families in general position of mappings of the (y, z)-plane into the base $(\mathbf{R}^2 = (\lambda_1, \lambda_2)$ of the universal family $x^3 + \lambda_1 x + \lambda_2 = 0$, taking O into O, only those singularities occur that are w-equivalent to the singularities at O of the following four mappings:

$$\begin{cases} \lambda_1 = y, \\ \lambda_2 = z, \end{cases} \quad \begin{cases} \lambda_1 = \pm y^2, \\ \lambda_2 = z, \end{cases} \quad \begin{cases} \lambda_1 = y^3 + yz, \\ \lambda_2 = z. \end{cases}$$

These four singularities are mutually inequivalent and are irremovable in the sense that a small perturbation of the two-parameter family in general position of the mappings $(\mathbf{R}^2, O) \to (\mathbf{R}^2, O)$ cannot change the first singularity for a general value of the parameters, the two middle ones on a curve in the parameter plane, and the last one for individual values of the parameter.

The proof proceeds by standard methods [1]; calculations can be avoided by noting that the pair consisting of a semicubical parabola and a line passing through its vertex and transversal to the tangent, or a pair of two semicubical parabolas with transverse tangents at the vertex, has no moduli (which follows immediately from the normal forms of the singularities D_5 and $Y_{5,5}$ of functions of two variables).

The projections of surfaces corresponding to these for singularities of λ-mappings, are equivalent to the four projections indicated in Theorem 3, with $\mu = 3$ $(z = x^3 + yx,\ z = x^3 \pm y^2 x,\ z = x^3 + y^3 x)$. However, the λ-mapping constructed from the last formula is not equivalent to the mapping $\lambda_1 = y^3 + yz$, $\lambda_2 = z$ in the sense of w-equivalence. Thus, the converse assertion of Lemma 4 is not true: *equivalent projections can have inequivalent λ-mappings.*

REMARK 3. Apparently, Lemma 5 generalizes to diffeomorphisms: each diffeomorphism of the base that can be lifted to Γ seems to be the projection onto the base of some diffeomorphism of the ambient space $\mathbf{R}^\mu$ that preserves the hypersurface Γ. In this context it would be useful to compute the group of connected components of the group of w-liftable diffeomorphisms (in the real and in the complex case). Besides, the analogous proposition generalizing Lemma 5 apparently holds for a wide class of hypersurfaces Γ.

When the manuscript of this paper was completed, I learned that O. V. Lyashko has solved the problems raised here (see [7]), and that V. V. Goryunov has found all the simple singularities of projections of hypersurfaces onto spaces of any codimension.

10.3. Vector fields tangent to projections and sections of the discriminant varieties. The theorem on w-stability of curves can be regarded as a dual (in the sense of the informal duality of §9) to the theorem proved in [1] on the w-stability of hypersurfaces. This duality also appears in the more delicate properties of liftable fields described below.

We consider the variety of orbits B of an irreducible reflection group in Euclidean space, acting on its complexification $\mathbf{C}^\mu$. This variety of orbits B is diffeomorphic to $\mathbf{C}^\mu$.

We fix a general line $\mathbf{C}^1$ passing through O in B (transversal to the variety of non-regular orbits of Σ at O). We consider the projection $\pi: B \to \tilde{B}$ of B along $\mathbf{C}^1$ onto $\tilde{B}$ (diffeomorphic to $\mathbf{C}^{\mu-1}$). We denote by $\tilde{\Sigma}$ the projection of the set of singular points of Σ onto $\tilde{B}$. The complex hypersurface $\tilde{\Sigma}$ in $\tilde{B}$ is, in terms of singularity theory, the bifurcation diagram of the functions. As a rule, it is reducible. From Theorem 2 we obtain the very strong stability of $\tilde{\Sigma}$ relative to the choice of the $\mathbf{C}^1$ (the projection may even be curvilinear).

THEOREM 4.[1] *The germs of vector fields on $\tilde{B}$ tangent to $\tilde{\Sigma}$ are the projections of germs of vector fields on B tangent to Σ, at least in the cases A_2, A_3, B_2, B_3.*

The dual theorem consists in the following. We fix a general hyperplane passing through O in B (transversal to the "worst direction" on Σ at O). We denote it by $\tilde{\tilde{B}}$, the embedding of $\tilde{\tilde{B}}$ in B by i, and the intersection of $\tilde{\tilde{B}}$ and Σ by $\tilde{\tilde{\Sigma}}$. From [1] we obtain the very strong stability of $\tilde{\tilde{\Sigma}}$ relative to the choice of hyperplane (or hypersurface) $\tilde{\tilde{B}}$.

THEOREM 5. *The germs of vector fields on $\tilde{\tilde{B}}$ tangent to $\tilde{\tilde{\Sigma}}$ can be extended to germs of fields on B tangent to Σ, at least in the cases A_2, A_3, B_2, and B_3.*

REMARK. Theorems 4 and 5 use the special properties of the discriminant varieties. It is easy to construct a plane curve and a general projection of it onto a line so that not every field on the line that (the field) vanishes at the projection of a singular point lifts to a field on the plane tangent to the curve. However, in the case of curves it suffices for fields to be liftable that the curve is quasi-homogeneous and that the projection is along an axis. In the case of surfaces quasi-homogeneity is already insufficient, as is shown by the example of the surface

$$z^4 - x^2z^2 + x^3y = 0,$$

[1] While the manuscript of this paper was being prepared, I learned that O. V. Lyashko (see [7]) and V. M. Zakalyukin have found far-reaching generalizations of Theorem 4. In particular, the theorem is true for all simple Lie algebras ($A_k, B_k, C_k, D_k, E_k, F_4, G_2$).

which can be projected onto the (x, y)-plane along the z-axis. The field

$$(4xy - x^2)\partial/\partial x$$

on the (x, y)-plane is tangent to the discriminant variety $xy(4y - x) = 0$, but is not liftable to a field tangent to a hypersurface in 3-space.[1]

For a section of a quasi-homogeneous hypersurface of the coordinate hyperplane the vector fields in the hyperplane tangent to the intersection extend to fields in the ambient space tangent to the hypersurface whenever the intersection has an isolated singularity at the origin. For let

$$F \equiv \sum \alpha_k x_k \partial F/\partial x_k = 0$$

be the equation of the hypersurface, $x_1 = 0$ the equation of the hyperplane, $f \equiv F|_{x_1=0} = 0$ the equation of the section, and $v_2 \partial_2 + \ldots + v_n \partial_n$ the field tangent to a section. Then

$$\sum v_k \, \partial f/\partial x_k = hf,$$

that is,

$$\sum (v_k - h\alpha_k x_k)\, \partial f/\partial x_k = 0.$$

If the singularity of f at O is isolated, the Koszul complex for $\{\partial f/\partial x_k\}$ is acyclic. From the acyclicity in dimension 1 it follows that the module of fields in the hyperplane tangent to the section is generated by the fields

$$\omega = \sum \alpha_k x_k \partial/\partial x_k \text{ and } \omega_{k,\,l} = (\partial f/\partial x_k)\partial_l - (\partial f/\partial x_l)\partial_k \qquad (k,\ l \geqslant 2).$$

These fields extend to the fields

$$\Omega = \sum \alpha_k x_k \partial_k \quad (k \geqslant 1) \text{ and } \Omega_{k,\,l} = (\partial F/\partial x_k)\partial_l - (\partial F/\partial x_l)\partial_k \quad (k,\ l \geqslant 1),$$

tangent to the hypersurface $F = 0$ in the ambient space. Hence, in particular, we can obtain Theorem 5 for the cases A_2, A_3, B_2, and B_3. However, already in the case A_4 the module of fields tangent to the section $\tilde{\tilde{\Sigma}}$ is not generated by ω and $\omega_{k,l}$ (example: the liftable fields $i^* v_{k,l}$, $v_{k,l} = d_l \lambda_l v_k - d_k \lambda_k v_l$, where $d_p = p + 1$, and $v_k = v_{\lambda_k}$ are basis liftable fields).

PROOF OF THEOREMS 4 and 5. The bases of the modules of vector fields tangent to Σ, $\tilde{\Sigma}$, and $\tilde{\tilde{\Sigma}}$ in the cases A_2, A_3, and B_3, and also the equations of $\tilde{\Sigma}$ and $\tilde{\tilde{\Sigma}}$ are written down in the following Table 1. We use coordinates in B such that

$\Sigma(A_2) = \{\lambda_1, \lambda_2\colon x^3 + \lambda_1 x + \lambda_2 = 0$ has a multiple root$\}$,
$\Sigma(A_3) = \{\lambda_1, \lambda_2, \lambda_3\colon x^4 + \lambda_1 x^2 + \lambda_2 x + \lambda_3 = 0$ has a multiple root$\}$,
$\Sigma(B_3) = \{\lambda_1, \lambda_2, \lambda_3\colon x^3 + \lambda_1 x^2 + \lambda_2 x + \lambda_3 = 0$ has a multiple or a zero root$\}$.

The direction of projection of π is the λ_2-axis for A_2, the λ_3-axis for A_3 and B_3. The coordinates in $\tilde{B}$ are λ_1 for A_2, (λ_1, λ_2) for A_3 and B_3. The plane $\tilde{\tilde{B}}$

[1] O. V. Lyashko (see [7]) has proved liftability under the condition that the set of points of the discriminant variety corresponding to roots of multiplicity greater than 2 has codimension at least 2.

is $\lambda_1 = 0$, the coordinates in $\widetilde{B}$ are λ_2 for A_2, (λ_2, λ_3) for A_3 and B_3. We denote the field $\partial/\partial\lambda_k$ by ∂_k. The projection of B onto $\widetilde{B}$ is denoted by π, and the embedding of $\tilde{\tilde{B}}$ in B by i.

The bases $\{v_k\}$ of the modules of fields tangent to Σ are obtained by the method of [1] (see §8). The bases of the modules of fields tangent to $\widetilde{\Sigma}$ and $\tilde{\tilde{\Sigma}}$ can be found by direct computations. With the use of the natural gradings the calculations become very short (the degree of λ_k is equal to the coefficient of $\lambda_k \partial_k$ in v_1).

Each of the basis fields in Table 1 tangent to $\widetilde{\Sigma}$ can be represented as the projection $\pi_* v$ of a vector field v on B tangent to Σ, and each of the basis fields in Table 1 that is tangent to $\tilde{\tilde{\Sigma}}$ can be represented as the restriction i^*v of a vector field v on B and tangent to Σ. For fields tangent to $\widetilde{\Sigma}$ (respectively, $\tilde{\tilde{\Sigma}}$), which are indicated first in Table 1, we have $v = v_1$. For A_3 and B_3 there is a second field. For it v has the form of a linear combination of basis fields v_k with coefficients p_k (for tangency with $\widetilde{\Sigma}$) or q_k (for tangency with $\tilde{\tilde{\Sigma}}$), as given in Table 2:

TABLE 1

Type	Tangent to Σ	Tangent to $\widetilde{\Sigma}$	Tangent to $\tilde{\tilde{\Sigma}}$
A_2	$v_1 = 2\lambda_1\partial_1 + 3\lambda_2\partial_2$ $v_2 = 3\lambda_2\partial_1 - \frac{2}{3}\lambda_1^2\partial_2$	$2\lambda_1\partial_1$ $\widetilde{\Sigma}$: $\lambda_1 = 0$	$3\lambda_2\partial_2$ $\tilde{\tilde{\Sigma}}$: $\lambda_2 = 0$
A_3	$v_1 = 2\lambda_1\partial_1 + 3\lambda_2\partial_2 + 4\lambda_3\partial_3$ $v_2 = 3\lambda_2\partial_1 + (4\lambda_3 - \lambda_1^2)\partial_2 - $ $- \frac{1}{2}\lambda_1\lambda_2\partial_3$ $v_3 = 4\lambda_3\partial_1 - \frac{1}{2}\lambda_1\lambda_2\partial_2 +$ $+ \left(2\lambda_1\lambda_3 - \frac{3}{4}\lambda_2^2\right)\partial_3$	$2\lambda_1\partial_1 + 3\lambda_2\partial_2$ $(8\lambda_1^3 + 27\lambda_2^2)\,\partial_1$ $\widetilde{\Sigma}$: $(8\lambda_1^3 + 27\lambda_2^2)\,\lambda_2 = 0$	$3\lambda_2\,\partial_2 + 4\lambda_3\,\partial_3$ $64\lambda_3^2\,\partial_2 + 9\lambda_2^3\,\partial_3$ $\tilde{\tilde{\Sigma}}$: $27\,\lambda_2^4 = 256\,\lambda_3^3$
B_3	$v_1 = \lambda_1\partial_1 + 2\lambda_2\partial_2 + 3\lambda_3\partial_3$ $v_2 = 2\lambda_2\partial_1 + (3\lambda_3 + \lambda_1\lambda_2)\partial_2 +$ $+ 2\lambda_1\lambda_3\partial_3$ $v_3 = 3\lambda_3\partial_1 + 2\lambda_1\lambda_3\partial_2 +$ $+ \lambda_2\lambda_3\partial_3$	$\lambda_1\partial_1 + 2\lambda_2\partial_2$ $\lambda_2(4\lambda_2 - \lambda_1^2)\,(3\partial_1 + 2\lambda_1\partial_2)$ $\widetilde{\Sigma}$: $\lambda_2\,(4\lambda_2 - \lambda_1^2)\,(3\lambda_2 - \lambda_1^2) = 0$	$2\lambda_2\partial_2 + 3\lambda_3\partial_3$ $(4\lambda_2^3 + 27\lambda_3^2)\,\partial_2$ $\tilde{\tilde{\Sigma}}$: $\lambda_3\,(4\lambda_2^3 + 27\lambda_3^2) = 0$

TABLE 2

Type	$\{p_k\}$	$\{q_k\}$
A_3	$4\lambda_1^2-12\lambda_3,\ 9\lambda_2,\ 6\lambda_1$	$0,\ 16\lambda_3,\ -12\lambda_2$
B_3	$\lambda_1\lambda_2-9\lambda_3,\ 6\lambda_2-2\lambda_1^2,\ 3\lambda_1$	$2\lambda_2^2,\ 9\lambda_3,\ -6\lambda_2$

The coefficients p_k and q_k in Table 2 were calculated by means of the grading in the ring of polynomials in λ described above. However, these calculations are not needed in the proof: it can be checked directly that for the fields v constructed from these coefficients the fields $\pi_* v$ and $i^* v$ coincide with those in Table 1. This proves Theorems 4 and 5.

REMARK. Theorems 4 and 5 were discovered as by-products of The proof of Theorem 3 (which they do not use).

Warning: In spite of the fact that the surfaces in B that are of interest to us in Lemmas 6 and 7 of Theorem 3, and also their normal forms, turn out to be vertical (cylindrical with generating direction $\partial/\partial\lambda_\mu$), their reduction to normal forms, generally speaking, requires the whole group of Σ-preserving diffeomorphisms: the groups of diffeomorphisms preserving Σ when projected onto $\widetilde{B}$ (that is, taking vertical lines into vertical lines) need not be sufficient.

Theorems 1 and 2 were proved as preliminaries to Theorem 3. The question of the normal form of singular projections of smooth surfaces, which is solved by Theorem 3, arose in O. A. Platonova's study of the singularities in general position in the problem of the fastest way of avoiding obstructions. The proofs of Theorems 1–5 required the study of the Lie algebras of the fields w_β, and hence the operation of linearized convolution φ introduced in §8. We arrived at the proof of non-degeneracy (Theorem B of §8), and for this, Theorem C. Theorem A, which is equivalent to Theorem B, in combination with the formula of Levine–Eisenbud [3] and Khimshiashvili [5] suggests an interpretation of the signatures of the quadratic forms ψ_q generated by the displacement as the indices of singular points of vector fields. For the singularities A_μ, D_μ and E_μ this immediately reduces to Theorem D of §8, and then, as for the boundary singularities B_μ, C_μ, F_4 introduced in [2], the resulting index turned out not to be entirely usual. An analysis of this index led to all the results of §§1–8 on the boundary singularities of 1-forms. A comparison of both signature formulae for the index required the study of the duality between quadratic forms on Q and on T^*, which is carried out in §9. Thus, all the results of this paper arose from the problem of removing obstructions.

References

[1] V. I. Arnold, Wave fronts evolution and equivariant Morse lemma, Comm. Pure Appl. Math. **29** (1976), 557–582 (in English).

[2] V. I. Arnol'd, Critical points of functions on a manifold with boundary, the simple Lie groups B_k, C_k, F_4 and singularities of evolutes, Uspekhi Mat. Nauk **33**:5 (1978), 91–107.
= Russian Math. Surveys **33**:5 (1978), 99–116.

[3] D. Eisenbud and H. I. Levine, The topological degree of a finite C^∞ map germ, Ann. of Math. (2) **106** (1977), 19–39. MR **57** # 7651.

[4] E. Lloijenga, The complement of the bifurcation variety of a simple singularity, Inventiones Math. **32** (1974), 105–116.

[5] G. N. Khimshiashvili, The local degree of a smooth mapping, Sakharth. SSR Mecn. Akad. Moambe **85** (1977), no. 2, 309–311.

[6] A. G. Khovanskii, The index of a polynomial vector field, Funktsional. Anal. i Prilozhen. **13** (1979), 49–58.
= Functional. Anal. Appl. **13** (1979).

[7] O. V. Lyashko, The geometry of bifurcation diagrams, Uspekhi Mat. Nauk **34**:3 (1979), 205–206 (added by author, 18 November 1979).
= Russian Math. Surveys **34**:3 (1979).

Received by the Editors 15 October 1978

Translated by J. S. Joel

UDC 531.36:534

STEADY OSCILLATIONS WITH POTENTIAL ENERGY HARMONIC IN SPACE AND PERIODIC IN TIME

PMM Vol. 43, No. 2, 1979, pp. 360 - 363

V. I. ARNOL'D
(Moscow)

(Received March 7, 1978)

It is shown that the answer to the question posed by A. Iu. Ishlinskii about whether the equilibrium position $x = 0$ of the system $x^{\cdot\cdot} = -\partial U(x, t)/\partial x$ can be stable, if function U is harmonic with respect to x, is positive. A symmetric time-periodic matrix $A(t)$ whose trace is zero, such that the zero solution of system $x^{\cdot\cdot} = A(t)x$ is stable, is constructed. (We recall that the quadratic form $(Ax, x)/2$ is harmonic then and only then, when the trace of A is zero). The configuration space dimension, i.e. the number of coordinates of vector x, is assumed greater than unity.

Although the proposed here example is somewhat artificial, the method of its construction may prove to be of some interest in the investigation of Lie semigroups and in the analysis of the attainable set boundary in controllable systems. This method is presented here in Sect. 3, where the origin of this example formally defined in Sect. 1, is explained. The equilibrium stability of the obtained system with harmonic potential is proved in Sect. 2.

1. Definition of the example. Let N and ε be positive numbers (below, N is selected fairly large and ε fairly small). We specify the symmetric matrix S of order n with zero trace so that the eigenvalues of matrix S^2 are positive and different. Such matrix exists for $n > 2$. (It is possible, for instance, to take a diagonal matrix that has the indicated properties). We shall assume that $n > 2$.

We subdivide segment $[0, (4/N) + \varepsilon)$ of the t-axis into six consecutive segments whose length Δ appears in the first row of the table below, and specify matrix $A(t)$ which in each of these segments is equal to the constant matrix shown in the second row of the table (the segment left-hand ends, but not the right-hand ones are assumed to be part of each segment).

Δ	$1/N$	$\varepsilon/3$	$1/N$	$1/N$	$2\varepsilon/3$	$1/N$
$A(t)$	$3NS/2$	0	$-3NS/2$	$3NS/4$	0	$-3NS/4$

Having thus constructed $A(t)$ on segment $[0, T)$, $T = (4/N) + \varepsilon$, we continue that function over the whole t-axis periodically with period T. Note that the trace of $A(t)$ is zero for all t, so that the potential which corresponds to $A(t)$ is harmonic.

Theorem 1. If we set ε fairly small and N fairly large, the equilibrium position $x = 0$ of the equation $x^{\cdot\cdot} = A(t)x$ with periodic coefficients is stable.

Note that in the example defined in Theorem 1 the dependence of function $A(t)$ on time is discontinuous. This defect can be easily eliminated by smoothing that

Originally publ. in: Prikl. Mat. Mekh. 43(2), 360-363, © Nauka, Moscow 1979
English transl.: J. Appl. Math. Mech. 43(2), 385-389, © Elsevier, Oxford 1979
Reprint by Springer-Verlag Berlin Heidelberg 2016 with kind permission.

function. Let us fix function A (i.e. ε and N in Theorem 1), and complement its curve by segments that link the limit values to the left and right at points of discontinuity. Let us consider the smooth periodic function A_* whose curve lies entirely in the δ-neighborhood of the supplemented curve of function A, and such that for any t the trace of $A_*(t)$ is zero. Such functions can be readily defined explicitly.

Theorem 2. If δ is fairly small, the equilibrium position $x = 0$ of the equation $x^{\cdot\cdot} = A_*(t)x$ with periodic coefficients is stable. Here $A_*(t) = -\partial U(x, t) / \partial x$ and the potential U is harmonic.

2. Proofs. Let us examine the monodromy operator of the equation $x^{\cdot\cdot} = A(t)x$ which is equivalent to the system $x^{\cdot} = y$, $y^{\cdot} = Ax$ with matrix

$$Q = \begin{Vmatrix} 0 & E \\ A & 0 \end{Vmatrix}$$

Hence the transformation of the phase space with matrix $\exp \Delta Q$ corresponds to a time segment of length Δ along which A is constant. The monodromy operator of the considered equation is the product of six of such exponents that correspond to six segments of constant A in a period.

First, let us consider segments of length $1/N$ in each of which A is of the form NB, where $B = 3S/2$, $-3S/2$, $3S/4$ and $-3S/4$, respectively. Hence as $N \to \infty$ the related operator of the phase space transformation in the time $1/N$ has the following finite limit:

$$\lim_{N\to\infty} \exp \frac{1}{N} \begin{Vmatrix} 0 & E \\ NB & 0 \end{Vmatrix} = \exp \begin{Vmatrix} 0 & 0 \\ B & 0 \end{Vmatrix} = \begin{Vmatrix} E & 0 \\ B & E \end{Vmatrix}$$

Consequently the product of the three exponents that correspond to the first three segments has the following finite limit as $N \to \infty$:

$$\exp \left[\begin{Vmatrix} E & 0 \\ -B & E \end{Vmatrix} \Delta \begin{Vmatrix} 0 & E \\ 0 & 0 \end{Vmatrix} \begin{Vmatrix} E & 0 \\ B & E \end{Vmatrix} \right], \quad B = \frac{3S}{2}, \quad \Delta = \frac{\varepsilon}{3}$$

Multiplication of the matrices yields

$$\exp \Delta \begin{Vmatrix} B & E \\ -B^2 & -B \end{Vmatrix}$$

The product of transformations of the last three segments is of the same form, except that in this case $\Delta = 2\varepsilon/3$, and $B = 3S/4$.

Finally, as $N \to \infty$ the matrix of the monodromy operator has the limit

$$M(\varepsilon) \doteq \exp \frac{2\varepsilon}{3} M_2 \exp \frac{\varepsilon}{3} M_1, \quad M_n = \begin{Vmatrix} 3S/2^n & E \\ -9S^2/4^n & -3S/2^n \end{Vmatrix}, \quad n = 1, 2$$

When $\varepsilon \to 0$ we have

$$M(\varepsilon) = E + \varepsilon M_0 + \ldots, \quad M_0 = \begin{Vmatrix} S & E \\ -9S^2/8 & -S \end{Vmatrix}$$

We shall prove that for fairly small ε all eigenvalues of $M(\varepsilon)$ are different with their modulus equal unity.

The formula for M (ε) implies that for any fixed τ

$$\lim_{k\to\infty} M(\tau / k)^k = \exp \tau M_0 \qquad (2.1)$$

The transformation in the right-hand side has eigenvalues the modulus of all of which is equal unity, while being pairwise different (if τ is not too large). In fact, this is the transformation in time τ of the phase stream of the system

$$x^{\cdot} = Sx + y, \quad y^{\cdot} = -{}^{9}/_{8}S^2x - Sy$$

Eliminating y we obtain $x^{\cdot\cdot} = -{}^{1}/_{8}S^2x$, i. e. the equation of small oscillations with positive potential energy (the eigenvalues of S^2 are positive) and pairwise different natural frequences ω_j (the eigenvalues ω_j^2 of matrix S^2 are pairwise different). The eigenvalues of matrix $\exp \tau M_0$ are equal, hence $\exp(\pm i\tau\omega_j)$. These numbers are pairwise different for small τ.

Note that all of the considered here differential equations are Hamiltonian equations. Hence the operators of phase space transformation are simplicial. Simplicial operators whose eigenvalue moduli are all equal unity, are different from ± 1, and pairwise different, have the property of strong stability: they are stable together with all simplicial operators that are close to them (see [1]).

Thus the simplicial operator $\exp \tau M_0$ is highly stable. If follows now from formula (2.1) that when k is fairly large the operator $M(\tau / k)$ is highly stable, which means that for a fairly small ε the operator M (ε) is highly stable. But then when N is fairly large but finite, the monodromy operator $M(N, \varepsilon)$ close to M (ε) is also stable.

This proves Theorem 1. Theorem 2 follows from that the operator $M(N, \varepsilon)$ is highly stable, and the monodromy operator for A_* is fairly close to it.

R e m a r k s. 1°. The above reasoning provides also an example of the one-dimensional system $x^{\cdot\cdot} = A(t)x$ with periodic coefficients, which remains stable when the sign of the right-hand side is changed. Moreover, the function with such properties can be selected odd (or even) with respect to t. Formulas in Sect. 1 in which $S = 1$ provide such an example with a discontinuous function A; the same properties can be obtained for a smooth function A by the smoothing procedure.

2°. It is now possible to construct an example of stable equation with a potential that is harmonic in space and time periodic in the case of $n = 2$. For this it is necessary to consider a system which decomposes into the direct product $x_1^{\cdot\cdot} = A(t)x_1$, $x_2^{\cdot\cdot} = -A(t)x_2$, where function A is defined in Remark 1°.

3. L i e s e m i g r o u p s a n d t h e a t t a i n a b i l i t y l i m i t. 1°. This problem can be considered to be a particular case of the problem of Lie semigroups generated by the convex cones in Lie algebras. The Lie semigroup generated by a cone represents the closure of products of exponents of Lie algebra elements that lie in the cone.

In the considered problem the Lie group is a group of simplicial matrices of order $2n$, and the cone directrix consists of matrices of the form

$$\begin{Vmatrix} 0 & E \\ S & 0 \end{Vmatrix}$$

where S is a symmetric matrix whose trace is zero.

It was shown above that for $n \geqslant 3$ the semigroup generated by such cone contains only highly stable simplicial matrices. It would be interesting to try to classify the semigroups generated by convex cones belonging to classical Lie groups. The reasoning presented below makes possible the first steps in that direction. The example in Sect. 1 has been formulated on that basis.

2°. The problem of Lie semigroups is in turn a particular case of the problem of determination of the attainable set in controlled systems. The latter problem is generally formulated as follows: in the space tangent to a manifold a subset (called the indicatrix of possible velocities) is specified at every point. It is required to determine the set of points that can be reached from a given point in motion whose velocity at every instant of time belongs to the indicatrix, or to determine at least the closure of such set.

One of the ways of solving this problem consists of successive extension of the indicatrix, which is carried out as follows. First, instead of the indicatrix we can consider the cone drawn over it, since only the direction of the curve reaching it is essential, not the velocity of motion along it. Second, the obtained cone can be closed, hoping that points attainable with the new cone will lie either inside or at the boundary of the set of points attainable prior to the closure. Third, the obtained closed cone can be extended to its convex shell, considering the possible mixed strategies (i. e. of motions with rapidly changing sections along which the velocity is selected differently).

Let us assume that as the result of these operations we obtain such cones in the tangent spaces such that each of them contains in the tengent space some subspace. A fourth possibility of extending the cone-indicatrix exists in that case. Let us consider any vector field whose vector at each point lies in the considered subspace of the tangent space.

Let us consider some point on the phase curve passing through the selected reference point. Transformation invers to the transformation of the phase stream, which transforms the reference point to the end one, maps the end point indicatrix into the tangent space at the reference point. The shifted indicatrix can be joined to the reference point indicatrix.

It is actually possible to move for a certain time N over the vector field to the end point, then for some short time to follow the indicatrix of the end point, and then during time N move back along the phase curve of the vector field. As the result we shift from the reference point by a distance of order ε by following the shifted indicatrix (within smalls of higher order with respect to ε).

The combination of all indicatrices shifted in this manner can be again subjected to convexification, closure, etc., until these operations no longer alter the indicatrices.

3°. The example in Sect. 1 was derived as follows. The initial indicatrix in the unit of a simplicial group consists of matrices of the form

$$\begin{Vmatrix} 0 & E \\ A & 0 \end{Vmatrix}$$

where A is any symmetric matrix with zero trace. At the remaining points of the group the indicatrix is obtained from the latter by a right-hand shift.

Thus at each point the indicatrix is an affine plane, and the corresponding cone

is an open half-plane of the number of dimensions greater by one. Its closure represents a closed half-plane, and the boundary (at point unity) consists of matrices of the form

$$\begin{Vmatrix} 0 & 0 \\ A & 0 \end{Vmatrix}$$

where A is a symmetric matrix with zero trace.

In this way we have obtained a right-invariant field of planes lying in the extended indicatrices. The multiplication of the group elements (from the right) by

$$\exp t \begin{Vmatrix} 0 & 0 \\ A & 0 \end{Vmatrix}$$

specifies the phase stream, whose velocity vectors belong to the derived planes. Using the extension of the indicatrix by shifts along that field we join to the indicatrix in the unity all matrices of the form

$$\begin{Vmatrix} E & 0 \\ A & E \end{Vmatrix} \begin{Vmatrix} 0 & E \\ B & 0 \end{Vmatrix} \begin{Vmatrix} E & 0 \\ -A & E \end{Vmatrix} = \begin{Vmatrix} A & E \\ B-A^2 & -A \end{Vmatrix}$$

where A and B are symmetric matrices with zero trace, and the first matrix in the left-hand side is one of matrices of phase stream transformation, and the last one is its inverse matrix.

We convexificate the set of all indicatrices shifted in this manner, and obtain the set of all matrices of the form

$$\begin{Vmatrix} S & E \\ C & -S \end{Vmatrix}$$

where S and C are symmetric matrices with the trace of S equal zero and $\operatorname{tr} C \leqslant -\operatorname{tr} S^2$. Among these matrices there are matrices of the form

$$\begin{Vmatrix} S & E \\ -\lambda S^2 & -S \end{Vmatrix}, \quad \lambda > 1$$

to which corresponds the system $x^{\cdot} = Sx + y$, $y^{\cdot} = -\lambda S^2 x - Sy$ or the equation $x^{\cdot\cdot} = -(\lambda - 1)S^2 x$ which defines stable oscillations. It is now not difficult to plot the path that leads from unity to a highly stable simplicial matrix, as was done in Sect. 1.

REFERENCES

1. A r n o l ' d, V. I., Mathematical Methods of Classical Mechanics. Moscow, "Nauka", 1974.

Translated by J.J. D.

The Loss of Stability of Self-Induced Oscillations near Resonances

V.I. Arnold

(Translation of *Nonlinear Waves*, A.V. Gaponov-Grekhov (Ed.), Nauka, Moscow, 1979, 116–131; summary on p. 359)

The results of studying the loss of stability of self-induced oscillations are presented on the "physical" level of rigor. The problem of choosing the necessary number of bifurcational parameters is discussed, and all the resonances are divided into two classes: weak ones and strong ones. Poincaré's normal form method and the results of its application to resonance studies are described.

1. Choosing the number of parameters in bifurcation theory

The loss of stability of stationary or self-induced oscillating regimes is typically regarded as a one-parameter phenomenon. Results of the one-parameter study are well known. Upon change of the parameter (e.g. the Reynolds number) the eigenvalues of the linearized problem, for a certain critical value of the parameter, cross the imaginary axis (or the multipliers of a periodic motion cross the unit circle). After this one can observe either a hard loss of stability (a transition from the initial regime of motion to a remote one) or a soft loss of stability. The latter occurs as the appearance of auto-oscillations, for the case of the loss of stability of an equilibrium, and as doubling of the period or the appearance of a two-frequency quasi-periodic regime, for the case of the loss of stability of a periodic motion. The difference between the initial motion and the one appearing for a soft loss of stability has the order of magnitude of the square root of the over-criticality (i.e. the difference of the parameter from the critical value).

This picture of the stability loss of equilibria and auto-oscillations has been known from the times of Poincaré and was thoroughly studied by A.A. Andronov and his school in the thirties.

Originally publ. in: Nonlinear Waves, A.V. Gaponov-Grekhov (Ed.), 116-131, © Nauka, Moscow 1979
Translated by Boris Khesin and Andrey Khesin.

It turns out that many phenomena that result from the loss of stability of auto-oscillations become clear only upon examination of two-parameter, rather than one-parameter, families of oscillating systems.

The choice of the natural number of parameters in one problem or another is a challenge of the same type as choosing the number of degrees of freedom while describing a particular physical phenomenon. Taking additional parameters into account does not lead to significantly different results if the initial parameters were correctly chosen.

A one-parameter point of view is natural when examining degenerations of codimension one, i.e. phenomena occurring on hypersurfaces (surfaces defined by one equation) in the functional (generally speaking, infinite-dimensional) space of all the systems under consideration. These hypersurfaces can divide the functional space and they occur naturally as the boundaries of regions of different behavior of the system (e.g. the boundaries of stability regions, etc.).

By a small perturbation each individual system can be moved off the degeneration hypersurface, and in this sense, the degeneration does not occur in typical systems.

However, in a one-parameter family, simplest degenerations can occur in a way unavoidable with respect to a small perturbation of the family. Indeed, a one-parameter family of systems is a curve in the functional space of all systems. This curve can transversally intersect the hypersurface corresponding to a certain degeneration (say, the loss of stability) in the functional space.

In that case any nearby curve also intersects our hypersurface, at a point close to the intersection point of the initial curve and the hypersurface. Thus even though one can remove the degeneration for every fixed value of the parameter, it will still inevitably occur for some nearby value of the parameter.

While studying generic one-parameter families, we can neglect submanifolds of codimension two in the functional space of all systems. Indeed, if in the functional space the curve representing our one-parameter family intersects a submanifold of codimension two, then by an arbitrarily small deformation of this curve one can move it off of the submanifold.

Manifolds of codimension two in the functional space correspond to double degenerations. The study of such degenerations should be conducted in two-parameter families. While investigating two-parameter families, double degenerations appear as unavoidable at certain points of the parameter plane. In this plane, the traces of the hypersurfaces corresponding to simple degenerations divide a vicinity of the point of a double degeneration into sectors. This division is called a *bifurcation diagram*. In each sector of the bifurcation diagram, the phase portrait of the system in question is of a certain kind. Upon moving from one sector to another, a certain rearrangement occurs, determined by the simple degeneration on the sector boundary. In the singular point itself, the phase portrait degenerates most of all, but this degenerate portrait is actually of least real interest to us: complicated degenerations should be examined not on their own, but in families with the number of parameters with which the degeneration becomes unavoidable. The answer should be given in the form of a combination of the bifurcation diagram ("clock-face") and phase portraits corresponding to the sectors, the rays, and the center of the diagram.

Knowledge of the bifurcations in a two-parameter family for a double degeneration allows us to learn essential information about phenomena occurring in one-parameter families. Indeed, on the plane of parameters of a two-parameter family, let us examine a curve in general position passing near the special point of double degeneration. This curve will cross, generally speaking, several sectors of the bifurcation diagram. This means that in the one-parameter family corresponding to the given curve, several subsequent rearrangements will occur, determined by which boundaries of the sectors the curve intersects. This way a local study of bifurcations in two-parameter families allows us to obtain information about some global rearrangements in one-parameter families (for example, information about the bifurcations following the first loss of stability, which is rather difficult to acquire in any other way).

2. Strong and weak resonances

In the problem of stability loss of a periodic motion, one can immediately see one parameter that can be roughly imagined as the absolute value of the multiplier[1] (the loss of stability occurs when it turns one).

The second parameter in this problem is the argument of the multiplier: the multiplier can cross the unit circle at various points along it. If the argument of the multiplier at the moment of crossing the unit circle is commensurable with 2π, then one encounters a *resonance* at the loss of stability. This resonance is a double degeneration and must be studied in a two-parameter family. The corresponding two parameters have a simple physical interpretation: one of them is the decrement and the second is the deviation of the frequency from the resonance (the difference between the argument of the multiplier and its resonance value).

On the plane of these two parameters there appears, as described above, a bifurcation diagram whose sectors correspond to different types of behavior of the system. If in a one-parameter family, the loss of stability occurs near a resonance then this one-parameter family is portrayed by a curve on our bifurcation diagram, passing near the central point of the diagram. In this case our diagram can be used to read off the sequence of rearrangements of motion types, occurring near the loss of stability. By drawing different curves on the diagram, we will get a finite collection of sequences of rearrangements, implemented in generic one-parameter systems losing stability near the resonance.

Calculations show that resonances of orders $q = 1, 2, 3, 4$ have a particularly strong influence on the picture of stability loss. (A resonance has order q if the

[1] The multipliers of a periodic motion are the eigenvalues of the linearization of the Poincaré recurrence function at the point corresponding to this periodic motion.

multiplier is equal to $\exp(2\pi ip/q)$.) These resonances are said to be *strong*. Thus for strong resonances the arguments of the multipliers are 0, ±90, ±120, 180°.

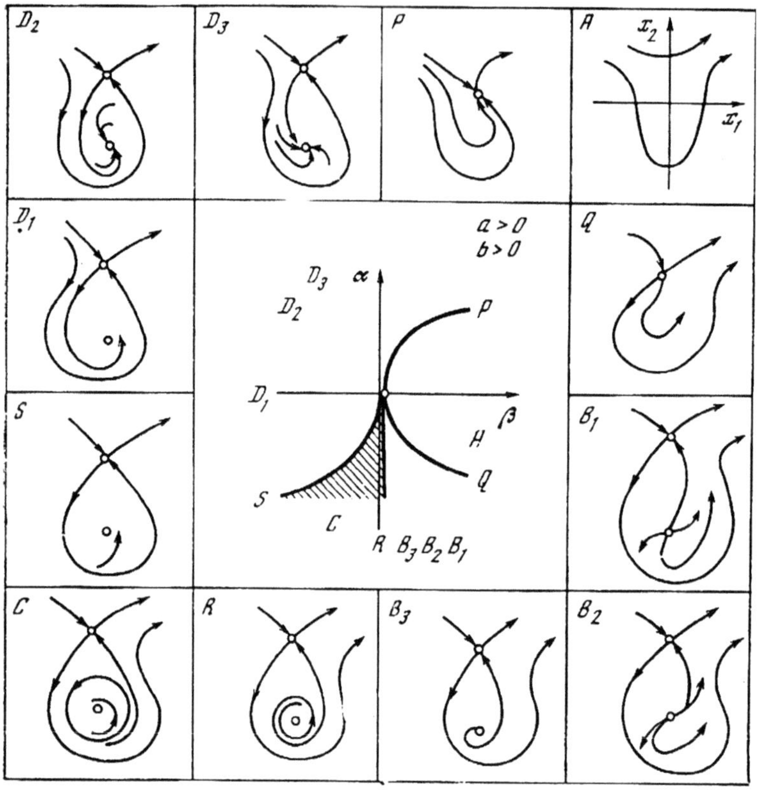

Fig. 1 Bifurcation diagram (in the center) and rearrangement of the phase portrait in the case of $q = 1$

Resonances of orders $q > 4$ are said to be *weak*. Near a weak resonance the loss of stability occurs almost in the same way as in the non-resonance case. From this point of view, only the angles close to 0, ±90, ±120, 180° should be regarded as angles "physically commensurable" with 360°, while other angles are "practically incommensurable" with 360°.

The description of bifurcation diagrams and rearrangements of the motion types near the loss of stability in a neighborhood of strong resonances is presented in the sections below.

It turns out that after some simplification (roughly speaking, upon averaging and "discarding almost inessential terms"), the problem is reduced to the study of the bifurcations of the phase portraits of planar vector fields that are invariant under rotation by the angle $2\pi/q$ and generically depend on two parameters.

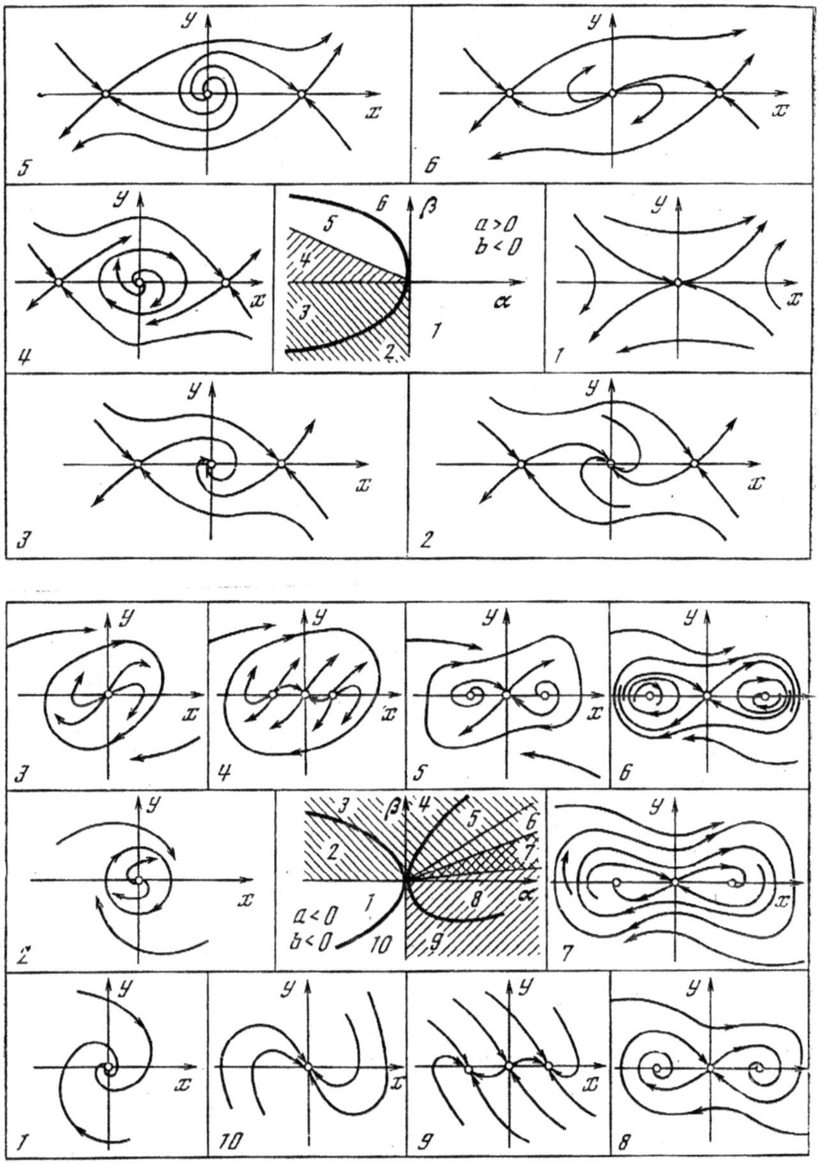

Fig. 2 Bifurcation diagram and rearrangement of the phase portrait in the two possible cases of the resonance $q = 2$

The reduced differential equations corresponding to strong resonances have the form

$$\begin{aligned}
\ddot{x} &= \alpha x + \beta + ax^2 + bxy, & q &= 1;\\
\ddot{x} &= \alpha x + \beta y + ax^3 + bx^2 y, & q &= 2;\\
\dot{z} &= \varepsilon z + Az|z|^2 + \bar{z}^2, & q &= 3;\\
\dot{z} &= \varepsilon z + Az|z|^2 + \bar{z}^3, & q &= 4.
\end{aligned}$$

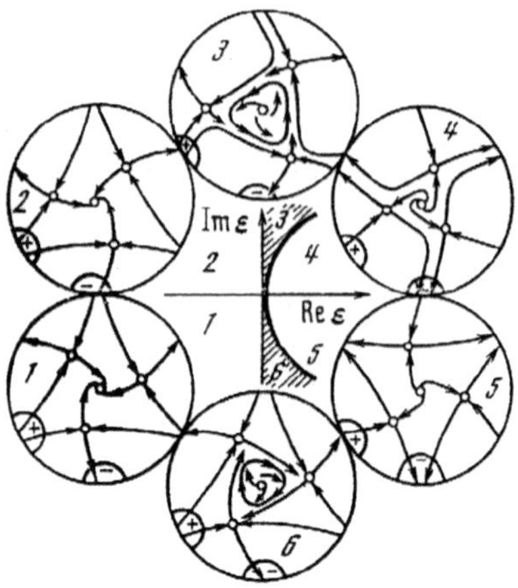

Fig. 3 Bifurcation diagram and rearrangement of the phase portrait in the case of $q = 3$ (Re$A < 0$)

Here $y = \dot{x}$; x, y, a, b are real and z, ε, A are complex numbers; the parameters of deformation are α, β, and ε; the values a, b, ReA, ImA are non-zero.

The picture of bifurcation diagrams and phase portraits is provided in Figs. 1–5. In the cases $q = 1, 2$, or 3, the answers depend only on the signs of a, b, ReA, ImA.

In the case of $q = 4$, the plane of the complex variable A is divided into a number of regions, in each of which the answer (the type of bifurcation diagrams and rearrangements) is distinct. The values of parameter A on the curves separating these regions do not need to be examined: they correspond already not to a double but to a triple degeneration (for the same reason we do not need to explore the cases $a = 0$, $b = 0$, Im$A = 0$, Re$A = 0$ for $q \neq 4$).

It should be emphasized that despite the simplicity of the reduced equations, their analysis requires overcoming serious mathematical difficulties (mainly related to the study of questions about the number of limit cycles).

The case $q = 4$ has not been yet completely studied; the pictures of rearrangements provided below correspond to just several of the regions on the A plane.

3. Poincaré's method of normal forms

Poincaré's method of normal forms is the basis of bifurcation theory. It plays the same role in the study of non-linear problems as Fourier's method (the method of expansion into the oscillation modes) plays in the linear theory. Although this method, already developed in Poincaré's thesis, has been known for over one hundred years, it became really widely applied to non-linear problems only in the last decade. Now there is quite a few works where this method is applied to systems with a finite or an infinite number of degrees of freedom. (Unfortunately, by now there has not yet been developed an analogue of Poincaré's method of normal forms for systems with a continuous spectrum.)

Fig. 4 Some possibilities of rearrangements for $q = 4$

The essence of Poincaré's method is that instead of searching for solutions of differential equations, we look for diffeomorphisms (changes of variables) that reduce the equation to a form as simple as possible.

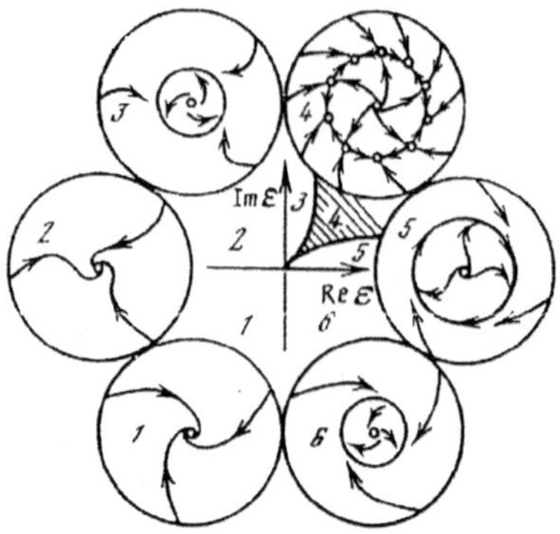

Fig. 5 Bifurcation diagram and rearrangement of the phase portrait for $q \geq 5$ (the equation is $\dot{z} = \varepsilon z + Az|z|^2 + \bar{z}^{q-1}$)

Let us study, for example, a system with an equilibrium in a neighborhood of this point. Choose the origin at the equilibrium. Then the system of equations can be written in the form

$$\dot{x} = A(x), \quad A(x) = \Lambda x + B(x), \tag{1}$$

where Λ is a linear operator, $x = (x_1, \ldots, x_n)$, and B contains only terms of degrees higher than one in x.

For simplicity, let us suppose that all eigenvalues λ_s of the operator Λ are distinct. Then we can choose a coordinate system such that the operator Λ is diagonalized, and in the linear approximation the system splits into one-dimensional ones.

We will look for a change of coordinates $x = y + h(y)$, where h stands for terms of degrees greater than one, to eliminate, if possible, the terms of degrees greater than one in the right-hand side of system (1). It turns out that in the general ("non-resonance") case, one can use a suitable change of coordinates to completely eliminate all the non-linear terms in the Taylor series of the right-hand side of the system at the origin.

Let us compute the right-hand side of (1) in the new coordinate system. We get

$$(E + \partial h/\partial y)\dot{y} = A(y + h(y)),$$
$$\dot{y} = (E + \partial h/\partial y)^{-1} A(y + h(y)).$$

Thus, $\dot{y} = \Lambda y + B(y) + C(y) + \ldots$, where $C(y) = \Lambda h(y) - \dfrac{\partial h}{\partial y}\Lambda y$ (the dots indicate terms of degrees greater than one in B and h).

Thus, *the effect of a deformation of the coordinate system is, in the first approximation, the change of the right-hand side of the equation by the value of the Poisson bracket of the linear part of the initial field and the deformation field.*

Consider the terms of the second degree in the Taylor series of the right-hand side

$$B(x) = B_2(x) + \dots$$

We will look for terms of the second degree in the Taylor series of the change of coordinates

$$h(y) = h_2(y) + \dots$$

so that in the new coordinate system one could get rid of the terms of the second degree in y. We obtain the following equation for h_2:

$$\Lambda h_2(y) - \frac{\partial h_2}{\partial y} \Lambda y = -B_2(y).$$

Denote by $h_2^{k,m}$ (respectively, $B_2^{k,m}$) the coefficient of the expansion of the kth component of h (respectively, of B) at $y^m = y_1^{m_1} \dots y_n^{m_n}$. The solution of the equation for h_2 is given by

$$h_2^{k,m} = -\frac{B_2^{k,m}}{\lambda_k - (m, \lambda)}.$$

The vanishing of one of the denominators $\lambda_k - (m, \lambda)$ is called a *resonance*. The number $|m| = m_1 + \dots + m_n$ is called the *order of the resonance.*

If a system does not have second order resonances, then the terms of the second degree can be eliminated by an appropriate change of coordinates.

Consider the resulting system with the right-hand side $\Lambda y + B_3(y) + \dots$ and perform a coordinate change $y = z + h_3(z)$. If there are no third order resonances, one can exclude all third degree terms.

Continuing this way, *one can eliminate all the non-linear terms in the Taylor series of the right-hand side, provided that the system does not have resonances of any order.* In the new coordinate system, the equation reduces to the simplest linear form $\dot{w} = \Lambda w$.

It should be mentioned that the formal procedure described above does not always lead to a converging series.

In applications to bifurcation theory, a primary role is played by several first approximations according to the method, and the question of convergence is not very important.

If the system has resonances, then certain non-linear terms cannot be rid of by changes of variables. More specifically, in the presence of the resonance $\lambda_k = (m, \lambda)$ it is impossible to kill x^m in the kth component of the field. This term is said to be *resonant.*

In this case Poincaré's method allows us to eliminate all non-linear terms in the Taylor series of the right-hand side of the equation, except for the resonant ones.

Let us remark that a resonance is one of the types of degenerations, thus, it must not be present in generic systems and should be studied not for an individual system but for a family of systems depending on parameters.

In a general family of systems depending on a parameter, resonances will be observed for isolated values of the parameter. In this case, for values of the parameter close to a resonant one, Poincaré's method allows us to exclude all the non-linear terms.

However, the resulting changes of variables will discontinuously depend on the parameter, since in the formulas of the change there appears a pole at the resonant value of the parameter. Thus, while studying the passage through a resonance, we should keep the resonant terms not only at the resonant value of the parameter (when it is impossible to eliminate them in any case), but also for near-resonant values of the parameter. In this case the change of variables can be made smoothly dependent on the parameters at the moment of passing through the resonance as well.

4. Normal forms near a periodic motion

Poincaré's method, described above for the case of an equilibrium, can be carried over to periodic motions as well. There are two essentially equivalent ways to perform this extension.

One of these ways consists in studying the first recurrence map or Poincaré map, i.e. the mapping of a piece of a hypersurface ("platform") transversal to the cycle in question into itself, defined by the phase curves of the given system. Let us choose a coordinate system on the transversal hypersurface so that the origin would be at the point of our cycle.

The recurrence function will be written in the form

$$x \mapsto \Lambda x + \dots, \quad x = (x_1, \dots, x_n),$$

where Λ is a linear operator and the dots indicate non-linear terms. The eigenvalues of the operator Λ are called the *multipliers* of the periodic motion under consideration.

By a change of coordinates

$$x = y + (\text{non-linear terms})$$

one can get rid of the non-linear terms of the recurrence function. In this case the resonances are relations $\lambda_k = \lambda^m$ where $\lambda^m = \lambda_1^{m_1} \dots \lambda_n^{m_n}$.

If there are no resonances then one can eliminate all non-linear terms in the Taylor series of the recurrence function. If there are some resonances then one should keep the resonant terms (y^m in the kth component). While studying families near resonances, one should retain the resonant terms.

Within the second approach to the problem of periodic motions, the phase curves close to a given cycle are described by an equation with periodic coefficients, in which the number of the phase coordinates is one less than in the initial one, while the time variable is now the phase of motion along our cycle.

In this case, Poincaré's method produces changes of variables that are periodic in time. In the non-resonant case, these coordinate changes allow one to kill all non-linear (in the new phase coordinates) terms in the Taylor expansion of the right-hand side in powers of the components of deviation from the cycle in question (the coefficients of these series are Fourier series in the time variable).

In the resonant case, there remain the resonant terms. Suppose that our equation has the form

$$\dot{x} = \Lambda x + \ldots, \quad x = (x_1, \ldots, x_n),$$

where the non-linear terms have the period 2π in t. Then the resonance condition has the form $\lambda_k = \lambda^m + im_0$, and the corresponding resonant term contains $x^m e^{im_0 t}$ in the kth component.

While studying families, one should keep the latter term in the normal form not only at the moment of resonance but also in its vicinity.

5. Preliminary analysis of the problem of stability loss of a periodic motion

Now we apply the above to the problem of stability loss. Consider the functional space of systems possessing periodic motions. The boundary of the region of systems with stable periodic motions is a hypersurface consisting of three components, each of which has codimension 1 in the functional space.

Indeed, on the boundary there are systems that have a multiplier whose absolute value is equal to 1. But since the characteristic equation is real, there is exactly one multiplier on the unit circle only if it is equal to 1 or -1. If 1 and -1 are not multipliers, then on the circle there must be at least two complex conjugate multipliers.

It is easy to see that each of the three cases (a multiplier equal to 1, a multiplier equal to -1, and a complex pair) corresponds to a hypersurface in the functional space, i.e. it is given by one constraint on the system. For systems on the boundary of the stability region, all other multipliers lie strictly inside the unit circle. In accordance to section 1, each of the three cases is encountered in an unavoidable way in generic one-parameter families, and we have to study the bifurcations as the parameter passes through its critical value.

In the case where one of the multipliers is equal to 1, the corresponding bifurcation is a merger of a stable cycle with an unstable one, past the critical value both cycles disappear; the loss of stability here is hard.

When one of the multipliers is equal to -1, the cycle losing stability does not disappear. At the critical value of the parameter, the cycle losing stability is approached by a stable cycle that goes twice around it. This double cycle either is born when passing through the critical value, in which case the loss of stability is soft, or it dies, in which case the radius of attraction of the stable cycle becomes small even before the hard loss of stability.

All of these results easily follow from Poincaré's theory presented above. Indeed, the Poincaré recurrence function with several initial resonant terms has the form

$$x \mapsto (1+\varepsilon)x + Cx^2 + \dots \quad (\lambda = +1),$$
$$x \mapsto (-1+\varepsilon)x + Cx^3 + \dots \ (\lambda = -1).$$

Here x stands for the coordinate in the eigendirection corresponding to the eigenvalue λ with the modulus equal to 1. We have denoted the parameter of deformation by ε; the coefficient C, generally speaking, depends on ε. The sign of $C(0)$ regulates the birth and death of cycles. In the directions of the remaining eigenvalues there is a compression, so that they do not affect the bifurcations (see Shoshitaĭshvili's general theorem [1]).

Now let us address the third case of stability loss, when a pair of multipliers crosses the unit circle.

The corresponding hypersurface in the functional space intersects the previous hypersurfaces (corresponding to the cases $\lambda = 1$ and $\lambda = -1$) by surfaces of codimension 2. The intersection points on these surfaces correspond to systems with a double eigenvalue $\lambda = 1$ (or $\lambda = -1$) and with a Jordan block of order 2. These two cases occur as unavoidable in the presence of two parameters and must be examined separately.

For the remaining cases where a pair of eigenvalues of the linearization of the recurrence function crosses the unit circle, Poincaré's method is applicable. Let us assume for simplicity that there are only two eigenvalues, i.e., that the initial phase space, where the cycle in question lies, is three-dimensional.

The computations according to this method lead to the following conclusions.

1. The recurrence function can be reduced to a normal form that is the time-one-map of the phase flow of a planar vector field (at least at the level of the Taylor expansions of finite order).

2. The resulting planar vector field is invariant under the group of all rotations of the plane in the non-resonant case and under the finite group of rotations by angles that are multiples of $2\pi/q$ in the case of a resonance of order q.

3. The first two conclusions hold also in the case where the system depends on a parameter. The change of coordinates giving the normal form for a finite part of the Taylor series smoothly depends on the parameter without resonances as well as near them, provided that one keeps the resonant terms for the exact resonance as well as for nearby values of the parameter. Then the symmetry of the normal form with respect to rotations by angles that are multiples of $2\pi/q$ holds not only for the exact resonance but also for close values of the parameter.

4. The equation of the normal form near a resonance of order $q > 2$ is as follows:

$$\dot{z} = \varepsilon z + A(|z|^2)z + B\bar{z}^{q-1},$$

provided that one neglects the terms of degrees $q+1$ and higher in $|z|$.

The same conclusions are obtained by applying the averaging method; in that case the normal form is interpreted as the averaged system describing the slow evolution of phase curves in a vicinity of the closed one.

While applying all of these results, one should keep in mind that the exact recurrence function is not, generally speaking, a map of the phase flow of any planar vector field and is not invariant under the action of any finite symmetry group. The fact that these properties hold for the normal form is one of indications that the series taking the system to a normal form diverge in the general case.

Nevertheless, many conclusions that can be drawn from the qualitative picture on the phase plane of the normal form allow us to make conclusions about the behavior of the phase curves of the initial system. For example, the equilibria of the reduced system (the truncated normal form or the averaged system) correspond to closed phase curves of the initial system, the cycles of the reduced system correspond to invariant tori in the initial system, etc. At the same time, finer effects, like separatrix splitting and appearance of homoclinic and heteroclinic trajectories, do not take place in the reduced system.

When the dimension of the initial phase space is greater than three, while studying the loss of stability with two eigenvalues crossing the unit circle, one can confine oneself to the study of the three-dimensional submanifold in the phase space on which the phenomena described above occur. Over the other eigendirections the recurrence function is a contracting mapping.

The foundation for discarding these eigendirections as irrelevant is obtained by a reference to the aforementioned general theorem due to Shoshitaĭshvili.

6. Study of the normal forms

From the above one can see how and in what sense the investigation of loss of stability near resonances reduces to the study of bifurcations of the phase portraits in two-parameter families of planar vector fields invariant under rotation by the angle $2\pi/q$, where q is the order of the resonance. This study is to be carried out in a small neighborhood of the origin, independent of the parameters.

The size of the region in which the system is being examined is the main distinction between bifurcation theory and local qualitative theory. Indeed, for an individual system, as well as for a system with parameters, one can study the system in a neighborhood of an equilibrium by means of the methods of local qualitative theory. This way we can learn, for example, for which value of the parameter the equilibrium becomes unstable, and we can find the phase portrait in a certain neighborhood of the equilibrium for each value of the parameter, both bifurcational and not.

However, the size of such a neighborhood tends to zero as the parameter approaches the bifurcational value, even though it is non-zero at the very moment of bifurcation.

For this approach the most important phenomenon escapes our attention: branching out of a new equilibrium or a cycle at the instant of bifurcation. Indeed, at the critical moment itself there is no branching object yet, and at the next moment it is already off the scope of applicability of local theory.

Thus, while studying bifurcations it is necessary to examine a neighborhood in the phase space not shrinking to zero as the parameter tends to the critical value. In other words, we have to consider not a parameter-dependent neighborhood in the phase space, but a neighborhood in the direct product of the phase space and the parameter space.

For instance, the phase pictures shown in Figs. 1–5 are observed near resonances in neighborhoods of the origin that do not depend on the parameters. A convenient tool for formulating the answers is the language of so-called versal deformations (or versal unfoldings).

A *deformation* of a vector field is a family of fields depending on parameters, that becomes the given field for the zero values of the parameters. Two deformations are said to be *equivalent* if there exist (1) a homeomorphic (continuous and bijective) change of parameters and (2) a family of homeomorphisms of the phase space, continuously depending on the parameters, that take the parameters and the family of phase portraits of the first deformation into the parameters and the corresponding phase portraits of the second one.

If one fixes a deformation with a certain parameter space, then any mapping of another parameter space into this one induces a new deformation.

A deformation of a vector field is said to be *topologically versal* if any other deformation of the same field is equivalent to a deformation induced from the given one. Thus a versal deformation is, in some sense, the largest deformation, containing in itself all others.

A versal deformation with a finite number of parameters does not necessarily exist for any field, but if it does, if it has been found and studied, then it provides a rather complete information about all phenomena and rearrangements occurring for any deformations of the given field.

Versal deformations are defined in a similar way for other objects, for example for vector fields with certain symmetries (meaning that both deformations and phase homeomorphisms in the definition of equivalence are symmetric).

As explained above, the problem about loss of stability near a resonance of order q requires the study of deformations of planar vector fields invariant under rotation by the angle $2\pi/q$.

Theorem. *Vector fields on the plane, invariant under rotation by the angle $2\pi/q$, $q \neq 2, 4$, and having an equilibrium at the origin with the zero linear part and* *generic non-linear terms, admit a two-parameter versal deformation in the class of symmetric fields. The corresponding system of differential equations can be written in the form*

$$\dot{z} = \varepsilon z + Az|z|^2 + \bar{z}^{q-1},$$

where ε is a complex parameter.

In the case of $q = 4$ the theorem is apparently also true, but has not yet been proven. The cases of strong resonances of orders $q = 1$ and 2 are studied separately and lead to the formulas pointed out above (in section 2).

The proofs of the formulated results are not simple: the difficulties are related to the study of limit cycles. In the simplest case of $q = 1$ these difficulties were overcome by Bogdanov in 1970–1971 [2]. In 1974, a preprint by Takens appeared where conjectures about types of rearrangements were announced.

Results of studies of the order 3 resonance were also published by Gavrilov [3]. However, it is unclear from these works how the difficulties related to the analysis of cycles of the reduced system were overcome. Complete proofs for all cases with $q \neq 4$ and for some cases with $q = 4$ were given by Horozov (1976–1977). The proofs were based on non-trivial considerations in algebraic geometry and the theory of Riemann surfaces, developed by Il′yashenko [4].

A more detailed exposition of the question of versal deformations can be found in the author's survey paper [2] cited above, in [5], and in the book "Supplementary Chapters of the Theory of Ordinary Differential Equations" currently being prepared for publication.[2]

References

[1] *Shoshitaĭshvili A.N.*, Trudy Sem. Im. Petrovskogo, no. 1 (1975), 279–309 (in Russian). English transl.: Amer. Math. Soc. Transl., Ser. 2, v. 118 (1982), 91–121.
[2] *Arnold V.I.*, Russian Math. Surveys, v. 27 (1972), no. 5, 54–123.
[3] *Gavrilov N.K.*, In: Studies on Stability and Oscillation Theory. Yaroslavl′ State University, Yaroslavl′, 1977 (in Russian).
[4] *Il′yashenko Yu.S.*, Funct. Anal. Appl., v. 11 (1977), no. 2, 105–113.
[5] *Arnold V.I.*, Funct. Anal. Appl., v. 11 (1977), no. 2, 85–92.

[2] See the book: V.I. Arnold, *Geometrical Methods in the Theory of Ordinary Differential Equations*, Grund. der Math. Wiss., v. 250, Springer-Verlag, New York, 1983, xi+334 pp.

Catastrophe Theory

V.I. Arnold

(Translation of Priroda, 1979, no. 10, 54–63)

Vladimir Igorevich Arnold, Dr. Sc. (physical and mathematical sciences), is a professor at the Moscow State University. He studies the theory of differential equations, classical and celestial mechanics, singularity theory, real algebraic geometry, and more. He is the author of *Ordinary Differential Equations* (1971), *Mathematical Methods of Classical Mechanics* (1974), *Supplementary Chapters of the Theory of Ordinary Differential Equations* (1978), and *Ergodic Problems in Classical Mechanics* (jointly with A. Avez) (1968). Arnold is a member of the National Mechanics Committee of the USSR, a vice-president of the Moscow Mathematical Society, a honorary member of the London Mathematical Society, and a Lenin Prize laureate.

Catastrophe theory first appeared in the West about ten years ago. Newsweek-level magazines reported a revolution in mathematics comparable to the invention of Newton's calculus. It was claimed that this new science, catastrophe theory, was much more useful to humanity than mathematical analysis: while Newton's theory only allowed for the study of smooth, continuous processes, catastrophe theory provided a universal method of exploring all abrupt transitions, shocks, and sudden qualitative changes. There appeared hundreds of scientific and scientific-like publications which applied catastrophe theory to a huge variety of subjects, such as heart beats, geometric and physical optics, embryology, linguistics, experimental psychology, economics, hydrodynamics, geology, and the theory of elementary particles. Among published works on catastrophe theory there are studies of the effect of alcohol on drivers, stability of ships, prisoner rebellions, the behavior of stock market players, modeling brain activity and mental disorders, and the politics of censorship of erotic literature.

In the early 1970s, catastrophe theory quickly became a fashionable, broadly advertised theory. Its universal claims reminded those of the pseudo-scientific theories of the 19th century.

Originally publ. in the popular scientific magazine "Priroda" 10, 54-63, © Priroda. Russian Academy of Sciences 1979
Translated by Boris Khesin and Andrey Khesin.
Springer-Verlag Berlin Heidelberg 2016 with kind permission.

Mathematical papers of the founder of catastrophe theory, the French mathematician R. Thom, were republished as pocket editions and widely circulated. That had not happened in the mathematical world since the appearance of cybernetics, from which catastrophe theory borrowed many self-advertising tricks.

After the wave of praise for catastrophe theory passed, there appeared more sobering, critical papers. Some of these were also printed in publications intended for laymen, bearing eloquent titles such as "But he isn't wearing anything at all!". By now there have already appeared surveys of papers that are critical of catastrophe theory[1]. Today there is no unanimity about the value of catastrophe theory. Below one will find the author's personal opinion on the subject.

H. Whitney's Singularity Theory

In 1955, the American mathematician Hassler Whitney published his paper "Mappings of the plane into the plane"[2] which laid the foundation of a new mathematical theory, singularity theory for smooth mappings.

To grasp what catastrophe theory is, one has to first get acquainted with elements of Whitney's singularity theory. Singularity theory is a grandiose generalization of the study of maxima and minima of functions. In Whitney's theory, functions are replaced with mappings, which are, in the simplest case, mappings of surfaces into a plane.

A mapping of a surface into a plane is a matching of every point on the surface with a point on the plane. If a point in the surface is given by coordinates (x_1,x_2) and a point in the plane is given by coordinates (y_1,y_2), then the mapping is determined by a pair of functions:

$$y_1 = f_1(x_1,x_2), \quad y_2 = f_2(x_1,x_2).$$

Mappings of smooth surfaces into a plane are all around us. Indeed, most bodies that surround us are bounded by smooth surfaces. The visible contours of bodies are just projections of the surfaces bounding those bodies onto the eye's retina. By taking a closer look at the bodies that surround us, for example, the faces of people, we can study singularities of the visible contours.

Whitney noticed that in cases of "general position", there occur only two types of singularities. All more complicated singularities are destroyed at a slightest change of the bodies or the direction of projection, while singularities of these two types are stable and persist under small deformations of the mapping.

An example of a singularity of the first type, which Whitney named "a fold", is the singularity of the projection of a sphere into a plane at points of its equator (see

[1] Guckenheimer J., *The catastrophe controversy*, The Mathematical Intelligencer, 1978, v. 1, no. 1, 15–20.

[2] Whitney H., *On singularities of mappings of Euclidean spaces. I. Mappings of the plane into the plane*, Ann. Math. (2), 1955, v. 62, no. 3, 374–410.

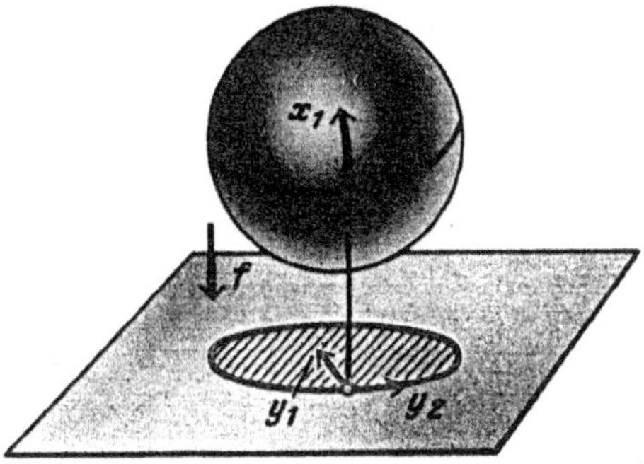

Fig. 1 Projection of the surface of a sphere into a horizontal plane. A singularity called a fold occurs on the equator of the sphere

Fig. 1). In local coordinates this mapping is given by the following formulas:

$$y_1 = x_1^2, \quad y_2 = x_2.$$

Unsurprisingly, projections of the surfaces of smooth bodies onto the eye's retina yield such a singularity in generic points. What is surprising is that in addition to this singularity, a fold, everywhere we encounter exactly one other singularity, but hardly ever notice it.

Whitney named this second singularity "a cusp". It results from projecting a surface shown in Fig. 2 into a plane. This surface is given by the formula $y_1 = x_1^3 + x_1x_2$ in the space with coordinates (x_1, x_2, y_1) and is projected into the horizontal plane (y_1, y_2). Thus, the mapping is determined in local coordinates by the formulas

$$y_1 = x_1^3 + x_1x_2, \quad y_2 = x_2.$$

On the horizontal plane of projection, one can see a semicubical parabola with a cusp point (a spike) at the origin. This curve divides the horizontal plane into two distinct parts. Points in the smaller part each have three inverse images (three points of the surface are mapped into each of them). Points in the larger part each have one inverse image only. Upon approaching the curve from the smaller part of the projection, two inverses merge and disappear (such a point on the curve corresponds to a fold singularity); upon approaching the cusp point, all three inverses merge.

Whitney proved that a cusp is stable, that is, any nearby mapping has a similar singularity in a suitable nearby point (i.e., a singularity such that the deformed mapping in appropriate coordinates near the indicated point is determined by the same formulas as the original mapping near the original point). Whitney also proved that any singularity of a smooth mapping of a surface into a plane, if slightly perturbed in a suitable way, "splits" into folds and cusps.

Thus, the visible contours of generic smooth bodies have cusp points in those places where the projections have cusps and do not exhibit other singularities: if we look closely, we can find these cusp points in the visible contours of every face and

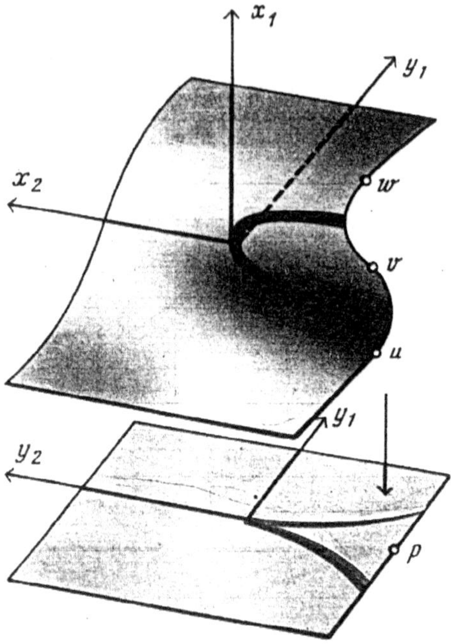

Fig. 2 The projection of the surface $y_1 = x_1^3 + x_1 x_2$ into a horizontal plane. There is a singularity called Whitney's cusp at point 0. Three inverses of point p (u, v, and w) are shown

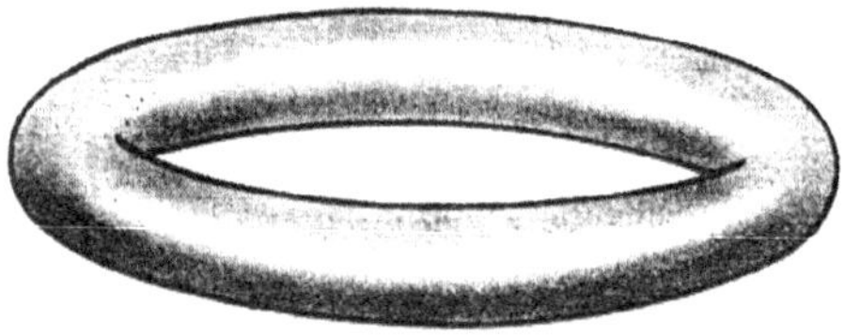

Fig. 3 The standard picture of a torus

body. Let us examine, for example, the surface of a smooth torus (e.g., an inflated tire). A torus is normally drawn as depicted in Fig. 3. If the torus were transparent we would see its contour as depicted in Fig. 4: the corresponding mapping of the torus into a plane contains four cusps. Thus, the ends of the line of the visible contour in Fig. 3 are cusp points. In these points the curve of the visible contour has a semicubical singularity.

Transparent tori are hard to come by. Let us examine another transparent body, a bottle. Fig. 5 depicts a bottleneck; two cusp points are visible. By rocking the bottle we can see for ourselves that a cusp is stable. Thus we obtain a convincing experimental confirmation of Whitney's theorem.

After Whitney's pioneering work, singularity theory vigorously developed further. Today, it is one of the central branches of mathematics where several paths intersect, linking the most abstract areas of mathematics (differential and algebraic

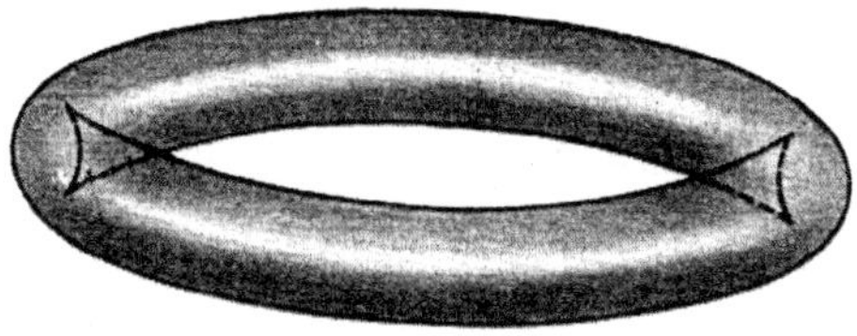

Fig. 4 Singularities of projection of a torus into a plane: one can see 4 cusp points

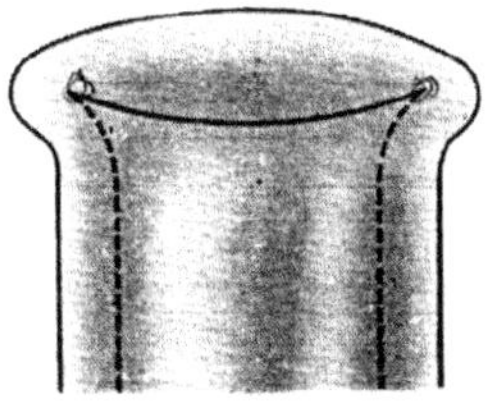

Fig. 5 A bottleneck. There are two visible cusp points

geometry and topology, the theory of groups generated by reflections, commutative algebra, the theory of complex spaces, etc.) with the most applied areas (the theory of stability of motion of dynamical systems, the bifurcation theory of equilibria, geometric and wave optics, and so on).

Applications of Whitney's Theory

Since smooth mappings are ubiquitous, their singularities are everywhere to be found as well. But since Whitney's theory provides a lot of information about the singularities of generic mappings, one can try to use this information to study a large number of various phenomena and processes in all natural sciences. This simple idea is the very essence of catastrophe theory.

In the case where the mapping under consideration is rather well known, we are talking about a more or less direct application of the mathematical singularity theory to various natural phenomena. Such an application does lead to useful results, for example, in elasticity theory and geometric optics (singularity theory of caustics and wave fronts, which we will discuss later).

However, in most works on catastrophe theory, one encounters a much more controversial situation when not only the mapping under study is unknown, but its very existence is rather problematic.

Applications of singularity theory in those cases are of speculative nature. To provide an idea of such applications, below we reproduce an example by the British mathematician C. Zeeman of a speculative application of Whitney's theory to the study of activities of a creative personality[3].

Let us characterize a creative character (e.g., a scientist) by three parameters: "skill", "enthusiasm", and "achievements". Apparently, there has to be a correlation between these parameters. Thus we get a surface in the three-dimensional space with coordinates (S,E,A).

Let us project this surface into the plane (S,E) along the A axis. For a generic surface, singularities can be either folds or cusps (according to Whitney's theorem). One claims that a cusp positioned as shown in Fig. 6 adequately describes the observed phenomena.

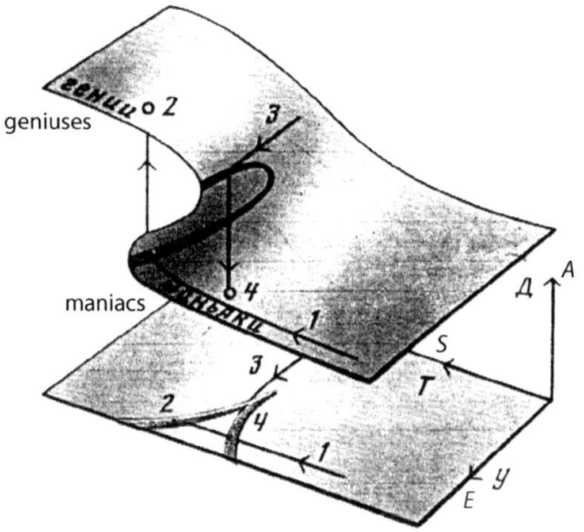

Fig. 6 A scheme of change of achievements (A) of a creative personality as a function of skill (S) and enthusiasm (E). With sufficient enthusiasm, an increase in skill leads to a jump from "maniacs" to "geniuses", but a growth of enthusiasm that is not supported by an increase in skill yields the "genius–maniac" catastrophe

Indeed, let us consider how under these assumptions, the scientist's achievements will change based on his/her skill and enthusiasm. If enthusiasm is low, then achievements monotonously and fairly slowly rise with skill. If enthusiasm is sufficiently high, then qualitatively new phenomena occur. In that case achievements can rise with a jump as skill increases (such a jump will happen if, for example, skill and enthusiasm change along curve 1 in Fig. 6 at point 2). The area of high achievements, which we finish in this situation, is labeled by the word "geniuses" in Fig. 6.

On the other hand, an increase in enthusiasm that is not supported with a corresponding increase in skill leads to a catastrophe (on curve 3 at point 4, Fig. 6), where achievements fall with a jump and we finish in the area labeled by the word "mani-

[3] Zeeman E.C., *Catastrophe theory: a reply to Thom*, In: Dynamical Systems (Warwick, 1974), Lecture Notes in Mathematics, v. 468 (A. Manning, ed.), Springer, Berlin, 1975, pp. 373–383.

acs". It is instructive that jumps from the state of "genius" to the state of "maniac" and back happen on different curves. Thus, at a sufficiently high enthusiasm level, a genius and a maniac may have equal levels of skill and enthusiasm but different achievements (and the preceding history).

The shortfalls of the described model and of a multitude of similar speculations in catastrophe theory are too obvious to discuss them in detail.

Catastrophe Machine

Unlike the example described above, applications of singularity theory to the study of bifurcations of equilibrium points in elasticity theory are perfectly justified.

In many elastic structures, there can be several points of equilibrium under the same external forces. Let us examine, for instance, a horizontal ruler whose ends are fixed with hinges, that is weighted down by a weight placed on the midpoint of the ruler.

Along with the point of equilibrium where the ruler is bent downwards by the weight, another point is possible where the ruler is bent upwards like a bridge.

As the weight is increased, there comes an instant when a "catastrophe" or a "clap" occurs: the ruler moves abruptly from one state to the other. Singularity theory is applicable to the study of such claps and its predictions are perfectly supported by experiments.

For a visual illustration of applications of this type there exist a number of inventions. One of the simplest devices is called Zeeman's catastrophe machine and is shown in Fig. 7.

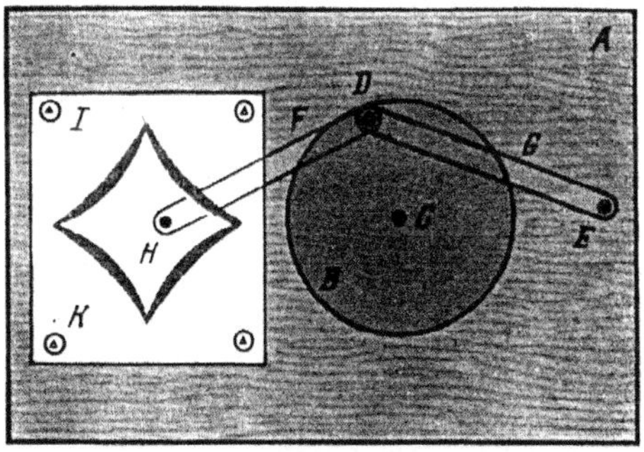

Fig. 7 Zeeman's catastrophe machine. *A* is the board; *B* is the cardboard disc; *C* is the needle that the disc can revolve around; *D* is the needle inserted into the disc only; *E* is the needle inserted into the board only; *F* and *G* are the elastic bands; *H* is the pencil; *I* is the sheet of paper; *K* is the catastrophe curve. As the pencil *H* moves, the disc *B* makes jumps when the pencil *H* crosses the catastrophe curve *K*

Everyone can easily make his/her own catastrophe machine. For this, one needs to take a board (*A* in Fig. 7) and, having cut out a cardboard disc (*B*), to insert a needle (*C*) both into the center of the disc and into the board so that the disc can rotate about this axis. Another needle (*D*) is inserted only into the disc near its edge, and a third needle (*E*) only into the board. To complete the assembly of the machine, one only needs two rubber bands (*F*, *G*), a pencil (*H*), and a sheet of paper (*I*).

After the needle at the edge of the disc has been attached with elastic bands to the fixed needle and to the pencil, the tip of the pencil is placed at some point on the sheet of paper, thus stretching the elastic bands. The disc is affixed in a certain position. Now, with movement of the tip of the pencil, the disc will rotate. It turns out that at certain positions of the tip of the pencil, a small change in this position can produce a "catastrophe", i.e., the disc will jump to a new position. Marking the places of all such "catastrophes" on the sheet of paper, we get the "catastrophe curve" (*K*).

It turns out that the catastrophe curve itself contains four cusp points. Upon crossing the catastrophe curve a jump may or may not occur, depending on how the path of the tip of the pencil goes around the cusp points of the catastrophe curve.

If the reader were to experiment with this machine and to try finding a rule that determined if there is a jump, the reader would easily be convinced of the necessity of a mathematical theory of the phenomenon and would better appreciate the contribution of singularity theory to its explanation.

The state of the catastrophe machine is determined by three numbers. Indeed, the position of the tip of the pencil is given by two coordinates (called the control parameters). The position of the disc is determined by one more number (the angle of rotation), which is also called the intrinsic parameter of the system. If all three numbers are known, the amounts of strain of the elastic bands are determined. Consequently, the potential energy of the entire system is determined as well. The disc rotates in such a way as to minimize this energy (at least locally). For a fixed position of the pencil, the potential energy is a function of the position of the disc, i.e., a function defined on a circle. This function can have, depending on the values of the control parameters, one or several minima (Fig. 8a). If upon changing the control parameters, the minimum changes smoothly, then a jump does not occur. A jump occurs at the values of the control parameters for which the local minimum disappears, having merged with a local maximum (Fig. 8b). After the jump, the disc ends up in a position corresponding to a different local minimum (Fig. 8c).

Consider the three-dimensional space of the possible states of this machine. The states at which the disc is in equilibrium form a smooth surface in this space. Let us study the projection of this surface into the plane of the control parameters along the axis of the intrinsic parameter (Fig. 9). This projection exhibits folds and cusps. The projection of the fold points is the catastrophe curve. From Fig. 9 we see why a passage of the control parameters across the catastrophe line sometimes causes, and sometimes does not cause, a jump (this depends on to which sheet of our surface the current position of the disc belongs). Using this figure, we can move from one place on the equilibrium surface to another without any jumps.

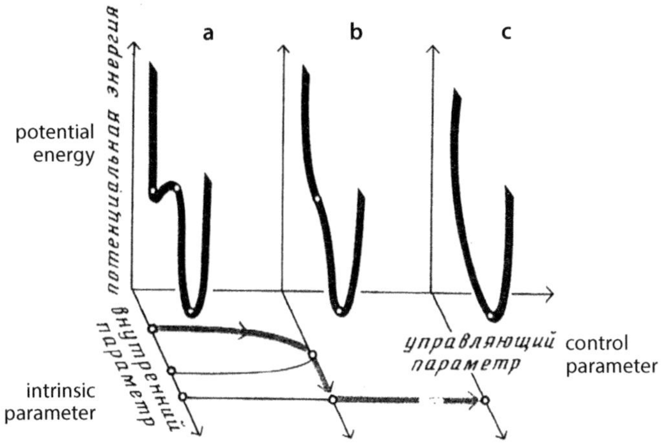

Fig. 8 Dependence of the potential energy of the system on the control and intrinsic parameters. A jump from one minimum of the potential energy to another occurs when one minimum gets "eaten" by a neighboring maximum

The outline of most applications of catastrophe theory is the same as in the described examples. One supposes that the process in question can be characterized by several control and inner parameters. The equilibrium states of the process form a surface of certain dimensions in this space. The projection of the equilibrium surface into the "plane" of control parameters can have singularities. One may assume that these singularities are generic. In this case, singularity theory predicts the geometry of the "catastrophes", i.e., jumps from one equilibrium state to another when changing the control parameters.

Applications of this type can be more or less justified depending on how justified the initial assumptions are. For example, in the theory of claps of elastic structures and in the theory of capsizing ships, the predictions of the theory are fully confirmed by experiments. On the other hand, in biology, psychology, and social sciences (say, in applications to the theory of stock brokers' behavior or the study of nervous disorders), both the initial assumptions and conclusions of the theory have rather a heuristic value.

Caustics and Wave Fronts

One of the most important conclusions of singularity theory is the universality of several simple patterns, such as a fold, cusp, and cusp point on a curve, that should be ubiquitous, and it is useful to be able to recognize them. In addition to the listed singularities, one often encounters several more patterns that have also received their own names. Some of them are depicted in Fig. 10.

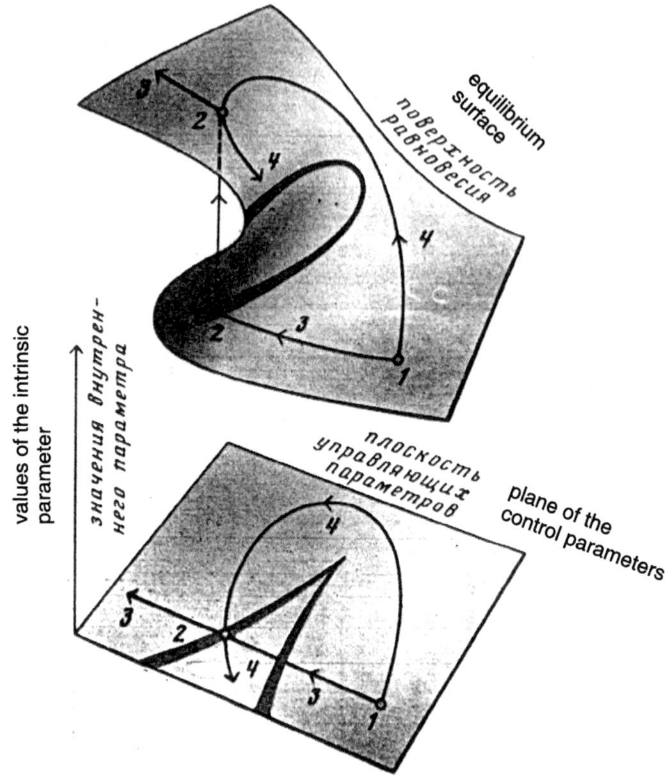

Fig. 9 Scheme of the three-dimensional space of the possible states of the catastrophe machine. If we move from point 1 along path 3, then there will be a jump at point 2 as we intersect the catastrophe curve; if we move along path 4, there will be no jump at point 2

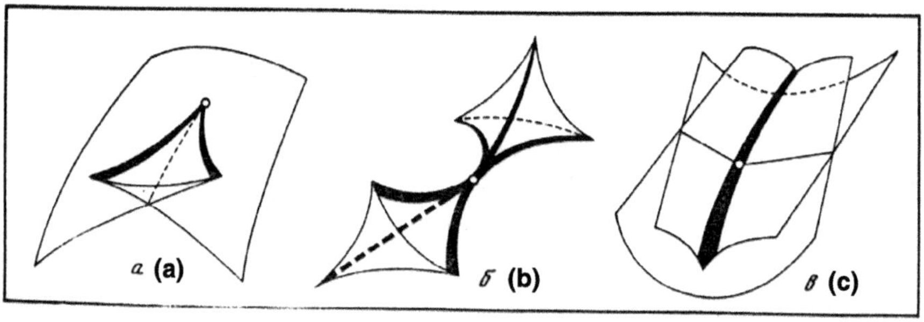

Fig. 10 Standard singularities of generic caustics in a three-dimensional space: swallowtail (*a*), pyramid (*b*), and purse (*c*)

Suppose that a perturbation is propagating in some medium (for example, a shock wave, light, or an epidemic).

For simplicity, let us start with the planar case. Suppose that in the initial moment the perturbation existed on curve "*a*" (Fig. 11) and that its speed of propagation is equal to 1. To figure out where the perturbation will be after time *t*, we should measure a segment of length *t* on every normal to the curve. The resulting curve is called a wave front.

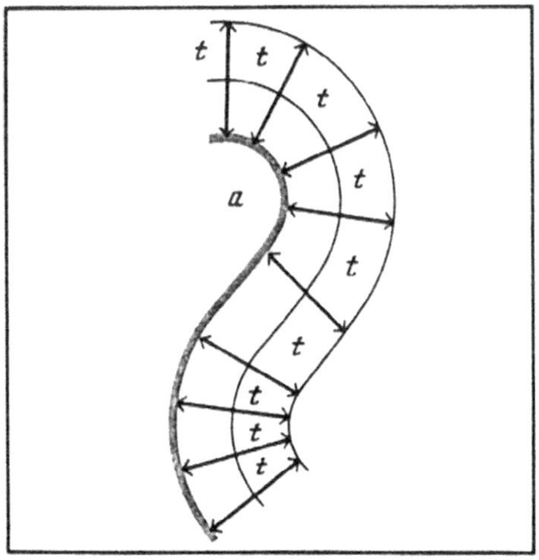

Fig. 11 Wave front propagating from curve "*a*"

Even if the initial wave front did not contain any singularities, after some time singularities will start appearing. For example, for propagation of a perturbation inside of an ellipse, the singularities shown in Fig. 12 appear. These singularities are stable (they cannot be removed by a small deformation of the initial front). Given a generic smooth initial front, only standard singularities of the same type will form after some time.

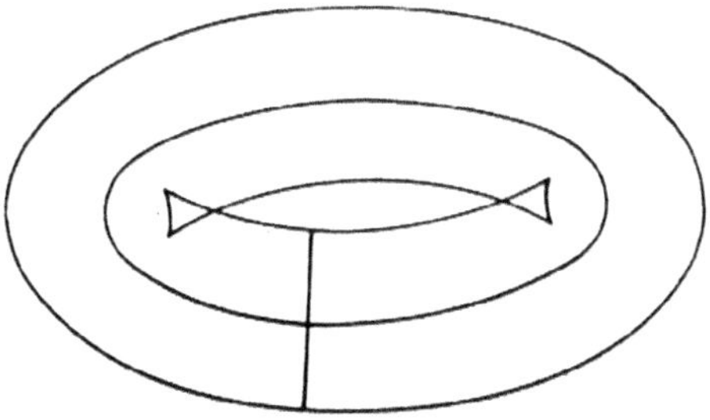

Fig. 12 Wave front propagating inside of an ellipse

In a three-dimensional space on a generic smooth wave front, as time passes by, there appear only cusp edges and standard "swallowtail"-type singularities shown in Fig. 10a. But in certain moments, the front experiences other bifurcations, also of standard types. These bifurcations are depicted in Fig. 13.

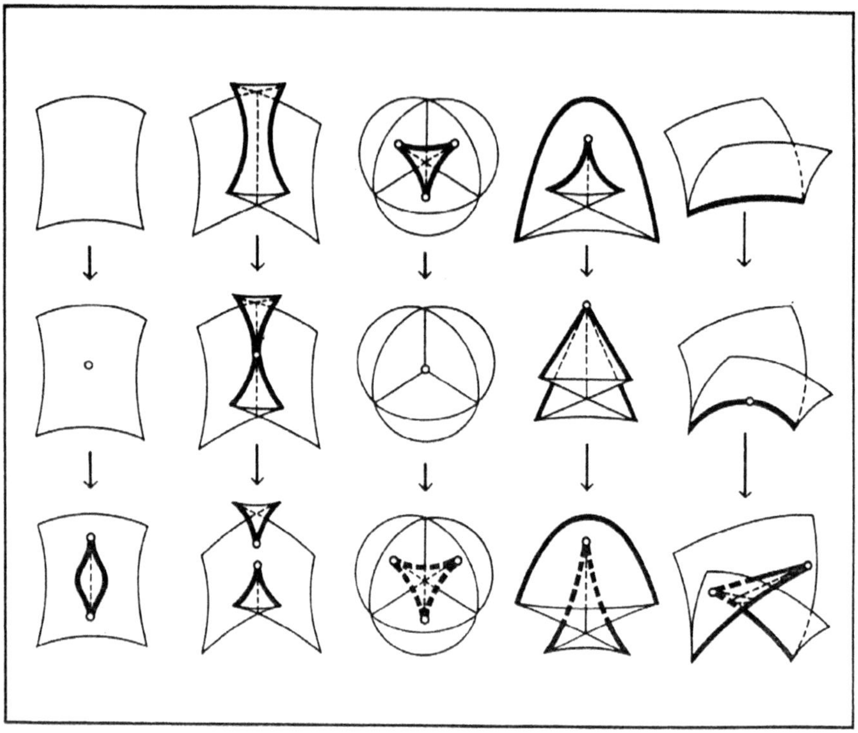

Fig. 13 Five standard types of bifurcations of a generic wave front as it moves in a three-dimensional space

Along with wave fronts, the propagation process can be described with the help of ray systems. For example, the propagation of perturbations inside of an ellipse can be described with the help of the family of internal normals to the ellipse (Fig. 14). This family admits an envelope. The envelope of a family of rays is called a caustic (i.e., "burning", since the light is concentrating at those points). A caustic is easy to see on the inside surface of a cup lit by sunlight. A rainbow, which we see in the sky, is also explained by the caustic of the system of rays that passed with total internal reflection through a droplet of water (Fig. 15).

The caustic of an elliptic front contains 4 cusp points. These singularities are stable: a front close to an ellipse will determine a caustic with the same singularities. Generic caustics in a three-dimensional space also have only standard singularities. These singularities are called the swallowtail, the pyramid, and the purse, and they are shown in Fig. 10.

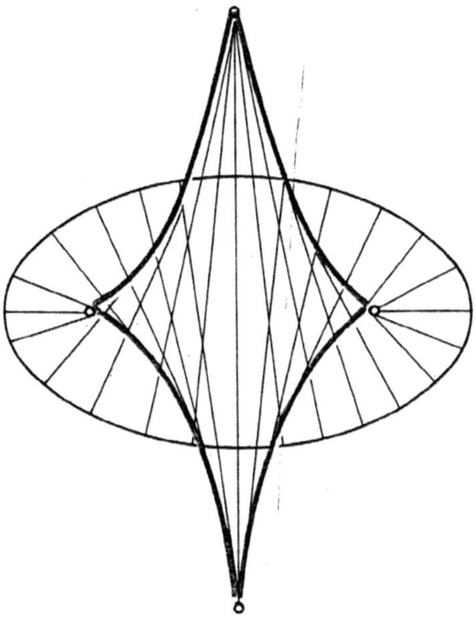

Fig. 14 Internal normals to an ellipse

The singularities of a propagating wave front slide along the caustic. This phenomenon is easily observable when comparing Fig. 12 and Fig. 14, as well as Fig. 10 and Fig. 13.

The singularities in caustics, fronts, and their bifurcations (Fig. 13 and Fig. 16) are well observed in experiments, and today it seems odd why this theory was not constructed, say, two hundred years ago. The reason is that the corresponding mathematical apparatus is non-trivial and is connected to such areas of mathematics as the classification of simple Lie algebras and Coxeter's crystallographic groups; it is also related (in a fairly mysterious way) to the classification of regular polyhedra in the three-dimensional Euclidian space.

Mysticism of Catastrophe Theory

Applications of singularity theory in natural sciences do not exhaust all the directions in catastrophe theory: in addition to concrete studies such as Zeeman's works there are more philosophical essays of the mathematician R. Thom, who was the first to realize the unifying character of Whitney's works on singularity theory (as well as that of the preceding works by H. Poincaré and A.A. Andronov on bifurcation theory), introduced the term "catastrophe theory", and launched its widespread propaganda.

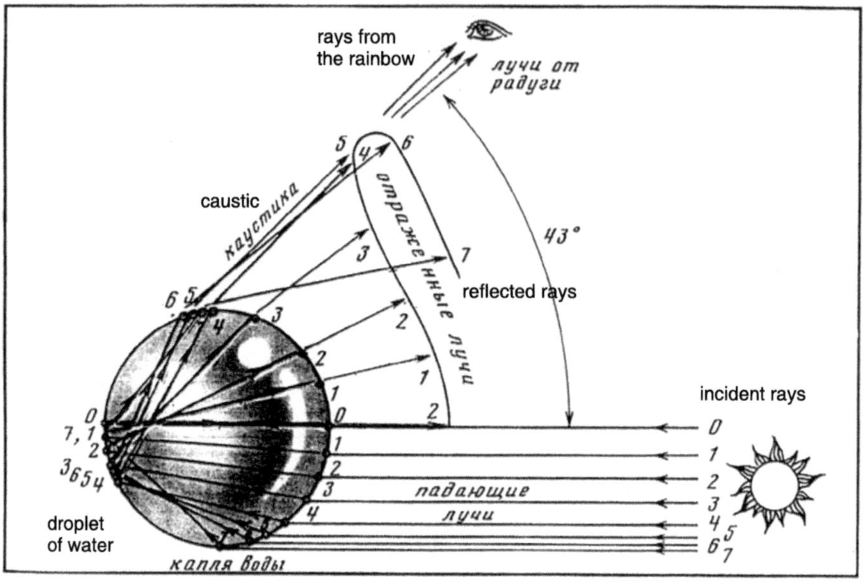

Fig. 15 Mechanism for the formation of a rainbow. Rays 1–7 after two refractions and one reflection in a droplet of water form angles from 0° to 43° with respect to the direction to the Sun. The density of the rays deflected at angles close to the maximal possible one is higher than the density of the rays deflected at other angles. The caustic (outside the droplet) is the line bounding the system of rays; it is very close to ray 5

The characteristic feature of Thom's works on catastrophe theory is their peculiar style. Zeeman, a passionate follower of this style, noted that the meaning of Thom's words only becomes clear when between every pair of Thom's lines one inserts 99 of his/her own lines.

To give the reader an opportunity to make his/her own opinion on this style, I include here an excerpt from Thom's 1974 overview of perspectives of catastrophe theory:

"In the philosophical and metaphysical aspects the catastrophe theory certainly cannot give answers to major problems worrying the mankind. But it encourages a dialectic, Heraclitian vision of Universe, a vision of the world as of the theatre of a permanent battle between 'logoi', between archetypes. Catastrophe theory leads us to a deeply polytheistic view: everywhere we should recognize the hands of gods. And here, catastrophe theory might find inevitable limits of its practical applicability. Perhaps it will share the fate of psychoanalysis. There is no doubt that Freud's main psychological discoveries are true. And yet knowledge of these facts has yielded very little use in practice (for the treatment of mental disorders, in particular). Just like the hero of the Iliad could not oppose the will of a god [say, Poseidon, without invoking the power of an opposite deity, such] as Athena, we also cannot restrict the actions of an archetype without counterposing it with an an-

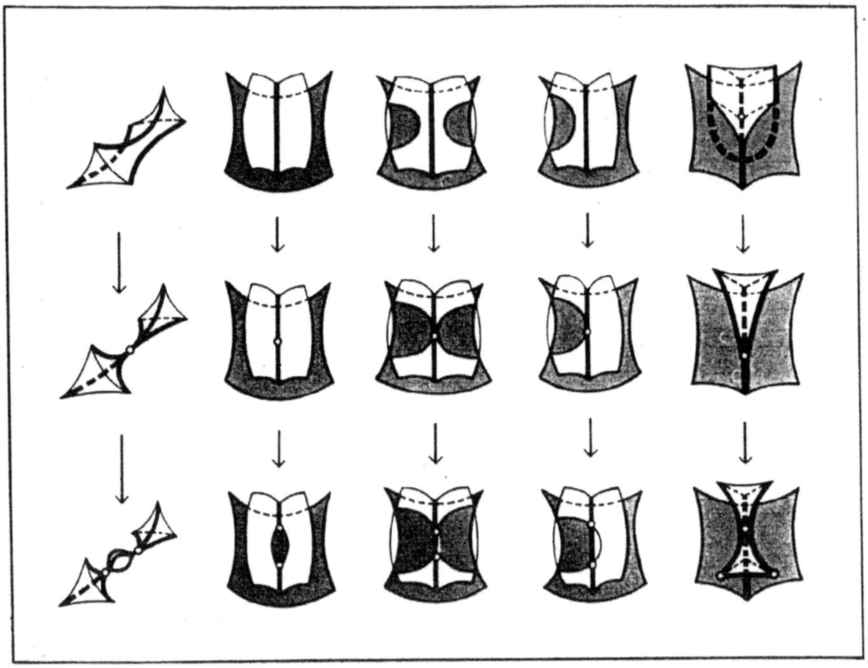

Fig. 16 Five standard types of bifurcations of a caustic. (This figure is missing in works on catastrophe theory and is apparently being published for the first time in this paper.)

tagonist archetype, in an ambiguous battle with an uncertain outcome.[4] The same reasons that allow us to command our actions in some situations, will condemn us to impotence in others. Maybe we will be able to prove the inevitability of some catastrophes, such as plague or death. Knowledge will no longer be necessarily a promise of success or survival; it could equally be a certainty of our failure, our end."[5]

Luckily, singularity theory's excellent results do not depend on catastrophe theory's gloomy mysticism.

Recommended Literature

[1] Whitney H.: *On singularities of mappings of Euclidean spaces. I. Mappings of the plane into the plane*, Ann. Math. (2), 1955, v. 62, no. 3, 374–410.

[2] Poston T., Stewart I.: *Catastrophe Theory and Its Applications*, Pitman, London, 1978.

[4] In the Russian original the bracketed part of the phrase was missing, which skewed the meaning of the sentence. (Translators' remark)

[5] Thom R., *La théorie des catastrophes: état présent et perspectives*, ibid., pp. 366–372.

[3] Thom R.: *Stabilité Structurelle et Morphogénèse*, W.A. Benjamin, Inc., Reading, Mass., 1972.
[4] Zeeman E.C.: *Catastrophe Theory. Selected Papers, 1972–1977*, Addison-Wesley Publishing Co., Reading, Mass., 1977.
[5] Bröcker T., Lander L.: *Differentiable Germs and Catastrophes*, Cambridge University Press, Cambridge, 1975.
[6] Golubitsky M., Guillemin V.: *Stable Mappings and Their Singularities*, Springer, New York, 1973.

SuperpositionofAlgebraicFunctions

V.I. ArnoldandG.Shimura

VII. Superposition of algebraic functions (V. Arnold, G. Shimura). Hilbert asked, as his 13th problem, whether or not an arbitrary analytic function of three variables can be obtained by finitely many substitutions of (continuous) functions of two variables. This was settled affirmatively, in fact against Hilbert's negative conviction, by V. I. Arnol'd and A. N. Kolmogorov.[1] Hilbert posed this question especially in connection with the solution of a general algebraic equation of degree 7. It is reasonable to presume that he formulated it in terms of continuous functions partly because he had an interest in nomography and partly because he expected a negative answer. Now that it is settled affirmatively, one can ask an equally fundamental, and perhaps more interesting, question with algebraic functions instead of continuous functions.

To be specific, take an algebraically closed basic field k and n independent variables $x_1, \ldots, x_n$ over k. Let K be an algebraic closure of $k(x_1, \ldots, x_n)$. For each positive integer $r \leqslant n$, we can define the subfield M_r of K consisting of all the elements of K obtained by successive compositions of algebraic functions of r variables. More precisely, let $[t_1, \ldots, t_r]$ denote, for any $t_1, \ldots, t_r \in K$, the algebraic closure of $k(t_1, \ldots, t_r)$ in K. Let $M_r^1 = k(x_1, \ldots, x_n)$, and let M_r^{s+1} be the composite of $[t_1, \ldots, t_r]$ for all choices of $t_1, \ldots, t_r \in M_r^s$. Then M_r is defined to be the union $\bigcap_{s=1}^{\infty} M_r^s$. Obviously $M_1 \subset M_2 \subset \cdots \subset M_n = K$. Now we can ask:

Can it happen that $M_r = K$ for some $r < n$?

Or more generally:

How many distinct fields are there among $M_1, M_2, \ldots, M_n$, or among the M_r^s? It is also interesting to ask the smallest M_r containing a given element of K. For example:

Determine the smallest M_r or M_r^s that contains an element f satisfying

$$f^n + x_1 f^{n-1} + \cdots + x_{n-1} f + x_n = 0.$$

One can also formulate and ask the questions within the algebraic closure of $k(x_1, \ldots, x_n, \ldots)$ with infinitely many variables x_i. This may or may not make a difference in the nature of the problem.

[1] V. I. Arnol'd, Dokl. Akad. Nauk SSSR **114** (1957), 679–681; Mat. Sb. **48** (90) (1959), 3–74; English transl., Amer. Math. Soc. Transl. (2) **28** (1963), 51–54, 61–147. MR **22** #2668; #12191; **27** #3758; #3759.

A. N. Kolmogorov, Dokl. Akad. Nauk SSSR **114** (1957), 953–956; English transl., Amer. Math. Soc. Transl. (2) **28** (1963), 55–59. MR **22** #2669; **27** #3760.

Originally publ. in: Mathematical Developments Arising from Hilbert Problems (1974) (Proc. Symp. Pure Math. 28), 45-46,

The *A*-*D*-*E* Classifications

V.I. Arnold

VIII. The *A-D-E* classifications (V. Arnold). The Coxeter-Dynkin graphs A_k, D_k, E_k appear in many independent classification theorems. For instance:

(a) Classification of the platonic solids (or finite orthogonal groups in the euclidean 3-space).

(b) Classification of the categories of linear spaces and maps (Gabriel, Gelfand-Ponomarev, Roiter-Nasarova) (see Sém. Bourbaki, no. 444, 1974).

(c) Classification of the singularities of algebraic hypersurfaces with a definite intersection form of the neighboring smooth fiber (Tjurina).

(d) Classification of critical points of functions, having no modules (see Sém. Bourbaki, no. 443, 1974).

(e) Classification of the Coxeter groups generated by reflections, or of Weyl groups with roots of equal length.

The problem is *to find a common origin of all the A-D-E classification theorems, and to substitute a priori proofs to a posteriori verifications of the parallelism of the classifications.*

Originally publ. in: Mathematical Developments Arising from Hilbert Problems (1974) (Proc. Symp. Pure Math. 28), 46,

Real Algebraic Geometry (the 16th Hilbert Problem)

V.I. Arnold

XIII. Real algebraic geometry (the 16th Hilbert problem) (V. Arnold). It is still unknown *how many connected components can have the complement of an algebraic hypersurface of degree n in the real projective k-dimensional space, even for* $k = 3$.

It follows from a theorem of Herrmann on the nodes of vibrating membranes (mentioned in Courant-Hilbert, *Methoden der math.*, Physik, Vol. I, Chapter VI, §6) that this number of components does exceed $1 + \binom{n+k-2}{k}$. This maximum is attained for plane curves (for n straight lines) and for surfaces of degree 4 (some smooth ones). However, the general theorem on membranes is not true, so the above estimation is only proved for curves.

Another still open problem is *to find all possible configurations of ovals of a plane curve of degree n with the maximum number of ovals* (i.e. $1 + (n - 1)(n - 2)/2$).

There is only one configuration of the 4 ovals of a curve of degree 4 (all exterior to one another), and 3 configurations of the 11 ovals of a curve of degree 6 (only one oval contains others, and it can contain 1, 5, or 9 of them). For the 22 ovals of a curve of degree 8, there are 145 configurations which verify all the known restrictions, and algebraic realizations are known for 9 configurations only.

In contrast to the recent progress with the algebraic part of the 16th Hilbert problem (due to Goudkov, Rohlin, Harlamov and others), there is not much progress with the second part, dealing with limit cycles: *it is still unknown whether a plane vector field, given by two polynomials of degree* 2, *can have more than* 3 *limit cycles.*

Profound ideas of I. G. Petrovskiĭ, concerning the relation of real limit cycles to the topology of the corresponding 2-dimensional foliation of the complex projective plane have not yet found their realizations.

Originally publ. in: Mathematical Developments Arising from Hilbert Problems (1974) (Proc. Symp. Pure Math. 28), 50-51,

StudyofSingularities

V.I. Arnold

XVI. Study of singularities.

(D) *Analytical properties of the thermodynamical functions at critical states* (V. Arnold). The thermodynamical functions are defined, with the help of some limit procedure, by the interaction potential of particles. *To find the singularities of these functions, knowing the interaction potential* is a very difficult analytical problem. The solution is not known for generic potential, nor for such simple models as the Ising lattice model in 3 dimensions.

(E) *Calculation of the order of the wildest generic singularities in l-parametric families of functions* (V. Arnold). For every integer $l > 0$ there is a rational number β_l, describing the asymptotics of the l-dimensional generalization of the Airy function, and the monodromy of the wildest singularity, generic in the l-parametric families of functions.

One can find an exact definition of the order of the wildest singularity β_l, in *Remarks on the stationary phase method and Coxeter numbers,* Russian Math. Surveys, 1973, no. 5, 17–44. The first 10 values of β_l are 1/6, 1/4, 1/3, 3/8, 5/12, 1/2, 1/2, 13/24, 9/16, 2/3. The optimistic formualtion of the problem: *to find all the* β_l; the pessimistic one: *to find* β_{1000}.

This problem is one of the ways to formulate precisely the fundamental but vague "problem of classification of singularities".

(F) *Topological properties of the bifurcation subsets* (V. Arnold). The bifurcation subsets are subsets in spaces of smooth or holomorphic functions on jets. There are two bifurcation subsets. A function belongs to the first (resp. second) subset, iff it is not a Morse function (resp. iff 0 is a critical value).

It seems very interesting *to study the topological properties of the complement of the bifurcation subsets* (for functions and also for differentiable mappings in general).

Near holomorphic functions with simple critical points of types A_k, D_k, E_k the complements to both bifurcation subsets have locally the homotopy type of a $K(\pi, 1)$ space (Brieskorn, Deligne, Ljaschko, etc.). The cohomology of the complement in the simplest case A_k is that of the second loop space of the 3-sphere (G. Segal, D. Fuks). On the other hand, in the real case there exist interesting relations of the topology of the bifurcation subsets to pseudo-isotopy and algebraic K-theory (Cerf, Wagoner, Volodin, etc.). See also the talk of Thom in *Manifolds,* Amsterdam 1970, Lecture Notes no. 197, Springer 1971.

Originally publ. in: Mathematical Developments Arising from Hilbert Problems (1974) (Proc. Symp. Pure Math. 28), 58-59,

DynamicalSystemsandDifferentialEquations

V.I. Arnold

XVII. Dynamical systems and differential equations.

(B) *Is the stability problem for stationary points algorithmically decidable?* (V. Arnold). The well-known Lyapounov theorem solves the problem in the absence of eigenvalues with zero real parts. In more complicated cases, where the stability depends on higher order terms in the Taylor series, there exists no *algebraic* criterion.

Let a vector field be given by polynomials of a fixed degree, with rational coefficients. Does an algorithm exist, allowing to decide, whether the stationary point is stable?

A similar problem: Does there exist an algorithm to decide, whether a plane polynomial vector field has a limit cycle?

A model problem: Does there exist an algorithm to decide, whether a period of a given abelian integral is positive (the equation of the curve and the integrand have rational coefficients and are of fixed degree). The simplest model problem is that of positivity on the real line of sums of products of exponentials and of polynomials. For this problem there exists an algorithm. This follows from the solution of Hilbert's problem on transcendental numbers (Matijasevich).

(C) *To prove the instability of equilibrium, which is not a minimum point for the potential energy* (V. Arnold). The equation of motion is that of Newton, $x'' = -\partial U/\partial x$ with an analytical potential energy U defined in the n-dimensional real space. The problem goes back to Lagrange and Dirichlet, who proved that points of minimum are stable. The instability seems obvious, but is not always the case for infinitely differentiable potentials.

Originally publ. in: Mathematical Developments Arising from Hilbert Problems (1974) (Proc. Symp. Pure Math. 28), 59-60,

FixedPointsofSymplecticDiffeomorphisms

V.I. Arnold

XX. Fixed points of symplectic diffeomorphisms (V. Arnold). The problem goes back to the "last geometric theorem" of Poincaré. The simplest case is the following problem: *Does every symplectic diffeomorphism of a 2-dimensional torus, which is homologous to the identity, have a fixed point*?

A symplectic diffeomorphism is a diffeomorphism which preserves some nondegenerate closed 2-form (the area in the 2-dimensional case). It is homologous to the identity iff it belongs to the commutator subgroup of the group of symplectic diffeomorphisms homotopical to the identity. With coordinates, such a diffeomorphism is given by $x \longrightarrow x + f(x)$, where x is a point of the plane and f is periodic. It is symplectic iff the Jacobian $\det(D(x + f(x))/Dx)$ is identically 1, and it is homologous to the identity iff the mean value of f is 0.

The "last geometric theorem" of Poincaré (proved by G. D. Birkhoff) deals with a circular ring. The existence of 2 geometrically different fixed points for symplectic diffeomorphisms of the 2-sphere is also proved (A. Shnirelman, N. Nikishin). In the general case, one may conjecture that the number of fixed points is bounded from below by the number of critical points of a function (both algebraically and geometrically).

Originally publ. in: Mathematical Developments Arising from Hilbert Problems (1974) (Proc. Symp. Pure Math. 28), 66,

PartialDifferentialEquations

V.I. Arnold

XXI. Partial differential equations.

(E) *What is a mathematical equivalent to physical "turbulence"*? (V. Arnold). The well-known (at least since 1962) conjecture is that one can define turbulence of viscous incompressible fluids with the help of such finite dimensional attractors, in the infinite dimensional phase space of initial conditions, that the dynamical systems on these attractors are themselves exponentially unstable ("hyperbolic"); the dimension of an attractor is finite for every finite Reynolds number and grows with it. However there exist other points of view, e.g. that mathematical expression of turbulence is the nonexistence or nonuniqueness of solutions of the Navier-Stokes equations, and that these equations are not an adequate model of physical motion. I hope the last is not true; *to find good existence and uniqueness theorems for the 3-dimensional Navier-Stokes equations* is only one of the aspects of the problem (which can as well be considered for the 2-dimensional case, where existence and uniqueness are known).

Originally publ. in: Mathematical Developments Arising from Hilbert Problems (1974) (Proc. Symp. Pure Math. 28), 67-68,

The Beginning of a New Style in the Scientific Literature

Professor V.I. Arnold and Academician Ya.B. Zeldovich

(A review of V.V. Beletsky's book "Essays on the Motion of Celestial Bodies", Moscow, Nauka Publishing House, 1972, 360 pp.[1])

"Dear Fagotto, show us something simple for the start" — it is with this epigraph from M.A. Bulgakov's "The Master and Margarita" that the first essay of the book under review begins. And the reader is not being lied to — the book does begin with some well known, and for that reason simple, classical results. Appreciating the style of the exposition, which is rather unusual in the scientific literature, from the very first lines (more precisely, from the brilliant dedication to Private Ferri Bren, a soldier in the Austro-Hungarian army, later a fighter in the International Red Guard regiment), the reader will only gradually begin to be captivated by the unexpected novelty of the results. And after turning the last page, he will certainly agree with the reviewers that the appearance of books such as this, and of writers such as its author, is a reflection of the amazing revival that celestial mechanics is currently experiencing in response to the demands of the theory of space flight.

Here the rules encountered more than once in other fields of knowledge repeat themselves. One would have thought that the problem of the motion of a body in the gravitational field of the Earth or in the Solar System has been very well studied. However, the practical implementation of the process has necessarily led to demands for more detailed and precise information, for the development of optimal strategies. At first glance, when one tries to answer those demands, the clarity and beauty of the simplified theory are lost. But in the next stage new points of view as well as theories that describe the results of more detailed calculations and new beauty arise. It is precisely this stage that is reflected in V.V. Beletsky's book.

[1] Translation of Priroda, 1973, no. 10, 115–117; cf. "A science book of a new style" by Ya.B. Zeldovich, Uspekhi Fiz. Nauk, **112** (1974), no. 3, 541–542.

English translation of the book: V.V. Beletsky, "Essays on the Motion of Celestial Bodies", Basel, Birkhäuser–Springer, 2001, xviii + 372 pp. The present translation of the review partially uses the material from that edition and its preface.

Originally publ. in the popular scientific magazine "Priroda" 10, 115-117, © Priroda. Russian Academy of Sciences 1973
Translated by Boris Khesin and Andrey Khesin.
Springer-Verlag Berlin Heidelberg 2016 with kind permission.

The author of the book uses an integral sign to pull dynamic trajectories out of the "Chest of Oblivion". The image contains an allusion to S.Ya. Marshak's verse "Chest" (Image by I.V. Novozhilov, Dr. Phys.-Math. Sc.)

As the author rightly notes, the names of the specialists working in the applied areas, whose successes in the development of space technology and the conquest of cosmic space have stimulated the progress of many exact sciences, are not commonly found in the pages of scientific journals; these people, as a rule, do not write books and only rarely defend dissertations. But without their efforts the development of mechanics, as reflected in many dissertations, would have been unthinkable, and many books would have remained unwritten, among them the present "Essays on the Motion of Celestial Bodies". For that reason, from the mouth of its author, a well-known specialist in the mechanics of space flight, the ordinary-sounding words "This book is a modest attempt to pay tribute to my professors, friends, and colleagues" are filled with a special importance.

To give a very rough idea of the richness of the book's content and its character, let us mention the names of the essays (briefly explaining them, where especially necessary) and the epigraphs to several of them.

1. On the unperturbed and perturbed motion of a satellite, with a digression on asymptotic methods of nonlinear mechanics.

2. On the rebirth of an old problem, or what happens if two masses are placed at a purely imaginary distance from one another.

> ... and the more he looked at the bell-rope, the more he felt that he had seen something like it, somewhere else, sometime before.
>
> A.A. Milne, *The World of Pooh*

The struggles of learning. Quoting the book: “Intelligent beings emerging on a planet belonging to the system of a double star would be placed by Nature in worse conditions, compared with the humans inhabiting the Earth, when it comes to understanding its laws.” (Image by I.V. Novozhilov, Dr. Phys.-Math. Sc.)

3. Yet another reincarnation of an old problem.

This essay describes an extreme case of the problem of attraction of two fixed centers (where one of the centers is shifted to infinity and is generating a homogeneous field).

4. Motion of the worlds.

> The Old Grey Donkey, Eeyore, stood by himself in a thistly corner of the forest, ... and thought about things. Sometimes he thought sadly to himself, “Why?” and sometimes he thought, “Wherefore?” and sometimes he thought, “Inasmuch as which?”
>
> A.A. Milne, *The World of Pooh*

The essay is devoted to problems of the stability of motion, to Laplace’s theorem on stability of the Solar System, to the results obtained in that area after the fundamental work by A.N. Kolmogorov.

5. The restricted three-body problem and a flight to the Moon.

> *Mephisto*: ... O Moon, for you — my kiss. (Spreading his dark red cape on the ground, he sits down on it, and flies away with a whir through the window, which opens widely with a noise.)
>
> A.V. Lunacharskiĭ, *Faust and the City*

The donkey Eeyore is downcast by the "pathetic view" of an invariant torus, which so frequently appears in the works of one of the reviewers (Image by I.V. Novozhilov, Dr. Phys.-Math. Sc.)

6. They are waltzing in orbits.
The essay talks of several notable effects in the rotation and orientation of satellites.[2]

7. In a spiral to space.

> Imagine a dazzling, elastic spiral, whirling away into the sky!...
> A.A. Voznesenskiĭ, *The Triangular Pear*

This essay is about acceleration of a satellite under the action of the traction force directed along its orbit.

8. The Sun fills the sails.
This essay presents a scientific commentary to Arthur C. Clarke's short story "The Wind from the Sun" describing a race of space yachts with solar sails. The solar wind is actually taken into account when calculating orbits of some satellites.

[2] By the way, in this essay the author suggests making space apparatuses in the shape of animals that move around on their snouts: *Rhinogradentia otopteryx volitans*, discovered by H. Stümpke and described in his monograph "Body Structure and Life of Rhinogradentia" (Stuttgart, 1967). A more accessible exposition on the subject of rhinogradentia can be found in the journal "Nauka i Zhizn′", 1963, no. 4. Quoting H. Stümpke, V.V. Beletsky indicates that "the existence of literature about rhinogradentia is not a sufficient condition for the existence of rhinogradentia" (pp. 171 and 348).

9. The gravity flyer.

> ... He swam, although that was extremely stupid — swimming in space, where there is nothing to push away from. He threw his arms forward and moved his legs the way a swimmer does...
>
> V.D. Mikhaĭlov, *Amid the Stars*

In this essay, while discussing the effect of variations of gravity forces resulting from changes in the shape of the attracted body, the author writes, in passing, "Speaking in principle, an astronaut that was left behind by a satellite and lost his individual rocket engine may catch up with this satellite by modifying his orbit using purposeful pulsations. Let him swim breaststroke! Thus he will have a chance of survival. But do not ask me how much time this will take" (p. 254).

Attached to this essay correspondence between the author and the readers of the magazine "Tekhnika — Molodezhi", where an article about a gravity flyer was published, makes a strong impression. One of the readers, not understanding the concept of a gravity flyer, writes indignantly, "The discussion provided in this paper is based on plain ignorance and testifies to the wrong practice of awarding scientific titles... The cunning fabrications of Professor V.V. Beletsky, Dr. Phys.-Math. Sc., came about as a result of his complete lack of understanding of the physical laws of celestial mechanics, in which he, apparently, 'drifts with breaststroke' " (p. 254).

10. Interplanetary flights: low thrusts for high goals.

> We must carry to other planets
> The gospel of our little Earth!
>
> V.Ya. Bryusov, *Childhood Hopes*

11. Relative motion of orbiting bodies.

12. Cosmic pinwheel.

> — Please, señor, — countered Sancho Panza, — those are not so much giants as windmills.
>
> Cervantes, *Don Quixote*

In the final essay the author presents a problem which would have been unthinkable in old, pre-satellite mechanics. It is about an unusual evolution of the rotation and orientation of "Proton"-type satellites related to small aerodynamic forces.

For a solid scientific monograph, the style of V.V. Beletsky's book is out of ordinary in many respects. Without exaggeration one can say that it marks the beginning of a new style in the scientific literature. The author explains in a frank and detailed manner the reasons behind each calculation, its difficulties, and the psychological side of the research. The book contains no attempts to look important pompously, to present results while hiding the methods used to obtain them. The book is adorned and brightened up by humorous illustrations by I.V. Novozhilov, Dr. Phys.-Math. Sc.; some of them have been reproduced here. Not a boring lecture, but a discussion

The Sun fills the sails! (a space flight by means of a solar sailboat) (Image by I.V. Novozhilov, Dr. Phys.-Math. Sc.)

with a brilliant, knowledgeable, and witty companion — that is the general impression from the "Essays". Even a person remote from space problems would enjoy going through this book, skipping, possibly, the calculations.

Now, obviously, it is necessary to consider the book's shortcomings.

Being a physicist, one of the reviewers sees a major flaw in that the book contains little physics. Right from the start the problems are stated as systems of Newton's equations in three Cartesian coordinates. Probably, for computer calculations this is, indeed, the most straightforward approach. However, the physicists have developed a whole set of concepts such as energy, momentum, resonance, parametric resonance. Each concept has its own (precise or approximate) mathematical meaning. Computer calculations automatically satisfy the energy conservation law for forces independent of time. Having a good command of mathematics makes it possible in certain cases to avoid using physics notions. But should one do that, especially in the matter of presenting and explaining results?!

The other reviewer has discovered that in the captivatingly interesting problem "Leonov and the lens cap" (which, by the way, would decorate any course on differential equations), the cap that is thrown towards the Earth by the astronaut falls behind the satellite, while in fact the cap has to overtake the satellite, as follows from the momentum conservation law (pp. 306–309). That same reviewer must confess that he was not able to understand the section "Quantization of stable motions".

These remarks, of course, are not that crucial, but, perhaps, could come useful while preparing the second edition of this book, which is absolutely necessary as there were only four thousand copies of V.V. Beletsky's "Essays" printed!

A parody of V.V. Beletsky's peculiar imaginary spaceship — "the gravity flyer" (Image by I.V. Novozhilov, Dr. Phys.-Math. Sc.)

The famous 19th century Frenchman Camille Flammarion published in succession "Astronomy", "Astronomy for All", and "Astronomy for Ladies". Shouldn't V.V. Beletsky and the "Nauka" publishing house follow that same path when publishing the next editions of "Essays"?

A Science Book of a New Style

Ya.B. Zeldovich

(A review of V.V. Beletsky's book "Essays on the Motion of Celestial Bodies", Moscow, Nauka Publishing House, 1972, 360 pp.[1])

In my opinion the book under review is outstanding because of its take on the following three principal benchmarks: firstly, this is an important topic, secondly, it exposes novel and often unexpected results, and finally, thirdly, its exposition style is bright and quite unusual.

Half of mankind was born after the launch of the Sputnik, the first artificial Earth satellite. In the space age there is no need to prove the importance of the problem of motion of objects in space, and most of all, the flight of artificial space objects. Therefore I am not going to discuss here the importance of this topic. However, the richness of the content and abundance of new results in the book require some comments.

One might think that the problem of body motion in the gravity field of the Earth or in the Solar System is so simple and so thoroughly studied that it is difficult to look for anything new in this problem. However, the following rule often observed in other sciences is also working here: practical applications require more detailed and precise knowledge and pose a problem of finding optimal modus operandi. In the beginning one only encounters complications in a given problem and loses the clarity and beauty of simplified theories. However, at the next step there appear both new points of view and novel theories describing the results of more detailed computations, and new beauty arises as well. To avoid being proofless I will give the titles of chapters (or essays, as the author named them): 1. On unperturbed and perturbed motion of satellites with a digression on asymptotic methods of nonlinear

[1] Translation of Uspekhi Fiz. Nauk, **112** (1974), no. 3, 541–542; cf. "The beginning of a new style in the scientific literature" by V.I. Arnold and Ya.B. Zeldovich, Priroda, 1973, no. 10, 115–117.

English translation of the book: V.V. Beletsky, "Essays on the Motion of Celestial Bodies", Basel, Birkhäuser–Springer, 2001, xviii + 372 pp.

Originally publ. in: Usp. Fiz. Nauk 112(3), 541-542, © Lebedev Physical Institute, Russ. Acad. Sciences 1974
Translated by Boris Khesin and Andrey Khesin.
Springer-Verlag Berlin Heidelberg 2016 with kind permission.

mechanics; 2. On the second birth of an old problem, or what happens if two masses are located at a purely imaginary distance from each other? 3. Another old problem returned to a new life; 4. Motion of the worlds; 5. The restricted three-body problem and a flight to the Moon; 6. They waltz in orbits; 7. To the space along a spiral; 8. Sails are full of the Sun; 9. Gravity flyer; 10. Interplanetary flights: low thrusts for high goals; 11. Relative motions of orbital bodies; 12. Space turntable.

Here one examines, e.g., the weak forces of light pressure as well as the problem of motion in the gravity field of several bodies. What is very interesting is the consideration of resonance phenomena in a system of several bodies.

Of particular interest is the possibility to change the trajectory of a celestial body by means of the internal motions, namely thanks to a periodic change of the quadrupole moment ("gravity flyer"). I have to warn the review's reader that the effect exists due to the fact that the gravity field of the Earth depends on the coordinates and cannot be reduced to a constant one. In other words, it contains tidal forces. Without this remark it might have seemed that a gravity flyer is impossible, but actually this effect is small but nonzero (it involves the square of the ratio of the body size to the Earth radius). The author cites several letters from angry but not sufficiently thoughtful readers of previously published papers, where a gravity flyer was described. There are also letters with strong statements about "ignorance and bad practices", "it is long overdue to expose some science doctors", etc.

Now let us discuss the book's style. The style is quite unusual for a solid scientific monograph. The author explains frankly and in detail the motives for each calculation, its difficulties, and every time emphasizes the psychological side of the study. There is no trace of ranting importance in the book or of an attempt to present results while hiding the corresponding methods. Among beautiful and lively decorations of the book there are humorous drawings and illustrations by I.V. Novozhilov, Doctor of Phys.-Math. Sciences, as well as unexpected but relevant epigraphs, from R.I. Rozhdestvensky to M.A. Bulgakov, in verse and prose. The general impression after reading the book is that this is not a dull lecture but rather a chat with a bright, knowledgeable, and witty companion. Even a person quite remote from space problems would browse through the book with pleasure, possibly skipping the calculations.

The famous XIX century French author, Camille Flammarion, published successively "Astronomy", "Astronomy for All", and "Astronomy for Ladies". Wouldn't the book's author V.V. Beletsky and its publisher like to consider urgently publishing the second edition of the book, as well as, probably, the Soviet equivalent of "Celestial Bodies for Ladies"? Anyway, now it is not easier to acquire the book by Beletsky than "Dangerous Liaisons" by P. Choderlos de Laclos, a volume by B.L. Pasternak, or "Airport" by A. Hailey. Even for this review I am using a copy specifically lent to me by the publisher.

Evidently, one has to move to the discussion of the drawbacks of the book, at least to prove that the reviewer has read it (in difficult conditions!).

Being a physicist the reviewer sees the main shortcoming of the book in the shortage of physics there. From the very beginning, the problems are formulated as systems of Newton's equations in three Cartesian coordinates: $\ddot{x} = f(x,y,z,t),\dots$ Apparently for computer calculations this is the most straightforward way. However, the physicists have developed a whole bunch of notions such as energy, momentum, resonance, parametric resonance, etc. Each notion has a particular (precise or approximate) mathematical meaning. Computer calculations automatically satisfy the energy conservation law for time-independent forces. Having mastered mathematics one can possibly get by without physical notions. But is it worth doing that when one discusses and explains the results?!

One specific suggestion: it would be worth to pay particular attention to the fact that in the central field of Newton's gravity of a single point mass, the orbits are closed ellipses. As is well known, this fact does not follow from the energy and momentum conservation laws, but is a feature of the potential $1/r$.

In the hydrogen atom and in the Solar System, small deviations of the orbit from closedness (the precession of Mercury's perihelion and the Lamb energy shift) become of great, principal importance. Further developments of the atomic theory (quantum electrodynamics) and the theory of gravity (Einstein's theory of general relativity) are confirmed by these small effects.

Thus, one needs the second revised and extended edition of the remarkable "Essays" by V.V. Beletsky, marking the beginning of a new style of scientific literature (as the reviewer very much hopes for). And until then let the readers of this review envy the readers and owners of the first edition of that book.

On the First All-Union Mathematical Student Olympiad

V.I. Arnold, A.A. Kirillov, V.M. Tikhomirov, M.A. Shubin

(Translation of Uspekhi Mat. Nauk, **30** (1975), no. 4, 281–288)

On October 23–24, 1974, there occurred the third and final round of the All-Union olympiad "The student and progress of science and technology". This was the first All-Union student olympiad; it was on several different disciplines, among them, mathematics.

The third round of the subject olympiad in mathematics was held at the Department of Mechanics and Mathematics of the Moscow State University. To hold the competition, a jury chaired by P.S. Alexandrov was created from the faculty of the Department. Many professors and graduate students of the Department proposed problems and graded the solutions. The final selection of the problems was done by the authors of this article.

There were 18 teams at the olympiad, namely, two teams from the RSFSR,[1] one team from each of the other Union republics, and one team from each of Moscow and Leningrad. In total, 89 students participated in the olympiad. Note that all of the proposed problems were solved by the olympiad's participants.

Here is a list of the top 10 participants:

1. S.V. Konyagin (Moscow, Moscow State Univ., Mech.-Math., 2nd year),
2. A.B. Alexandrov (Leningrad, Leningrad State Univ., Math.-Mech., 4th year),
3. V.V. Sologubov (RSFSR-2, Moscow Phys.-Tech. Inst., 2nd year),
4. A.N. Drobotenko (Moldavia, Kishinev Polytech. Inst., 3rd year),
5. S.E. Belkin (Moscow, Moscow State Univ., Mech.-Math., 3rd year),
6. E.M. Matveev (Moscow, Moscow State Univ., Mech.-Math., 3rd year),
7. A.L. Figotin (Ukraine, Kharkov Univ., Mech.-Math., 4th year),
8. E.V. Sallinen (Leningrad, Leningrad State Univ., Math.-Mech., 4th year),
9. N.V. Kartashev (Ukraine, Kiev Univ., Mech.-Math., 4th year),
10. A.M. Shteĭnberg (RSFSR-1, Ural Univ., Mech.-Math., 4th year).

[1] Russian Soviet Federative Socialist Republic (Translators' note)

Originally publ. in: Usp. Mat. Nauk 30(4), 281-288, © Russ. Acad. Sciences 1975
Translated by Boris Khesin and Andrey Khesin.
Springer-Verlag Berlin Heidelberg 2016 with kind permission.

In the team standing, Moscow and Leningrad shared the first and second places, with Ukraine coming in third.

Below we present the problems of the third round of the All-Union mathematical olympiad, as well as their solutions. In total there were 11 problems: five on the first day and six on the second.

Day 1

1. *Prove that functions* $f_1, \dots, f_n$ *are linearly independent if and only if there exists a set of points* $t_1, \dots, t_n$ *such that* $\det \|f_i(t_j)\|_{i,j=1}^n \neq 0$.

Solution. If $\det \|f_i(t_j)\|_{i,j=1}^n \neq 0$, then it is clear that $f_i(t)$ are linearly independent. Let us prove the converse by induction. For $n = 1$ it is evident. Suppose that the statement is true for $n = k$ and prove it for $n = k+1$. Let us find k points $t_1, \dots, t_k$ such that $\det \|f_i(t_j)\|_{i,j=1}^k \neq 0$. Suppose that there is no $(k+1)^{\text{th}}$ point with the required property. This means that

$$\det \begin{Vmatrix} f_1(t_1) & \dots & f_{k+1}(t_1) \\ \dots\dots & \dots & \dots\dots\dots \\ f_1(t_k) & \dots & f_{k+1}(t_k) \\ f_1(t) & \dots & f_{k+1}(t) \end{Vmatrix} \equiv 0.$$

Expanding the determinant along the last row we get

$$f_1(t)C_1 + \dots + f_{k+1}(t)C_{k+1} \equiv 0,$$

where

$$C_{k+1} = \det \|f_i(t_j)\|_{i,j=1}^k \neq 0,$$

by the inductive hypothesis. We have arrived at a contradiction with our assumption of linear independence. Thus, the statement of the problem is true for $n = k+1$.

This problem was fully solved by 25 people. 12 people did not attempt it. The rest gave incomplete or incorrect solutions.

2. *The function* $f(t)$ *is defined over* $[0, +\infty)$ *and has a continuous derivative.*

a) *It is known that* $f(t) + f'(t) \to A$ *as* $t \to +\infty$. *Prove that* $f(t) \to A$ *as* $t \to +\infty$.

b) *It is known that* $f(t) - f'(t) \to A$ *as* $t \to +\infty$. *Is it true that* $f(t) \to A$ *as* $t \to +\infty$?

Solution. a) Subtracting a constant, we can simplify the problem to $A = 0$. Let $f(t) + f'(t) = g(t)$. Then

$$f(t) = Ce^{-t} + e^{-t} \int_0^t e^{\tau} g(\tau)\,\mathrm{d}\tau = Ce^{-t} + e^{-t} \left[\int_0^T e^{\tau} g(\tau)\,\mathrm{d}\tau + \int_T^t e^{\tau} g(\tau)\,\mathrm{d}\tau \right].$$

Thus

$$|f(t)| \leqslant |C|e^{-t} + e^{-t+T} \max_{t\geqslant 0} |g(t)| + \max_{t\geqslant T} |g(t)|,$$

from which the problem's statement a) follows.

b) It is not true that necessarily $f(t) \to A$. Indeed, if $f(t) = e^t$, then $f(t) - f'(t) = 0$, but e^t does not approach 0 as $t \to +\infty$.

This problem was fully solved by 19 people. 45 people managed to solve part b), while 11 people did not attempt the problem.

3. *The function $f(x)$ is defined over $[0, +\infty)$, has a continuous derivative, and satisfies the inequality $\int\limits_0^{+\infty} |f'(x)|\,\mathrm{d}x < +\infty$. Prove that the series $\sum\limits_{n=0}^{\infty} f(n)$ converges if and only if the integral $\int\limits_0^{\infty} f(x)\,\mathrm{d}x$ converges.*

Solution. The Newton–Leibniz formula and the mean-value theorems imply that for $0 \leqslant \theta \leqslant 1$ one has

$$\left| \int\limits_n^{n+\theta} f(x)\,\mathrm{d}x - \theta f(n) \right| \leqslant \theta \int\limits_n^{n+1} |f'(x)|\,\mathrm{d}x.$$

Therefore for $0 < u < v$

$$\left| \int\limits_u^v f(x)\,\mathrm{d}x - \sum_{n=[u]+1}^{[v]-1} f(n) - \theta_1 f([u]) - \theta_2 f([v]) \right| \leqslant \int\limits_{[u]}^{[v]+1} |f'(x)|\,\mathrm{d}x,$$

$$0 \leqslant \theta_j \leqslant 1 \qquad (j = 1, 2).$$

Whence the required statement is obtained by applying the Cauchy criterion.

This problem was solved by 14 participants. 31 participants did not even attempt the problem.

4. *Evaluate $\int\limits_1^{10} x^x\,\mathrm{d}x$ with a relative error of no more than 1% (it is required to give the answer, not to prove the estimate of the error;* $\ln 10 \approx 2.3026$).

Answer. $\int\limits_1^{10} x^x\,\mathrm{d}x \approx \dfrac{10^{10}}{1 + \ln 10} \approx 0.30 \cdot 10^{10}.$

Here is the reasoning:

$$\int_1^{10} x^x\,\mathrm{d}x = \int_1^{8} x^x\,\mathrm{d}x + \int_8^{10} x^x\,\mathrm{d}x = A + B; \qquad A < \int_1^{8} 8^x\,\mathrm{d}x < \frac{8^8}{\ln 8} < 10^7;$$

$$B = \int_8^{10} x^x\,\mathrm{d}x = \int_8^{10} \frac{\mathrm{d}(e^{x\ln x})}{1+\ln x} = \left.\frac{x^x}{1+\ln x}\right|_8^{10} + \int_8^{10} \frac{x^x\,\mathrm{d}x}{(1+\ln x)^2 x}$$

$$\approx \frac{10^{10}}{1+\ln 10} + \left.\frac{x^x}{(1+\ln x)^3 x}\right|_8^{10} \approx \frac{10^{10}}{1+\ln 10} \approx 0.30 \cdot 10^{10}$$

(in the computation above, we have omitted values giving an error of under 1%).

Only two participants solved this problem: A.B. Alexandrov (Leningrad) and P.È. Normak (Estonia). 50 people did not attempt to solve the problem.

5. *Let coordinate axes be chosen in a three-dimensional space, and let V_1 be a convex polyhedron, all vertices of V_1 lying on the integer lattice. Denote by V_k the polyhedron obtained from V_1 by multiplying by k the radii vectors of all of the points of V_1. Let $N(V)$ denote the number of lattice points that lie inside of polyhedron V or on its surface, and let $\mu(V)$ denote the volume of polyhedron V. Prove that*

$$N(V_3) - 3N(V_2) + 3N(V_1) - 1 = 6\mu(V_1). \tag{1}$$

Solution will be carried out by induction over the dimension. On a line the following formula is true:

$$N(V_1) - 1 = l(V_1), \tag{2}$$

where $l(V_1)$ is the length of the line segment V_1 whose endpoints have integer coordinates. Let us show that on a plane for a convex polygon V_1 with vertices in the integer lattice points, the following formula holds:

$$N(V_2) - 2N(V_1) + 1 = 2S(V_1), \tag{3}$$

where $S(V_1)$ is the area of V_1. From equation (2) it follows that (3) holds for degenerate two-dimensional polygons, namely, line segments. This immediately implies that if polygon V_1 were cut by a diagonal into two pieces V_1' and V_1'', then

$$N(V_2) - 2N(V_1) + 1 = \left[N(V_2') - 2N(V_1') + 1\right] + \left[N(V_2'') - 2N(V_1'') + 1\right].$$

Thus one only needs to prove (3) for triangles.

Let us put the origin O at a vertex of the lattice triangle V_1 (Fig. 1) and let A and B be the triangle's two other vertices. Let $OA'B'$ be the triangle V_2. From Fig. 1 it is clear that the left-hand side of equation (3) is equal to $N(\Pi)$, the number of lattice points in the shape Π, which is the parallelogram $OACB$ without sides AC

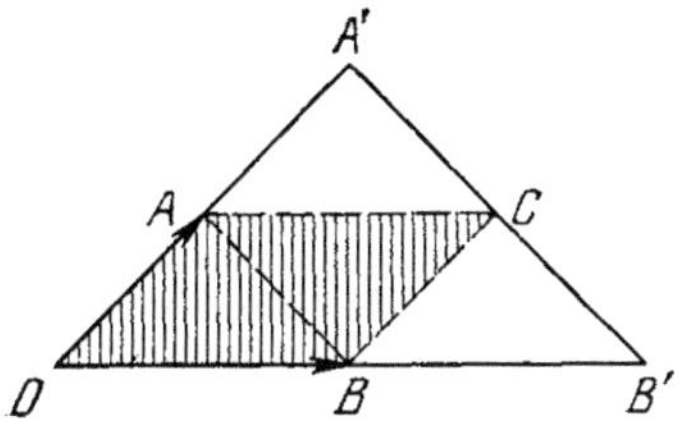

Fig. 1

and BC. If $S(\Pi)$ is the area of parallelogram $OACB$ (equal to $2S(V_1)$), then to prove (3) it remains to verify that $N(\Pi) = S(\Pi)$. Note that shifting Π to lattice points we can tessellate the entire plane (without intersections). Moreover, a square with side length M intersects with $\dfrac{M^2}{S(\Pi)} + O(M)$ Π-parallelograms. The number of lattice points in these parallelograms is equal to $N(\Pi)\dfrac{M^2}{S(\Pi)} + O(M)$. On the other hand, the number of lattice points in the square under consideration is equal to $M^2 + O(M)$. Thus, $N(\Pi) = S(\Pi)$, which proves formula (3).

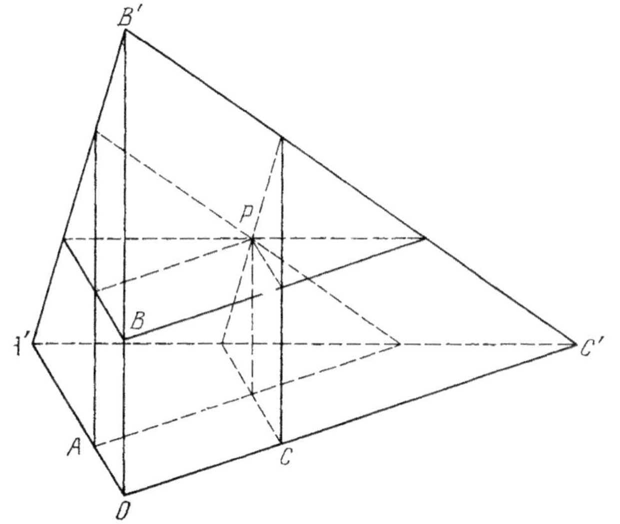

Fig. 2

Now consider the three-dimensional case. From (3) it follows that it is sufficient to prove equation (1) for tetrahedra only, where the proof is analogous to the two-dimensional case but is based on a more complicated drawing (Fig. 2). From the polyhedron $V_3 = OA'B'C'$ we exclude three polyhedra of type V_2, located "at the corners" A', B', C', then we add their pairwise intersections (of type V_1), and exclude one extra point P. The remainder is the parallelepiped Π with a volume of $6\mu(V_1)$ without a part of the surface. These parallelepipeds can tessellate all space without intersections. Therefore, $N(\Pi) = \mu(\Pi)$, which proves the desired statement.

This problem was solved only by È.P. Turkevich (Ukraine). Several participants were on the right track. 67 people did not attempt the problem.

Day 2

1. *Let $f(x)$ be a continuous function over $[0,1]$ and let $\varphi(x)$ be a continuous periodic function over $(-\infty,+\infty)$ with a period T. Prove that*

$$\lim_{n\to+\infty}\int_0^1 f(x)\varphi(nx)\,\mathrm{d}x = \frac{1}{T}\int_0^T \varphi(t)\,\mathrm{d}t \cdot \int_0^1 f(x)\,\mathrm{d}x. \tag{4}$$

Solution. Let $\chi_{[\alpha,\beta]}(x)$ be the function equal to 1 for $x \in [\alpha,\beta]$ and to 0 for $x \notin [\alpha,\beta]$. Let us prove equation (4) for $f(x) = \chi_{[\alpha,\beta]}(x)$, where $[\alpha,\beta] \subset [0,1]$. We get

$$\int_0^1 \chi_{[\alpha,\beta]}(x)\varphi(nx)\,\mathrm{d}x = \int_\alpha^\beta \varphi(nx)\,\mathrm{d}x = \frac{1}{n}\int_{\alpha n}^{\beta n} \varphi(x)\,\mathrm{d}x \to (\beta-\alpha)\frac{1}{T}\int_0^T \varphi(x)\,\mathrm{d}x$$

as $n \to +\infty$.

Now, to prove (4) in the general case we need to uniformly approximate the function $f(x)$ by piecewise constant functions.

This problem was solved by 17 people. 25 students did not attempt it.

2. *What is the largest number of normals that one can draw to an ellipse from a point outside of it? Describe and draw the set of points outside of the ellipse with the maximum number of normals.*

Solution. Write the equation for an ellipse in a parametric form: $x = a\cos t$, $y = b\sin t$. The normal passing through the point $(a\cos t, b\sin t)$ on the ellipse has the equation

$$ax\sin t - by\cos t = (a^2-b^2)\sin t\cos t.$$

The envelope of this family of normals can be determined parametrically by combining this equation with the relation obtained by differentiating this equation in t:

$$ax\cos t + by\sin t = (a^2-b^2)\cos 2t.$$

Excluding the parameter t, we arrive at the equation for an astroid:

$$(ax)^{2/3} + (by)^{2/3} = (a^2-b^2)^{2/3}.$$

Note that the transition from one number of normals to another can only happen on this astroid. From this it is easy to verify that one can draw 4 normals to the ellipse from any point that lies inside the astroid; there are 2 normals from any point outside of the astroid, as well as from its vertices; while there are 3 normals from points on the astroid itself (except for the vertices).

If $a^2 > 2b^2$, then the astroid goes beyond the ellipse (Fig. 3). In Fig. 3, the domain to be drawn in the problem is shaded.

This problem was fully solved only by S.E. Belkin (Moscow). 39 students did not attempt the problem.

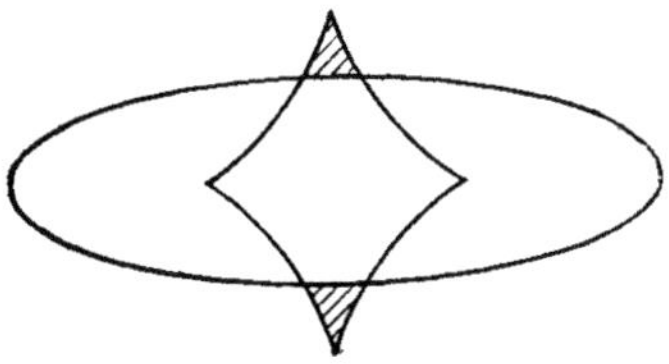

Fig. 3

3. *Given two trigonometric polynomials $p(t)$ and $q(t)$, prove that there exists a non-zero polynomial $R(x,y)$ such that $R(p(t),q(t)) \equiv 0$.*

Solution. Let $p(t)$ and $q(t)$ be algebraic polynomials in e^{it} and e^{-it} of a degree no greater than m. Let $R(x,y)$ be an algebraic polynomial of degree N in two variables.

Then $R(p(t),q(t))$ is a trigonometric polynomial, which is an algebraic polynomial in e^{it} and e^{-it} of a degree no greater than mN. If we were to take a polynomial $R(x,y)$ with undefined coefficients and to set the coefficients of e^{ikt} to zero in the expression for $R(p(t),q(t))$, then we would get $2mN+1$ linear homogeneous equations on $\frac{1}{2}(N+1)(N+2)$ coefficients of the polynomial R. Consequently, for a large N the number of unknowns will outgrow the number of equations, from which the existence of a desired polynomial R follows.

This problem was solved by 11 people. 51 people did not attempt to solve it.

4. *Consider a sequence of balls whose radii tend to 0, but whose total volume is infinite. Prove that it is possible to fit a finite number of these balls inside a cube so that they would fill 99% of the cube's volume.*

Solution. It is sufficient to prove the following statement: there exists a number α, $0 < \alpha < 1$, such that for any "vessel" G (a region with a piecewise smooth boundary) it is possible to insert a finite number of balls of the given sequence into the vessel in such a way that they would fill a volume no less than $\alpha\mu(G)$, where $\mu(G)$ is the volume of vessel G. Indeed, if this is true, then we can place balls into the initial cube in such a way that there will remain an empty space of volume $\leqslant (1-\alpha)\mu_0$ (μ_0 being the volume of the initial cube). Regarding the remaining unfilled part of the cube as a new vessel, we once again fill an αth part of it and so on. After n steps, the unfilled volume will be $\leqslant (1-\alpha)^n \mu_0 \to 0$ as $n \to +\infty$.

Since any vessel can be filled with cubes with an arbitrarily small unfilled volume, we see that it suffices to learn how to fill an αth part of any cube. Here is one of the possible algorithms for such a filling. Suppose we want to fill a cube with side length 2 with balls. Take a cube with side length 1 whose center is at the center of the initial cube. Sort the balls in the order of decreasing radius, and let their radii be $r_1 \geqslant r_2 \geqslant \ldots$ Choose a number N such that $r_N < 1/2$ and start placing the consecutive balls no. $N, N+1, \ldots$ by positioning their centers inside of the unit cube. The process will inevitably terminate on some M^{th} step as the total volume of all of the balls is infinite. So, suppose that we have managed to fit the balls of radii $r_N, r_{N+1}, \ldots, r_{N+M}$. Let us show that the balls with the same centers and doubled radii cover the unit cube in its entirety. Indeed, if we could find a point in the unit cube that did not get covered by the balls with doubled radii then this means that the distance from this point to any ball is $\geqslant r_{N+M}$, which implies that we could have fit one more ball with its center at that point and with radius $r_{N+M+1} \leqslant r_{N+M}$. All constructed balls obviously lie within the cube with side length 2. Thus, if V is the total volume of the balls we embedded then $8V > 1$ and so $V > 1/8$, which comprises $1/64$ of the volume of the cube with side length 2.

Three people solved this problem: S.E. Belkin (Moscow), S.V. Konyagin (Moscow), and V.V. Sologubov (RSFSR-2). 60 people did not attempt the problem.

5. *A vertex of a curve is a point where the curvature has an extremum. Prove that any smooth (infinitely differentiable) closed convex curve admits at least four vertices.*

Solution. If $(x_1(s), x_2(s))$ is the equation for the given curve Γ with an arc-length parameter s, then, as is well known,

$$\ddot{x}_1(s) = -k(s)\dot{x}_2(s), \qquad \ddot{x}_2(s) = k(s)\dot{x}_1(s).$$

It follows that for any numbers a_0, a_1, and a_2

$$\int\limits_\Gamma (a_0 + a_1 x_1 + a_2 x_2)\frac{\mathrm{d}k}{\mathrm{d}s}\,\mathrm{d}s = a_0 \int\limits_\Gamma \mathrm{d}k - a_1 \int\limits_\Gamma \ddot{x}_2\,\mathrm{d}s + a_2 \int\limits_\Gamma \ddot{x}_1\,\mathrm{d}s = 0. \tag{5}$$

The curvature $k(s)$ has a maximum and minimum on Γ. If they were the only vertices then by drawing the straight line $a_0 + a_1 x_1 + a_2 x_2 = 0$ through them we would arrive at a contradiction with equation (5), since $(a_0 + a_1 x_1 + a_2 x_2)\dfrac{\mathrm{d}k}{\mathrm{d}s}$ does not change sign in this case.

This problem also allows the following simple geometric solution (suggested by S.M. Guseĭn-Zade).

Let us inscribe a circle S of maximum possible radius r into the curve Γ (Fig. 4). Let A and B be points of tangency of the curve Γ and the circle S so that there are no other tangency points on the arc AB. It is easy to see that $\smile AB \leqslant \pi$. Moreover, the

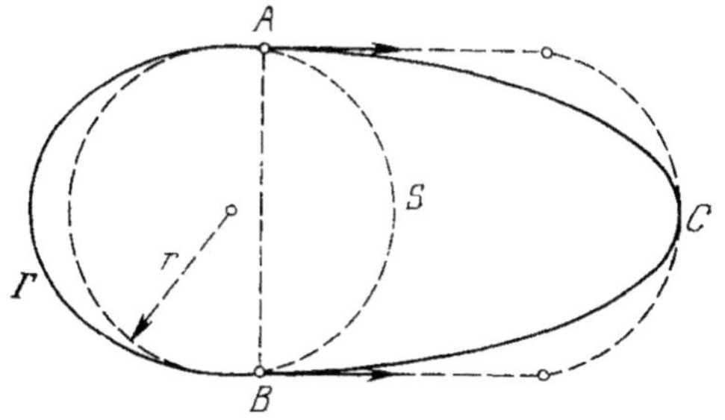

Fig. 4

curvature of Γ at points A and B is no larger than r^{-1} and is strictly less than r^{-1} at some points near A and B. Let us shift the arc AB of the circle S orthogonally to the line segment AB in the direction of its convexity by the maximum possible distance so that this arc still intersects the curve Γ (at point C). In this case C is a point of tangency of the arc and the curve Γ, and the curvature of Γ at the point C is no less than r^{-1}. This immediately implies the existence of at least two local maxima of the curvature and hence the existence of two minima.

Only one participant managed to solve this problem: A.N. Drobotenko (Moldavia), while 68 people did not attempt it.

6. *Does there exist a convex polyhedron with* 6 *vertices and* 15 *edges in the four-dimensional space*?

Solution. Such a polyhedron does exist. It is sufficient to take any two triangles in the planes $(x,y,0,0)$ and $(0,0,z,u)$ containing the origin and then to consider their convex hull. As an example let us take the points

$$A_1 = (1,0,0,0), \quad A_2 = (-1,1,0,0), \quad A_3 = (-1,-1,0,0),$$
$$A_4 = (0,0,1,0), \quad A_5 = (0,0,-1,1), \quad A_6 = (0,0,-1,-1).$$

To prove that these are the points we were looking for, it suffices for any pair of points A_i and A_j to find a linear function $\alpha x + \beta y + \gamma z + \delta u$, which attains larger values at the points A_i and A_j than at any others. Obviously, it is sufficient to examine just the pairs A_i and A_j with $i \leqslant 3$, $j > 3$. The required linear forms are easy to find. For instance, $-x+y+2z$ for A_2 and A_4; $-x+y-z+u$ for A_2 and A_5; $-x+y-z-u$ for A_2 and A_6.

This problem was only solved by A.B. Alexandrov (Leningrad) and 59 people did not attempt it.

Remark. The last problem reflects a rather particular case of an interesting phenomenon: *in the four-dimensional space for any n it is possible to find a convex polyhedron with n vertices such that from any vertex one can "see" any other vertex, that is the line segments connecting any two vertices are edges of this polyhedron.*

As an example of such a polyhedron, one can take the convex hull of any n distinct points lying on the curve

$$x(t) = (t, t^2, t^3, t^4), \qquad -\infty < t < +\infty.$$

This curve possesses the following remarkable property: if $x_1 = x(t_1)$ and $x_2 = x(t_2)$ are two distinct points on the curve then there exists a three-dimensional hyperplane passing through them such that all the other points lie strictly on one side of it. Indeed, let us draw the hyperplane of type $l(x) = c$ (l is a linear form) through points x_1 and x_2, and parallel to the tangent vectors to the curve at both points (it is easy to verify that a vector parallel to the chord x_1x_2 and these tangent vectors are linearly independent). Then $l\big(x(t)\big) - c$ is a polynomial of degree four with two multiple roots, t_1 and t_2, and therefore $l\big(x(t)\big) - c$ preserves its sign.

ХРОНИКА

РЕГИОНАЛЬНАЯ МАТЕМАТИЧЕСКАЯ ШКОЛА В СЫКТЫВКАРЕ

В. И. Арнольд, А. М. Вершик, Д. Б. Фукс, Я. М. Элиашберг

С 15 по 25 марта 1976 г. в Сыктывкаре состоялась Региональная математическая школа по глобальному анализу, дифференциальной топологии и теории особенностей. Школа была организована Сыктывкарским государственным университетом имени 50-летия СССР и привлекла многих специалистов. Кроме сыктывкарцев в работе школы приняли участие 39 математиков из Москвы (14 чел.), Ленинграда (18 чел.), Горького (3 чел.), Новосибирска (3 чел.) и Петрозаводска (1 чел.)

Программа школы была очень насыщенной. Ежедневно читались три, а иногда и четыре двухчасовые лекции.

В школе были прочитаны циклы обзорных лекций по особенностям гладких отображений (В. И. Арнольд, 16 часов), по теории вариации слоений (Д. Б. Фукс, 6 часов), по комбинаторным классам Понтрягина (А. М. Габриэлов, 4 часа), по взвешенно-проективным алгебраическим многообразиям (И. В. Долгачев, 4 часа), по теории узлов (О. Я. Виро).

Кроме того, были прочитаны доклады о многогранниках Ньютона и их приложениях (Д. А. Бернштейн), о топологии комплексных алгебраических поверхностей (Ф. А. Богомолов), о римановых многообразиях неотрицательной кривизны и выпуклых множествах (Ю. Д. Бураго), о классификации целочисленных решеток в небольших размерностях (Б. Б. Венков), о представлениях групп гладких функций со значениями в группах Ли и об асимптотической теории представлений симметрических групп (А. М. Вершик), о лагранжевых и лежандровых особенностях (В. М. Закалюкин), о фильтрации Ходжа и ее применениях (А. Н. Кириллов). Доклады сыктывкарцев Н. М. Мишачева и Я. Б. Элиашберга были посвящены хирургии особенностей слоений; в докладе Я. Б. Элиашберга было рассказано о решении давно стоявшей проблемы о неподвижных точках преобразований, сохраняющих площадь. Доклад Д. Б. Фукса содержал доказательство гипотезы Адамса, в докладе В. И. Арнольда о бифуркациях инвариантных многообразий дифференциальных уравнений и о жесткости окрестностей эллиптических кривых описаны неожиданные связи между этими двумя теориями.

Параллельно с работой школы некоторые участники прочли лекции для студентов и преподавателей университета. Так, В. И. Арнольд прочел лекции «Теория качелей» и «О построении курса дифференциальных уравнений», А. М. Вершик прочел спецкурс «Выпуклый анализ» и сделал обзор по эргодической теории.

Небольшое оставшееся от заседаний и обсуждений время участники стремились использовать для лыжных тренировок. Погода этому весьма благоприятствовала.

Безусловно, школа оказалась очень полезной для всех ее участников. Оргкомитет решил сделать в дальнейшем такие школы традиционными и проводить их в Сыктывкарском университете каждые два года.

Originally publ. in: Usp. Mat. Nauk 31(4), 301, © Russ. Acad. Sciences 1976
Reprint by Springer-Verlag Berlin Heidelberg 2016 with kind permission.

Kolmogorov's School

(Published in the newspaper "Trud", May 27, 1978, p. 2)

Recently, Andrey Nikolaevich Kolmogorov turned seventy-five. In the last several years he no longer risks dips in an ice-hole, only allowing himself to bath in snow. It is doubtful that anyone suspects that the elderly, gray-haired man, skiing in nothing but shorts along the forty kilometer trail between the Skalba and Vorya rivers or working his way barefoot to the lonely Chyornoe[1] lake through a forest near Fryazino of the Moscow region in the springtime is one of the most prominent mathematicians of modern times, a Hero of Socialist Labour, and a member of six European and American academies.

[1] meaning "Black" in Russian

Originally publ. in the newspaper "Trud", May 27, p. 2, © Newspaper "Trud", Moscow 1978
Translated by Boris Khesin and Andrey Khesin.
Springer-Verlag Berlin Heidelberg 2016 with kind permission.

Andrey Nikolaevich's road into big mathematics was not easy. He began making a living himself early on, worked as a train conductor. Kolmogorov graduated from school as an external student. He continued working as an undergraduate student.

...We are constantly surrounded by random events. Some examples are the motion of molecules and particles suspended in fluid, genetic mutations, failures of components of complicated systems, weather changes, and the load degree of telephone station channels. Probability theory plays a major role in forecasting all of these processes.

The works by Academician A.N. Kolmogorov are fundamental to the science of chance. Applications of mathematics to all of these areas allowed for the turn from a purely descriptive approach to well-defined scientific methods. In some cases, mathematical explanations of complicated processes allowed for discoveries of the patterns in many phenomena typical for microscopic objects.

Andrey Nikolaevich Kolmogorov was a pioneer and the first explorer in many areas of mathematics. He is the author of many outstanding achievements in probability theory, the theory of functions, functional analysis, topology, the theory of differential equations, the theory of dynamical systems, the turbulence theory for fluids, etc. It is difficult to indicate an area of mathematical analysis to which A.N. Kolmogorov did not make a significant contribution or where he did not solve (often two century) old problems. For probability theory and the theory of stochastic processes A.N. Kolmogorov accomplished what Euclid had done for geometry: he created a rigorous axiomatic theory, providing a reliable mathematical foundation for this area of knowledge.

A.N. Kolmogorov was always working on a wide range of applied problems. Among his approximately three hundred published works, one can find studies in mechanics, geology, biology, crystallization of metals, etc.

In our epoch of large scientific teams, it seems that the age of major individual advances in science is a thing of the past. For A.N. Kolmogorov, however, major individual accomplishments are typical. This had not prevented Andrey Nikolaevich from creating a very large scientific school in Moscow State, the university inseparable from all of his activities since 1920. Later, many of A.N. Kolmogorov's students themselves became the leaders of entire scientific fields and created new scientific schools, became academicians and corresponding members of the Academy of Sciences. Andrey Nikolaevich's students, having defended their dissertations under his supervision, number over sixty, while the number of his scientific "grandchildren" is in the hundreds, and thousands of scientists have experienced the influence of his ideas, lectures, books, and works.

Andrey Nikolaevich puts a lot of time and effort into the improvement of mathematical education in this country. "Mathematics can be taught truly well," says A.N. Kolmogorov, "only by one who is passionate about it and perceives it as a living, developing science." Having possessed the highest degree of this contagious enthusiasm, Andrey Nikolaevich became one of the initiators of creating physical/mathematical boarding schools. The only kids who can get there are those living in small towns and villages of the country who were winners of mathematical olympiads. Many of these "provincials" have already grown up to become big scientists. For many years A.N. Kolmogorov has taught the classes himself to the students there and his pedagogical load can sometimes be as high as twenty-eight hours a week. Many new programs and textbooks have been written with direct contribution from Andrey Nikolaevich.

...Today, like yesterday and long ago, the academician is full of creative vigour and aspirations. He does not give himself any easy time. Scientific work, teaching, all follows the old rhythm. It seems that Kolmogorov has discovered the recipe to eternal youth.

V. ARNOLD,
Professor of the Lomonosov Moscow State University,
Lenin Prize laureate.

Preface to the Collection "Singularities of Differentiable Mappings" of Russian Translations of Papers in English and French

V.I. Arnold

(Singularities of Differentiable Mappings. A collection of Russian translations of papers in English and French. Editor: V.I. Arnold. Mir Publishing House, Moscow, 1968. The editor's preface is on pp. 5–7.)

Singularity theory of differentiable mappings is a young branch of mathematics situated at the crossroads of differential topology, algebraic, analytic, and differential geometry, functional analysis, and theory of differential equations.

The work by H. Whitney on mappings of a plane into a plane (1955) is considered to be the birth of this new theory, although certain separate results appeared noticeably earlier (the Morse theory of critical points of functions, Whitney's theorems on singularities of embeddings, the works of L.S. Pontryagin and others on the relation between singularities and characteristic classes).

The main contributions into the modern development of singularity theory were made by H. Whitney, R. Thom, H.I. Levine, B. Malgrange, J.M. Boardman, B. Morin, J.N. Mather, J.-C. Tougeron. The present book is a collection of translations of their papers, mostly so far unavailable to the Soviet reader.

One can single out three parts in singularity theory of differentiable mappings: differential, topological, and homological ones. Presently, the most advanced is the first part, centered around the studies of differentially stable singularities. The singularity of the map $y = x^2$ at the origin is an example of such singularity; any nearby map has the same type of singularity near the origin. On the other hand, the singularity of the map $y = x^3$ at the origin is unstable, since it is destroyed by a small perturbation ($y = x^3 - \varepsilon x$).

It turns out that in low dimensions any mapping can be approximated by a stable one, while stable mappings admit a comprehensible algebraic classification. For

Originally publ. in: "Singularities of Differentiable Mappings", V.I. Arnold (Ed.), 5–7, © Mir, Moscow 1968
Translated by Boris Khesin and Andrey Khesin.
Springer-Verlag Berlin Heidelberg 2016 with kind permission.

instance, for $\mathbb{R}^2 \to \mathbb{R}^2$ there are three such mappings:

$$\begin{array}{lll} \text{I} & y_1 = x_1, & y_2 = x_2, \\ \text{II} & y_1 = x_1, & y_2 = x_2^2, \\ \text{III} & y_1 = x_1, & y_2 = x_1 x_2 + x_2^3. \end{array}$$

However, a mapping of one nine-dimensional space to another cannot always be approximated by a stable one. This is related to the existence of continuous invariants of the singularity with respect to diffeomorphisms of the source and target spaces. Under a deformation of the mapping such an invariant can change so that the mapping will be unstable along with all mappings close to it.

One may say that the set of different types of singularities from the differential viewpoint becomes a continuum. In this case it is natural to consider a more robust, topological classification of singularities.

The study of topological stability of singularities is the second important part of the theory. Here one encounters fewer specific results; the solution of this problem is, in principle, announced at the Bombay talk by R. Thom, whose translation is given in this book, on pp. 164–178, but the proofs have not been published yet.

The third aspect of singularity theory is the "homological" one. Here one discusses the relation between topology of manifolds M and N and the singularities of mappings $f: M \to N$. An example is the Morse theory (where N is a line), connecting the homology of M and the critical points of a function f. Another example is delivered by the Whitney theorem, stating that a mapping of the projective plane into the plane has an odd (and hence non-vanishing) number of cusps of type (III). The homological side of singularity theory is hardly present in this collection. One can get acquainted with the results of this research by reading the papers "Les singularités des applications différentiables" by R. Thom, *Ann. Inst. Fourier, Grenoble*, **6** (1956), 43–87; "The singularities, S_1^q" by H.I. Levine, *Illinois J. Math.*, **8** (1964), no. 1, 152–168.

Connections of singularity theory with other areas of mathematics are numerous and multifaceted. Sometimes the Jacobians vanish in analysis, topology, and physics.[1] The value of papers in this collection is not only in specific results on the structure of singularities and their classification. Equally important is the development of general approaches to various problems on degenerations, a precise definition of "general position", codimensions, transversality, etc.

Presently singularity theory is on the rise. It is far from completion, however; one may hope that the readers of the present collection will take part in further developments of singularity theory and its applications.

[1] An instructive example of interaction of physics and singularity theory is provided by the work of F. Pham on Feynman diagrams, see the book "Homology and Feynman Integrals" by R.C. Hwa and V.L. Teplitz whose Russian translation will appear soon.

Preface to the Russian Translation of "Introduction à l'étude topologique des singularités de Landau" by F. Pham

V.I. Arnold

(The French original: F. Pham. Introduction à l'étude topologique des singularités de Landau. Mémorial des Sciences Mathématiques, Fasc. 164. Gauthier-Villars Éditeur, Paris, 1967, ii + 142 pp.

The Russian translation: Mir Publishing House, Moscow, 1970, 184 pp. Translator: M.V. Yakobson. Editor: V.I. Arnold. The editor's preface is on pp. 5–6.)

The book by F. Pham is a survey of studies on ramifications of functions defined by integrals over manifolds. It also contains a lucid exposition of the necessary topological toolbox, accessible to both mathematicians and physicists.

The interest of physicists in the ramifications of integrals is related to the study of singularities of Feynman integrals in perturbation theory. The mathematical problem is as follows. One considers the integral of an algebraic differential form, algebraically depending on parameters, over a cycle of an algebraic manifold. The integral value depends analytically on parameters outside of a certain variety of the complex parameter space, which is called a Landau variety. A study of the singularity type near the Landau variety boils down to examining the action of the fundamental group of the complement to the Landau variety in the complex parameter space, the action being on the homology of the manifold whose cycles are used in the integration.

In the simplest situation the problem reduces to the Picard–Lefschetz formula and the Leray residue theory; to determine the fundamental group one may use Zariski's results.[1]

The calculations carried out by the French physicists turned out to be useful in purely mathematical studies as well. In particular, E. Brieskorn[2] derived from

[1] Zariski's formulas [O. Zariski, Algebraic Surfaces, Springer, Berlin, 1935] were applied to Feynman integrals by G. Ponzano and T. Regge [The monodromy group of one loop relativistic Feynman integrals (preprint)].

[2] Beispiele zur Differentialtopologie von Singularitäten, *Invent. Math.*, **2** (1966), no. 1, 1–14.

Originally publ. as preface to the Russian translation of the book: F. Pham. Introduction à l'étude topologique des singularités de Landau. V.I. Arnold (Ed.), 5-6, © Mir, Moscow 1970. Translated by Boris Khesin and Andrey Khesin.
Springer-Verlag Berlin Heidelberg 2016 with kind permission.

F. Pham's results that 28 manifolds

$$z_1^2 + z_2^2 + z_3^2 + z_4^3 + z_5^{6k-1} = 0, \quad |z_1|^2 + \ldots + |z_5|^2 = 1, \qquad k = 1, \ldots, 28,$$

represent J.W. Milnor's 28 differentiable structures on the seven-dimensional sphere.

The book by F. Pham is intended for mathematicians not familiar with Feynman integrals and for physicists not familiar with topology. The author does not assume from the reader any preliminary knowledge exceeding the standard analysis course. A major part of the book can be used as a topology textbook geared towards analytical applications.

Preface to the Russian Translation of "Singular Points of Complex Hypersurfaces" by J. Milnor

V.I. Arnold

(The English original: J.W. Milnor. Singular Points of Complex Hypersurfaces. Annals of Mathematics Studies, No. 61. Princeton Univ. Press, Princeton, N.J.; Univ. of Tokyo Press, Tokyo, 1968, iii + 122 pp.

The Russian translation: Mir Publishing House, Moscow, 1971, 128 pp. Translator: V.M. Buchstaber. V.I. Arnold's preface is on pp. 5–6.)

There exists a remarkable relation between the theory of singular points of planar algebraic curves, on the one hand, and the theory of knots, links, and braids, on the other hand. The knots and links arise as the intersections of the three-dimensional sphere which bounds a small neighborhood of a singular point and the (real two-dimensional) surface of complex points of our curve. The algebraic invariants of the curve (such as Puiseux exponents, etc.) turn out to be connected with the topological invariants of the obtained knot or link.

The study of this relation and the corresponding higher-dimensional generalizations is the subject of the present book. An example of achievements in this area is the theorem of E. Brieskorn according to which the 28 manifolds

$$z_0^{6k-1} + z_1^3 + z_2^2 + z_3^2 + z_4^2 = 0, \quad |z_0|^2 + \ldots + |z_4|^2 = 1, \qquad k = 1, 2, \ldots, 28,$$

represent 28 Milnor's spheres (which are all homeomorphic to the standard seven-dimensional sphere, but pairwise non-diffeomorphic).

As another example one can mention the following construction of a non-smoothable topological manifold (M. Kervaire, E. Brieskorn, N. Kuiper).

Consider the subset of the seven-dimensional complex space given by

$$z_1^2 + \ldots + z_5^2 + z_6^3 + z_7^5 = \varepsilon, \qquad \sum z_i \bar{z}_i \leqslant 1,$$

where ε is sufficiently small. This set is a 12-dimensional smooth manifold with a boundary homeomorphic to the topological 11-dimensional sphere. By contracting

Originally publ. as preface to the Russian translation of the book: J. Milnor. Singular Points of Complex Hypersurfaces. V.I. Arnold (Ed.), 5-6, © Mir, Moscow 1971. Translated by Boris Khesin and Andrey Khesin.
Springer-Verlag Berlin Heidelberg 2016 with kind permission.

the boundary to a point, one obtains a 12-dimensional topological manifold without boundary. It does not admit any differentiable structure.

As an example of very recent results, it is worth mentioning the work by H.A. Hamm, who extended the results of J. Milnor and E. Brieskorn to isolated singularities of complete intersections.

A feature of the book by J. Milnor is that algebraic geometry is regarded here not as a branch of algebra or number theory, but as a part of geometry and analysis. In the world literature, this book is one of the first attempts of the modern geometric exposition of some basic notions of algebraic geometry.

One should mention that the recently increased interest in the geometry of singularities of algebraic manifolds is due not only to the mathematical appeal of this area, but also to its applications in physics (ramification of integrals, singularities connected to Feynman diagrams, etc.). In particular, the Brieskorn theorem mentioned above is based on the calculations of the French physicist F. Pham.

Judging by its expository style, the book by J. Milnor can be used as a textbook for beginners: an abundance of examples, the presence of figures, and a visual and geometric character of the author's arguments make the book accessible to the broad community of mathematicians of all areas of expertise.

Preface to the Russian Translation of "Differentiable Germs and Catastrophes" by T. Bröcker and L. Lander

V.I. Arnold

(The English source: Th. Bröcker. Differentiable Germs and Catastrophes. Translated from the German by L. Lander. Last chapter and bibliography by L. Lander. London Mathematical Society Lecture Note Series, No. 17. Cambridge Univ. Press, Cambridge, 1975, vi + 179 pp.

The Russian translation: Mir Publishing House, Moscow, 1977, 208 pp. Translator: A.G. Kushnirenko. Editor: V.I. Arnold. The editor's preface is on p. 5.)

The subject of the book by T. Bröcker and L. Lander is critical points of families of smooth functions depending on parameters. One often encounters the study of such families in various applications. The book by Bröcker and Lander is the first textbook in the world literature devoted to such questions and written with a great attention to the reader, of whom very little mathematical knowledge is assumed. For a more fundamental study of the Whitney–Thom–Mather general theory of singularities of smooth mappings we address the reader to the textbook "Stable Mappings and Their Singularities" by M. Golubitsky and V. Guillemin, which recently was translated into Russian.[1]

[1] The Editors cannot resist providing here the epigraph of the English edition:

> There once lived a man
> who learned how to slay dragons
> and gave all he possessed
> to mastering the art.
>
> After three years
> he was fully prepared but,
> alas, he found no opportunity
> to practise his skills.
>
> Dschuang Dsi.
>
> As a result he began
> to teach how to slay dragons.
>
> René Thom.

Originally publ. as preface to the Russian translation of the book: Th. Bröcker, L. Lander. Differentiable Germs and Catastrophes. V.I. Arnold (Ed.), 5, © Mir, Moscow 1977. Translated by Boris Khesin and Andrey Khesin.
Springer-Verlag Berlin Heidelberg 2016 with kind permission.

Preface to the Russian Translation of "Stable Mappings and Their Singularities" by M. Golubitsky and V. Guillemin

V.I. Arnold

(The English original: M. Golubitsky and V. Guillemin. Stable Mappings and Their Singularities. Graduate Texts in Mathematics, Vol. 14. Springer, New York, 1973, x + 209 pp.

The Russian translation: Mir Publishing House, Moscow, 1977, 292 pp. Translator: A.G. Kushnirenko. Editor: V.I. Arnold. The editor's preface is on p. 5.)

The book by M. Golubitsky and V. Guillemin is the first textbook on the singularity theory of smooth mappings in the world literature. The authors do not assume of the reader any preliminary knowledge outside of the standard courses of calculus and linear algebra: the necessary facts on manifolds, bundles, jets, etc. are given in detail in the first two chapters of the book.

After that there are chapters devoted to stable mappings, the Weierstrass–Malgrange preparation theorem, and an introduction to the classification of singularities. The authors always confine themselves to the simplest cases, particularly trying to make sure that the exposition of the main ideas would not be obstructed by technical details. This style of exposition makes the book very attractive for a first acquaintance with singularity theory.

Singularity theory is one of the most actively developing branches of modern mathematics, and it is located at the crossroad of analysis, topology, differential and algebraic geometry, and the theory of differential equations.

Numerous applications of singularity theory in physics and other natural sciences (such as caustics and wave fronts in optics, bifurcations and stability loss in elasticity theory, oscillating integrals and the method of stationary phase in quantum mechanics, etc.) lately acquired a unifying name of "catastrophe theory" and are actively pursued in hundreds of publications.

One may hope that the book by Golubitsky and Guillemin written with great pedagogical skills will allow the Soviet readers to enter the scope of ideas and methods of the singularity theory of smooth mappings, where a lot remains to be done.

Originally publ. as preface to the Russian translation of the book: M.Golubitsky, V.Guillemin. Stable Mappings and Their Singularities. V.I. Arnold (Ed.), 5, © Mir, Moscow 1977. Translated by Boris Khesin and Andrey Khesin.
Springer-Verlag Berlin Heidelberg 2016 with kind permission.

Acknowledgements

Springer-Verlag thanks the publishers of V.I. Arnold's papers for having granted permission to reprint them in these Collected Works.
The numbers behind the original sources correspond to the numbers given in the Table of Contents. We have made every effort to identify the copyright owners for all articles included in this book in order to obtain reprint permission. Some of our requests however remained unanswered. The following list includes all sources.

Reprinted from Comm. Pure Appl. Math.	15, 16
Reprinted from Funct. Anal. Appl.	1, 2, 4, 7, 8, 9, 20, 21, 25
Reprinted from Invent. Math.	12
Reprinted from J. Appl. Math. Mech.	28
Reprinted from Moscow Univ. Math. Bull.	19
Reprinted from Proc. Symp. Pure Math.	13, 31, 32, 33, 34, 35, 36, 37
Reprinted from Russ. Math. Surv.	3, 5, 11, 17, 26, 27
Reprinted from Selecta Math. Sov.	10
Reprinted from Usp. Mat. Nauk	24, 40

GPSR Compliance

The European Union's (EU) General Product Safety Regulation (GPSR) is a set of rules that requires consumer products to be safe and our obligations to ensure this.

If you have any concerns about our products, you can contact us on ProductSafety@springernature.com

In case Publisher is established outside the EU, the EU authorized representative is:

Springer Nature Customer Service Center GmbH
Europaplatz 3
69115 Heidelberg, Germany

Batch number: 10755398

Printed by Printforce, the Netherlands